Essential Mathematics for Undergraduates

Simon G. Chiossi

Essential Mathematics for Undergraduates

A Guided Approach to Algebra, Geometry, Topology and Analysis

 Springer

Simon G. Chiossi (iD)
Departamento de Matemática Aplicada
Instituto de Matemática e Estatística
Universidade Federal Fluminense
Niterói
Rio de Janeiro, Brazil

ISBN 978-3-030-87176-5   ISBN 978-3-030-87174-1   (eBook)
https://doi.org/10.1007/978-3-030-87174-1

Mathematics Subject Classification: 15-01, 03-01, 51-01, 54-01, 97N70

This Springer imprint is published by the registered company Springer Nature Switzerland AG
The registered company address is: Gewerbestrasse 11, 6330 Cham, Switzerland

*To Peter J. Hilton (1923–2010),
eternally grateful for the 1996
masterclass on algebraic topology
and the lectures on Bletchley Park.
It was a privilege.*

Preface

Books on the so-called rudiments of mathematics typically fall into two categories. There are pragmatic texts, which above all emphasise computing strategies for solving problems; albeit effective, their deliberate lack of accuracy unwittingly reinforces the lethal misconception that mathematics boils down to memorising and manipulating formulas. In the other group belong books presenting elementary material from a more penetrating angle, which is usually adequate—in terms of background knowledge and level of sophistication—for post-graduate students. In 2015 I had the idea of writing a manual interpolating, so to speak, between those two extremes and thus organise the proverbial pile of loose notes accumulated over the years. Clearly, there was no way I could write an introductory-level book of that kind. So here it is. Whoever writes a book of this kind must, in some degree, rely on his predecessors' labours, so there is nothing particularly revolutionary in the contents, apart perhaps from the assortment of subjects. But the driving idea and the exposition's ensuing structure are completely novel. In this, the present book distinguishes itself from the majority of other texts on similar matters.

Contents and Layout Nowadays, alas, several undergraduate courses are subsidiary to the infamous Calculus, as if they were a sort of prologue to it—at best a companion. Here I strived to stop this bad habit, and focused on material that is not subject to whims of fashion but on the contrary preps for abstract algebra, topology, analysis and geometry. It is regrettable that many of the topics studied here are absent from many current syllabi. I am convinced they should be part of the baggage of knowledge that any student concluding a Bachelor in mathematics should take away with them.

Structure The book is made of morsels of theory—some bite-size, while other mouthfuls—selected from a broad spectrum. To harmonise this matter and organise it didactically, the book has been divided into five parts consisting of 19 chapters, plus two appendices. *Part 1* (Chaps. 1–5) presents the language, its rules and the elementary bricks. It concentrates on the idea of formalisation and provides 'meta-mathematical' methods and the tools of the trade. Chapter 5 has the specific role to provide the algebraic background to logic and set theory.

Chapters 6–7 form the core of *Part 2* and deal with the number systems of our daily practice. The more abstract Chapter 8 furnishes a pleasant contextualising complement.

The material necessary to solve equations and inequalities (Chaps. 9–13) is in *Part 3*, disguised under a more general framework. I appended a little chapter (Chap. 14) on enumerative combinatorics, the binomial formula and how they tie in with functions.

Part 4 elaborates on Felix Klein's revolutionary approach (Chaps. 17–19) and unequivocally attests the wedlock between algebra and geometry imbuing the book. This pervasive point of view, also taken up in Chaps. 8, 13 and elsewhere, exemplifies an archetype of Mathematics without the intellectual and disciplinary barriers of many academic curricula. Chapter 15 is included for convenience, to flesh out the Euclidean structure of $\mathbb{R}^n$, while Chaps. 16 and 19 stress the key idea of symmetry transformation.

The final part (*Part 5*) consists of two appendices: 'Etymologies' lists the Greek origin of a selection of terms of common use; 'Author Index' is an anthology of people cited throughout.

Added Features   Detailed proofs and much scattered material is made accessible for the first time in textbook form. Apart from offering the essential theory, chapters are disseminated with abundant and poignant examples. A careful selection of exercises, in the stead of endless routine lists, accompanies the learning process and encourages the student to become independent. This is all the more crucial as we move into a completely skills-based society. Many simple proofs are left as homework, and of course any example can be used for training.

☕ marks sections and chapters that require a bit more of background. Albeit absolutely standard, these subjects somehow deviate from the usual path and reflect my own geometrical-topological taste and my tendency towards abstraction. If judged detrimental to the narrative thrust, they may be skipped without affecting the rest. But the incorporation of this material is intended to open the door on a contemporary scenery and point out interdisciplinary connections.

To enrich the offering, several figures and a number of historical comments and trivia are thrown in (a selection is on p.xiii).

What Is Not Here   With the exception of a few non-essential parts, I purposely made the effort to fight off both the angel of topology and the devil of abstract algebra [97], so that the model structures of universal algebra (e.g. groups, rings, fields, lattices) and point-set topology are absent. There is nothing on infinitesimal calculus, nor analytic geometry. Occasionally, and only in optional parts, a minimum familiarity with those subjects might be useful. Prerequisites are indicated at the beginning of sections.

Readership   The book is targeted at undergraduate students and instructors alike. To begin with, students who need to consult a reference text to find, say, a proof they have never been through, a never-heard-of theorem, or the broader context of a concept. It works as a guide for young undergraduates taking the first steps through the thick foliage of the maths jungle, by building up the maturity to read and write proofs and expanding their skillsets.

Teachers will capitalise on supporting material, snippets and indications, in view of their classes. They might zig-zag through the contents, jump back and forth and across, skip certain parts and dig deeper into others. This starts with the terminology, some of which is excessive for a first-year class, and solecistic in later parts of the text. The book's embryo was, at least initially, devised to solidify and fill in the gaps of some Brazilian curricula that, for lack of time, do not address certain content or cannot answer the current demand (one such existed at my institution). It subsequently grew into a homogeneous companion, designed to be flexible and allow for a non-sequential reading (p.xv). For these reasons—and the fact it adapts to a variety of pedagogical choices—it serves well a large number of basic disciplines, both is Mathematics and other STEM courses, and is not specific to a particular area, country or region.

Anticipating the natural protest that the book is too difficult for young students to understand, I'll paraphrase Gödel, who attributed such objection to the prejudice against everything that is abstract, and to the ensuing anxiety. Reading it leisurely, without attempting to understand everything immediately, will make it less unintelligible.

Main Intents For starters, I wished to familiarise the novice with the language of mathematics. Beginning with abstract logic permits to provide rigorous formulations and demonstrate the majority of results, because I believe in Richard Dedekind's tenet that 'what can be demonstrated, should not be believed without proof in the sciences' [24, Preface]. Doing so also allows to correct any wrong notion that students might have picked up in school. I am a self-confessed etymological stickler, and I like to keep in mind that 'education' comes from the Latin *educĕre*, meaning 'to conduct, lead away (from mistakes)'—in contrast to what one reads in certain poorly informed websites. It goes without saying that certain facts will not have a formal justification, for the latter would require a more robust knowledge on the reader's side, which definitely goes beyond the present scope.

Secondly, and related to the above, I sought to establish clear and unambiguous theoretical bases on which to build a curriculum. This 'essential mathematics' is crucial and cannot in any way be presented in an approximate fashion, lest we leave indelible scars in the education of students or, even worse, future high-school teachers. You cannot build a house without solid foundations.

Lastly, readers will be exposed to a broad and modern perspective, much in the same spirit as [20]. After all, the ancient Greek noun μαθεματικὴ means learning, and μαθεματικός is someone who is eager to learn. In order to tickle the curiosity and fire up the scientific appetite, I pointed out subjects and areas (☞) that the reader might wander into at a later stage.

Acknowledgements My scientific and personal development has been deeply influenced by my mentors Simon Salamon, Fabio Podestà and Thomas Friedrich, to whom I will be everlastingly beholden. I am particularly indebted to my dear friend and colleague Letterio Gatto, a refined soul and well-read scholar. His relentless encouragement and support have always been a lifeline to me.

My deepest gratitude goes to Robinson Nelson dos Santos (Editor Mathematics at Springer Brazil) for his adroit and patient guidance and the savvy tips. I also wish to thank Francesca Bonadei (Executive Editor Mathematics and Statistics at Springer Italy) for the enduring trust over the years.

This edition undoubtedly contains mistakes and many omissions. I would be glad to receive feedback at simongc@id.uff.br.

Niterói, Brazil   Simon G. Chiossi
June 2021

Chapter Flow

Either white box below is recommended as a starting point. That is because a very efficient way to feed logic to students is to set it up in parallel to the more digestible set theory. The cryptomorphism between algebras of sets and propositional calculus can be taught at a basic level, and makes it easier to convince students of the validity of certain methods, such as the proof by contradiction, the syllogism, or the explosion principle.

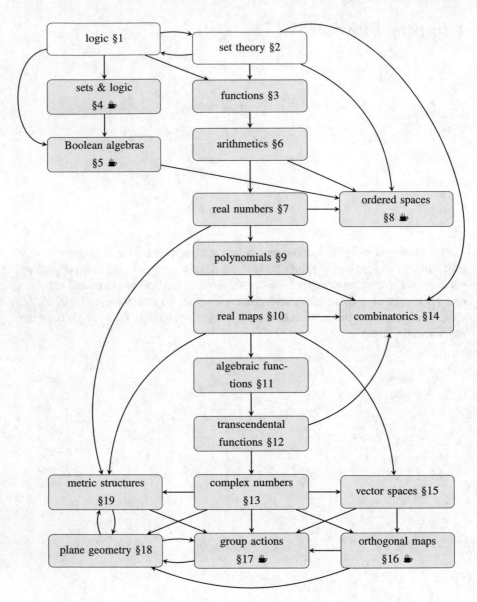

Hot Topics

Here is a list of topics that are not usually found in elementary books. These occasional digressions take the reader down winding corridors and new alleys.

- Soundness and completeness of first-order logic   Sects. 1.5, 4.1, 4.4
- Boolean nature of logic   Sect. 5.3
- Mathematical structures (algebraic, order, incidence etc.)   Sects. 4.3, 3.3
- Order topology   Chap. 8
- Galois connections   Sect. 4.2
- Equivalents of the axiom of choice   Sect. 3.4
- Primitive recursion   Sect. 6.3
- Cardinal arithmetics   Sects. 6.6, 7.7
- Real numbers as Dedekind cuts   Sect. 7
- Algebraic and transcendental numbers   Chap. 13.3
- Characterisation of the exponential map   Sect. 12.1
- Compactifications, stereographic projections, projective spaces Sects. 7.6, 18.2.1
- Actions of matrix groups   Sects. 16, 17
- Metric completions   Sect. 19.6
- Generating functions   Sect. 14.4
- Kinematics and electromagnetism in a nutshell   Sect. 15.7
- Greek origin of major terms   Appendix: Etymologies
- Complex structures   Remark 13.6, Sect. 15.3
- Projective axiomatics   Remark 18.7
- Birkhoff's axiomatisation of plane geometry   Sect. 18.1
- The Cantor set   Example 3.20

A selection of historical comments and critical remarks has been disseminated throughout:

- Undecidable theorems or theories   Remark 1.24
- Non-standard models of arithmetics   Example 4.16
- (Non-)solubility of polynomial equations by radicals   Remark 9.6
- The RSA cryptosystem   Remark 6.27

Devising Courses

The pedagogical undertaking of teaching is massive, as it balances two antagonistic needs. On one side, the longer the students are exposed to vague and equivocal lectures, the more complex they will find it to adapt to abstract reasoning and rigour and tap into their talents. That is why it is highly desirable to teach the present material sooner rather than later. On the other, instructors must not give in to the temptation of teaching what green students are not in the condition to assimilate properly. But I strongly believe that the majority of sections, provided that they are suitably taught, are appropriate for early undergraduates. That said, each lecturer will decide what to present and what not.

This book provides much material to accompany existing courses, but was also envisaged towards creating entirely new disciplines. Several single chapters lend themselves to introductory courses, such as Chap. 19 (*'Metric spaces'*), Chap. 15 (*'Vector spaces'*), Chap. 18 (*'Euclidean and non-Euclidean geometry'*), Chap. 5 (*'Boolean algebras'*), Chap. 8 (*'Ordered spaces'*), and Chaps. 1–4 (*'Mathematical Logic'*).

The table below attempts to match chapters (or sections thereof) to the syllabi of some more-or-less standard courses that spring to mind:

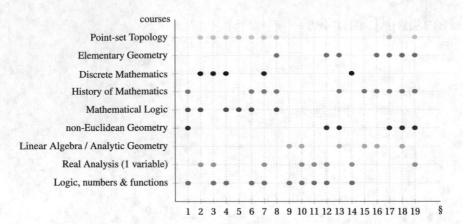

The following sequence is for a 60 hour course on, for lack of a better name, *'Logic, numbers & functions'* (table, bottom row). This model, excluding ☕ sections, was tested at Politecnico di Torino (2010–2011) and Universidade Federal Fluminense (2015–2016, 2019), where it has now become part of the undergraduate degree.

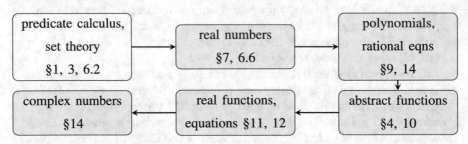

Below is some material I put to use when lecturing on *'History of Mathematics'* for a teaching degree (UFF, 2019). Due to the modular nature of the class, it looks more like a web than a directed graph.

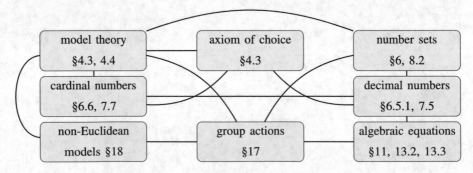

The next chart could be cherry-picked to set up a non-standard course on *'Geometry and algebra'*, in which one would revise and cement Group Theory, Linear Algebra and Analytic Geometry and prepare for tackling Differential Geometry, PDEs or Lie Theory.

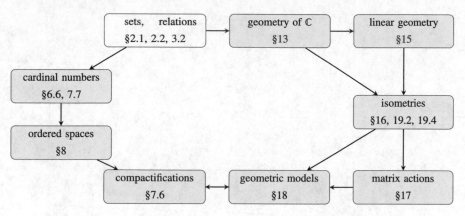

Special symbol legend:

:=	Definition	⚡	Contradiction (typically within proofs)
□	End of proof	✗	Non-acceptable solution
✓	Acceptable solution	☞	Pointer
☕	Optional material		

Contents

Part II   Numbers and Structures

Part I
Basic Objects and Formalisation

Chapter 1
Round-Up of Elementary Logic

This first chapter is an incomplete summary of classical intuitive logic. The importance of tackling this at the very beginning—of a book and of a curriculum—cannot go amiss. A satisfactory treatment deserves a specific course, that clearly cannot be taught in the first semesters, nor in one semester only. Therefore here we shall just present the more practical aspects. Details can be found in a number of references, for instance [81, 92, 95].

Before we get going, recall that propositional logic is better assimilated jointly with set theory (Chap. 2).

There are two basic stops along the way towards a formal mathematical theory (see Definition 1.15). The first is to adopt a formal language, the second is to classify statements as true or false, in an exclusive way, in a given context.

1.1 First-Order Languages

A formal language may be thought of in a simple way as a model that describes certain aspects of a natural language (Danish, Yoruba, Māori, . . .). It is a translation that works as a more-or-less satisfactory symbolic approximation of statements, that is to say the natural language's declarative sentences. In this book the word logic refers to so-called **classical logic**, which only deals with the formalisation of declarative sentences: more complex logics, e.g. deontic, modal, temporal etc., are discussed in [40].

Any formal language is made of three constitutive elements, defined shortly: an alphabet $\mathscr{A}$ of symbols, the words $\mathscr{T}$ of the vocabulary, and a set $\mathscr{F}$ of statements.

Definition Such a triple $\mathscr{L} = (\mathscr{A}, \mathscr{T}, \mathscr{F})$ is called a **(predicative) language of first order**.

S. G. Chiossi, *Essential Mathematics for Undergraduates*,
https://doi.org/10.1007/978-3-030-87174-1_1

We shall not distinguish between object language (the one about which we speak) and metalanguage (the one we speak), although it is important to keep them separate lest we incur in liar's-type paradoxes. The convention will be that in the metalanguage we shall speak about the truth and falsity of sentences of the object language.

So let's introduce the ingredients of a language.

Definition An **alphabet** is a finite or countable collection of primitive logical symbols, namely:

- a finite or countable set V of **variables**: x y z t w $\cdots$
- 5 (Boolean) **connectives** $\neg$ $\wedge$ $\vee$ $\rightarrow$ $\leftrightarrow$
- 2 **quantifiers** $\forall$ $\exists$
- punctuation symbols , : | and brackets: ( ) [ ] { }
- the logical predicate $\doteq$, called **equality**

and primitive extra-logical symbols:

- **constants** a b c $\cdots$
- n-ary **predicates** P Q R $\cdots$
- n-ary **relations / functions** f g h $\cdots$

These eight sets are disjoint. Extra-logical symbols depend strongly on the situation considered, so we shall not define them more precisely. For example, in set theory we would include $\in, \varnothing$, in elementary arithmetics $s, +, \cdot, \leqslant, 0$ etc.

The definition of alphabet we gave is far from being minimal: it would be enough to take the connectives $\neg, \vee$ and the quantifier $\exists$ to define the others, cf. Exercise 1.2 and Theorem 1.9.

Finite concatenations of symbols make up words, like $x\exists)fa \doteq (\doteq\rightarrow))Pq \wedge \forall$. Among these we wish to find meaningful expressions, the analogues to those in the natural language.

Definition The set of **terms** $\mathcal{T}$ is the smallest set of words such that

i) $\mathcal{T}$ contains variables and constants;
ii) for every $t_1, \ldots, t_n \in \mathcal{T}$ and every n-ary relation f, the expression $ft_1t_2t_3 \cdots t_n$ is a term (conventionally indicated $f(t_1, \ldots, t_n)$).

The language's formal grammar is built recursively starting from the equality of terms and predicates (a bit like the atoms of matter), and combining the latter by connectives and quantifiers (the chemical bounds that form molecules). Cf. Fig. 1.1.

Definition The set of **(well-formed) formulas** (or **propositions**, or **statements**) is the smallest set $\mathcal{F}$ of expressions such that

i) if $t, s \in \mathcal{T}$ then $(t \doteq s)$ belongs to $\mathcal{F}$ (a formula of this kind is called **equation**);
ii) $Pt_1t_2 \cdots t_n \in \mathcal{F}$, written $P(t_1, \ldots, t_n)$, for every n-ary predicate P and terms $t_1, t_2, \ldots, t_n \in \mathcal{T}$;
iii) if $A, B \in \mathcal{F}$ then $\neg A$, $A \wedge B$, $A \vee B$, $A \rightarrow B$, $A \leftrightarrow B \in \mathcal{F}$;
iv) if $A \in \mathcal{F}$ and x is a variable, then $(\exists x)(A) \in \mathcal{F}$ and $(\forall x)(A) \in \mathcal{F}$.

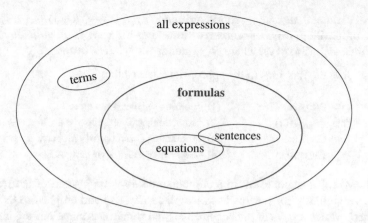

Fig. 1.1 Types of words of a language

Regarding the formulas in (iii) and (iv), the standard nomenclature is:

$\neg A$	**negation**	(read 'not A');
$A \wedge B$	**conjunction**	('A and B');
$A \vee B$	**disjunction**	('A or B');
$A \rightarrow B$	**conditional**	('A implies B', 'if A then B', 'A only if B'),

A is called **hypothesis**, **premise**, or **sufficient condition** for B,
B is the **thesis**, **consequence**, or **necessary condition** for A;

$A \leftrightarrow B$	**biconditional**	('A if and only if B'),

A is the **necessary and sufficient condition** for B (and vice versa).

Terminology: we shall tweak the dictionary and refer to conditional propositions as 'implications', and to biconditionals as 'equivalences'.

Furthermore,
$\forall$ is called **universal** quantifier: $\forall x$ is read 'for every x';
$\exists$ is the **existential** quantifier: $\exists x$ reads 'there exists x such that', or 'for some x'.

Examples

i) The proposition '*some real numbers are rational*' is understood as '*there exists (at least) one real number that is rational*'. This is written $(\exists x)(R(x) \wedge Q(x))$, where $R(x)$: '*x is a real number*' and $Q(x)$: '*x is a rational number*'. We can also read it as '*there exists a real number x such that x is rational*' or '*for some real number x, x is rational*', for example.

ii) *'Every rational number is real'* is written $(\forall x)\big(Q(x) \to R(x)\big)$.
iii) The proposition *'Brexiteers do exist'* translates in ' *there is an individual that is a Brexiteer'*: $(\exists y)B(y)$, where $B(y)$ stands for '*y is a Brexiteer*'.

Exercise Join the formulas to the appropriate description

(a)	$(\forall x)A(x)$	(1)	nothing satisfies property A
(b)	$(\forall x)\neg A(x)$	(2)	something satisfies property A
(c)	$(\exists x)A(x)$	(3)	something doesn't satisfy property A
(d)	$(\exists x)\neg A(x)$	(4)	everything satisfies property A

Already at this stage, to avoid a cumbersome notation instead of $(\forall x)(A)$ one uses $\forall x\ A$; similarly $\exists x \mid A$ or $\exists x : A$ replace $(\exists x)(A)$ and emphasise the words 'such that'. Moreover, $\doteq$ simply becomes $=$, and sometimes negations are indicated by "crossing out" the negified symbol:

$$\neg(a \doteq b) \rightsquigarrow a \neq b \qquad \neg(a \in A) \rightsquigarrow a \notin A$$
$$\neg(B \subseteq A) \rightsquigarrow B \nsubseteq A \qquad \neg(P \to Q) \rightsquigarrow P \nrightarrow Q$$

just like $\nexists x \mid P(x)$ is shorthand for $\neg(\exists x)P(x)$.

Note how quantifiers only act on variables (this is the technical meaning of the language being of 'first' order). Variables that aren't quantified are called **free**. A formula without free variables is called a **sentence** (example: $3^2 + 4^4 - 5^2 = 0$). In practice when a variable is free in a formula, every time we attribute to predicates a meaning, that variable behaves as an 'indeterminate', since the statement speaks about the variable itself. For instance in the formulas $x < 7$ and $(\exists y)(y > x)$ the variable x is free, whereas y is not. On the other hand, in formula $(\exists y)(y < x) \wedge (\forall x)(x > 0)$ the first occurrence of x is free but the second one isn't. This shows that, as regards the meaning of a sentence, free and non-free occurrences of the same variable have nothing to do with one another.

In Sects. 1.2 and 1.3 we shall discuss the semantics of propositions and predicates (with further material in Sect. 4.3), whilst Sect. 1.4 is about syntax.

1.2 Propositional Calculus

The locution 'propositional calculus' refers to the manipulations meant to determine whether a formula is true or false. This is formalised by the next notion.

Definition 1.1 A **truth function** is a map $\upsilon\colon \mathscr{F} \to \{0, 1\}$ with the following propositional values

for any proposition $A, B \in \mathscr{F}$. The values 0, 1 are called 'false', 'true', and a proposition P is **false** or **true** if there exists a function υ such that $\upsilon(P) = 0$ or $\upsilon(P) = 1$ respectively.

A	B	$\neg A$	$A \wedge B$	$A \rightarrow B$	$A \leftrightarrow B$	$A \vee B$
1	1	0	1	1	1	1
1	0	0	0	0	0	1
0	0	1	0	1	1	0
0	1	1	0	1	0	1

In our daily life it's common to use a disjunction $\vee$ with a different meaning to what the above table prescribes. In mathematics when we say $A \vee B$ is true three things can happen: only A is true, only B is true, or A, B are both true. In other words the logical disjunction $\vee$ is *inclusive*. The *exclusive or* (one, and only one of A, B is true) is expressed by $\nleftrightarrow$, which corresponds to the Latin *aut*:

$$A \text{ aut } B \quad \dot{=} \quad \neg(A \leftrightarrow B).$$

On the contrary, the conjunction $\wedge$ has the everyday meaning: simultaneous occurrence.

Another crucial aspect that must be clearly laid out concerns the truth of conditional statements, lest we overlook it. A formula $A \rightarrow B$ is false only when a false conclusion B follows from a true premise A. In particular, whenever A is false the implication is true, irrespective of B (look at the table). This at first seems rather surprising. It suggests that when we assume something false then any statement becomes true, as if the mechanism governing the rules of thought exploded, spitting out anything: truth, falsity, absurdity, triviality... We will explain why it "has" to be so, and the importance of this fact, in Remark 1.23.

Exercises

i) Read out loud the sentences

$$(0 < a < b) \rightarrow (a^2 < b^2) \quad \text{and} \quad (a^2 < b^2) \leftrightarrow (-b < a < b),$$

and convince yourself they are true.

ii) Construct the truth table for $\big((P \vee Q) \rightarrow (R \leftrightarrow \neg S)\big) \wedge Q$.

iii) Consider the following argument: *If prices are high, wages are high. Prices are high or capped. In presence of price caps there's no inflation. But there is inflation.* Can you conclude that wages are high?

Exercise 1.2 Prove that any truth function υ satisfies the following properties

i) $\upsilon(\neg A) = 1 - \upsilon(A)$

ii) $\upsilon(A \wedge B) = \min\big\{\upsilon(A), \upsilon(B)\big\}$

iii) $\upsilon(A \vee B) = \max\big\{\upsilon(A), \upsilon(B)\big\}$

iv) $\upsilon(A \rightarrow B) = \upsilon(\neg A \vee B) = \upsilon(\neg B \rightarrow \neg A)$

v) $\upsilon(A \rightarrow B) = 1$ if and only if $\upsilon(A) \leqslant \upsilon(B)$

vi) $\upsilon(A \leftrightarrow B) = \upsilon\big((A \rightarrow B) \wedge (B \rightarrow A)\big)$

Properties iv) and vi), in particular, come in very handy during proofs, see p. 20.

Formula $\neg B \rightarrow \neg A$ is called **contrapositive** to $A \rightarrow B$. By item iv) above, the statement '*if m is a multiple of* 3 *then m is a multiple of* 9' has the same exact meaning as its contrapositive proposition '*if m isn't a multiple of* 9 *then m isn't a multiple of* 3'. Notice the difference with the 'reciprocal' statement '*if m is a multiple of* 9 *then m is a multiple of* 3' (where hypothesis and thesis are swapped).

Because of this exercise one could define truth functions in a simpler way, by imposing only the values of negation and disjunction:

Proposition 1.3 *A map* $\upsilon : \mathscr{F} \rightarrow \{0, 1\}$ *is a truth function if and only if*

$$\upsilon(\neg A) = 1 - \upsilon(A), \qquad \upsilon(A \vee B) = \max\big\{\upsilon(A), \upsilon(B)\big\}.$$

Proof Assign values 0, 1 to A and B and use the table on page 7. $\qquad\qquad$ □

Definition Two statements A, B are called **(tauto)logically equivalent**, written

$$A \models\mid B,$$

if $\upsilon(A) = \upsilon(B)$ for any truth function υ.

Exercise 1.2 leads to the following result, which is fundamental in indirect proofs such as the proofs by contradiction (Sect. 1.4).

Proposition 1.4 *The following are tautological equivalences:*

i) $\neg(A \wedge B) \models\mid (\neg A \vee \neg B), \quad \neg(A \vee B) \models\mid (\neg A \wedge \neg B) \qquad$ *(De Morgan laws)*
ii) $(A \rightarrow B) \models\mid (\neg A \vee B) \models\mid (\neg B \rightarrow \neg A)$
iii) $(A \leftrightarrow B) \models\mid \big((A \rightarrow B) \wedge (B \rightarrow A)\big)$

Proof This can be shown using the table on page 7, so we leave it as an exercise. $\qquad\qquad$ □

Definition A statement T is called a **tautology** if $\upsilon(T) = 1$ for any truth function υ, and one writes

$$\models T$$

read 'T is true' (in the model under exam, cf. Sect. 4.3).

Since tautologies are all logically equivalent, we may take formula $A \vee \neg A$ as the standard tautology, for any $A \in \mathscr{F}$, and indicate it by $\top$.

The notion of tautology depends on that of truth function, in particular on its definition on negations and disjunctions. We are implicitly assuming of working in a logic system where at least one between A and $\neg A$ is always true:

$$\models A \vee \neg A \qquad \textbf{(law of excluded middle).} \tag{1.1}$$

This is related in a technical way to the fact our logic is bivalent, or two-valued: any formula A is either true or false (*aut–aut*, of Wittgenstein fame), and no other option is allowed. There would be much more to elaborate on regarding this, but we shall leave it to that.

On the other hand

Definition A statement C whose truth value is 0 independently of the truth function chosen, is said to be a **contradiction**:

$$\vDash \neg C.$$

By duality with (1.1) we have that $A \wedge \neg A$ is always a contradiction, whatever A is. That's to say, a statement and its negation cannot be concurrently true:

$$\vDash \neg(A \wedge \neg A) \qquad \textbf{(principle of non-contradiction)}.$$

As any contradiction is logically equivalent to $A \wedge \neg A$, we'll denote it by $\bot$.

A straightforward consequence is that $A \equiv\mathrel{\mathop{\mid}} \top$ is the same as saying $\vDash A$, or $\neg A \equiv\mathrel{\mathop{\mid}} \bot$.

Theorem 1.5 *For any formula A, B*

i) $\vDash A \leftrightarrow B$ *if and only if* $A \equiv\mathrel{\mathop{\mid}} B$.
ii) $\vDash A$ *and* $\vDash A \to B$ *imply* $\vDash B$.

Proof Exercise. □

Example Let's show that

$$\vDash \neg((\neg P \vee Q) \vee (Q \wedge (\neg Q \vee \neg P))) \leftrightarrow (\neg Q \vee P).$$

Here we go (in each passage we suggest the reader detect which rules were employed):

$$\neg((\neg P \vee Q) \vee (Q \wedge (\neg Q \vee \neg P))) \equiv\mathrel{\mathop{\mid}} \neg(\neg P \vee Q) \wedge \neg(Q \wedge (\neg Q \vee \neg P))$$

$$\equiv\mathrel{\mathop{\mid}} (\neg\neg P \wedge \neg Q) \wedge (\neg Q \vee (\neg\neg Q \wedge \neg\neg P))$$

$$\equiv\mathrel{\mathop{\mid}} (P \wedge \neg Q) \wedge (\neg Q \vee (Q \wedge P))$$

$$\equiv\mathrel{\mathop{\mid}} (P \wedge \neg Q) \wedge ((\neg Q \vee Q) \wedge (\neg Q \vee P))$$

$$\equiv\mathrel{\mathop{\mid}} (P \wedge \neg Q) \wedge (\neg Q \vee P)$$

$$\equiv\mathrel{\mathop{\mid}} ((P \wedge \neg Q) \wedge \neg Q)) \vee ((P \wedge \neg Q) \wedge P))$$

$$\equiv\mathrel{\mathop{\mid}} (P \vee \neg Q) \vee (\neg Q \vee P)$$

$$\equiv\mathrel{\mathop{\mid}} \neg Q \vee P$$

Also note that in the course of the simplification we managed to eliminate all connectives except $\vee$ and $\neg$.

Exercises 1.6 Prove the following formulas are tautologies, for any formula A, B, C

$A \wedge (A \to B) \to B$	$A \to (B \to A \to B)$
$(A \to B) \wedge (B \to C) \to (A \to C)$	$(A \to B \wedge \neg B) \to \neg A$
$(A \wedge B \to C) \to (A \to (B \to C))$	$(A \to (B \to C)) \to (A \wedge B \to C)$
$(A \to B) \to (A \vee C \to B \vee C)$	$(A \to B) \to (A \wedge C \to B \wedge C)$
$(A \to B) \wedge (C \to B) \leftrightarrow (A \vee C \to B)$	$(A \to B) \wedge (A \to C) \leftrightarrow (A \to B \wedge C).$

In the next proposition the savvy reader might recognise a pattern, that bespeaks an algebraic structure of sorts (the lines might be called 'commutativity', 'associativity', 'distributivity', 'neutral elements' and 'idempotence' ... More will be said in Sect. 5.3)

Proposition 1.7 *For any statement P and Q*

$P \vee Q \rightrightarrows Q \vee P \qquad P \wedge Q \rightrightarrows Q \wedge P$

$P \vee (Q \vee R) \rightrightarrows (P \vee Q) \vee R \qquad P \wedge (Q \wedge R) \rightrightarrows (P \wedge Q) \wedge R$

$P \wedge (Q \vee R) \rightrightarrows (P \wedge Q) \vee (P \wedge R) \qquad P \vee (Q \wedge R) \rightrightarrows (P \vee Q) \wedge (P \vee R)$

$P \vee \perp \rightrightarrows P \qquad P \wedge \top \rightrightarrows P \qquad P \wedge \perp \rightrightarrows \perp \qquad P \vee \top \rightrightarrows \top$

$P \vee P \rightrightarrows P \qquad P \wedge P \rightrightarrows P$

Proof Exercise.   □

Definition A formula B is **tautological consequence** of another formula A, written

$$A \vDash B,$$

in case $\upsilon(B) = 1$ for any truth function υ such that $\upsilon(A) = 1$ (i.e. when $\vDash A$ implies $\vDash B$).

Theorem (Tautological Consequence) *For any formulas A, A_1, A_2, B*

i) $A \vDash B$ *if and only if* $\vDash A \to B$
ii) $A_1, A_2 \vDash B$ *if and only if* $\vDash A_1 \wedge A_2 \to B.$

Proof Exercise.   □

Exercise Prove that $\vDash \big((P \to Q) \to P\big) \to P$ (classically known as Peirce law).

Then take a contradiction as Q, and deduce principle (1.1). This guarantees the law of excluded middle can be phrased using just one connective, namely: $\to$.

1.3 Predicative Calculus

In contrast to the formulas seen thus far, which are always true of false for a given truth function, predicative statements—or **predicates**, those containing variables—can be both, depending on the value of the variables. If we consider $P(x)$: '$x^2 > 0$', it's clear that $\vDash P(1)$ but $\nvDash P(0)$.

This naive idea can be made formal by use of the notion of *substitution of variable*. Put succinctly, given a formula $A \in \mathscr{F}$ containing a variable x, and a term $t \in \mathscr{T}$, we write $A(t)$ for the formula A in which every free occurrence of x is replaced by t (under the proviso that no variable becomes captured by a quantifier during the substitution).

Examples

i) Suppose $A(x, y)$ indicates $\big(\exists x \mid \neg(x \doteq y)\big)$. The substitution of y by x produces $A(x, x)$: $\big(\exists x \mid \neg(x \doteq x)\big)$, with completely different meaning from the original. The reason is that y becomes quantified after becoming x.

ii) In the game of chess consider these constants (pieces): $\Bbbk$ (white king), $\mathbf{q}$ (black queen), $\mathbf{p}$ (black pawn), and these formulas:

$$F(x) : \text{'}x \text{ is a piece'} \quad G(x) : \text{ '}x \text{ is a pawn'} \quad B(x) : \text{ '}x \text{ is black'}$$

$$C(x, y) : \text{ '}x \text{ can capture } y\text{'}.$$

In the familiar interpretation of the rules $F(\Bbbk)$, $G(\mathbf{p})$, $C(\mathbf{p}, \Bbbk)$ are trues sentences, whilst $\neg B(\mathbf{q})$, $G(\Bbbk)$, $C(\mathbf{q}, \mathbf{p})$ are false.

Definition Given an implication $A(x) \to B(x)$, an **example** is a term e that makes the hypothesis true, $\vDash A(e)$, and the thesis true, $\vDash B(e)$. In particular, therefore, $\vDash A(e) \to B(e)$.

A **counterexample** is a term c that makes the hypothesis true and the thesis false: $\vDash A(c)$, $\vDash \neg B(c)$, so that $\nvDash A(c) \to B(c)$.

Example Consider the statement

$$\underbrace{\text{if } x \text{ is a positive integer}}_{A(x)}, \text{ then } \underbrace{x^2 + x + 41 \text{ is a prime number}}_{B(x)}.$$

It's not hard to see 1 is an example (both $A(1)$ and $B(1)$ are true), 40 is a counterexample ($A(40)$ is true, but $B(40)$ is false since $40^2 + 40 + 41 = 41^2$). As a matter of fact all of $B(1)$, $B(2), \dots, B(39)$ are true, so fathoming that $B(40)$ might be false is unexpected, and highly counter-intuitive.

Theorem *Let $A(x)$ be a formula and y a free variable in $A(x)$. Then*

$$\vDash (\forall x)A(x) \to A(y), \qquad \vDash A(y) \to (\exists x)A(x).$$

A proper proof of this fact requires the study of variable substitution, so we'll leave it to logic textbooks. But now we know that formula $A(x) \to B(x)$ is false if it admits at least one counterexample.

Example The formula $(\exists x)P(x) \to (\forall x)P(x)$ is false. If we assume x belongs to a set with at least two elements $a \neq b$, we may find a truth function υ for which $P(a)$ is true and $P(b)$ false. Hence $\upsilon\big((\exists x)P(x)\big) = 1$ and $\upsilon\big((\forall x)P(x)\big) = 0$, showing that the initial formula is false.

This tells us that formula $A(x) \to B(x)$ is true if there exist no counterexamples, since in this case any object satisfying the hypothesis must necessarily also satisfy the thesis. Alas, most of the time this is just a nice theoretical observation, for showing the non-existence of counterexamples is a tall order, typically more difficult than proving the implication directly.

Theorem *Let $A(x)$ be a formula, and B a formula where x does not appear as a free variable. Then*

$$\models B \to A(x) \quad implies \quad \models B \to (\forall x)A(x),$$

$$\models A(x) \to B \quad implies \quad \models (\exists x)A(x) \to B.$$

Corollary $\models A(x)$ *if and only if* $\models (\forall x)A(x)$.

Example 1.8 We wish to prove $(\forall x \in \mathbb{R})(\sin^2 x + \cos^2 x = 1)$. To do that we begin by viewing x as an unknown, but fixed, real number, say $\overline{x}$. After proving that $\sin^2(\overline{x}) + \cos^2(\overline{x}) = 1$ for this specific value $\overline{x}$, we argue that since is $\overline{x}$ arbitrary, the statement is proved in general. This passage typically leaves the novice flummoxed, and with good reason: it secretly involves switching from considering the variable free $(\overline{x})$ to non-free (x). Once this is understood, there's no need to introduce any artificial symbol $\overline{x}$ just for the argument's sake, so one sticks with x throughout.

Theorem 1.9 *Let x, y be distinct variables, $A(x), B(x), E(x, y)$ arbitrary formulas, C a formula where x is not free. Then*

i) $(\forall x)(\forall y)E(x, y) \models\mid (\forall y)(\forall x)E(x, y), \ (\exists x)(\exists y)E(x, y) \models\mid (\exists y)(\exists x)E(x, y)$

ii) $\neg(\exists x)A(x) \models\mid (\forall x)(\neg A(x)), \qquad \neg(\forall x)A(x) \models\mid (\exists x)(\neg A(x))$

iii) $(\exists x)(A(x) \vee B(x)) \models\mid (\exists x)A(x) \vee (\exists x)B(x),$
$(\forall x)(A(x) \wedge B(x)) \models\mid (\forall x)A(x) \wedge (\forall x)B(x)$

iv) $\models (\forall x)A(x) \vee (\forall x)B(x) \to (\forall x)(A(x) \vee B(x))$

v) $\models (\exists x)(A(x) \wedge B(x)) \to (\exists x)A(x) \wedge (\exists x)B(x).$

vi) $(\exists x)(C \vee B(x)) \models\mid C \vee (\exists x)B(x), \qquad (\forall x)(C \wedge B(x)) \models\mid C \wedge (\forall x)B(x).$

vii) $(\forall x)(C \vee B(x)) \models\mid C \vee (\forall x)B(x), \qquad (\exists x)(C \wedge B(x)) \models\mid C \wedge (\exists x)B(x).$

Proof Exercise. More relevant than proving these statements is understanding what they mean.   □

Exercise Find counterexamples to invalidate the converses to implications iv)-v) above.

Examples

i) In linear algebra we say that vectors $v_1, v_2, \ldots v_k \in \mathbb{R}^n$ are linearly independent when

$$(\forall \alpha_1) \cdots (\forall \alpha_k) \bigg(\big((\alpha_1 \in \mathbb{R}) \wedge \cdots \wedge (\alpha_k \in \mathbb{R}) \big) \wedge \sum_{i=1}^{k} \alpha_i v_i = 0$$

$$\rightarrow \big((\alpha_1 = 0) \wedge \cdots \wedge (\alpha_k = 0) \big) \bigg).$$

The writing is usually shorted to:

$$\forall \alpha_1, \cdots, \alpha_k \in \mathbb{R} \quad \sum_{i=1}^{k} \alpha_i v_i = 0 \rightarrow \alpha_1 = \cdots = \alpha_k = 0.$$

Hence the v_i aren't linearly independent when

$$\exists \alpha_1, \cdots, \alpha_k \in \mathbb{R} \quad \text{such that} \quad \sum_{i=1}^{k} \alpha_i v_i = 0 \quad \text{and} \quad \exists j \text{ such that } \alpha_j \neq 0.$$

ii) In topology a set $A \subseteq \mathbb{R}$ is called open if

$$(\forall p) \bigg(p \in A \rightarrow \big((\exists \epsilon)(\epsilon > 0 \wedge (p - \epsilon, p + \epsilon) \subseteq A) \big) \bigg).$$

Hence A is not open if $\exists p \in A$ such that $\forall \epsilon, (p - \epsilon, p + \epsilon) \nsubseteq A$.

iii) In analysis a real map f of real variable is called continuous at a if

$$(\forall \epsilon) \bigg(\epsilon > 0 \rightarrow \big((\exists \delta)(\delta > 0 \wedge (\forall x)(|x - a| < \delta \rightarrow |f(x) - f(a)| < \epsilon)) \big) \bigg).$$

Using standard shortcuts, we may say a map is not continuous at a if

$$\exists \epsilon > 0 \text{ such that } \forall \delta > 0 \, \exists x \in \mathbb{R} \text{ such that } |x - a| < \delta \text{ and } |f(x) - f(a)| > \epsilon.$$

The order of the quantifiers in a formula is essential. While '$(\forall a \in \mathbb{R})(\exists b \in \mathbb{R}) \, b > a$' is true, the statement '$(\exists b \in \mathbb{R})(\forall a \in \mathbb{R}) \, b > a$' is patently false. With regard to this, we have

Proposition *If x, y are distinct variables and $A(x, y)$ is any formula,*

$$\vDash (\exists x)(\forall y)A(x, y) \rightarrow (\forall y)(\exists x)A(x, y)$$

Proof Exercise.   □

Example The placing of quantifiers distinguishes two well-known types of convergence for sequences of functions $\{f_n \colon X \rightarrow \mathbb{R}\}_{n \in \mathbb{N}}$ to some f:

pointwise convergence:

$$\forall \epsilon > 0, \underbrace{\forall x \in X\ \exists N}_{N \text{ depends on } x} \text{ such that } |f_n(x) - f(x)| < \epsilon \text{ for all } n > N;$$

uniform convergence:

$$\forall \epsilon > 0 \underbrace{\exists N \text{ such that } \forall x \in X}_{N \text{ doesn't depend on } x} |f_n(x) - f(x)| < \epsilon \text{ for all } n > N.$$

Another convention, to simplify the writing, is to use another existential quantifier, $\exists !$, whose meaning is 'there exists one, and only one', formally defined as follows:

$$(\exists ! x)P(x) \vDash\!\!\dashv (\exists x)P(x) \wedge (\nexists y)\big(P(y) \wedge \neg(x \doteq y)\big)$$

$$\vDash\!\!\dashv (\exists x)P(x) \wedge (\forall y)\big(P(y) \rightarrow (x \doteq y)\big).$$

Example Consider this version of Euclid's famous V Postulate: *for any line r and any point $P \notin r$ on some plane π, there exists a unique line s parallel to r through P*. Its negation reads

$$\forall r \subseteq \pi, \forall P \in \pi \text{ such that } P \notin r \begin{cases} \text{there exists another line } s \parallel r \text{ through } P \\ \qquad \text{or} \\ \text{there exists no line } s \parallel r \text{ through } P \end{cases}.$$

Taking one of the above options as axiom opens the door on non-Euclidean geometries, respectively on *hyperbolic geometry* and *spherical geometry*, see Chap. 18.

Definition 1.10 A set of formulas Γ is called **unsatisfiable** if $\Gamma \vDash \bot$. In other words there are no truth functions for which all formulas in Γ are true.

It's called **satisfiable** if any element in it is true for one (the same) truth function.

Corollary 1.11 *Suppose Γ is a collection of formulas, and A a given formula. Then $\Gamma \vDash A$ if and only if $\Gamma \cup \{\neg A\}$ is unsatisfiable.*

In particular, $\vDash A$ if and only if $\{\neg A\}$ is unsatisfiable.

The idea behind the notion of truth function υ is that of a model of a language. Here it will suffice to say informally that a model $\mathcal{M}$ is a structure that examines semantic elements (meaning and truth) by means of syntactic elements (formulas and proofs). See Sect. 4.3 for more information.

1.4 Deduction ☕

It's a well-known fact that mathematics establishes its truths not like the other sciences, through experiments, but with implications or equivalences. Every mathematical result is always formulated following the same pattern, namely as a theorem, proposition, lemma or corollary (more on the distinction among these later). The notion of theorem is well formalised, see Definition 1.12, just like the notion of proof. Any theorem is made by

- **hypotheses/premises**: for which objects and under which conditions the theorem holds;
- **thesis/conclusion**: which property is true if the hypotheses are satisfied.

This structure agrees with the standard form of any implication $P \to Q$.

A theorem without proof is just a conjecture—hence not a theorem—and therefore it cannot be accepted as true, even though mathematicians would take inspiration from conjectures, 'false' theorems and counterexamples. A proof is a process meant to detect true statements, and indicates how to reach the thesis starting from the hypotheses (cf. the Latin root *de-monstro*). In the choice of steps in a proof it is important to omit irrelevant information, i.e. all hypotheses that are not necessary to get to the thesis. Mathematicians prefer to economise when formulating hypotheses and theorems, and don't like to assume premises so strong that from them would follow much more than what is needed.

The present section aims to present a global point of view on the demonstration process. The latter is based on an axiomatic system (Definition 1.15), i.e. axioms and inference rules. Axioms are propositions we accept without proof nor questioning, as if they were universal truths, and without enquiring about their possible interpretation.

Definition The set $\Lambda \subseteq \mathscr{F}$ of **logical axioms** is formed by these formulas:

i) $x \doteq x$ for every variable x
ii) every tautology $\top$
iii) $A(t) \to (\exists x)A(x)$ for all $A \in \mathscr{F}$
iv) $t \doteq s \to \big(A(t) \leftrightarrow A(s)\big)$ for every variable t, s and every formula $A \in \mathscr{F}$

(We shall not address more technical things such as the dependence of Λ on the language $\mathscr{L}$.)

Since axioms, also known as postulates, cannot (nor should!) be proved, they play a completely different role from theorems.

One also needs **inference rules** $\mathscr{I}$, the 'instructions' prescribing how to prove and allowing to generate theorems starting from propositions one already knows are true. The basic inference rule, called **modus ponens**, asserts that starting from P and $P \to Q$ we can deduce Q, for any formula $P, Q \in \mathscr{F}$:

(**MP**) if the hypothesis P and the implication $P \to Q$ are true, then the thesis Q is true.

Modus ponens is actually a ternary relation with elements $(P, R, Q) \in \mathscr{F} \times \mathscr{F} \times \mathscr{F}$ where $R \doteq (P \to Q)$. One says Q is **obtained by MP from** P, R.

Definition 1.12 A (logical, formal) **proof** is a finite sequence of formulas $A_1, \ldots A_n \in \mathscr{F}$ such that every A_k either belongs to the set of logical axioms Λ or is obtained by MP from A_i, A_j, $j, i < k$.

We call **theorem** any formula A for which there exists a proof $A_1, \ldots, A_n$ ending in $A \doteq A_n$, and we write

$$\vdash A.$$

(The symbol $\vdash$ is called entailment.) If $\Gamma \subseteq \mathscr{F}$, we call **deduction from Γ** a proof $A_1, \ldots, A_n$ where $A_i \in \Lambda \cup \Gamma$ or A_i is obtained by MP from Γ. The last formula in a proof is said to be **deducible** from Γ (or **logical consequence** of Γ, or a **theorem** of Γ):

$$\mathfrak{Thm}_\Gamma = \Gamma \cup \Lambda \cup \{A \mid \Gamma \vdash A\}.$$

Hence, a theorem is a formula deducible from $\varnothing$. In these terms MP rephrases as

$$\{A, A \to B\} \vdash B, \quad \text{or} \quad A \wedge (A \to B) \vdash B. \tag{1.2}$$

Exercises (Other Inference Rules) Prove that

i) $\Gamma \vdash A$, $A \vdash B$ imply $\Gamma \vdash B$ *(transitivity of deduction)*
ii) if $\Gamma' \subseteq \Gamma$ and $\Gamma' \vdash A$ then $\Gamma \vdash A$ *(inflation of hypotheses)*
iii) $\vdash A$ implies $\Gamma \vdash A$ for any Γ
iv) $\vdash A \to B$ implies $\vdash (\exists x)(A \to B)$ for any $x \in \mathscr{V}$ that is not free in B.
                                                          (existence of examples)

Throughout the text the reader will encounter several proof techniques, both direct and indirect. A direct proof reaches the conclusion by means of a chain of steps, as was mentioned above. An indirect method is usually based on contradiction (Corollary 1.22).

At any rate, any proven statement is a theorem. Therefore the difference among a theorem, a proposition, a lemma and a corollary is not clear-cut, and left to the taste of the single writer. Traditionally (mainly historically) there exist 'fundamental' theorems, thus called for the central role they play in certain theories: for example, the *fundamental theorem of calculus*, the *fundamental theorem of algebra* (Theorem 13.8), the *fundamental theorem of arithmetics* (Theorem 6.23), the *fundamental theorem of Riemannian geometry* etc.

Among all theorems one calls **lemma** an auxiliary proposition, customarily needed in the proof of another proposition, or a preliminary fact. The most famous examples are the lemmas attached to the names of *Zorn, Schur, Urysohn, Lindelöf, Margulis, Morse, Hartogs, Riesz, Schwarz, Abel, Nakayama, Poincaré, Fatou, Yoneda*, the *'five' lemma*, the *'snake' lemma*, and at least 3 *Gauß lemmas* (!).

The aforementioned are so important that it's universally agreed upon it's actually reductive to name them lemmas, yet tradition is strong. In this book we shall use **proposition** to denote a theorem slightly less crucial, albeit equally deserving of mention. A **corollary** is a statement that follows, sometimes immediately, from another fact.

We shall discuss now a few 'metatheorems' (theorems in the metalanguage) that are extremely valuable to be able to work.

The reciprocal assertion to modus ponens (1.2) is

Theorem 1.13 (Deduction Theorem) *For every subset* $\Gamma \subseteq \mathscr{F}$ *and formulas* $A, B \in \mathscr{F}$ *we have*

$$\Gamma, A \vdash B \text{ implies } \Gamma \vdash A \to B.$$

In case Γ *is empty:* $A \vdash B$ *implies* $\vdash A \to B$.

Proof We shall postpone the justification to p. 77, after we have introduced tools to simplify arguments. It must be said that there are proofs not involving Moore operators, which are though rather lengthy and technical, and wouldn't add much to the present understanding. □

More generally,

Corollary *For any formulas* $A_1, \ldots, A_n, B$, *we have that* $\{A_1, \ldots, A_n\} \vdash B$ *if and only if* $\vdash A_1 \to (A_2 \to (\cdots (A_n \to B) \cdots))$.

Proof By induction on n. □

Demonstrations may also be classified as 'constructive' or 'non-constructive'. A paramount feat of mathematics is proving the existence of an object without describing it explicitly, without a formula so to speak. For instance, it is important to know that every bijection has an inverse map, although this same statement is too general to provide an expression for the inverse to a generic map. Proofs that use the axiom of choice are in principle non-constructive, because it's not possible to get hold of the required choice function explicitly in most cases. One only knows such a map exists, and proceeds with the proof. Hence in practical applications non-constructive proofs have their limitations, simply because often it becomes necessary to describe an object concretely to carry out a serious study. On the other hand it might be instructive and useful to know beforehand that a certain equation admits at least one solution, to compute estimates by numerical methods. This is perfectly exemplified by the *theorem of existence of zeroes of continuous maps*. And even when a formula does exist, as in the case of *Cramer's rule* for $n \times n$ linear systems, implementing the solving formula in the computer can be complicated if the system is large, say $n > 10^{10}$ (these cases do occur, and are actually the most frequent ones). If so, the theory guarantees the existence of a solution, which for practical reasons we may want to approximate and compute in an easier way.

There are several methods to prove certain kinds of statements. One is the so-called *proof by induction*, which we'll encounter in Sect. 6.2. A second type is the *case-by-case proof* 1.14, whereby one proves a statement about a (perhaps long) list of objects. The argument can proceed by examining every possible case occurring in the list. For example, a property relative to real, associative division algebras of finite dimension can be proved by verifying it on the three possible instances, i.e. $\mathbb{R}, \mathbb{C}, \mathbb{H}$.

Proposition 1.14 (Case-by-Case Proof) *If $\Gamma \vdash A_1 \vee \ldots \vee A_n$, and $\Gamma \vdash A_i \to B$ for all $i = 1, \ldots, n$, then $\Gamma \vdash B$.*

Example Let's show there don't exist natural numbers n, m such that $n^2 - m^2 = 10$. Note that the list of naturals numbers is infinite, so any attempt to consider all pairs $(m, n) \in \mathbb{N}^2$ would never end. Hence, let's decompose $n^2 - m^2 = (n + m)(n - m)$ and consider the possible cases for $n + m, n - m$ giving 10 as a product. This is easy (and ends in finite time) since the only divisors of 10 are $\pm 10, \pm 5, \pm 2, \pm 1$. For example, if $n + m = -10$ then $n - m = -1$, resulting in $n = -11/2 \notin \mathbb{N}$, thus excluding this possibility. The other cases are completely analogous.

This sort of proof is not elegant and very often quite heavy going, especially when the list is long. (However, remember than a complicated and inelegant proofs is preferable to no proof at all.) Certain results required years, or centuries, before they got a proof that wasn't based on the examination of all possible cases. The math community is still awaiting a theoretical argument for the four-colour Theorem 18.27.

Theorem (Generalisation) *Assume $\Gamma \vdash A$ and x is a non-free variable in every formula of Γ. Then $\Gamma \vdash (\forall x)A$.*

Example 1.8 will give an application of the generalisation theorem.

Other methods exist that apply to specific situations or theorems, and don't have the same importance of the processes mentioned so far. Every branch of mathematics has developed over time its own proving techniques, which are learnt automatically upon studying the corresponding fields. At the end of the day, refining one's own taste and developing the perception of which strategies are appropriate in which situations is one of the soft skills of a researcher in mathematics.

Definition 1.15 A **formal theory of first-order** $\mathbb{T}$ consists of

- a first-order language $\mathscr{L}$
- a set $\Lambda_{\mathbb{T}}$ of axioms containing the logical axioms Λ
- a set of inference rules $\mathscr{I}$
- a non-empty collection of theorems $\mathfrak{Thm}_{\mathbb{T}} \supseteq \Lambda_{\mathbb{T}}$ that is closed under $\mathscr{I}$, meaning:

$$A \in \mathfrak{Thm}_{\mathbb{T}} \text{ if and only if } \mathfrak{Thm}_{\mathbb{T}} \vdash A.$$

Besides the logical axioms, a theory typically requires other specific axioms ($\Lambda \subsetneq \Lambda_{\mathbb{T}}$). Notable examples are Zermelo–Fraenkel set theory (Chap. 2), plane geometry (Chap. 18), Dedekind–Peano arithmetics (Sect. 6.1), non-standard arithmetics (p. 90), and any algebraic theory (groups, rings, fields, lattices ...). See Sect. 4.3 for more details.

There also are theories of higher order, such as topological theories, category theory, topos theory... Actually, any branch of mathematics is made of various interacting theories.

Definition 1.16 A set $\Gamma \subseteq \mathscr{F}$ is **inconsistent** or **contradictory** when $\Gamma \vdash \bot$. Otherwise it's called **consistent**, or **sound**. ($\bot \notin \mathfrak{Thm}_\Gamma$).

Corollary 1.17 *The singleton $\{A\}$ is consistent if and only if $\neg A$ is not a theorem, i.e. $\neg A$ cannot be deduced (in formulas: $\nvdash \neg A$).*[1] *More generally, if Γ is consistent and $C \in \Gamma$, then $\Gamma \nvdash \neg C$.*

When a set Γ is contradictory, the set of its theorems $\mathfrak{Thm}_\Gamma$ coincides with the whole $\mathscr{F}$, because we can prove any statement whatsoever from Γ, see Remark 1.23.

1.5 Soundness and Completeness ☕

The heart of this section are two theorems that explain the relationship between the notions of **soundness** (or logical validity) and **completeness**. Soundness conveys the idea of correctness, in other words that any provable statement is true. Completeness is the opposite, namely that a truth can be proved. It justifies phrases like *'it's true that X'* to mean *'X can be proved'* i.e. *'X is a theorem'*. The great power of the completeness theorem resides in the fact it's often easier to establish whether a formula is true or false, than to prove it.

We'll defer the proof of the soundness theorem to p. 77.

Theorem 1.18 (Soundness of First-Order Logic) *In a language of first order, any proven statement A is true:*

$$\vdash A \quad implies \quad \vDash A.$$

Regarding the completeness theorem, although it is fundamental in first-order logics, its formal proof (by Gödel in 1922, with earlier significant contributions by Skolem) is not essential in order to formulate many related concepts. Hence we shall not provide a proof.

Theorem 1.19 (Completeness of First-Order Logic) *In a language of first order, any true statement A is a theorem (it can be proved):*

$$\vDash A \quad implies \quad \vdash A.$$

---

[1] Beware not to confuse $\nvdash$ with $\forall$ when hand-writing.

Using Corollaries 1.17 and 1.11 we may reformulate the completeness theorem by saying that any consistent set of formulas is satisfiable. Or, in one word, first-order logic is **complete**. This means that the syntax captures and controls the semantics (if we accept the contents of Γ). If a formula A isn't a theorem, we can produce an inconsistent theory by adding, as axiom, any formula coming from substituting arbitrary formulas in place of the variables of A.

Corollary *Predicative calculus and propositional calculus (of first order) are consistent theories.*

Corollary $\Gamma \subseteq \mathscr{F}$ *is unsatisfiable if and only if it is inconsistent, if and only if it contains a contradiction.*

Observe that the existence of incomplete formal systems is, on its own, not that surprising. A system might be incomplete simply because not all the necessary axioms have been discovered. For instance, Euclidean geometry without the V postulate is incomplete: it's not possible to prove or disprove the postulate starting from the other axioms (i.e., within the theory).

With regard to (1.2), now we have

Theorem 1.20 (Modus Ponens) $\vdash \big(A \wedge (A \to B)\big) \to B.$

Proof The implication $\big(A \wedge (A \to B)\big) \to B$ is tautological. Hence Theorem 1.19 allows to conclude.   □

Theorem 1.21 (Principle of Non-contradiction) *The formula $A \wedge \neg A$ is never a theorem, for any A.*

Only within inconsistent theories, therefore, we simultaneously have $A, \neg A \in \mathfrak{Thm}$ (so $\vdash A$ and $\vdash \neg A$), cf. (1.1). Using the tautology $(\neg A \to \bot) \to A$ (cf. Definition 1.16) we may rephrase the principle of non-contradiction as follows:

Corollary 1.22 (Proof by Contradiction) *Let $A \in \mathscr{F}$, $\Gamma \subseteq \mathscr{F}$. Then $\Gamma \vdash A$ if and only if $\Gamma \cup \{\neg A\}$ is inconsistent, that is*

$$\Gamma, \neg A \vdash \bot \quad \text{implies} \quad \Gamma \vdash A.$$

Now, recall we showed in Proposition 1.4, ii) that

$$\underbrace{P \to Q}_{\text{implication}} \; \dashv\vdash \; \neg P \vee Q \; \dashv\vdash \; \underbrace{\neg Q \to \neg P}_{\text{contrapositive}} \tag{1.3}$$

are logically equivalent formulas. Then on the one hand we have

Corollary (Proof by Contraposition) *Formula $(P \to Q)$ is deducible if and only if $(\neg Q \to \neg P)$ is deducible.*

Proof Using soundness (Theorem 1.18) and completeness (Theorem 1.19) we may equivalently show that $P \rightarrow Q$ is true iff $\neg Q \rightarrow \neg P$. But this is precisely (1.3). □

On the other hand, Corollary 1.22 delivers at our feet yet another method of proof, one that essentially adapts Theorem 1.20 to the contrapositive statement $\neg Q \rightarrow \neg P$. The recipe is referred to, in fancy terms, as *reductio ad absurdum* [reduction to the impossible]

$$\vdash \big((P \rightarrow Q) \wedge \neg Q\big) \rightarrow \neg P \qquad \text{(technically called } modus\ tollens\text{)}$$

and is based on **refutation**: the implication is initially assumed true, $\vDash P \wedge \neg Q$. Then using valid arguments one reaches facts that contradict one another, say $P \wedge \neg P$ or any deducible contradiction. (We opted to signal this situation with the *non sequitur* symbol ⚡.) Now, since logical fallacy is not allowed by Theorem 1.21, we conclude it's incorrect to presume $P \rightarrow Q$ was true. Therefore $\vDash P \rightarrow Q$. (Due to the completeness/soundness theorems we may interchange $\vDash$ and $\vdash$ as we please.)

Examples

1) Prove that *If m is an integer and m^2 is even, then m is even.* Let's suppose, by contradiction, the statement is false. Then there must be a counterexample, hence an m satisfying the hypothesis and not the thesis. In other words there is an integer m such that m^2 is even but m is odd. Now, since m is odd, there is a k such that $m = 2k + 1$ and so $m^2 = (2k + 1)^2 = 2(2k^2 + 2k) + 1$. Consequently m^2 is odd ⚡: an integer number cannot be even and odd at the same time (contradiction). Therefore it's wrong to suppose the initial proposition is true, and we must conclude it is false.

ii) Let's show that *among $n \geqslant 2$ random people hobnobbing at a dinner party, there are at least two with the same number of acquaintances.* The possibilities for the number of acquaintances are $0, 1, 2, \ldots, n - 1$ (so n possibilities in total). Suppose, by contradiction, none of the n people knows the same number of acquaintances. This means every person has a different number of acquaintances. Since there are n participants and n possibilities for the number of acquaintances, necessarily person p_0 must know no one (0 acquaintances), person p_1 must know 1 other person only, p_2 knows 2 other people etc. Then person p_{n-1} has $n - 1$ acquaintances, meaning he knows everybody, including p_0. But this contradict the fact that p_0 knew nobody ⚡. An alternative argument will be provided in Example 6.37.

These examples show the usefulness of Definition 1.10 in proving a set Γ of formulas is unsatisfiable. Such an argument follows the same model of a straightforward proof in all aspects except for one: when proving a certain formula holds, the last one in the sequence, the conclusion, is known beforehand. In the general proof of unsatisfiability the final formula can be any contradiction.

Remark 1.23 (Explosion Principle) If $A \models\mid \bot$ is a contradiction we know that $\vdash A \to B$ for any formula B. Hence we can prove any statement starting from a set containing contradictions. The Latin *ex falso quodlibet* [from the false anything follows] leaves no doubt:

$$\bot \to Q \text{ for any formula } Q.$$

The statement '*if* $x \in \mathbb{R}$ *and* $x^2 < 0$, *then* $x = 42$', for instance, is true (!), because its premise is false. In fact, there exists no x satisfying the hypothesis, and we can conclude anything we want, even that global warming doesn't exist.

Exercises Prove the following statements by the method(s) of your choosing:

i) Let n be a positive integer. If $2^n - 1$ is prime, n is prime.
ii) If a is an odd number, the quadratic equation $x^2 - x - a = 0$ doesn't have integer solutions.
iii) For $m, n \in \mathbb{N}$, if $mn = 100$ then $m \leqslant 10$ or $n \leqslant 10$.
iv) If $n \in \mathbb{Z}$ can be written as sum of two odd integers, then n is even.
v) For any real $\epsilon > 0$, there exists an $N \in \mathbb{N}$ such that $\frac{1}{n} < \epsilon$ for every $n \geqslant N$.
vi) Suppose $m, n \in \mathbb{Z}$ are such that mn is even. Then m is even or n is even.
vii) What can we say about the subset $X \subseteq \mathbb{N}$ knowing that:

   – for every $x \in X, x > 10$;
   – if there exists a $y \in X$ such that $y > 20$, then $5 \in X$;
   – for every $x \in X$ we have: x odd $\iff x > 25$?

To summarise, and speaking loosely, the completeness theorem for first-order theories (hence, consistent ones) establishes a thesaurus between the world of grammar and the world of meanings:

syntax	$\Gamma \vdash A$	deducible	$\vdash \neg A$	inconsistent	consistent
semantics	$\Gamma \models A$	true	$\models \neg A$	contradiction	satisfiable

Exercises Prove

i) $A \wedge (A \vee B) \models\mid A \models\mid A \vee (A \wedge B)$ (*absorption*)
ii) $A \wedge A \models\mid A$, $A \models\mid A \vee A$, $A \models\mid \neg\neg A$ (*idempotence*)
iii) $P, Q \vdash (P \wedge Q)$ $(P \wedge Q) \vdash Q$ $P, Q \vdash (P \vee Q)$
iv) $(B \to C) \models\mid (\neg C \to \neg B)$
v) $A \models\mid B \vee A$ if and only if $\vdash B \to A$ if and only if $A \wedge B \models\mid B$
vi) $\neg(P \wedge Q) \models\mid (\neg P \vee \neg Q)$ $\neg(P \vee Q) \models\mid (\neg P \wedge \neg Q)$ (*De Morgan laws*)
vii) $(P \leftrightarrow Q) \models\mid (Q \leftrightarrow P) \models\mid ((P \wedge Q) \vee (\neg P \wedge \neg Q))$
viii) $A \wedge B \models\mid \bot$ if and only if $\vdash A \to \neg B$ if and only if $\vdash B \to \neg A$
ix) $\{P \to Q, Q \to P\} \vdash (P \leftrightarrow Q)$ $(P \leftrightarrow Q) \vdash (P \to Q), (Q \to P)$
x) $((P \to Q) \wedge (Q \to R)) \vdash (P \to R)$ (*syllogism*)

Remark 1.24 Establishing whether a formula F is a theorem in a given theory is a very hard problem, and normally one cannot decide it beforehand, by some 'higher' theoretical argument. At the same time there exist formulas that cannot be proved in any way. Even worse, for a certain viewpoint, there exist formulas F such that neither F nor $\neg F$ can be proved. A statement of the latter kind, which is not provable nor refutable, is said to be **undecidable**. The are countless examples, in every area of maths. Some rather famous ones are:

- the parallel postulate, see Chap. 18
- the axiom of choice in ZF theory, see Sect. 3.4
- the continuum hypothesis in ZFC theory (Theorem 7.29)
- the existence of Whitehead group (proved by Shelah in 1974) ☞ *algebra/algebraic topology*
- deciding whether a polynomial $p \in \mathbb{Z}[x_1, \ldots, x_n]$ has integer roots (*10th Hilbert problem* [55], proved by Matiyasevich in 1970) ☞ *Diophantine geometry*
- deciding whether a given formula is provable in a theory (*Hilbert's Entscheidungsproblem / Church–Turing thesis* (1936)) ☞ *recursion theory* (see p. 90)
- deciding whether two topological manifolds of dimension ≥ 4 are homeomorphic ☞ *topology*
- the stopping problem for Turing machines (1936) / Rice theorem (1951) ☞ *computability theory*.

There exist entire theories that are undecidable, the most prominent probably being arithmetics (*Church's undecidability theorem*, p.90). The references [21, 78] are a good source of undecidable propositions.

To finish, and counterbalance the emphasis on formalisation of this book, there's no better way than quoting the great René Thom, according to whom in any creative activity *"whatever is rigorous is insignificant."*

Notation: from this point onwards we shall write

$\implies$ to denote the conditional connective $\to$, or any if-then implication
$\iff$ for the biconditional connective $\leftrightarrow$, or any if-and-only-if equivalence.

The symbol $\to$ will be reserved for functions.

Chapter 2
Naive Set Theory

All basic subjects rest heavily on the intuitive theory elaborated by Frege and Cantor known as 'naive' set theory. It's called naive because not long after its foundation it underwent a profound crisis, caused by the discovery of famous antinomies which showed its inadequacy. This fact led to the creation of **ZF theory**, after Zermelo and Fraenkel (and major contributions by often forgotten Skolem). ZF theory is a deductive system with 8 specific axioms, called (just out of curiosity) *extensionality, infinity, pairing, union, power set, regularity, specification schema, replacement schema*. We shall work within the theory called **ZFC**, which is ZF theory augmented with the *axiom of choice*, dealt with in Sect. 3.4). These axioms allow to describe sets either by explicitly listing their members

$$X = \{ ⚔, ♉, ☿, ♠, ⚓, ♇, \dots \},$$

or through a predicate

$$X = \{ x \mid p(x) \}.$$

The latter should be read as '*X is the set of all elements x such that* $\models p(x)$ *(property p holds/is true)*'. For example:

$$\{ x \mid x \in \mathbb{Z} \wedge (-8 < x^3 \leqslant 35) \} = \{-1, 0, 1, 2, 3\}.$$

A **set** is a primitive notion, just like the relation $\in$ ('belongs to', 'is an element of'[1]) and $\doteq$ ('equals'). 'Primitive' means undefined, since these concepts are justified only informally, usually by appealing to the intuition and day-to-day experience. Since definitions are stated so that each new object depends on

---

[1] The symbol $\in$ is a calligraphic variant of the Greek letter ε. Peano chose it because it reminded him of the verb ἐστί[to be], and he read '$x \in y$' as 'x is a y'.

© The Author(s), under exclusive license to Springer Nature Switzerland AG 2021
S. G. Chiossi, *Essential Mathematics for Undergraduates*,
https://doi.org/10.1007/978-3-030-87174-1_2

previously introduced objects, and the latter depend on objects introduced before them and so forth, to define anything we would need to go backwards along the ancestral line without end. This is the 'infinite regress' much disliked by mathematicians and many philosophers alike. Lest we risk incurring in it, no formal theory (Definition 1.15) can shun primitive notions. Other examples of primitive concepts include the notions of 'point' and 'lying between' in Euclidean geometry (Chap. 18), or 'zero' and 'successor' in Peano's arithmetics (Sect. 6.1).

In practice a semi-heuristic approach will suffice for us. Axioms and primitive notions will be used implicitly, because here we shan't give a rigorous axiomatic presentation, but intuitive and simplified, including merely the material used in the sequel. Once we accept a naive theory, we'll also go along with having a 'universal' set $\mathscr{U} := \{x \mid x \doteq x\}$ (technically, a Grothendieck universe). This is not really necessary for a treatise of our level, but it simplifies life when writing certain expression (and didactically speaking it's no big leap). Remark 6.44 will provide a little more information in this respect.

2.1   The Algebra of Sets

The foundations of set theory as we present them here were discovered and developed mainly by Cantor and Dedekind, around 1870. Whilst Cantor's work is well known, the influence of Dedekind, here and in other areas, is less so [73].

Definition Let A, B be sets. One says A is a **subset** of B, written $A \subseteq B$, if $x \in A \implies x \in B$ for any element x.

The symbol $A \subset B$, or $A \subsetneq B$, indicates A is a **proper** subset of B: $(A \subseteq B) \wedge (A \neq B)$, or equivalently:

$$A \subset B \iff A \subseteq B \wedge \exists t \in B \text{ such that } t \notin A.$$

Proposition 2.1 *The relation $\subseteq$, called **inclusion**, satisfies three properties:*

$$A \subseteq A, \qquad A \subseteq B \wedge B \subseteq A \implies A = B, \qquad A \subseteq B \wedge B \subseteq C \implies A \subseteq C$$

for all sets A, B, C. (In the language of Definition 2.9, the relation $\subseteq$ is a partial order.)

Proof Exercise (think about the properties of logical connectives).   □

To represent a set 'without elements' we'll use the following artifice: we call $\varnothing := \{x \mid x \neq x\}$ the **empty set**. The idea is that a putative element of $\varnothing$ would have to be different from itself, and hence should not exist. An equivalent characterisation, actually making for a proper definition, is the content of the next statement:

Proposition *The empty set $\varnothing$ is the unique set with the property that $\varnothing \subseteq A$ for any set A.*

Proof First of all let's prove that $\varnothing$ is contained in any set. To do that suppose, by contradiction, $\varnothing$ is not contained in one particular set A. This means $\varnothing$ contains some element x that doesn't belong in A. But $\varnothing$ does not contain elements, so there cannot be any such x.

Now let's prove uniqueness: suppose E is another set such that $E \subseteq A$ for every set A. Taking $A = \varnothing$ in particular, we have $E \subseteq \varnothing$. But $\varnothing \subseteq E$ by what we have proved above. Hence Proposition 2.1 implies $E = \varnothing$. $\qquad\square$

If we are given two sets A, B we can generate other sets by means of set-theoretical operations:

Definition 2.2 We define the following sets/operations between sets:

$A \cup B := \{ x \mid x \in A \lor x \in B \}$ is the **union** of A and B

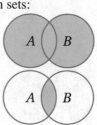

$A \cap B := \{ x \mid x \in A \land x \in B \}$ is the **intersection** of A and B

$A \setminus B := \{ x \mid x \in A \land x \notin B \}$ is the **difference** of A and B

$A \triangle B := (A \setminus B) \cup (B \setminus A)$ is the **symmetric difference** of A and B, a.k.a. **Boolean sum**.

Venn diagrams (the pictures on the right) are an effective way to represent these sets visually.

Two sets A, B with empty intersection $A \cap B = \varnothing$ are called **disjoint**. In that case their union is called **disjoint union** $A \sqcup B$. Another way of saying this is that A, B form a **partition** of $A \cup B$, cf. Definition 2.6.

Definition The **set-complement** to A is the set

$$\overline{A} := \mathscr{U} \setminus A = \{ x \mid x \notin A \}$$

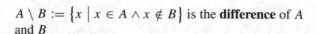

(sometimes written $\complement A$).

Proposition 2.3 *The operations union, intersection, difference defined above satisfy the following properties:*

i) $A \cap B \subseteq A \subseteq A \cup B$
ii) $A \cup B = B \cup A, \qquad A \cap B = B \cap A$
iii) $(A \cup B) \cup C = A \cup (B \cup C) \qquad (A \cap B) \cap C = A \cap (B \cap C)$
iv) $A \cup (B \cap C) = (A \cup B) \cap (A \cup C) \qquad A \cap (B \cup C) = (A \cap B) \cup (A \cap C)$

$v)$ $A \cup \varnothing = A,$ $A \cap \mathcal{U} = A$
$vi)$ $A \cup \overline{A} = \mathcal{U},$ $A \cap \overline{A} = \varnothing$ *(that is, $A \sqcup \overline{A} = \mathcal{U}$)*

for every A, B, C.

Proof We'll only prove the first formula in iv), leaving the rest as an exercise. For any x we have

$$x \in A \cup (B \cap C) \overset{\text{defn}}{\Longleftrightarrow} (x \in A) \vee (x \in B \cap C)$$

$$\overset{\text{defn}}{\Longleftrightarrow} (x \in A) \vee \big((x \in B) \wedge (x \in C)\big)$$

$$\overset{1.7}{\Longleftrightarrow} \big((x \in A) \vee (x \in B)\big) \wedge \big((x \in A) \vee (x \in C)\big)$$

$$\overset{\text{defn}}{\Longleftrightarrow} \big(x \in A \cup B\big) \wedge \big(x \in A \cup C\big)$$

$$\overset{\text{defn}}{\Longleftrightarrow} x \in (A \cup B) \cap (A \cup C).$$

$\square$

The associative property 2.3, iii) permits to understand $A \cup B \cup C$ and $A \cap B \cap C$ without ambiguity.

Exercises Show that for any set $A, B, C,$

i) $A \cap (A \cup B) = A = A \cup (A \cap B)$ ii) $A \cap A = A,$ $A = A \cup A,$ $A = \overline{\overline{A}}$

iii) $A \setminus \varnothing = A,$ $A \setminus A = \varnothing$ iv) $B \subseteq C \Longrightarrow A \setminus C \subseteq A \setminus B$

v) $A = B \cup A \iff B \subseteq A \iff A \cap B = B$ vi) $\overline{A \cup B} = \overline{A} \cap \overline{B},$ $\overline{A \cap B} = \overline{A} \cup \overline{B}$

vii) $A \setminus B = A \setminus (A \cap B) = A \cap \overline{B}$ viii) $A \cap B = \varnothing \iff A \subseteq \overline{B} \iff B \subseteq \overline{A}$

ix) $A \bigtriangleup B = \varnothing \iff A = B$ x) $A \bigtriangleup B = (A \cup B) \setminus (A \cap B)$

xi) $A \bigtriangleup B = B \bigtriangleup A$ xii) $(A \bigtriangleup B) \bigtriangleup C = A \bigtriangleup (B \bigtriangleup C)$

xiv) $A \bigtriangleup \varnothing = A,$ $A \bigtriangleup A = \varnothing.$

This exercise shows, on purpose, that intersection and complementation are enough to define union and difference of sets.

Exercise Consider the three sets X (red), Y (blue) and Z (green) in the figure below. Using set-theoretical operations describe the regions:

$$A, \quad E, \quad I, \quad E \cup D, \quad E \cup D \cup F, \quad E \cup D \cup F \cup I,$$

$$A \cup B \cup C, \quad A \cup B \cup E, \quad A \cup B, \quad C \cup D \cup F, \quad A \cup B \cup C \cup I$$

in terms of X, Y, Z only, and in the most 'economical' possible way.

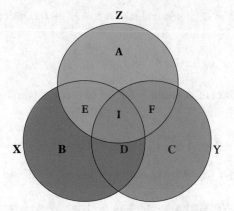

Definition 2.4 The collection of subsets of a set A is called **power set**

$$\mathscr{P}(A) := \big\{ B \subseteq \mathscr{U} \mid B \subseteq A \big\} =: 2^A.$$

Note that the power set isn't empty since $\varnothing, A \in \mathscr{P}(A)$. The reason for the funny notation 2^A will become clear later.

Example Let's take 4 billiard balls $\{①, ②, ③, ④\} = A$. The power set

$$\mathscr{P}(A) = \big\{ \varnothing, \{①\}, \{②\}, \{③\}, \{④\}, \{①, ②\}, \{③, ①\}, \{①, ④\}, \{②, ③\}, \{③, ④\},$$
$$\{②, ④\}, \{①, ②, ③\}, \{①, ②, ④\}, \{①, ④, ③\}, \{②, ④, ③\}, \{①, ②, ③, ④\} \big\}$$

has $16 = 2^4$ elements.

Exercises Prove that for any set A, B

i) $a \in A \Longrightarrow \{a\} \in \mathscr{P}(A)$ ii) $A \subseteq B \Longrightarrow \mathscr{P}(A) \subseteq \mathscr{P}(B)$

iii) $\mathscr{P}(A \cap B) = \mathscr{P}(A) \cap \mathscr{P}(B)$ iv) $\mathscr{P}(A) \cup \mathscr{P}(B) \subseteq \mathscr{P}(A \cup B)$

v) $\mathscr{P}(A) \cap \mathscr{P}(\overline{A}) = \{\varnothing\}.$

Find an example showing $\mathscr{P}(A) \cup \mathscr{P}(B) \not\supseteq \mathscr{P}(A \cup B)$.

Some of the previous properties generalise to a finite number of sets. Typically when a collection of sets $A_1, A_2, \ldots, A_n$ is finite, the results are morally the same as the above ones, and can be proved by induction on n. But when n is infinite, something very different might happen, see p. 69.

2.2   Binary Relations

In this section we shall not discuss the complete theory of binary relations (although this would probably help with functions, later) but we'll concentrate on two types of relation that are important in the daily practice: equivalence relations and order relations. We begin by defining the operation of product among sets.

Definition Let $A, B \neq \varnothing$ be sets and take $a \in A, b \in B$. One calls **(ordered) pair** the set

$$(a, b) := \{\{a\}, \{a, b\}\}.$$

The **Cartesian product** of A and B is the set of ordered pairs:

$$A \times B := \{(a, b) \mid a \in A, b \in B\} \subseteq \mathscr{P}(A \cup B).$$

Consequently the pair (a, b) is equal to the pair (c, d) if and only if $a = c$ and $b = d$.

In case $A = B$ the Cartesian product is denoted $A^2 := A \times A$. It is rather immediate to check that

Proposition 2.5 *The Cartesian product $A \times B$ is empty $\iff A = \varnothing$ or $B = \varnothing$.*

Proof Exercise (use contradiction).   □

Corollary *The Cartesian product is distributive with respect to the operations $\cap, \cup, \setminus$, i.e. $A \times (B \cup C) = (A \times B) \cup (A \times C)$ etc.*

Proof Exercise.   □

Exercises

i) Suppose $A, B \neq \varnothing$. Prove that $A \times B \subseteq C \times D$ if and only if $A \subseteq C, B \subseteq D$. Explain with an example why the hypothesis that A, B are non-empty cannot be removed.

ii) Show that the Cartesian product is not commutative: $A \times B \neq B \times A$, nor associative: $(A \times B) \times C \neq A \times (B \times C)$.

The Cartesian product is the simplest instance of an *incidence structure* (X, L), given by a set X and a subset $L \subseteq \mathscr{P}(X)$ of 'blocks'. Another important class of examples is

Definition 2.6 A **partition** of a set X is an incidence structure $\left(X, L = \{X_i\}_{i \in I}\right)$ whose blocks $X_i \neq \varnothing$ are non-empty, pairwise disjoint, and they exhaust X:

$$X = \bigsqcup_{i \in I} X_i.$$

Examples

i) The set $X = \{\female, \male, \female\male\}$ can be partitioned in 5 different ways:

$$\{\female\} \cup \{\male\} \cup \{\female\male\}, \quad \{\female\} \cup \{\male, \female\male\}, \quad \{\male\} \cup \{\female, \female\male\},$$
$$\{\female\male\} \cup \{\male, \female\}, \{\female, \male, \female\male\}$$

The first partition (made of singletons) and last one (made of one block) are considered 'trivial'.

ii) The plane $X = \mathbb{R}^2$ admits an infinite partition made of squares $X_n = [n, n + 1) \times [n, n + 1)$ labelled by $n \in \mathbb{Z}$.

iii) The punctured plane $\mathbb{R}^2 \setminus \{(0, 0\}$ is partitioned by all the rays emanating from the origin:

$$X_m = \{(x, mx): x \in \mathbb{R}^+\}, \quad m \in \mathbb{R}; \ X_\uparrow = \{(0, y): y \in \mathbb{R}^+\};$$
$$X_\downarrow = \{(0, -y): y \in \mathbb{R}^+\}.$$

Exercise

i) Let $X = \bigsqcup_{i \in I} X_i$ be the partition of a finite set. Prove by induction the addition formula $|X| = \sum_{i \in I} |X_i|$. (The generalisation of this is relation (6.5).)
The above formula holds in case I is infinite, too, and will be justified in Sect. 7.7.

ii) How many partitions can one define over a set with $n = 0, 1, 2, 3, 4$ elements?

The notion of partition allows us to lead a little astray and talk about counting problems. The number of partitions over a set with n elements is called *Bell number* B_n:

$$B_0 = 1, \ B_1 = 1, \ B_2 = 2, \ \ldots,$$
$$B_{42} = 35742549198872617291353508656626642567, \ \ldots$$

Bell numbers are important in ☞ *number theory* and *computer science*, for instance to determine all possible factorisations of an integer. The number 66 possesses three distinct prime factors 2, 3, 11, and can be written in $B_3 = 5$ different ways, up to rearranging factors: $2 \cdot 3 \cdot 11 = 2 \cdot 33 = 3 \cdot 22 = 11 \cdot 6 = 66$. The number 200 has, besides the prime factorisation $2^3 \cdot 5^2$, further three (essentially) distinct decompositions $4 \cdot 50 = 2 \cdot 100 = 40 \cdot 5$. But note $B_2 = 2$. In general, if N is a square-free integer with n prime factors, then B_n gives the number of prime factorisations for N (up to the order of factors). To determine B_n when n is small one can use Bell's triangle (Fig. 2.2), a table of integers $x_{i,j}$ arranged so that:

• $x_{1,1} = 1$;
• each row begins with the last number on the previous row $x_{i,1} := x_{i-1,i-1}$;
• after the first, each element is the sum of the number to its right and the one above the latter: $x_{i,j} := x_{i,j-1} + x_{i-1,j-1}$.

The numbers on the first column (or on the slanted side) are the $B_i := x_{i,1}$.

Exercise v), p. 328, furnishes a recursive relationship to compute the Bell numbers (essentially, a reformulation of the process just described). In example d), p. 334 there is a more abstract formula.

Here's an enticing application [37]. Take a deck of n playing cards, remove the top card and put it back in the deck randomly. Then shuffle the deck, take the top card and insert it back anywhere. Repeat this operation n times. Clearly the cards can be ordered in n^n different ways in the deck. Of these possibilities, the number of times the deck will be rearranged exactly as in the beginning is B_n. Therefore the odds the deck is in the original order after the shuffles equals B_n/n^n, which is significantly larger than the odds $1/n!$ of a completely random reshuffle. The Bell numbers represents certain moments of Poisson distributions (☞ *probability theory*).

Definition A **binary relation** R between sets X, Y is a subset of $X \times Y$. One writes $x \mathrm{R} y$ to mean $(x, y) \in \mathrm{R}$, read 'x is related to y'.

(In a more formal definition $(X, Y, \mathrm{R} \subseteq X \times Y)$ would be called a *relational structure*.)

Examples 2.7

i) $\mathrm{R} = \{(m, n) \in \mathbb{N} \times \mathbb{N} \mid n \neq 0\}$.

ii) Fix $n \in \mathbb{N}$ and consider $\mathrm{R} = \{(m, r) \in \mathbb{N}^2 \mid \exists q \in \mathbb{N} \text{ such that } m = qn + r\}$.
    This particular relation is indicated by $\equiv_n$, see 2.1.

iii) $\mathrm{R} = \{(p_1, p_2) \in \mathbb{N}^2 \mid \gcd(p_1, p_2) = 1\}$.

iv) $\mathrm{R} = \{(x, y) \in X^2 \mid x = y\}$, called **diagonal** of X.

v) $\mathrm{R} = \{(\{x\}, X) \in \mathscr{P}(X)^2 \mid \{x\} \subseteq X\}$. That is, $\{x\} \mathrm{R} X \iff x \in X$.

2.2.1 Equivalence Relations

When $X = Y$ the Cartesian product is the square X^2, and $\mathrm{R} \subseteq X^2$ is said to be a (binary) relation on X.

Definition A binary relation on X is called an **equivalence relation** (the customary symbol is $\sim$) if it is

reflexive	$a \sim a$
symmetric	$a \sim b \implies b \sim a$
transitive	$a \sim b \wedge b \sim c \implies a \sim c$

for every $a, b, c \in X$.

Examples

i) The *extensionality axiom* of ZF theory implies that $=$ is an equivalence relation (actually, the diagonal is the smallest equivalence relation on a set X).

ii) The relation 'have the same birthday' among human beings is an equivalence relation.

iii) $\{(n, 3^n + 2) \in \mathbb{N}^2\}$ is not an equivalence relation on $\mathbb{N}$.

iv) Let $f: X \longrightarrow Y$ be a map. The relation $x_1 \sim_f x_2 \iff f(x_1) = f(x_2)$ is an equivalence, cf. Exercise i) on p.57.

v) The relation $\leqslant$ on $\mathbb{R}$ is not an equivalence, because it fails the symmetry property. Analogously, $<$ is neither reflexive nor symmetric.

vi) Parallelism (Sect. 18.1.1) among lines in space is an equivalence relation.

Definition Let $\sim$ be an equivalence relation on $X \neq \emptyset$. The **equivalence class** or **coset** of the element $x \in X$ is the non-empty set

$$[x] := \{z \in X \mid z \sim x\}.$$

The set formed by the equivalence classes is called the **quotient** of X by $\sim$

$$X/_\sim := \{[x] \mid x \in X\} \subseteq \mathscr{P}(X).$$

Observe that an equivalence class doesn't depend on the representative chosen in it:

$$[x] = [y] \iff x' \sim y' \quad \forall x' \in [x], \ y' \in [y].$$

In fact, $t \in [x]_\sim = [y]_\sim \implies (x \sim t) \wedge (t \sim y) \implies x \sim y$. Conversely, $x \sim y \implies x \in [y]_\sim$, so $t \in [x]_\sim$ implies $t \in [y]_\sim$ by transitivity, i.e. $[x]_\sim \subseteq [y]_\sim$. By symmetry we may swap the roles of x, y and obtain $[y]_\sim \subseteq [x]_\sim$, so eventually the two classes coincide.

This sets up (see p. 61) a link between equivalence relational structures (that are 'algebraic', i.e. defined through the Cartesian product) and partitions (which are incidence structures, 'geometrical' so to speak):

Theorem 2.8 *The equivalence relations of a set $X \neq \emptyset$ and its partitions are cryptomorphic:*

- *an equivalence relation $\sim$ determines a partition $P_\sim = \bigsqcup_{x \in X} [x]_\sim$;*
- *a partition $P = \bigsqcup_{i \in I} X_i$ generates an equivalence relation: $x \sim_P y \iff \exists j \in I : x, y \in X_j$;*
- *the two constructions are inverse to each other: an equivalence relation defined by a partition induces exactly that partition: $P_{\sim_P} = P$. Vice versa, a partition defined by an equivalence relation induces precisely the relation we started from: $\sim_{P_\sim} = \sim$.*

Proof The first two claims are left as exercises. As regards the last bullet, it will be immediate once we establish the following.

We claim

$$[x]_\sim \cap [y] = \emptyset \iff [x] \neq [y],$$

meaning that equivalence classes are disjoint precisely when they differ by one element at least (and if they have one common element, they coincide). To prove it we'll show the equivalent statement

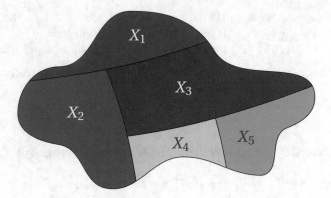

Fig. 2.1 A partition with 5 blocks

Fig. 2.2 Recursive
construction of Bell's triangle

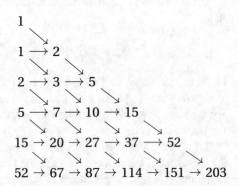

$$[x] \cap [y] \neq \emptyset \iff [x] = [y].$$

The sufficient direction ($\Longleftarrow$) is clear (one detail: no class $[x]$ is empty, for it
contains at least its representative x, by reflexivity). For the necessary condition
($\Longrightarrow$) suppose there exists $t \in [x] \cap [y]$, so $x \sim t \sim y$ and then $x \sim y$ by transitivity.
As we showed earlier, this implies $[x] = [y]$. $\square$

Examples

i) The relation 'have a T-shirt of the same colour' partitions the students in the
   classroom into equivalence classes, each one grouping all students wearing a
   certain colour. In Fig. 2.1 the quotient set

$$\frac{\{\text{students}\}}{\sim} = \left\{\mathbf{X}_1, \ \mathbf{X}_2, \ \mathbf{X}_3, \ \mathbf{X}_4, \ \mathbf{X}_5\right\}$$

   has five elements.

ii) Example 2.7 ii) defines on $\mathbb{Z}$ a relation $\equiv_n$ (with $n > 1$ fixed):

$$m \equiv_n r \iff \exists q \in \mathbb{N} \text{ such that } m = qn + r, \tag{2.1}$$

called **congruence modulo** n. Numbers m, r obeying (2.1) are said congruent modulo n, and the elements of the quotient

$$\mathbb{Z}_n := \mathbb{Z}/_{\equiv_n} = \{[0], [1], \dots, [n-2], [n-1]\}$$

are the congruence classes, or simply **integers modulo n**. In words, congruent numbers mod n give the same remainder when divided by n (☞ *modular arithmetics*). Each set $\mathbb{Z}_n$ is a commutative ring with unit, and being finite it's a field if and only if n is prime (*Dickson–Wedderburn little theorem*).

Clock arithmetics and musical scales are based on $\mathbb{Z}_{12}$, binary numbers are elements of $(\mathbb{Z}_2)^k$ (finite sequences of bits, i.e. elements of $\mathbb{Z}_2$), and bytes are elements of $(\mathbb{Z}_2)^8$.

Exercise Prove that $a^2 \equiv_8 0$ or 1 or 4, for any $a \in \mathbb{Z}$.

iii) Take the unit interval $I = [0, 1] \subset \mathbb{R}$ and define the partition $I = \{0, 1\} \sqcup \bigsqcup_{\substack{x \in I \\ x \neq 0,1}} \{x\}$. The induced equivalence relation, in practice, only identifies $0 \sim 1$.

If we imagine to perform this by bending I until we manage to 'glue' the endpoints together, we may view the quotient as if it were a circle

$$[0, 1]/_\sim = S^1.$$

A similar procedure applied to $\mathbb{R} = (-\infty, +\infty)$ produces the *one-point compactification* of $\mathbb{R}$, see Proposition 7.14.

iv) Let $a \sim b$ be the relation on $\mathbb{R}$ defined by $a - b \in \mathbb{Z}$. This is an equivalence, with classes $[a] = a + \mathbb{Z}$. The quotient $\mathbb{R}/_\sim = S^1$ arises by identifying the endpoints $x \sim x + 1$ of any real interval of length 1.

v) Suppose $\sim$ is the relation on the Cartesian square I^2 that identifies the opposite sides of the square:

$$(a, 0) \sim (a, 1), \quad (0, b) \sim (1, b), \quad \forall a, b \in I.$$

The quotient $I \times I/_\sim = T^2$ shown in Fig. 2.3 is called two-dimensional *torus*. The last three examples are paramount in ☞ *topology*.

Fig. 2.3 The 2-torus

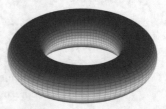

vi) The relation $\equiv_{n,m}$ on $\mathbb{N}$ given by

$$x \equiv_{n,m} y \iff \exists\, m, n \in \mathbb{N}: \begin{cases} x, y < n \ \wedge\ x = y \\ \text{or} \\ x, y \geqslant n \ \wedge\ x \equiv_m y \end{cases}$$

is an equivalence. It generalises case ii) above, because $\equiv_{0,m}$ coincides with the congruence $\equiv_m$, and it can be proved that any *congruence* on $\mathbb{N}$ is of the type $\equiv_{n,m}$, or the identity. This fact is crucial when studying models of Peano arithmetics, see Sect. 6.1.

2.2.2   Order Relations

Now we drop symmetry and replace it with skew-symmetry.

Definition 2.9 A binary relation R on a set X is called a **partial order** if it is

reflexive	$a\mathrm{R}a$
skew-symmetric	$a\mathrm{R}b \ \wedge\ b\mathrm{R}a \implies a = b$
transitive	$a\mathrm{R}b \ \wedge\ b\mathrm{R}c \implies a\mathrm{R}c$

for every $a, b, c \in X$. The set X is called **partially ordered** by R, or **poset**.

(As before, we won't insist on an 'order structure' being a particular relational structure $(X, \mathrm{R} \subseteq X^2)$.)

Example The relation $\leqslant$ ('smaller than or equal to') is a partial order on subsets of real numbers. That's why it's customary to denote by $\leqslant$ a generic order relation, even though it may have nothing to do with numbers.

Notation: we write $A \leqslant B$ to mean $a \leqslant b$ for all $a \in A, b \in B$. Similarly, $a \leqslant B$ stands for $a \leqslant b \ \forall b \in B$.

An effective way to represent an order relation on a finite set X is to use its **Hasse diagram**. This is a *graph* whose vertices are the elements of X and the edges $x\!-\!y$ join elements in immediate relation $x \leqslant y$. The graph is implicitly directed from bottom to top, so that x appears below y (and connected to it) if $x \leqslant y$: thus the skew-symmetry of $\leqslant$ becomes manifest, and transitivity can be read off consecutive edges that go only up or only down.

Examples (relative to Fig. 2.4)

i) The relation $\subseteq$ on $\mathscr{P}(X)$ is a partial order, as proven in Proposition 2.1.
   Take for instance $X = \{1, 2, 3\}$ and its subsets. The left figure explains $\varnothing \subseteq \{2\} \subseteq \{1, 2\} \subseteq X$, while there's no relation between $\{3\}$ and $\{2\}$, or $\{3\}$ and $\{1, 2\}$. Figure 5.1 represents a four-element case.

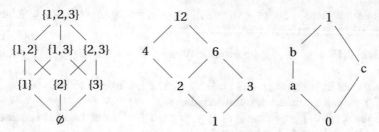

Fig. 2.4 Hasse diagrams

ii) $X = \mathbb{N}^*$ with relation 'being a divisor of' , cf. Definition 6.18,

$$b \preccurlyeq a \iff b|a \iff \exists n \in \mathbb{N} \text{ such that } a = nb$$

is partially ordered.

In the middle picture we have the divisors of 12. For example $1 \preccurlyeq 2 \preccurlyeq 6 \preccurlyeq 12$, and since $\gcd(2, 3) = 1$, then $2 \npreccurlyeq 3$ and $3 \npreccurlyeq 2$.

iii) $\big(\{1, a, b, c, 0\}, \leqslant \big)$ where the only relations among elements are $0 \leqslant a \leqslant b \leqslant 1$, $0 \leqslant c \leqslant 1$. The corresponding diagram is called 'pentagon' (right).

Exercises Construct Hasse diagrams for

i) $\big(\mathscr{P}(X), \subseteq \big)$ with $X = \{1, 2, 3, 4, 5\}$.

ii) $\{x \in \mathbb{N} \mid x \preccurlyeq 144\}$, where $\preccurlyeq$ means 'divides' (as in Definition 6.18).

Definition Let $(X, \leqslant)$ be a poset. An element $x \in X$ is called

- **minimal** if there is no smaller element $y \in X$: $\nexists y \in X : y < x$.
- **maximal** if there is no bigger element $y \in X$: $\nexists y \in X : y > x$.
- a **minimum** if $x \leqslant X$, written $x = \min X$.
- a **maximum** if $X \leqslant x$, written $x = \max X$.

Typically all these are clearly visible in the Hasse diagram.

Exercise Explain why 'minimum $\implies$ minimal', and 'maximum $\implies$ maximal', but not vice versa (saying that x is minimal is not the same as saying $x \leqslant X$).

Under which conditions x minimal $\iff$ $x \leqslant X$ certainly stands?

Taking the power set $\mathscr{P}(X)$ ordered by $\subseteq$, the set X itself is the only maximum element, and $\varnothing$ is the unique minimum element. In this situation the convention is to speak of minimal and maximal elements excluding these obvious elements. In Fig. 2.4 left, therefore, $\{1\}, \{2\}, \{3\}$ are minimal, whereas $\{2, 3\}, \{1, 3\}, \{1, 2\}$ are maximal in X.

Exercises

i) Show that the singletons $\{x\}$, $x \in X$, are minimal in $\mathscr{P}(X)$, and their set-complements $X \setminus \{x\}$ are maximal.

ii) Prove that if the singletons $\{x\}_{x \in X}$ are (minimal and) maximal in $\mathscr{P}(X)$, then X has only 2 elements.

iii) Prove that there is no set X in which two-element subsets $\{x, y\} \subseteq X$ are both minimal and maximal in $\mathscr{P}(X)$.

Definition 2.10 Let $(X, \leqslant)$ be a partially ordered set and $Y \subseteq X$ a non-empty subset.

If the set $\mathfrak{B}^+(Y) := \{r \in \mathbb{R} \mid r \geqslant y, \ \forall y \in Y\}$ of **upper bounds** is non-empty, we call Y **upper bounded** or **bounded above**.

If the set $\mathfrak{B}^-(Y) := \{l \in \mathbb{R} \mid l \leqslant y, \ \forall y \in Y\}$ of **lower bounds** is non-empty, Y is said **lower bounded** or **bounded below**:

$$\mathfrak{B}^-(Y) \leqslant Y \leqslant \mathfrak{B}^+(Y).$$

In particular, upper and lower bounds are $\leqslant$-related to any other element in Y.

Examples Taking $X = \mathbb{R}$,

i) $-3, -239, 0$ are lower bounds for $\mathbb{R}^+$, because $\mathfrak{B}^-(\mathbb{R}^+) = (-\infty, 0]$.
ii) 18 and π^{23} are upper bounds for the interval $[-7, 3]$, and actually $\mathfrak{B}^+([-7, 3]) = [3, +\infty)$.

Definition Let $Y \subseteq (X, \leqslant)$ be a non-empty subset.

- If Y is upper bounded, we call **least upper bound** or **supremum** of Y the element

$$\sup Y := \min \mathfrak{B}^+(Y).$$

- If $\mathfrak{B}^+(Y) = \varnothing$ (Y is **unbounded above**), we write $\sup Y = +\infty$.
- If Y is lower bounded, we call **greatest lower bound** or **infimum** of Y the element

$$\inf Y := \max \mathfrak{B}^-(Y).$$

- If $\mathfrak{B}^-(Y) = \varnothing$ (Y is **unbounded below**), we set $\inf Y = -\infty$.

A further convention is $\inf \varnothing = +\infty$, $\sup \varnothing = -\infty$ (this is merely due to formal uniformity, in order to include $\varnothing$ when stating theorems, cf. Sect. 7.6).

Exercise Suppose $Y \subseteq (X, \leqslant)$ is a non-empty bounded subset. Show that

$$\sup Y \in Y \implies \max Y = \sup Y, \qquad \inf Y \in Y \implies \min Y = \inf Y.$$

Examples Still using $X = \mathbb{R}$,

i) $\sup \{q \in \mathbb{Q} \mid q^2 < 2\} = \sqrt{2} = \inf \{q \in \mathbb{Q}^+ \mid q^2 > 2\}$; in reality this is the definition of the number $\sqrt{2}$.
ii) $\min[5, +\infty] = 5 = \sup[-2, 5] = \max[-2, 5]$ and $\sup[5, +\infty) = +\infty$.
ii) $\sup \mathbb{R} = +\infty$, $\inf \mathbb{R} = -\infty$.

iv) Since $k > h > 0 \iff 0 < \dfrac{1}{k} < \dfrac{1}{h}$, the set $Y = \left\{ \dfrac{1}{n} \,\middle|\, n \in \mathbb{N}^+ \right\}$ is bounded:

$\max Y = 1$ while $\nexists \min Y$ and $\inf Y = 0$.

v) $\{2^n + 1 \mid n \in \mathbb{Z}\}$ is unbounded above, but $\inf\{2^n + 1 \mid n \in \mathbb{Z}\} = 1$.

Exercises 2.11

i) Find (in case they exist!) sup and inf for the sets

$$\left\{ \frac{n}{n+1} \,\middle|\, n \in \mathbb{N} \right\}, \quad \left\{ \frac{3n^2}{4n+1} \,\middle|\, n \in \mathbb{N} \right\}, \quad \left\{ \frac{y}{y-1} \,\middle|\, y \in \mathbb{R} \setminus \{-1\} \right\},$$

$$\left\{ \frac{y}{y-1} \,\middle|\, y \in [-2, 1) \right\}.$$

ii) Let $A, B \subseteq \mathbb{R}$ be subsets, and prove:

$$A \subseteq B \implies \inf B \leqslant \inf A \leqslant \sup A \leqslant \sup B$$

(assuming the supremum and infimum exist). This shows sup is a non-decreasing function, and inf is non-increasing.

iii) Consider upper bounded sets $\varnothing \neq A, B \subseteq \mathbb{R}$. Show $A \cup B$ is upper bounded. Then prove

$\sup(A \cup B) = \max\big(\sup A, \sup B\big);$

if $A \cap B \neq \varnothing$ then $\sup(A \cap B) \leqslant \min\big(\sup A, \sup B\big)$.

iv) Take real numbers $a_{ij} \in \mathbb{R}$, $i = 1, \dots, n$, $j = 1, \dots, m$. Show that

$$\max_i \big(\min_j a_{ij}\big) \leqslant \min_j \big(\max_i a_{ij}\big).$$

v) Consider upper bounded sets $\varnothing \neq A, B \subseteq \mathbb{R}$ and prove

$\sup\{a + b : a \in A, b \in B\} = \sup A + \sup B$
(if $A, B \subseteq \mathbb{R}^+$) $\sup\{ab : a \in A, b \in B\} = \sup A \cdot \sup B$
$\inf\{-a : a \in A\} = -\sup A$
$\sup\left\{ \dfrac{1}{a} : a \in A \right\} = \dfrac{1}{\inf A}, \quad \inf\left\{ \dfrac{1}{a} : a \in A \right\} = \dfrac{1}{\sup A}$ (assuming
$\sup A, \inf A > 0$).

The existence of a least upper bound and an greatest lower bound is not guaranteed, nor is their uniqueness.

Examples

i) The set $\{x \in \mathbb{Q} \mid x^3 > 2\}$ does not have least upper bound in $\mathbb{Q}$, because it's unbounded.

ii) Consider $X = \{1, 2, 3, 12, 18\}$ ordered by divisibility: $a \preccurlyeq b \iff a|b$. The subset $Y = \{2, 3\}$ has upper bounds $\mathfrak{B}^+(Y) = \{12, 18\}$ in X, but there is no least upper bound because 12 and 18 don't divide one another.

iii) To understand better, take $\mathbb{N}$ ordered by divisibility. As 1 divides any number, $\min(\mathbb{N}, \preccurlyeq) = 1$. Perhaps surprisingly, $\max(\mathbb{N}, \preccurlyeq) = 0$ because 0 is the only common multiple of all natural numbers. An infinite subset $Y \subseteq \mathbb{N}$ will always have $\sup Y = 0$, but the infimum might be bigger than 1. On the contrary, if we take a finite subset $Y = \{n_1, \ldots, n_p\} \subset \mathbb{N}$, then

$$\sup Y = \operatorname{lcm}\{n_1, \ldots, n_p\}, \quad \inf Y = \gcd\{n_1, \ldots, n_p\}.$$

Questions of this sort are object of the ☞ *theory of lattices*.

The supremum and infimum always exist and are unique in families of subsets. (This fact is what makes power sets $\mathcal{P}(Y)$ stand out among Boolean algebras, see Theorem 5.8.)

Proposition 2.12 *Let Y be a non-empty set. Any upper bounded collection $\varnothing \neq \mathcal{I} \subseteq \mathcal{P}(Y)$ admits a unique least upper bound, namely the family's union*

$$\sup \mathcal{I} = \bigcup_{T \in \mathcal{I}} T.$$

Similarly, every lower bounded collection $\varnothing \neq \mathcal{W} \subseteq \mathcal{P}(Y)$ admits a unique greatest lower bound, i.e. the intersection

$$\inf \mathcal{W} = \bigcap_{S \in \mathcal{W}} S.$$

Proof Let's see why the union is the supremum, the intersection the infimum. The union $\bigcup_{T \in \mathcal{I}} T$ is larger than any element $T \in \mathcal{I}$ (it contains it), and if there exists another upper bound T', then $T' \supseteq T$ for every $T \in \mathcal{I}$, and necessarily $T' \supseteq \bigcup_{T \in \mathcal{I}} T$. This makes the union the smallest upper bound. Moreover, the supremum is unique: if there were another element $S \in \mathcal{P}(Y)$ with the same properties, any $T \in \mathcal{I}$ would be a subset of the upper bound $S \supseteq T$, hence $S \supseteq \bigcup_{T \in \mathcal{I}} T$; but since S is the smallest among upper bounds, $\bigcup_{T \in \mathcal{I}} T = S$.

The argument for the infimum is completely analogous, and left as an exercise. $\square$

Proposition 2.12 is what makes the family $\mathcal{P}(Y)$ (or a subfamily) special: it turns it into a so-called *algebra of sets*. This is related to Exercise 5.1 (iii) and Theorem 5.16.

Definition A partial order relation $\leqslant$ on X is a **total order** (or **linear** order) if for any $a, b \in X$ one always has either $a \leqslant b$ or $b \leqslant a$ (any two elements are related). Then $(X, \leqslant)$ is a **(totally) ordered space**.

Example Number sets such as $\mathbb{R}, \mathbb{Q}, \mathbb{Z}, \mathbb{N}$ are totally ordered by the standard relation $\leqslant$. This fact is not straightforward and will be proved in due course.

None of the lattices in Fig. 2.4 defines a total order. Relatively to the first diagram, it's easy to show that $(\mathscr{P}(X), \subseteq)$ is not totally ordered in general: any two disjoint subsets $A, B \subseteq X$ will not be related.

The structure in Fig. 2.5, left, is not totally ordered.

Definition 2.13 Let $X \neq \varnothing$ be partially ordered by $\leqslant$. One calls **chain** of X a totally ordered subset $C \subseteq X$ under the induced order relation, i.e. the restriction $\leqslant |_{C \times C}$ of $\leqslant$ to C.

Exercise Determine all chains of length 3, 4 and 5 in the diagram of Fig. 2.5.

A chain C is called **ascending** if it admits minimum, and **descending** in case there is a maximum. These notions are mainly interesting for infinite posets X. In that case, an **infinite descending chain** is an infinite, strictly decreasing sequence of elements $x_1 > x_2 > x_3 > \cdots$.

Example In $X = (\mathbb{Z}, \leqslant)$ the set $\mathbb{Z}^- = \{-1, -2, -3, \ldots\}$ is an infinite descending chain.

Theorem 8.12 proves that $(\mathbb{N}, \leqslant)$ does not have infinite descending chains, because any descending chain of natural numbers has a minimum.

Our discussion on total order relations will continue in Chap. 8.

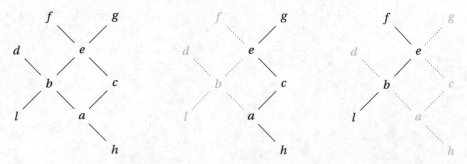

Fig. 2.5 The Hasse diagram on the left contains several chains: the right and middle pictures show two maximal chains

Chapter 3
Functions

This chapter introduces the general dictionary of functions, some properties and examples. Although many aspects (e.g., Sect. 3.2) holds for general binary relations, we shall concentrate on functions, probably the most important notion in mathematics. The theory of real functions of one real variable deserves separate treatment (Chap. 10).

Prerequisites: Sect. 3.5 uses a rough idea of finite, countable and uncountable set, and the axiom of choice. The proof of property 3.18 depends on Theorem 19.19.

3.1 Lexicon

Definitions A **function** $f : X \to Y$ is a relation $R_f \subseteq X \times Y$ such that

i) for every $x \in X$ there exists an element $(x, y) \in R_f$ (f is defined everywhere)

ii) if $(x, y) \in R_f$ and $(x, \tilde{y}) \in R_f$ then $y = \tilde{y}$.   (f is uniquely defined)

The function f associates each element $x \in X$ with a unique element $y =: f(x) \in Y$ called the **image** of x under f:

$$f : X \longrightarrow Y$$
$$x \longmapsto f(x)$$

The words function, **map**, **mapping**, or **transformation** are normally used interchangeably, as synonyms. Some are traditionally preferred in certain areas (the first is more common in analysis, the others in geometry/topology). In the writing $y = f(x)$, x is called **independent variable**, y **dependent variable** (dependent on the variable of which it is function). Let's say, clearly, that a function is *not* a law associating x with y.

S. G. Chiossi, *Essential Mathematics for Undergraduates*,
https://doi.org/10.1007/978-3-030-87174-1_3

Definition The largest subset where the relation R_f becomes a function is called (natural, or maximal) **domain**

$$\mathrm{Dom}(f) := \{x \in X \mid f(x) \in Y\} \subseteq X.$$

This doesn't contradict (i) in the first definition. The fictional equivocation between X and $\mathrm{Dom}(f)$ depends on the habit of not indicating the domain in concrete cases, since the latter should be easily computable when not given explicitly (which students hate): by writing $f(x) = \sqrt{x-4}$, for example, we are tacitly taking $X = \mathbb{R}$, but it's known that $\mathrm{Dom}(f) = [4, +\infty) \subsetneq X$ (see p. 224). As a matter of fact, the passage from $X \times Y$ to $\mathrm{Dom}(f) \times Y$ corresponds to a *cilindrification*[1] of the relation/function.

Definition The set Y is called **codomain** and

$$f(X) := \{y \in Y \mid y = f(x) \text{ for some } x \in \mathrm{Dom}(f)\} =: \mathrm{Im}(f)$$

is the **image** (or **range**) of the map f. Furthermore, one calls

- $f(B) := \bigcup_{x \in B}\{f(x)\}$ image of the subset $B \subseteq \mathrm{Dom}(f)$
- $f^{-1}(y) := \{x \in X \mid f(x) = y\}$ **pre-image** of $y \in Y$

  (in geometrical contexts the name **fibre** is much liked)

- $f^{-1}(A) := \{x \in X \mid f(x) \in A\}$ pre-image of the subset $A \subseteq Y$.

Exercise Prove that $f^{-1}(A) = \bigcup_{y \in A} f^{-1}(y)$, for any map $f: X \to Y$ and any $A \subseteq Y$.

The space of functions between X and Y is indicated with

$$Y^X := \{f: X \to Y \text{ function}\} \subseteq \mathscr{P}(X \times Y).$$

The reason for this 'exponential' notation (and why X^Y isn't used instead) will become manifest in Proposition 7.17) (Fig. 3.1).

Definition Two maps $f: X \to Y$, $g: Z \to W$ are said to be **equal** when $\mathrm{Dom}(f) = \mathrm{Dom}(g)$ (they have the same domain), $Y = W$ (same codomain) and $f(x) = g(x)\ \forall x \in \mathrm{Dom}(f)$ (same action).

---

[1] My colleague Petrúcio Viana pointed out this.

Fig. 3.1 Not a function

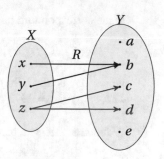

Examples

(1) Associating each place in the city of Rio de Janeiro with its temperature (°C) at a given moment defines a function: {place in Rio de Janeiro} → $\{n \in \mathbb{N} \mid n \geqslant 15\}$.

(2) The genetic code is a function: {human beings} → {nucleic acids} that maps each individual to their own DNA chain.

(3) $f : \mathbb{N} \to \mathbb{N}, n \mapsto f(n) = 3^n$ is a function: by virtue of the unique prime decomposition of natural numbers (Theorem 6.23), we have $3^n \neq 3^m$ if $n \neq m$.

Note that the functions $g : \{n \in \mathbb{N} \mid n \geqslant 4\} \to \mathbb{N}, g(n) = 3^n$ and $h : \mathbb{N} \to \{m \in \mathbb{N} \mid m < 10\}, h(n) = 3^n$ are different from f.

(4) The relation R on $X \times Y$ in Fig. 3.1, represented schematically via Venn diagrams with arrows, does *not* represent a function: the element $z \in X$ is related with two distinct elements $c, d \in Y$, which prevents the possibility of defining the image $R(z)$ univocally.

(5) The volume of the parallelepiped of sides $\mathbf{a}, \mathbf{b}, \mathbf{c}$ is a real function of 3 vector variables

$$\text{vol} : \mathbb{R}^3 \times \mathbb{R}^3 \times \mathbb{R}^3 \to [0, +\infty)$$

$$\text{vol}(\mathbf{a}, \mathbf{b}, \mathbf{c}) = \mathbf{a} \times \mathbf{b} \cdot \mathbf{c}$$

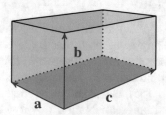

(6) Gödel's numbering is a function $\ddot{g} : \mathscr{F} \to \mathbb{N}$ mapping a formula of the language to its *Gödel number*. First one associates an integer $r_j \in \mathbb{N}^*$ with the jth symbol of a formula F in some pre-established way, so that to codify any $F \in \mathscr{F}$ by a finite sequence $(r_1, \ldots, r_n)$ in a unique manner. Then one defines the Gödel number of F as the product of the first n primes each raised

to the corresponding power in the sequence:

$$\ddot{g}(F) = \ddot{g}(r_1, r_2, r_3, \ldots, r_i, \ldots r_n) = 2^{r_1} \cdot 3^{r_2} \cdot 5^{r_3} \cdots p_i^{r_i} \cdots p_n^{r_n}$$

where p_i is the ith prime: $p_1 = 2, p_2 = 3, p_3 = 5, \ldots, p_{31} = 127, \ldots, p_{298} = 1973$ etc. (☞ *recursion theory*).

(7) A function defined piecewise: $f : \mathbb{R} \to \mathbb{R}$, $x \mapsto \begin{cases} 1 - x^3, & x < 0 \\ 0, & x = 0 \\ \sqrt{7 - x}, & x > 4 \end{cases}$. The

domain is a subset of the union of the sets to the right: $(-\infty, 0] \cup (4, 7]$.

(8) The complex-valued function of one complex variable

$$\zeta : \left\{ s \in \mathbb{C} \mid \mathrm{Re}(s) > 1 \right\} \longrightarrow \mathbb{C}, \quad \zeta(s) = \sum_{n=1}^{\infty} \frac{1}{n^s} \tag{3.1}$$

is the legendary *Riemann zeta function*.

A curiosity: this function, when evaluated on simple numbers (that should be forbidden), returns values that actually permeate the sciences. Here are some instances. In ☞ *number theory* it is possible to redefine $\zeta(0) = 1 + 1 + 1 + 1 + \cdots$ to be finite (and thus make the sum converge !). Similarly, $\zeta(-1)$ equals the divergent sum $1 - 2 + 3 - 4 + \ldots$, but in certain areas of theoretical physics (☞ *string theory*) it's important to give it a specific value. The sum of the harmonic series $\zeta(1)$ is infinite. Yet it can be proved that

$$\gamma := \lim_{\epsilon \to 0} \frac{\zeta(1 + \epsilon) + \zeta(1 - \epsilon)}{2} \approx 0.5772, \tag{3.2}$$

called the *Euler–Mascheroni constant*. Other examples include $\zeta(2) = \sum_{n=1}^{\infty} \frac{1}{n^2}$, which in 1734 Euler proved to equal $\frac{\pi^2}{6}$. The inverse $6\pi^{-2}$ are the odds for two random integers to be coprime. Or the *Apéry constant* $\zeta(3) \approx 1.202$, which plays a role in ☞ *quantum electrodynamics*.

It's known that $\zeta(-n) = (-1)^n \frac{\beta_{n+1}}{n + 1}$ for $n > 0$, where $(\beta_n)_{n \in \mathbb{N}} = (1, -\frac{1}{2}, \frac{1}{6}, 0, -\frac{1}{30}, \ldots)$ are the so-called *Bernoulli numbers* (see Example 14.11 (c)). Taking in account that $\beta_{2k+1} = 0$ when $k \geqslant 1$, one finds $\zeta(-2k) = 0$, and negative even integers are called the trivial zeroes of ζ. One of the hardest open problems in mathematics is to find the other zeroes. The 8th Hilbert problem [55], better known as *Riemann hypothesis* [15], predicts that every complex, non-real solution $s = x + iy$ to $\zeta(s) = 0$ has real part $x = \frac{1}{2}$. Today, more than a century after its initial formulation, the problem remains particularly inaccessible because it amalgamates both analytical and arithmetical aspects.[2] The person able to solve the conjecture will win a

---

[2] Apparently Hilbert himself said: *If I woke up after sleeping for a thousand years, my first question would be: has the Riemann hypothesis been proved?*

million-dollar prize, almost certainly the Fields Medal and guaranteed eternal
fame. The same can be said about the other *Millennium Problems* [101]: the
Birch and Swinnerton-Dyer conjecture (related to the value assigned to $\zeta(0)$),
the *Hodge conjecture*, the *P vs. NP problem* (see Remark 6.24), the solutions
to the *Navier–Stokes equations*, and *quantum Yang–Mills theory*.

(9) The function $\chi(S) = V - E + F \in \mathbb{Z}$, where V, E, F are the number of
vertices, edges and faces of a triangulation of a surface S, is called the Euler–
Poincaré characteristic of S, see Sect. 18.5.

(10) Fix $m, n \in \mathbb{N}^*$. A real $m \times n$ *matrix* is a function

$$A \colon \{1, 2, \ldots, m\} \times \{1, 2, \ldots, n\} \to \mathbb{R}$$

mapping the pair (i, j) to a number $A(i, j) = a_{ij}$. Conventionally one
identifies A with the collection $\{a_{ij} \mid i = 1, 2, \ldots, m; \ j = 1, 2, \ldots, n\}$
of mn elements arranged, for operational reasons, in a rectangular formation

$$A = (a_{ij}) = \begin{pmatrix} a_{11} & a_{12} & \cdots & a_{1n} \\ a_{21} & a_{22} & \cdots & a_{2n} \\ \vdots & \vdots & \ddots & \vdots \\ a_{m1} & a_{m2} & \cdots & a_{mn} \end{pmatrix}_{m \times n} \ .$$

The number i is the row index, j the column index, so element a_{ij} is placed at
the crossing of row i and column j. The set of real $m \times n$ matrices is indicated
by $\mathbb{R}^{m \times n}$.

(11) The number of roots of a real non-zero polynomial defines a map $\mathbb{R}[t] \setminus \{0\} \to$
$\mathbb{N}, \ p(t) \mapsto \operatorname{card} \{t_0 \in \mathbb{R} \mid p(t_0) = 0\}$. More on this in Sect. 9.2.

(12) $\pi_1 \colon \mathsf{Top} \to \mathsf{Grp}$ associates a path-connected topological space X with its
fundamental group $\pi_1(X)$. This π_1 is actually more than a function: it's an
instance of a functor (☞ *category theory, algebraic topology*).

(13) Functions defined on spaces that are not number sets are often called **operators**. Examples:

- matrices are **linear operators**, defined on vector spaces (Sect. 15.2);
- **differential operators**, defined on functions:

  - (*k*th derivative): $D^k \colon f(x) \mapsto f^{(k)}(x)$, for $k \in \mathbb{N}^*$
  - (*logarithmic derivative*): $f(x) \mapsto D\big(\log f(x)\big)$
  - the 'del', 'delbar' operators ∂, $\bar{\partial}$ (p.355);
  - grad, div, curl (p.370);
  - (*curvature*): $f(x) \mapsto \kappa_f(x) = \dfrac{f''(x)}{(1 + f'(x)^2)^{3/2}}$
  - (*Legendre transform*): $f(x) \mapsto f^*(x) = x\,(f')^{-1}(x) - f\big((f')^{-1}(x)\big)$

- **integral operators**:

  - (*arc length*): $f(x) \mapsto s(t) = \displaystyle\int_{t_0}^{t} \sqrt{1 + f'(x)^2}\,\mathrm{d}x$

Fig. 3.2 A curve defined by
many functions

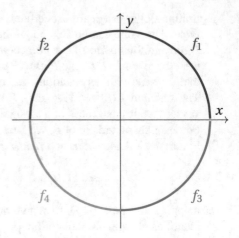

– **integral transforms** $f(t) \mapsto (Tf)(u) = \int_I f(t)\, K(t, u)\, \mathrm{d}t$, such as

- (*Laplace*): $K(t, u) = \mathrm{e}^{-ut},\ I = [0, +\infty)$
- (*Fourier*): $K(t, u) = \mathrm{e}^{-2\pi itu},\ I = \mathbb{R}$
- (*Mellin*): $K(t, u) = t^{u-1},\ I = [0, +\infty)$.

☞ *differential geometry, probability, complex geometry, harmonic analysis.*

When we consider functions it's easiest to think of explicit expressions between x
and y, like $f(x) = \sqrt{x^3 + 3\cos x}$. We soon realise, though, that the notion 'function
defined by a formula' is too restrictive, as the next example suggest.

Example 3.1 Equation $x^2 + y^2 - 1 = 0$ defines a relation (*not* a function) between
the real variables x, y.

Although we can't express $y \in [-1, 1]$ in terms of any $x \in [-1, 1]$
uniquely, it will be possible to do so by restricting the domain. In Fig. 3.2,
$f_1: [0, 1] \to \mathbb{R},\ f_1(x) = \sqrt{1 - x^2}$ has the red arc as graph. Not dissimilarly
$g: [-1, 1] \to \mathbb{R},\ g(x) = -f_1(x)$ describes the semi-circle $\mathrm{graph}(f_4) \cup \mathrm{graph}(f_3)$,
whilst $h: (-1, 1) \to \mathbb{R},\ h(x) = \begin{cases} -f_1(x) & -1 < x < 0 \\ f_1(x) & 0 < x < 1 \end{cases}$ gives opposite quarters
$\mathrm{graph}(f_1) \cup \mathrm{graph}(f_4)$.

The example shows the true reason for defining a function $f: D \to \mathbb{R}$ as a relation
(and not a 'law', whatever that might mean), i.e.: a subset $f \subseteq D \times \mathbb{R}$ such that
any choice of $x \in D$ determines a unique $y \in \mathbb{R}$ such that $(x, y) \in f$. It's only for
convenience that we abbreviate $(x, y) \in f$ with $y = f(x)$.

Exercises 3.2 Take subsets $P, Q \subseteq X$ and $A, B \subseteq Y$, and a map $f: X \to Y$.
Prove that

i) $f^{-1}(A) \cup f^{-1}(B) = f^{-1}(A \cup B)$, $f(P) \cup f(Q) = f(P \cup Q)$.

ii) $f^{-1}(A) \cap f^{-1}(B) = f^{-1}(A \cap B), \qquad f(P) \cap f(Q) \supseteq f(P \cap Q)$. Find an
example where $f(P) \cap f(Q) \not\subset f(P \cap Q)$.

iii) $f(f^{-1}(A)) = A, \quad P \subseteq f^{-1}(f(P))$. Find an example where
$f^{-1}(f(P)) \not\subset P$.

iv) $P \subseteq Q \Longrightarrow f(P) \subseteq f(Q), \quad A \subseteq B \Longrightarrow f^{-1}(A) \subseteq f^{-1}(B)$.

v) $f(P) \setminus f(Q) \subseteq f(P \setminus Q), \quad f^{-1}(Y \setminus B) = f^{-1}(Y) \setminus f^{-1}(B)$.

vi) $f(P \cap f^{-1}(A)) = f(P) \cap A$.

Definition 3.3 Fix a subset T in some set X. The **characteristic function** of T is
the map $\chi_T : X \to \{0, 1\}$

$$\chi_T(x) = \begin{cases} 1, & x \in T \\ 0, & x \notin T \end{cases}.$$

The characteristic function of $T \subseteq X$ therefore partitions the larger set

$$X = (X \setminus T) \sqcup T = \chi_T^{-1}(0) \sqcup \chi_T^{-1}(1),$$

and thus χ_T determines T completely as the pre-image of the value 1.

In logic, if $X = \mathscr{F}$ were the set of formulas, the characteristic function of $T = \{x \in \mathscr{F} \mid \vDash x\}$ (true formulas) would be a truth function υ.

Proposition 3.4 *Let* $B, C \in \mathscr{P}(X)$ *be subsets and* χ_B, χ_C *the corresponding characteristic functions. Then for all* $x \in X$

i) $\chi_{B \cup C}(x) = \max \{\chi_B(x), \chi_C(x)\}$

ii) $\chi_{B \cap C}(x) = \chi_B(x)\chi_C(x)$

iii) $B \subseteq C \iff \chi_B(x) \leqslant \chi_C(x)$

iv) *if, furthermore,* $X = B \sqcup C$ *is a partition, then* $(\chi_X(x) - \chi_B(x))(\chi_X(x) - \chi_C(x)) = 0$.

Proof Exercise.   □

Chapter 5 provides the general framework for all of this.

Besides, consider a subset $B \in \mathscr{P}(X)$, the partition $B \sqcup (X \setminus B) = X$ and the
piecewise-defined map $f : X \to Y$,

$$f(x) = \begin{cases} f_1(x) & x \in B \\ f_2(x) & x \notin B \end{cases}.$$

Because we can write $f = f_1 \chi_B + f_2 \chi_{X \setminus B}$ it's easy to see that

$$\mathrm{Dom}(f) = (\mathrm{Dom}(f_1) \cap B) \cup (\mathrm{Dom}(f_2) \cap X \setminus B).$$

3.2 Invertibility

In this section we'll address composition of maps and a number of issues surrounding invertibility: injectivity, surjectivity, and the obstructions to these properties.

Definition A function $f : X \to Y$ is **one-to-one** (or **1-1**, or **injective**) when

$$\forall a, b \in X \quad a \neq b \Longrightarrow f(a) \neq f(b).$$

By contraposition this is the same as demanding that, $f(a) = f(b) \Longrightarrow a = b$, $\forall a, b \in X$. A 1-1 map is also called an **injection**, and a shorthand notation is $f : X \hookrightarrow Y$.

Examples

i)   The map in Fig. 3.3 is not injective. If we eliminate the arrow departing from a, or the one from b, we do obtain a 1-1 map.
ii) The 'birthday' map: $\{$ humans $\} \to \{$ dates $\}$ is not injective.
iii) Suppose a set X has at least 2 elements. Then no constant map $X \to Y$, $x \mapsto y_0$ is 1-1.
iv) If $X \subseteq Y$, the injection $\iota : X \to Y$, $\iota(x) = x$ is called **inclusion** map of X in Y. It's very common to write $X \subseteq Y$ to indicate an inclusion.

Exercises

i)   Compute the pre-image $f^{-1}(81)$ for $f : \mathbb{R} \to \mathbb{R}$, $f(x) = x^4$ and conclude f isn't 1-1.
ii) Prove

$$\begin{aligned}
f : X \to Y \text{ is 1-1} \iff & \left(f^{-1}(y)\right) = 1 \quad \forall y \in \text{Im}(f); \\
\iff & f(P \cap Q) = f(P) \cap f(Q) \quad \forall P, Q \subseteq X; \\
\implies & f(P \setminus Q) = f(P) \setminus f(Q) \quad \forall P, Q \subseteq X.
\end{aligned}$$

Fig. 3.3 A non-injective map

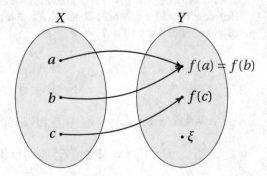

Injectivity is a property that talks about the domain of a function. The 'dual' condition, so to speak, on the codomain, is the following:

Definition A map $f: X \rightarrow Y$ is **onto** (or **surjective**, or a **surjection**) when $f(X) = Y$. An onto map is sometimes denoted by $f: X \twoheadrightarrow Y$.

Exercise Prove that $f: X \rightarrow Y$ is onto if and only if the pre-image $f^{-1}(y)$ is not empty for every $y \in Y$.

Examples 3.5

i) The map $f: \mathbb{N} \rightarrow 2\mathbb{N} := \{\text{even numbers}\}$, $f(n) = 2n$, is onto and 1-1.
ii) The function in Fig. 3.4 is not onto, because the element ω of the codomain doesn't belong to $\mathrm{Im}(f)$.
iii) The functions

$$\pi_X : X \times Y \longrightarrow X \qquad \pi_Y : X \times Y \longrightarrow Y$$
$$(x, y) \longmapsto x \qquad\qquad (x, y) \longmapsto y \qquad\qquad (3.3)$$

are called **projections** of the Cartesian product. They are onto if $X \neq \emptyset \neq Y$ (or in the trivial case $X = \emptyset = Y$).
iv) If $\sim$ is an equivalence relation on X, the surjective map

$$\pi : X \longrightarrow X/\!\sim, \qquad \pi(x) = [x]$$

is called the **canonical projection** associated with to $\sim$. We might interpret it as the map that collapses each coset in X to a single point in the quotient.

Exercises Explain why

i) $h : \mathbb{R} \rightarrow \mathbb{R}$, $h(s) = 3^s$ is 1-1, not onto.
ii) $f : \mathbb{R} \rightarrow \mathbb{R}$, $f(\xi) = \xi^2$ is neither onto, nor 1-1.
   Yet, $g : \mathbb{R} \rightarrow [0, +\infty)$, $g(\xi) = \xi^2$ is surjective.
iii) $f : X \rightarrow Y$ is onto $\iff Y \setminus f(P) \subseteq f(X \setminus P)$ for every $P \subseteq X$.
iv) Determine for which $a, b, c, d \in \mathbb{Z}$ the map $g : \mathbb{Z}^2 \rightarrow \mathbb{Z}^2$, $g(x, y) = (ax + by, cx + dy)$ is injective, or surjective.

Fig. 3.4 A non-surjective map

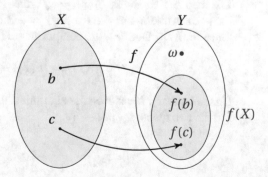

Establishing whether a function is 1-1, or onto is rather difficult in general. Think of the real map $f(x) = \arctan \sqrt[3]{x-1} + \dfrac{1}{\log(x^\pi + 1)}$ (luckily, in this particular case infinitesimal calculus comes to the rescue).

Definition A function $f : X \to Y$ is **bijective**, or a **bijection** if it is both 1-1 and onto.

Examples

i) The **identity** (map) on X: $\mathbb{1}_X : X \to X, x \mapsto x$;
ii) $f : \mathbb{R} \to \mathbb{R}$, $f(x) = mx + q$ for any real numbers q and $m \neq 0$.

Rather unfortunately, a bijection is also called a **1-1 correspondence**. The reader should have no doubts that a 1-1 map is not a 1-1 correspondence in general. The reason for the misnomer should emerge from the following exercise.

Exercises 3.6 Let $f : X \to Y$ be a function.

i) If f is 1-1, show that restricting the codomain to the image $\text{Im}(f)$, without altering the action of f, produces a bijection $f : X \to f(X)$ (we used the same letter for the obvious reason).
ii) Suppose $X = Y$ is finite. Prove that $f : X \to X$ is

$$\text{one-to-one} \iff \text{onto} \iff \text{bijective.}$$

(This is related to Propositions 6.38 and 6.36.)

Let's introduce an operation defined on functions, called **composition**.

Definition 3.7 Let $f : X \to Y$, $g : Y \to Z$ be functions such that $\text{Im}(f) \subseteq \text{Dom}(g)$. The **composite (map) of f and g** is the function

$$g \circ f : X \to Z, \quad (g \circ f)(x) = g\big(f(x)\big).$$

By definition, $\text{Dom}(g \circ f) = \big\{ x \in \text{Dom}(f) \mid f(x) \in \text{Dom}(g) \big\}$, see Fig. 3.5.

Examples

i) The map $h : \mathbb{R} \to \mathbb{R}$, $h(x) = \sqrt{x}$ has $\text{Dom}(h) = [0, +\infty)$ and $\text{Im}(f) = [0, +\infty)$, while $r : \mathbb{R} \to \mathbb{R}$, $r(x) = -3$ has $\text{Dom}(r) = \mathbb{R}$, $\text{Im}(r) = \{-3\}$. Since $\text{Im}(h) \subseteq \text{Dom}(r)$, the composite $r \circ h$ is well defined,

$$(r \circ h)(x) = r\big(h(x)\big) = r(\sqrt{x}) = -3.$$

On the contrary, $h \circ r$ doesn't exist: if it did, what should $h(-3)$ be?

ii) For the two maps $f, g : \mathbb{N} \to \mathbb{N}$

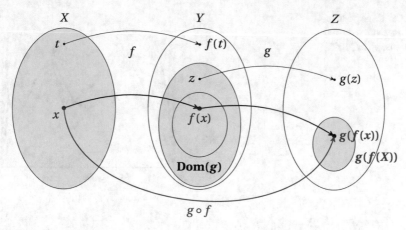

Fig. 3.5 How composition works

$$f(n) = n + 1, \quad g(n) = \begin{cases} 0 & n = 0 \\ n - 1 & n > 0 \end{cases}$$

we have $g \circ f = \mathbb{1}_\mathbb{N}$, but $f(g(0)) = 1 \neq 0$. Note that f is 1-1 (but not onto) and g is onto (and not 1-1).

iii) Since $f, g \colon \mathbb{R} \to \mathbb{R}$, $f(x) = x^2 + 3$, $g(x) = \sqrt{x}$ satisfy $f(\mathbb{R}) \subset \mathbb{R}^+ = \mathrm{Dom}(g)$, the composite exists:

$$(g \circ f)(x) = \sqrt{f(x)} = \sqrt{x^2 + 3}$$

Also $f \circ g$ exists and is given by $(f \circ g)(x) = f(\sqrt{x}) = x + 3$. This example shows that map composition is not a commutative operation: $g \circ f \neq f \circ g$, in general.

iv) Fix $r \in (0, 4]$. The logistic map $f \colon [0, 1] \to \mathbb{R}$, $f(x) = rx(1 - x)$ is related to demographic models with growth rate r. The iterated composites $f^k = \underbrace{f \circ \cdots \circ f}_{k \text{ times}}$ (Fig. 3.6) are the subject of ☞ *non-linear dynamics* and the entry point to *chaos theory*.

v) A map $p \colon X \to X$ such that $p \circ p = p$ is called **idempotent**. Convince yourself the identity map $\mathbb{1}_X$, any constant map, the absolute value $|\cdot| \colon \mathbb{R} \to \mathbb{R}$, and the two maps $\mathbb{R} \to \mathbb{Z}$:

$$\lfloor x \rfloor := \max \left\{ m \in \mathbb{Z} \mid m \leqslant x \right\} \quad (\textit{floor function / integer part})$$
$$\lceil x \rceil := \min \left\{ n \in \mathbb{Z} \mid n \geqslant x \right\} \quad (\textit{ceiling function}). \tag{3.4}$$

are all idempotent. A geometrical example are **projection** maps

Fig. 3.6 Logistic map
$f(x) = \frac{7}{2}x(1-x)$ and
iterations: f^2, f^3, f^4

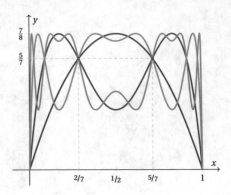

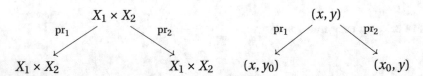

where $x_0 \in X_1$, $y_0 \in X_2$ are chosen elements.

Show that any idempotent map $p \colon X \to X$ has fixed points $x \in X$, i.e. solutions to $p(x) = x$. Describe the set of fixed points in the previous instances.

Exercises Prove the following facts.

i) Map composition is associative

$$f \circ (g \circ h) = (f \circ g) \circ h \tag{3.5}$$

(assuming both sides are well defined).

ii) Let $f \colon X \to Y$, $g \colon Y \to X$ be maps satisfying $g \circ f = \mathbb{1}_X$. Then f is 1-1 and g is onto:

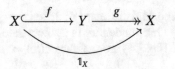

iii) The composite of injective/surjective maps is respectively injective/surjective.

Definition Suppose a map $g \colon Y \to X$ has the property that

$$g \circ f = \mathbb{1}_X \tag{3.6}$$

for some map $f \colon X \to Y$. Then f is called a **section** of g (and g is said to be right-invertible).

The existence of a section is not guaranteed in absolute terms. It is, as it turns out, equivalent to the AC, see p. 63.

Exercises Show that

i) any section $f : X \to Y$ (of some $g : Y \to X$) is injective;
ii) vice versa, a 1-1 map $f : X \hookrightarrow Y$ (with non-empty domain X) is a section of some map g.

Dually, if $f : X \to Y$ admits a map $g : Y \to X$ such that (3.6) holds, then f is called left-invertible and g a *retraction* of f, particularly in ☞ *topology*.

Exercise Prove that a 1-1 map (with Dom $\neq \varnothing$) admits a retraction.

The notion of section (and retraction) is categorical, and turns out to be paramount in algebra, geometry and topology.

Example The map $g : \mathbb{R} \to [0, +\infty)$, $g(x) = x^2$ is right-invertible. It admits two right inverses: $f : [0, +\infty) \to \mathbb{R}$, $f(x) = \sqrt{x}$ and $h : [0, +\infty) \to \mathbb{R}$, $h(x) = -\sqrt{x}$, because $g(f(x)) = \sqrt{x}^2 = x = \sqrt{(-x)^2} = g(h(x))$ for all $x \in [0, +\infty)$. There are exactly two inverses due to the uniqueness of the square root.

The same example shows that a left inverse may not be right inverse, and conversely: above, f is no left inverse to g since $f(g(-1)) \neq -1$, for instance.

Exercises

i) Write down other examples of maps and find their right or left inverses.
ii) Suppose f is both a right and left inverse for g. Prove f is unique.
iii) Plot the graph of $g(x) = x^2$ and its two right inverses.

Definition A map $f : X \to Y$ is said to be **invertible** if there exists a function, denoted by $f^{-1} : Y \to X$ and called **inverse** to f, such that

$$f \circ f^{-1} = \mathbb{1}_Y$$

and

$$f^{-1} \circ f = \mathbb{1}_X$$

$$X \xrightarrow{\; f \;} Y \xrightarrow{\; f^{-1} \;} X \xrightarrow{\; f \;} Y$$

Beware of the notation: $f^{-1}(x) \neq \left(f(x)\right)^{-1} = \dfrac{1}{f(x)}$. Also, recall that f^{-1} indicates the inverse function only when f is invertible, otherwise it simply denotes pre-images.

Exercises 3.8 Assume f is an invertible map. Prove that

i) the inverse map f^{-1} is unique;
ii) $\left(f^{-1}\right)^{-1} = f$;

iii) f^{-1} is the only right inverse and the only left inverse of f.

iv) Find a map admitting a right inverse and a left inverse, without being invertible.

As we said earlier, proving a map is bijective is quite hard. Thankfully, if we happen to know its inverse we are spared the task. More generally, actually

Lemma 3.9 *A function $f: X \to Y$ is bijective if and only if it is invertible.*

Proof ($\Longrightarrow$) As f is bijective, the pre-image of any $y \in Y$ contains only one element: $f^{-1}(y) = \{x\}$. Define a map $g: Y \to X$ by setting $g(y) = x$. Then $f(g(y)) = y$ and $g(f(x)) = x$ for all $x \in X, y \in Y$, meaning $g = f^{-1}$.

($\Longleftarrow$) f is 1-1 because $f(a) = f(b) \Longrightarrow f^{-1}(f(a)) = f^{-1}(f(b)) \Longrightarrow a = b$. To prove surjectivity, pick any $y \in Y$. Then $a := f^{-1}(y) \in X$ is an element with image $f(a) = y$. That's to say, f is onto. $\square$

The previous argument is instructive, in that it explains why bijectivity is a necessary and sufficient condition for the existence of the inverse map:

- as $\text{Im}(f^{-1}) = \text{Dom}(f)$, f must be 1-1 for f^{-1} to be uniquely defined;
- as $\text{Im}(f) = \text{Dom}(f^{-1})$, if f weren't onto then f^{-1} wouldn't be everywhere defined.

Corollary f is bijective $\Longleftrightarrow$ f^{-1} is bijective.

Proof Exercise. $\square$

Examples

i) The function $T : \mathbb{R} \to \mathbb{R}, T(x) = \dfrac{9}{5}x + 32$ has inverse $T^{-1} : \mathbb{R} \to \mathbb{R}, T^{-1}(x) = \dfrac{5}{9}(x - 32)$:

$$y = \frac{9}{5}x + 32 \iff x = \frac{5}{9}(y - 32).$$

T converts temperatures from Celsius degrees ($x\,°$C) to Fahrenheit degrees ($y\,°$F).

ii) The sine function $\sin : \left[-\frac{\pi}{2}, \frac{\pi}{2}\right] \to [-1, 1]$ is a 1-1 correspondence between the intervals $\left[-\frac{\pi}{2}, \frac{\pi}{2}\right]$ and $[-1, 1]$. Its inverse map $\arcsin : [-1, 1] \to \left[-\frac{\pi}{2}, \frac{\pi}{2}\right]$ is called arcsine, see p. 290.

Lemma *Let* $X \xrightarrow{g} Y \xrightarrow{f} Z$ *be bijections such that* $\mathrm{Im}(g) = \mathrm{Dom}(f)$. *Then*

$$(f \circ g)^{-1} = g^{-1} \circ f^{-1}.$$

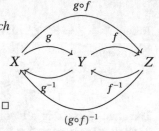

Proof Immediate, using Exercise 3.8. $\square$

Theorem *The set of bijections of a set* X

$$\mathfrak{S}(X) = \{\sigma : X \to X \text{ bijection}\}$$

is a group under composition, called the **symmetric group** *of* X.

Proof First of all, any composite is well defined since X is the domain and range of any $\sigma \in \mathfrak{S}(X)$. We know from (3.5) that $\circ$ is associative. Next, the identity $\mathbb{1}_X$ clearly satisfies $\mathbb{1}_X \circ \sigma = \sigma = \sigma \circ \mathbb{1}_X$, making it the neutral element. Finally, any $\sigma \in \mathfrak{S}(X)$ has an inverse element (the inverse map) by Lemma 3.9. $\square$

Aside, we also know that $\mathfrak{S}(X)$ won't be Abelian, except for very special sets X. Further information in Sect. 14.1.

Definition Take sets $X \subseteq Y$ and Z, and a map $f : Y \to Z$. The **restriction** (map) of f to X is the function $f\big|_X : X \to Z$ defined by $f\big|_X(x) = f(x), \forall x \in X$.

Using Definition 3.7 we may write $f\big|_X = f \circ \iota$ where ι is the inclusion $\iota : X \subseteq Y$. Sometimes this is expressed by saying f is an extension of $f\big|_X$.

In a similar way we'll speak (sloppily) of restriction of $f : Y \to Z$ to $W \subseteq Z$ to mean the composite map between the inclusion $W \hookrightarrow Z$ with f.

Exercises Consider a map $f : X \to Y$.

i) Prove the relation $\sim_f$ defined by

$$x \sim_f y \iff f(x) = f(y)$$

is an equivalence relation. Then show

$$\tilde{f} : X/\!\sim_f \longrightarrow f(X)$$
$$[x] \longmapsto f(x)$$

(3.7)

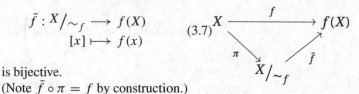

is bijective.
(Note $\tilde{f} \circ \pi = f$ by construction.)

- Taking $X = Y = \mathbb{R}^{n+1} \setminus \{0\}$ and defining $f(x) = \dfrac{x}{\|x\|}$, it's almost immediate to prove that

$$x \sim_f y \iff x = ty \quad \text{for some } t > 0.$$

The map f retracts all points x along lines through the origin to their normalised representative ($\|f(x)\| = 1$). That is to say

$$\frac{\mathbb{R}^{n+1} \setminus \{0\}}{\sim_f} \cong S^n$$

n-sphere (Fig. 3.7).

- If we now choose $X = Y = S^n$ and $g(x) = -x$, the cosets consist of antipodal points $[x] = \{x, -x\}$ on the sphere, and thus we obtain

$$\frac{S^n}{\sim_g} = \mathbb{R}\mathbb{P}^n,$$

the **real projective space**. Combining the two previous quotients leads back to the usual description of projective space as the set of lines through the origin

$$\mathbb{R}\mathbb{P}^n := \frac{\mathbb{R}^{n+1} \setminus \{0\}}{\mathbb{R}^*} := \frac{\mathbb{R}^{n+1} \setminus \{0\}}{\sim}$$

where $x \sim y \iff x \in \mathbb{R}^* y$.

Fig. 3.7 The n-sphere as retraction quotient of the punctured Euclidean space

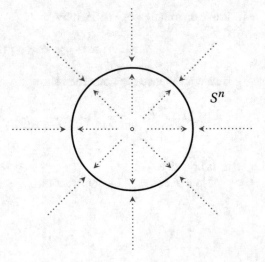

ii) Generalising (i), prove that for any map $f: X \to Y$ there exists a 1-1 function $\iota: X/{\sim_f} \to Y$ such that $\iota \circ \pi = f$. Conclude that every function can be expressed as the composition of a 1-1 map and an onto map.

iii) A map $f: X \to X$ such that

$$f \circ f = \mathbb{1}_X$$

is called an **involution** of X. Show that

- an involution f is bijective and self-inverse: $f = f^{-1}$.
- $f(x) = -x$ on $\mathbb{R}$, or $g(x) = \frac{1}{x}$ and $(f \circ g)(x) = (g \circ f)(x) = -\frac{1}{x}$ on $\mathbb{R}^*$, are involutions. A more exotic one is $s(x) = \log\left(\dfrac{e^x + 1}{e^x - 1}\right)$ defined on $\mathbb{R}^+$.

Changing fields, the following are involutions: reflections in the plane bout a line, or about a point; complex conjugation in $\mathbb{C}$; inversion in a group; taking complements of sets; negation in predicative calculus.

3.3 Operations

Operations form a distinguished class of functions of great interest.

Definition An (n-ary) **operation** on a set $X \neq \varnothing$ is a function $X^n \to X$.

We'll often suppose $n \geqslant 2$ because a 'unary' operation is simply a map: for instance the inversion $g \mapsto g^{-1}$ in a group, the negation $P \mapsto \neg P$ of logical formulas, or set-complementation $X \mapsto \complement X$.

The most relevant case is $n = 2$, i.e. **binary operations**. When X is finite, a binary operation may be read off the *Cayley table*. For instance, on $X = \{a, b, c\}$ we define $*: X \times X \to X$ by

$*$	a	b	c
a	a	a	a
b	b	c	a
c	a	c	b

That is, $a * a = a$, $b * a = b$, $c * b = c$ etc.
(using the convention $*(x, y) = x * y$).

Examples

i) the set-theoretical operations $\cap$, $\cup$, $\setminus$, $\triangle$, cf. Definition 2.2
ii) the logical connectives $\wedge$, $\vee$, $\to$, $\leftrightarrow$ between logical formulas, see Definition 1.1
iii) the multiplication $a \cdot b$ and addition $a + b$ of a ring

iv) the join $(x, y) \mapsto x \cup y$ and meet $(x, y) \mapsto x \cap y$ of a lattice
 v) the cross product $\mathbf{u} \wedge \mathbf{v}$ on $\mathbb{R}^3$ (Sect. 15.6)
vi) the connected sum # of compact surfaces (☞ *algebraic topology*)
vi) the composition between maps $f \circ g$, see Definition 3.7.

Definition An **algebraic structure** is a set $X \neq \varnothing$ equipped with one or more operations. (See Definition 4.11 for the formal definition.)

Algebraic structures should be viewed as fundamental objects in mathematics, alongside order structures and incidence structures (topological, projective, ...). But this point of view is relatively new. Since the Renaissance, mathematics was traditionally compartmentalised in well-separated fields: algebra, mechanics, astronomy, arithmetics, analytic geometry ... The astounding developments reached by these areas in the eighteenth century would not have sustained themselves hadn't mathematicians understood that although the many theories dealt with different objects and had distinct objectives, they were still linked by (un)canny similarities between their deductive systems, the respective theorems and properties (what this book is constantly trying to emphasise). Examples include the non-Euclidean revolution (Gauß, Lobachevskiĭ, Bolyai and Riemann), or the new forms of 'calculus' with non-arithmetic operations (Galois, Hamilton and Boole). Starting from that moment the properties of operations came to the fore, and at the same time the nature of the specific objects—on which one could act as on numbers—began to loose importance. Exemplary of this formalist stance is Russell's quip: "*mathematics may be defined as the subject in which we never know what we are talking about, nor whether what we are saying is true*", or Hilbert's approach. Around this time an idea of mathematics took shape as the study of formal relationships, from the first general and conscientious global perspectives (Graßmann, Riemann, Hankel) to the axiomatic systems in use today (Cantor, Dedekind, Hilbert, Peano and Klein). Jumping forward, in 1939 a group of mainly French mathematicians, known under the pseudonym Nicolas Bourbaki (H. Cartan, Chevalley, Coulomb, Delsarte, Dieudonné, Ehresmann, de Possel, Mandelbrojt, Weil and later Bass, Schwartz, Serre, Grothendieck, Koszul, Eilenberg, Lang and Godement), began fighting for more formal rigour and simplicity. Their aim was to stress the concept of structure and found the entire body of mathematics on set theory (they viewed sets as 'de-structured' objects). The Bourbaki tack has been ferociously critiqued in the last thirty so years, especially due to its more rigid ramifications in mathematical education. Nonetheless it heralded a systematic order amidst the jungle of theories and sub-disciplines, and created over the years new concepts and a useful terminology, all still in vogue.

The simplest example of an algebraic structure is a *magma* $(G, *)$, where G is a set together with a binary operation $* \colon G \times G \to G$.

Definition Let $(G, *)$ and $(\widetilde{G}, \widetilde{*})$ be magmas. A map $f \colon G \to \widetilde{G}$ such that

$$f(x * y) = f(x) \widetilde{*} f(y) \qquad \text{for every } x, y \in G$$

is called a **homomorphism**. The space of such is indicated by $\mathrm{Hom}\,(G, \widetilde{G})$ (the operations' symbols are dropped unless strictly necessary). The term **isomorphism** denotes a bijective homomorphism. In case $(G, *) = (\widetilde{G}, \widetilde{*})$ a homomorphism is called an **endomorphism**,

$$\mathrm{End}\,(G) := \mathrm{Hom}\,(G, G),$$

and a bijective endomorphism is an **automorphism** of $(G, *)$

$$\mathrm{Aut}\,(G) := \mathrm{End}\,(G) \cap \mathfrak{S}(G).$$

Exercise Consider magmas $(G, *), (H, \diamond), (K, \uplus)$ and composable homomorphisms f, g as in the diagram

$$(G, *) \xrightarrow{\ f\ } (H, \diamond) \xrightarrow{\ g\ } (K, \uplus)$$
$$g \circ f$$

Prove that $g \circ f \in \mathrm{Hom}\,(G, K)$.

Example The exponential function $\exp \colon (\mathbb{R}, +) \ \to\ (\mathbb{R}, \cdot)$ is a homomorphism between the additive and the multiplicative structures of $\mathbb{R}$, because $e^{x+y} = e^x e^y$ for all $x, y \in \mathbb{R}$ (Theorem 12.2). It is not an isomorphism, as not surjective: only positive real numbers are exponentials ($\mathrm{Im}(\exp) = \mathbb{R}^+$), or equivalently, only positive reals have logarithms (Sect. 12.2).

Two structures are called **isomorphic** if there exists an isomorphism between them (which preserves all operations), and we shall use the symbol $\cong$. **Universal algebra**, initiated by Garrett Birkhoff, targets the study of algebraic structures up to isomorphisms. From this point of view, the fact that $\exp \colon \mathbb{R} \to \mathbb{R}^+$ is bijective implies that $(\mathbb{R}, +) \cong (\mathbb{R}^+, \cdot)$ represent the same abstract magma. Similarly, one speaks of 'the' ring $\mathbb{Z}_2 = (\{1, 0\}, +, \cdot)$, where

+	0	1
0	0	1
1	1	0

$\cdot$	0	1
0	0	0
1	0	1

A *cryptomorphism*—another invention of Birkhoff's—denotes an isomorphism between structures of different nature. Roughly, a cryptomorphism is an invisible correspondence to the uninitiated eye, but with an extremely useful application to teaching. In the book we have made a big fuss about two cryptomorphic structures: the algebra of formulas $(\mathscr{F}, \wedge, \vee, \neg, \bot, \top)$ of a language and the algebra of subsets $(\mathscr{P}(X), \cap, \cup, \complement, \varnothing, X)$ of a non-empty set X.

3.4   Axiom of Choice ☕

Consider an arbitrary collection $\{X_i\}_{i \in I}$ of non-empty sets. For the present purposes we may suppose $X_i \cap X_j = \varnothing$, $\forall i \neq j$ without loss of generality (it suffices to take $\tilde{X}_i = X_i \times \{i\}$, which has the same cardinality as X_i). We wish to choose an element in each set X_i, that is to say, define a **choice function**

$$\psi : I \longrightarrow \bigsqcup_{i \in I} X_i$$
$$j \longmapsto \psi(j) \in X_j \quad .$$

To do that we start by picking at random an element $x \in \bigsqcup_{i \in I} X_i$, which will belong to some X_{i_1}; then we choose $x' \in \bigsqcup_{i \neq i_1} X_i$ inside some other X_{i_2} (with $i_1 \neq i_2$), and so on and so forth. If $\bigsqcup_{i \in I} X_i$ is finite the choosing process will end and eventually produce the choice function. But if the union is transfinite, on the other hand, without further information there is no elementary reason warranting we can make infinitely many choices and thus define ψ. To make any progress on the matter we must gear up, and include in our arsenal the uber-famous

Axiom of Choice (AC)   Every collection of non-empty sets $\{X_i\}_{i \in I}$ admits a choice function.

As mentioned at the beginning, at first sight the AC is an innocent-looking statement, because it seems self-evident that we can choose an element in each subset of a family, simultaneously. When I is finite or countable $\vdash_{\mathrm{ZF}}$ AC, since the axiom of choice can be proved by induction. In particular, for $|I| = 1$ (there's only one set X_i) the map ψ boils down to one choice only, so the AC reduces to the fact that a non-empty set possesses an element.

We'll now discuss three reformulations of the AC, which we'll prove are equivalent to it. The first one is so simple that many books consider it clearly equivalent to the AC (and rightly so).

Axiom of Choice (v2)   Any set $X \neq \varnothing$ admits a (choice) function $\tilde{\psi} : \mathscr{P}(X) \setminus \{\varnothing\} \to X$ such that $\tilde{\psi}(Y) \in Y$.

Proof (AC $\Longrightarrow$ v2) If X is non-empty, the elements of $\mathscr{P}(X) \setminus \{\varnothing\}$ form a collection (which we'll assume disjoint). The AC tells there exists a choice function $\psi : \mathscr{P}(X) \setminus \{\varnothing\} \to \bigsqcup_{Y \in \mathscr{P}(X) \setminus \{\varnothing\}} Y$, and so we can define $\tilde{\psi}(Y) := \psi(Y) \in Y$.

(v2 $\implies$ AC) Vice versa, any family of non-empty, disjoint sets $\{X_i\}_{i\in I}$ is contained in $\mathscr{P}\left(\bigsqcup_{i\in I} X_i\right)$. Then $\tilde{\psi}(X_j) \in X_j$ defines a choice map $\psi(j) := \tilde{\psi}(X_j)$. $\qquad\qquad\qquad\square$

Another less obvious avatar of the AC is the following:

Axiom of Choice (v3) Let $\{X_i\}_{i\in I}$ be a collection of sets. Then

$$\prod_{i\in I} X_i = \varnothing \iff \exists j \in I \text{ such that } X_j = \varnothing.$$

(In view of Proposition 2.5 and generalisations, the interest lies in the case card $I > \aleph_0$.)

Proof (v3 $\implies$ AC) Let $\{X_i\}_{i\in I}$ be a family of non-empty sets. By assumption $\prod_{i\in I} X_i$ is non-empty, so there exists an element $(x_i)_{i\in I}$ in the product. Take the jth projection

$$\mathrm{pr}_j \colon \prod_{i\in I} X_i \longrightarrow X_j, \quad \mathrm{pr}_j(\ldots, x_j, \ldots) = x_j.$$

We can define a choice function by $\psi(j) := \mathrm{pr}_j(\ldots, x_j, \ldots) = x_j \in X_j$.

(AC $\implies$ v3). If there exists $(x_i)_{i\in I} \in \prod_{i\in I} X_i$, by definition $x_j \in X_j$ for every index, so $X_j \neq \varnothing$ for every $j \in I$. Conversely, if $X_j \neq \varnothing$ for all $j \in I$, we may define a choice $\psi(j) \in X_j$ for every $j \in I$. Therefore the collection $(\psi(j))_{j\in I}$ is an element of the product $\prod_{i\in I} X_i$. $\qquad\qquad\qquad\square$

Eventually, the AC can be phrased in terms of sections, too, see (3.6):

Axiom of choice (v4) Every surjective map admits a section.

Proof (v4 $\implies$ AC). Suppose $\{X_i\}_{i\in I}$ is a family of non-empty sets and $X = \bigsqcup_{i\in I} X_i$ is their disjoint union. Consider the projections

$$\mathrm{pr}_X \colon X \times I \to X, \quad \mathrm{pr}_X(x, i) = x \qquad \text{and} \qquad \mathrm{pr}_I \colon X \times I \to I, \quad \mathrm{pr}_I(x, i) = i.$$

As pr_I is onto it has a section $f \colon I \to X \times I$, so $\mathrm{pr}_X \circ f \colon I \to X$ is a choice function.

(AC $\implies$ v4). Let $g \colon Y \to X$ be an onto map. Then $g^{-1}(x) \neq \varnothing$ for every $x \in X$, and $Y = \bigsqcup_{x\in X} g^{-1}(x)$ admits a choice function $\psi \colon X \to Y$, $\psi(x) \in g^{-1}(x)$. But that means exactly $g \circ \psi = \mathbb{1}_X$. $\qquad\qquad\qquad\square$

Concerning the role of the AC as an axiom in set theory, let's suppose ZF is consistent (this hypothesis is inoffensive, because if it were inconsistent by adding any other axiom the situation wouldn't change). Then it is possible to prove

Theorem 3.10 (Gödel) *The negation of the axiom of choice ¬AC is not deducible in ZF.*

This fact, in particular, ensures that ZFC is a consistent theory. Furthermore,

Theorem 3.11 (Cohen) *The AC is not deducible in ZF theory.*

The two facts together imply the AC is logically independent of ZF theory.

The other relevant feature of the AC is the flurry of situations in which it appears. For starters, there exist propositions that are logically equivalent to the AC in each and every field of mathematics, from logic to category theory, from measure theory to algebra, from set theory to topology. Other versions of the AC, some highly non-evident, include the following:

- *well-ordering theorem* (Theorem 3.13)
- *Zorn's lemma* (Lemma 3.12)
- *ordinal trichotomy* (Theorem 6.42)
- *Tarski's theorem* on the cardinality of products (Theorem 7.7)
- existence of bases in vector spaces (Theorem 15.2)
- *upward/downward Löwenheim–Skolem theorems*
- *Krull theorem* for rings with unit
- *Tikhonov theorem* on products of compact topological spaces

The fact the AC is an axiom naturally begs the question whether one should embrace it or make do without it, and what's the difference. Two arguments in favour of adopting the AC are:

1) possible simplifications: its independence (p. 23) implies many results can be proved in ZF if and only if they are provable in ZFC. For instance all theorems that can be formulated in Peano's arithmetics, like $P = NP$, or the Riemann hypothesis etc. [101] (clearly, there might exist a shorter proof in ZFC than in ZF!);
2) convenience: without AC certain results might not hold in categories with large cardinality. Below is a short list of classical results, whose proof is based on the AC:

  - countability of countable unions of countable sets (Theorem 7.22)
  - existence of subsets of $\mathbb{R}$ that aren't Lebesgue-measurable[3] [96]
  - existence of algebraic closures of fields
  - existence of transcendental bases for field extensions
  - *Stone representation for Boolean algebras* (Theorem 5.15)
  - *ultrafilter lemma* (Lemma 5.13), or the existence of free ultrafilters on any set
  - existence of skeletons for small categories
  - *Hahn–Banach theorem*

---

[3] If we replace the AC with axiom LM: "*every subset of $\mathbb{R}$ is Lebesgue measurable*" in ZF, the new system stays consistent, as shown by Solovay in 1960. Interestingly, ¬LM can be established only requiring that $\mathbb{R}$ is well ordered (just one set, rather than all).

- *Baire's category theorem*
- existence of orthonormal bases in Hilbert spaces
- *Hilbert's basis theorem*
- existence of Stone–Čech compactifications for Tikhonov spaces (cf. Example 5.19)
- completeness of first-order theories (Theorem 1.19)
- compactness of first-order theories.

Good references on the topic are [42, 50]. A singular and much cited consequence of the AC is the *Banach–Tarski paradox* (1924), according to which a solid ball in three-space can be decomposed in a finite number of pieces which can be reassembled in a different way to give two identical copies (same volume) of the original ball. Hausdorff had proved in 1914 a similar statement for the surface area of the two-sphere. The layman version goes by 'pea and sun paradox' (a pea can be chopped up and reassembled into the sun). Upon reflection, doubling a sphere or turning a pea into the sun is no less surprising than partitioning an infinite set into subsets of the same cardinality, see Sect. 7.7.

    A devil's advocate position against the AC is its non-constructive nature, already patent from the number of 'existence' results listed above. More to the point: defining a well order on $\mathbb{R}$ using a choice function is theoretically possible, but nonetheless a bit of a challenge.

    In the remaining part of the section we'll prove the equivalence between the AC, Zorn's lemma, and the well-ordering Theorem 3.13, which perhaps are the better known reformulations of the axiom of choice. To set the tone it's instructive to reproduce the pun [64] whereby the axiom of choice is obviously true, the well-ordering principle obviously false, and who can tell about Zorn's lemma? Although the three are equivalent (Theorem 3.14), what the quip means is that many mathematicians find the axiom of choice to be intuitive (of course we can choose elements), the well-ordering principle counterintuitive (how on earth is $\mathbb{R}$ well ordered?), and Zorn's lemma too complex for any insight.

    Recall that a chain in a poset $(X, \leqslant)$ is a (totally) ordered subset $C \subseteq X$ under the induced relation $\leqslant |_C$ (see Definition 2.13). We also remind the convention that $x \leqslant Y$ means $x \leqslant y$ for all $y \in Y$.

Lemma 3.12 (Zorn)   *Every partially ordered set $X \neq \varnothing$ whose chains are upper bounded contains a maximal element.*

Despite being attributed to Zorn (who proved it in 1935), Lemma 3.12 was known to Kuratowski already in 1922. Besides, the lemma is an alternative (and posterior) formulation of **Hausdorff's maximal principle** (1914): in a poset X, any ordered subset $Y \subseteq X$ is contained in a maximal ordered subset $Y_{\max} \supseteq Y$.

Theorem 3.13 (Well-Ordering Theorem (Zermelo)) *Every set $X \neq \varnothing$ possesses a well ordering.*

Theorem 3.14 *In ZF theory the following facts are logically equivalent:*

(AC)   *axiom of choice*
(ZL)   *Zorn's lemma*
(WO)   *well-ordering theorem.*

We shall follow [63, p. 619] to prove the theorem, and we'll need a few preliminaries.

Lemma 3.15 (Bourbaki–Witt) *Let $X \neq \varnothing$ be a poset whose chains $C \neq \varnothing$ admit least upper bound $\sup C \in X$. Any map $f \colon X \to X$ such that $x \leqslant f(x)$ for every $x \in X$ has a fixed point $x_0 \in X$, i.e. $f(x_0) = x_0$.*

Proof For the purposes of this lemma we'll call 'good' a subset E such that

$$f(E) \subseteq E \text{ and } \sup C \in E \text{ for every chain } C \subseteq E.$$

First, X is good, and the intersection I_x of all good sets containing $x \in X$ is also good. Moreover $\{y \in X \mid x \leqslant y\}$ is good and it contains x. Then $I_x \subseteq \{y \in X \mid x \leqslant y\}$, implying $x \leqslant I_x$.

Take an element $a \in X$ and define $I := I_a$ (a good set). If we prove that I is a chain, then $x_0 := \sup I \in X$ and hence $x_0 \in I$. But $f(x_0) \in f(I) \subseteq I$ because I is good. Therefore $x_0 \leqslant f(x_0) \leqslant x_0$ implies x_0 is a fixed point for f.

We need to show I is totally ordered. Consider the set $C \subseteq I$ of elements x for which there exists a chain $C_x \ni a, x$ such that

(1) $a \leqslant C_x \leqslant x$
(2) $f(C_x \setminus \{x\}) \subseteq C_x$
(3) the supremum of any non-empty subchain of C_x belongs in C_x.

Taking $C_a = \{a\}$ shows that $a \in C$. $\square$

Lemma 3.16 *If $x \in C$ then $I = I_x \cup C_x$.*

Proof By definition $C_x \subseteq I$, and $I \cap I_x$ is good so it contains I_x. Hence $I_x \subseteq I$, and therefore $I_x \cup C_x \subseteq I$.

For the opposite inclusion $I \subseteq I_x \cup C_x$ it's enough to prove that $I_x \cup C_x$ is good and contains a. Since $a \in C_x$ though, there remains to see to goodness. From $x \in I_x$ we have $I_x \cup C_x = I_x \cup (C_x \setminus \{x\})$, and by item (2) $f(I_x \cup C_x) = f(I_x) \cup f(C_x \setminus \{x\}) \subseteq I_x \cup C_x$.

To finish we must prove that the least upper bound u of any non-empty chain in $I_x \cup C_x$ belongs to $I_x \cup C_x$. It suffices to show $u \in I_x \cup C_x$. Now, u belongs to I, and observe that

$$C_x \leqslant x \leqslant I_x. \tag{3.8}$$

If a chain has an element in common with I_x, (3.8) forces $x \leqslant u$, so the chain's intersection with I_x is a non-empty chain in I_x with supremum u. Therefore $u \in I_x$. If there is no common element, the chain lies in C_x, and (3) implies $u \in C_x$. □

Aside observation: Lemma 3.16 and (3.8) imply $I_x \cap C_x = \{x\}$, so $C_x = (I \setminus I_x) \cup \{x\}$. In other words properties (1), (2), (3) actually characterise C_x.

Lemma *C is good and $a \in C$.*

Proof We already know $a \in C$. If we take an $x \in C$ Lemma 3.16 forces $C_x \cup \{f(x)\}$ to satisfy (1). Also (2) holds because $x \leqslant f(x)$. Any non-empty chain in $C_x \cup \{f(x)\}$ is contained in C_x and its supremum belongs to C_x, or it contains $f(x)$ and so its supremum is $f(x)$. Therefore $C_x \cup \{f(x)\}$ satisfies (3) as well. This confirms we can define $C_{f(x)}$ to be $C_x \cup \{f(x)\}$. Consequently $f(x) \in C$, in other words $f(C) \subseteq C$.

At last, pick a non-empty chain $(x_i) \subseteq C$ with $\sup_i \{x_i\} = u \in I$. The set $\bigcup_i C_{x_i} \cup \{u\}$ fulfils (1), (2), (3). □

Let's go back and finish off Lemma 3.15. Now $C \supseteq I_a = I$, so $C = I$. But taking $x, y \in I$, Lemma 3.16 forces $y \in I_x$ or $y \in C_x$. In the former situation (3.8) implies $x \leqslant y$, in the latter $y \leqslant x$. All-in-all I is totally ordered.

Proof of Theorem 3.14 We'll prove the three implications (AC) $\Longrightarrow$ (ZL) $\Longrightarrow$ (WO) $\Longrightarrow$ (AC).

(AC) $\Longrightarrow$ (ZL) Suppose all chains in the poset $(X, \leqslant)$ are upper bounded. Define the collection

$$\mathscr{C} := \{C_i \subseteq X \mid C_i \text{ non-empty chain}\}$$

ordered by $\subseteq$. An element $C \neq \varnothing$ of $\mathscr{C}$ is a collection of nested chains $C_\alpha \subseteq X$, and the chain $\bigcup_\alpha C_\alpha \subseteq X$ equals $\sup C$ (Proposition 2.12). By contradiction, suppose no chain in X is maximal. If we take a non-empty chain $C \subseteq X$ the AC gives us a map f such that $C \subset \psi(C)$. This choice function satisfies Lemma 3.15, so it must have a fixed point, which in turn means X contains a maximal chain ⨳. So necessarily there exists a maximal chain $C_0 \subseteq X$. By assumption, call $u_0 = \sup C_0$ its supremum. For any element $u \in X$ larger than u_0, $C_0 \cup \{u\}$ is a chain, so $C_0 \cup \{u\} = C_0$. Put differently, $u \in C_0$ and $u = u_0$ is maximal in X.

(ZL) $\Longrightarrow$ (WO) To define a well ordering on $X \neq \varnothing$ we introduce the family

$$\mathscr{Y} := \{(Y, \leqslant_Y) \mid Y \subseteq X, \ \leqslant_Y \text{ well order}\}.$$

Note $(\varnothing, \varnothing) \in \mathscr{Y}$, so the family is not empty. Let's equip $\mathscr{Y}$ with the partial order $\preccurlyeq$:

$$(Y, \leqslant_Y) \preccurlyeq (Z, \leqslant_Z) \iff \begin{cases} Y \subseteq Z \\ \leqslant_Y \text{ is a restriction of } \leqslant_Z \\ Y \leqslant_Z (Z \setminus Y) \end{cases}$$

By setting

$$e \leqslant_\cup f \text{ whenever } e \leqslant_{Y_k} f \quad (\text{supposing } e \in Y_j \subseteq Y_k \ni f),$$

it follows that every chain $\{(Y_i, \leqslant_i)\}$ in $\mathscr{Y}$ is bounded by the obvious 'union' $(\bigcup_i Y_i, \leqslant_\cup)$.

Lemma $\leqslant_\cup$ *is a well ordering on* $\bigcup_i Y_i$.

Proof Take $\varnothing \neq Z \subseteq \bigcup_i Y_i$, so any element in Z belongs to some Y_{i*}. As Y_{i*} is well ordered, there exists $z_0 = \min(Y_{i*} \cap Z)$. We claim $z_0 \leqslant_\cup z$ for every $z \in Z$. In fact, if $z \in Y_{i*}$ then clearly $z_0 \leqslant z$. If on the contrary $z \notin Y_{i*}$, then $z \in Y_{j*}$. Bus since $Y_{j*} \not\subseteq Y_{i*}$ necessarily $(Y_{i*}, \leqslant_{i*}) \preccurlyeq (Y_{j*}, \leqslant_{j*})$. Therefore $z \in Y_{j*} \setminus Y_{i*}$, and so $z_0 \leqslant_\cup z$. That means $z_0 = \min Z$. □

Now since $(\bigcup_i Y_i, \leqslant_\cup)$ is an upper bound of $\mathscr{Y}$, Zorn's lemma guarantees there is a maximal element $(M, \leqslant_M)$. We claim $M = X$. If not, pick $x \in X \setminus M$ and impose $M \leqslant x$. This would give a chain $M \cup \{x\} \succcurlyeq M$ ⨪⨪. Hence $X = M$ is well ordered.

(WO) $\implies$ (AC) Take a family $\{X_i\}_{i \in I}$ of non-empty sets. Its union $\bigcup_{i \in I} X_i$ has a well order $\preccurlyeq$ by hypothesis. Setting $\psi(i) = \min X_i$ with respect to the restriction $\preccurlyeq |_{X_i}$ we produce a choice function. □

One further consideration on Theorem 3.14. The implication (AC) $\implies$ (WO) is only valid in first-order logics, since already in second order it's no longer true.

Bertrand Russel [85, pp.125-127] made a (literally) fitting metaphor to explain why the axiom of choice is necessary to select a set from an infinite number of pairs of socks, but not an infinite number of pairs of boots. That's because we can distinguish right from left in boots, and therefore we can select one out of each pair (one possible choice function would pick all right boots, say). Without the axiom of choice one cannot assert that a choice function exists for pairs of socks, because left and right socks are indistinguishable, and we can't be sure to obtain a collection consisting of one sock from each pair.

3.5 Families of Sets ☕

We return to set-theoretical operations, this time defined on arbitrary numbers of sets. Handling families of any cardinality turns out to be important in many situations, first and foremost in matters involving ☞ *topology*. Whilst the finite (or countable) case can be treated inductively starting from what we have seen in Sect. 2.1, unions, intersections and products of uncountably many sets are deeply intertwined with the axiom of choice (Sect. 3.4).

Definition A **family** or **collection** of sets $A_i \subseteq \mathscr{U}$, written

$$\mathscr{A} = \{A_i \mid i \in I\} = \{A_i\}_{i \in I},$$

is a function $\mathscr{A} : I \to \mathscr{P}(\mathscr{U})$ with $I \neq \emptyset$. The family is called finite/countable if the index set I is finite/countable.

The familiar set-theoretical operations $\cup, \cap$ generalise to families in the form of unions/intersections of the images of the map. There are several notations to indicate them:

$$\bigcup \mathscr{A} = \bigcup_{A_i \in \mathscr{A}} A_i = \bigcup_{i \in I} A_i := \{x \mid \exists j \in I \text{ such that } x \in A_j\}$$

$$\bigcap \mathscr{A} = \bigcap_{A_i \in \mathscr{A}} A_i = \bigcap_{i \in I} A_i := \{x \mid x \in A_j \ \forall j \in I\}$$

The reason (essentially the only one) for imposing I be non-empty is that if we allowed an empty index set then, by contradiction, $\bigcap_{i \in \emptyset} A_i = \mathscr{U}$. So we'll assume henceforth every collection non-empty.

Theorem *Let $\mathscr{A} = \{A_i\}_{i \in I}$ be a family of sets and B another set. Then*

i) $B \cap \bigcup_{i \in I} A_i = \bigcup_{i \in I}(B \cap A_i), \quad B \cup \bigcap_{i \in I} A_i = \bigcap_{i \in I}(B \cup A_i),$ *(distributivity)*

ii) $\overline{\bigcup_{i \in I} A_i} = \bigcap_{i \in I} \overline{A_i}, \quad \overline{\bigcap_{i \in I} A_i} = \bigcup_{i \in I} \overline{A_i}$ *(De Morgan laws)*

iii) *If $J \subseteq I$ then* $\bigcup_{j \in J} A_j \subseteq \bigcup_{i \in I} A_i, \quad \bigcap_{j \in J} A_j \supseteq \bigcap_{i \in I} A_i.$ *(monotonicity)*

Proof Exercise. □

Exercises Take two families of sets $\mathscr{A} = \{A_i\}_{i \in I}$, $\mathscr{B} = \{B_j\}_{j \in J}$. Prove that

i) $\bigcup \mathscr{P}(\mathscr{A}) = \mathscr{A} \subseteq \mathscr{P}(\bigcup \mathscr{A})$;

ii) $\bigcup_{i \in I} A_i \cap \bigcup_{j \in J} B_j = \bigcup_{i,j}(A_i \cap B_j), \quad \bigcap_{i \in I} A_i \cup \bigcap_{j \in J} B_j = \bigcap_{i,j}(A_i \cup B_j),$

where $\bigcap_{i,j}$ indicates the intersection over the Cartesian product of indices $(i, j) \in I \times J$;

iii) $f(\bigcup_{i \in I} A_i) = \bigcup_{i \in I} f(A_i), \quad f(\bigcap_{i \in I} A_i) \subseteq \bigcap_{i \in I} f(A_i),$
for any map f defined on $\bigcup \mathscr{A}$.

When the family consists of a sequence of sets $A_1, A_2, \ldots$ (hence I is countable), the union and intersection are denoted by $\bigcup\limits_{i=1}^{\infty} A_i$ and $\bigcap\limits_{i=1}^{\infty} A_i$, while for a finite number we write

$$\bigcup_{i=1}^{n} A_i = A_1 \cup \cdots \cup A_n, \qquad \bigcap_{i=1}^{n} A_i = A_1 \cap \cdots \cap A_n.$$

Exercise 3.17 Consider n finite sets A_i with union $A = \bigcup\limits_{i=1}^{n} A_i$, and show

$$(\chi_A - \chi_{A_1})(\chi_A - \chi_{A_2}) \cdots (\chi_A - \chi_{A_n}) = 0.$$

Hint: generalise Proposition 3.4 (iv) by induction. (This will essentially prove formula (6.5).)

The next theorem handles the behaviour of **nested** collections $\mathscr{A} = \{A_i\}_{i \in \mathbb{N}}$, meaning $A_i \supseteq A_{i+1}$ for all i. We shall prove a special case regarding intervals, although the statement can be made slightly more general by fishing in the toolbox of metric spaces (Theorem 19.23).

Theorem 3.18 (Cantor's Intersection Property) *Let $\mathscr{A} = \{A_i\}_{i \in \mathbb{N}}$ be a nested family of non-empty, closed and bounded intervals $A_i = [a_i, b_i]$ in $\mathbb{R}$:*

$$\ldots \leqslant a_{i-1} \leqslant a_i \leqslant \overbrace{\underbrace{a_{i+1} \leqslant \ldots \leqslant b_{i+1}}_{A_{i+1}}}^{A_i} \leqslant b_i \leqslant b_{i-1} \leqslant \ldots$$

Then $\bigcap \mathscr{A} \neq \varnothing$.

Proof The proof relies crucially on the Bolzano-Weierstraß theorem. For every i pick any number $x_i \in [a_i, b_i]$. The sequence $\{x_i\}$ is contained in the largest (first) interval, A_1 say, which makes it a bounded set. Then Theorem 19.19 guarantees it accumulates at some ξ, and arguing along those lines it can be shown that $\xi \in \bigcap \mathscr{A}$. □

Definition 3.19 A **covering** (or **cover**) of a set X is a collection $\mathscr{A} = \{A_i\}_{i \in I}$ such that

$$X \subseteq \bigcup \mathscr{A}.$$

A **subcover** of $\mathscr{A}$ is a subcollection $\{A_j\}_{j \in J \subseteq I} \subseteq \mathscr{A}$ that still covers: $X \subseteq \bigcup\limits_{j \in J} A_j$.

Exercise

i) Show that any cover $\mathscr{A}$ of X can be 'trimmed', so that X is equal to $\bigcup \mathscr{A}$ (this doesn't mean we have a partition, as the subsets might overlap).
ii) Consider the cover $\{(n-1, n+1) \subseteq \mathbb{R} \mid n \in \mathbb{Z}\}$ of $\mathbb{R}$. Prove that any finite subcollection of it can't cover $\mathbb{R}$.

A specialised version of the *Heine–Borel theorem* states that if we take a closed bounded interval $[a, b] \subseteq \mathbb{R}$ and a covering $\{(a_i, b_i)\}_{i \in I}$ made of open subintervals, we may discard many subintervals and still cover the set with the rest. As a matter of fact there always exists a finite subcover:

$$\bigcup_{i \in I}(a_i, b_i) = [a, b] \implies \exists J \subseteq I \text{ finite}: \bigcup_{j \in J}(a_j, b_j) = [a, b]. \tag{3.9}$$

The theorem might seem trivial at first sight, but reveals its depth in multiple crucial applications, see Theorem 19.23. Related to this is the fact that a family of closed subintervals $\{[x_k, y_k]\}_{k \in K}$ of $[a, b]$ has the **finite-intersection property**:

$$\bigcap_{j \in J}[x_j, y_j] \neq \varnothing \text{ for every finite } J \subseteq K \implies \bigcap_{k \in K}[x_k, y_k] \neq \varnothing. \tag{3.10}$$

In ☞ *topology* a set such as $[a, b] \subseteq \mathbb{R}$ that satisfies either (3.10) or (3.9) would be called **compact**.

Example 3.20 (Cantor's Ternary Set) We won't be too fussy about the formalisation (read, induction) for the sake of the explanation. Let's trisect $[0, 1]$ in intervals of the same length, two closed and one open. Delete the middle open one, divide the others in three and erase the middle intervals in the second row:

$$[0, 1] = \qquad \left[0, \tfrac{1}{3}\right] \qquad \cup \left(\!\!\!\!\!\diagdown\diagdown\!\!\!\!\!\right) \cup \qquad \left[\tfrac{2}{3}, 1\right]$$

$$= \left[0, \tfrac{1}{9}\right] \cup \left(\!\!\!\diagdown\diagdown\!\!\!\right) \cup \left[\tfrac{2}{9}, \tfrac{1}{3}\right] \qquad \cup \qquad \left[\tfrac{2}{3}, \tfrac{7}{9}\right] \cup \left(\!\!\!\diagdown\diagdown\!\!\!\right) \cup \left[\tfrac{8}{9}, 1\right]$$

Now iterate the process. At the first stage we delete one subinterval of length $1/3$, at the second stage two intervals of length $1/9$, and by induction, at stage n we should erase $1 + 2 + 2^2 + \cdots + 2^{n-1} = \frac{1-2^n}{1-2} = 2^n - 1$ open intervals each of length $1/3^n$, as in Fig. 3.8. What's left is made of 2^n disjoint closed intervals, which we may stratify as a nested sequence following the above pattern:

Fig. 3.8 The Cantor set is like dust

$$C_1 = C_{1,1} \cup C_{1,2}$$

$$C_2 = C_{2,1} \cup C_{2,2} \cup C_{2,3} \cup C_{2,4}$$

Exercise To understands what is going on, write down the next level C_3 of the construction in the above format. Then prove by induction that

$$C_n = C_{n,1} \cup \cdots \cup C_{n,2^n}.$$

(To be pedantic, these sets are $C_n = \displaystyle\bigcup_{k=0}^{3^{n-1}-1} \left(\left[\tfrac{3k}{3^n}, \tfrac{3k+1}{3^n} \right] \cup \left[\tfrac{3k+2}{3^n}, \tfrac{3k+3}{3^n} \right] \right)$, but this is not relevant at the moment.)

Continuing the process indefinitely, a countable family of intervals will disappear from [0, 1], eventually leaving us with the **Cantor set**

$$C = \bigcap_{n=1}^{\infty} C_n.$$

In view of Theorem 3.18 we know this is not empty. Since each C_n has total length $2^n/3^n$, with a little leap of faith the length of C is seen to equal $\displaystyle\lim_{n\to\infty} \frac{2^n}{3^n} = 0$. This seems to indicate a peculiar structure resembling very fine dust. The Cantor set is used in set theory, ☞ *topology* and *fractal geometry* to exemplify a number of properties regarding cardinality (see Example 7.27), density (in itself, nowhere density), isolated points (any two points in C, as close as we want, are separated by an entire interval not contained in C), self-similarity (C is equal to two copies of itself shrunk by a factor of 3, hence the prototype of a *fractal*). Here are two features we can elaborate on, without invoking any topology whatsoever.

(1) We can show $C \neq \varnothing$ directly. The Cantor set certainly contains the endpoints of the intervals in every C_n, but not only. For instance, 1/4 is not an endpoint, since the denominator is not a power of 3. The base-three expansion of 1/4 reads

$$\frac{1}{4} = 2 \cdot \frac{1/9}{1 - 1/9} = 2 \sum_{n=1}^{\infty} \frac{1}{9^n} = \frac{0}{3^0} + \frac{0}{3^1} + \frac{2}{3^2} + \frac{0}{3^3} + \frac{2}{3^4} + \frac{0}{3^5} + \frac{2}{3^6} + \cdots$$

in other words $(1/4)_{\text{base }3} = 0.\overline{02} \in C$.

(2) The same circle of ideas allows us to describe C alternatively. Take the base-three representation of a number in [0, 1]

$$\sum_{n=1}^{\infty} \frac{c_n}{3^n} = 0.c_1 c_2 \ldots \qquad \text{where } c_i \in \{0, 1, 2\}.$$

For example:

$$\frac{1}{3} = 0.\overline{3} = \begin{cases} \dfrac{0}{3^0} + \dfrac{1}{3^1} + \dfrac{0}{3^2} + \dfrac{0}{3^3} + \ldots = _{\text{base 3}} 0.1 \\ \dfrac{0}{3^0} + \dfrac{0}{3^1} + \dfrac{2}{3^2} + \dfrac{2}{3^3} + \ldots = _{\text{base 3}} 0.0\overline{2} \end{cases}$$

With a little work it can be proved (we will not do it here) that only the rationals with digits $0, 2$ (not 1) belong to C. As a matter of fact, one could prove any number in C has a unique ternary expansion without any 1 (the idea is that a 1 can only arise if the number in question belongs to some middle interval).

Exercises

i) Write in base three: $\dfrac{2}{3} = 0.\overline{6}$ and $\dfrac{4}{9} = 0.\overline{4}$ (observe that $4/9$ has two expansions, both containing 1s).

ii) Show that $x \in C \implies x/3, \ 1 - x \in C$. Exploit these symmetries to find other elements of C.

To finish, we return to arbitrary families of sets, and consider products of any number of terms.

Definition The Cartesian product $\prod_{i \in I} A_i$ of the family $\mathscr{A} = \{A_i\}_{i \in I}$ is the set of maps

$$\prod_{i \in I} A_i := \left\{ f : I \to \bigcup_{i \in I} A_i \ \middle| \ f(i) \in A_i \ \forall i \in I \right\}.$$

Note how the definition is made possible by the axiom of choice.

Corollary *If in family $\{A_i\}_{i \in I}$ all sets are the same, $A_i = A$, then $\prod_{i \in I} A_i = A^I$.*

Proof Exercise. $\qquad\qquad\qquad\qquad\qquad\qquad\qquad\qquad\qquad\qquad\qquad$ □

As before, for countable collections the product is indicated $\prod_{i=1}^{\infty} A_i$, and for finite ones one writes $\prod_{i=1}^{n} A_i = A_1 \times \cdots \times A_n$. As usual, one uses the shorthand A^n in case $A_i = A$ for every i, tying in nicely with the previous corollary.

Corollary *For a finite collection $\{A_1, A_2, \ldots, A_n\}$*

$$\prod_{i=1}^{n} A_i = \varnothing \iff \exists j \in \{1, \ldots, n\} \text{ such that } A_j = \varnothing.$$

Proof We'll use induction on n. For $n = 1$ the statement is tautological, and $n = 2$ is proposition 2.5. Consider $\prod\limits_{i=1}^{k+1} A_i = \left(\prod\limits_{i=1}^{k} A_i\right) \times A_{k+1}$. By the aforementioned proposition, this is empty if and only if either $A_{k+1} = \varnothing$ or $\prod\limits_{i=1}^{k} A_i = \varnothing$ (which by induction hypothesis means one of the A_i is empty). $\qquad\square$

We stress once more that infinite products behave similarly, but the argument is now dictated by the AC (v3), p. 63.

Exercises Let $\{A_i\}_{i \in I}$, $\{B_j\}_{j \in J}$ be collections of sets, f a map with range containing the family $\{A_i\}_{i \in I}$. Prove that

i) $\displaystyle\bigcup_{i \in I} A_i \times \bigcup_{j \in J} B_j = \bigcup_{i,j}(A_i \times B_j)$ and $\displaystyle\bigcap_{i \in I} A_i \times \bigcap_{j \in J} B_j = \bigcap_{i,j}(A_i \times B_j)$;

ii) $\displaystyle f^{-1}\left(\bigcup_{i \in I} A_i\right) = \bigcup_{i \in I} f^{-1}(A_i), \quad f^{-1}\left(\bigcap_{i \in I} A_i\right) = \bigcap_{i \in I} f^{-1}(A_i).$

Chapter 4
More Set Theory and Logic ☕

This chapter develops set-theory topics that are useful in algebra and topology. It will give us the excuse to provide proofs to the Deduction and Soundness Theorems 1.13, 1.18, and introduce the first elements in model theory. The final part attempts to contextualise the scientific revolution unleashed by Gödel's theorems.

4.1 Moore Operators

Definition Let $S \neq \varnothing$ be a set. A **Moore (closure) operator** is a function $K : \mathscr{P}(S) \to \mathscr{P}(S)$ such that

i) $X \subseteq K(X)$
ii) K is increasing: $X \subseteq Y \implies K(X) \subseteq K(Y)$
iii) K is idempotent: $K\big(K(X)\big) = K(X)$

for all $X, Y \in \mathscr{P}(S)$.

Exercises Show that

i) K is increasing $\iff K(X \cup Y) \subseteq K(X) \cup K(Y)$;
ii) K is idempotent $\iff K\big(K(X)\big) \subseteq K(X)$;
iii) If K is a Moore operator, and $T \subseteq S$, the restriction $K\big|_T : \mathscr{P}(T) \to \mathscr{P}(T)$, $K\big|_T(Y) = K(Y) \cap T$ is a Moore operator.

Example 4.1 If (S, τ) is a topological space, the closure operator $K(X) = \overline{X}^\tau$ is a Moore operator.

A second example is the following:

© The Author(s), under exclusive license to Springer Nature Switzerland AG 2021
S. G. Chiossi, *Essential Mathematics for Undergraduates*,
https://doi.org/10.1007/978-3-030-87174-1_4

Definition 4.2 Let $\Gamma \subseteq \mathscr{F}$ be a set of formulas of a first-order language. The operator $\mathfrak{Con}\colon \mathscr{P}(\mathscr{F}) \to \mathscr{P}(\mathscr{F})$

$$\mathfrak{Con}(\Gamma) := \{X \in \mathscr{F} \colon \Gamma \vdash X\} =: \overline{\Gamma}$$

is called **logical consequence**.

Proposition $\mathfrak{Con}$ *is a Moore operator.*

Proof

i) If $A \in \Gamma$, trivially A alone is a deduction from Γ. Hence $A \in \mathfrak{Con}(\Gamma)$, which proves $\Gamma \subseteq \mathfrak{Con}(\Gamma)$.

ii) Take $\Gamma \subseteq \Delta$. Then $\Gamma \vdash A$ implies $\Delta \vdash A$, so $\overline{\Gamma} \subseteq \overline{\Delta}$.

iii) Consider $A \in \mathfrak{Con}(\mathfrak{Con}(\Gamma))$, i.e. $\mathfrak{Con}(\Gamma) \vdash A$, and a deduction $A_1, \ldots, A_n = A$ from $\mathfrak{Con}(\Gamma)$. We shall prove $\Gamma \vdash A_i$ by induction on i. Base: if $A_1 \in \Lambda$ then $\vdash A_1$ and $\Gamma \vdash A_1$; if $A_1 \in \overline{\Gamma}$ then $\Gamma \vdash A_1$. Inductive step: if $A_1 \in \Lambda \cup \mathfrak{Con}(\Gamma)$, as before, the claim is immediate. Suppose $A_k \doteq A_j \to A_i$, $k, j < i$ and $\Gamma \vdash A_k, A_j$ by hypothesis, i.e. $\Gamma \vdash A_k, \Gamma \vdash A_j \to A_i$. The sequence

$$\underbrace{\cdots, A_j,}_{\substack{\text{deduction} \\ \text{from } \Gamma}} \underbrace{\cdots, A_j \to A_i}_{\substack{\text{deduction} \\ \text{from } \Gamma}}$$

is a deduction from Γ. But A_i is obtained by MP from A_j and $A_j \to A_i$, so we can append A_i to the end and obtain a deduction $\cdots A_J, \cdots, A_j \to A_i, A_i$ from Γ. Therefore $\Gamma \vdash A_i$, for all i. In particular, $\Gamma \vdash A_n = A$, and so $A \in \mathfrak{Con}(\Gamma) = \mathfrak{Con}(\mathfrak{Con}(\Gamma))$.

□

Exercise Show that $\mathfrak{Con}(\varnothing) = \mathfrak{Con}(\Lambda) = \mathfrak{Thm}$.

Definition Γ is called **deductively closed** whenever $\Gamma \vdash A \implies A \in \Gamma$ for all formulas $A \in \mathscr{F}$. Equivalently, Γ is deductively closed if $\mathfrak{Con}(\Gamma) \subseteq \Gamma$ (closed under logical consequence).

Exercise Show that $\Gamma \subseteq \mathscr{F}$ is deductively closed if and only if $\Lambda \subseteq \Gamma$ and Γ is MP-closed ($A, A \to B \in \Gamma \implies B \in \Gamma$).

Definition We call **theory** a deductively closed set of formulas $\mathbb{T} \subseteq \mathscr{F}$. Any subset $\Gamma \subseteq \mathbb{T}$ such that $\mathfrak{Con}(\Gamma) = \mathbb{T}$ is called a **presentation** of the theory $\mathbb{T}$.

A **theorem** of theory $\mathbb{T}$ is a formula B that can be deduced from a presentation Γ:

$$\Gamma \vdash B \iff B \in \overline{\Gamma} \iff B \in \mathbb{T} \iff B \in \overline{\mathbb{T}}.$$

For that reason we also indicate theorems in $\mathbb{T}$ by $\mathfrak{Thm}_\mathbb{T} = \overline{\mathbb{T}}$. Note that any theory possesses at least one presentation (itself).

Definition A theory $\mathbb{T}$ is said to be **complete** if $X \in \mathbb{T} \iff \neg X \notin \mathbb{T}$ for every formula X. It is called **maximally non-contradictory** if it is a maximal element in the class of non-contradictory theories.

Exercise Show that the only contradictory theory is $\mathscr{F}$.

Moore operators give us the right language to provide simple proofs for two fundamental theorems of first-order logic: the Deduction Theorem 1.13 and the Soundness Theorem 1.18.

Deduction Theorem $\Gamma, A \vdash B \implies \Gamma \vdash A \to B$.

Proof Set $\Delta = \{X \in \mathscr{F} \mid \Gamma \vdash A \to X\}$. The goal is to show $\overline{\Gamma \cup \{A\}} \subseteq \Delta$. For this it suffices to prove Δ is deductively closed and that it contains both Γ and A, i.e.: $\Lambda \subseteq \Delta, \Gamma \cup \{A\} \subseteq \Delta$ and Δ is MP-closed.

So take an axiom $X \in \Lambda$. As the sequence $X, \underbrace{X \to A \to X}_{\text{tautology} \in \Lambda}, A \to X$ is a deduction, we have $\Gamma \vdash A \to X$ and $X \in \Delta$.

Similarly if $X \in \Gamma$ (same reasoning), so $\Gamma \subseteq \Delta$. Since $A \to A \in \Lambda$, we have $\Gamma \vdash A \to A$ and $A \in \Delta$. Therefore $\Gamma \cup \{A\} \subseteq \Delta$.

Suppose $\Gamma \vdash A \to X$ and $\Gamma \vdash A \to (X \to Y)$. The sequence

$$\cdots, A \to (X \to Y), \underbrace{\big(A \to (X \to Y)\big) \to (A \to X) \to (A \to Y)}_{\text{tautology}},$$

$$\underbrace{(A \to X) \to (A \to Y)}_{\text{by MP}}, \cdots, A \to X, \underbrace{A \to Y}_{\text{by MP}}$$

is a deduction from Γ. Hence $Y \in \Gamma$, and Δ is closed under modus ponens. $\quad\square$

Regarding the other announced result:

Soundness Theorem $\vdash A \implies \vDash A$.

Proof By setting $\Delta = \{X \in \mathscr{F} : \vDash X\}$ the proof reduces to showing $\overline{\Gamma} \subseteq \Delta$, that's to say: $\Gamma \cup \Lambda \subseteq \Delta$ and Δ is MP-closed.

If $X \in \Gamma$ then $\Gamma \vDash X$. If $X \in \Lambda$, it can be proved (we shall not) that every logical axiom is true: $\vDash \Lambda$. So again $\Gamma \vDash X$. In both cases $X \in \Delta$. To conclude, Δ is MP-closed by virtue of Theorem 1.5. $\quad\square$

Concerning the above observation that

$$\vDash \Lambda$$

let's just comment that axioms are not only true but **valid**: they are true in any model of the theory under exam. But since we are considering one model at a time, for us 'true' is a synonym of 'valid'.

Remark There are other types of 'closure' operators $\mathfrak{K} \colon \mathscr{P}(S) \to \mathscr{P}(S)$ that play a role in topology: *Kuratowski (closure) operators*. These are characterised by the four properties

$$\mathfrak{K}(\varnothing) = \varnothing, \qquad X \subseteq \mathfrak{K}(X), \qquad \mathfrak{K}(X \cup Y) = \mathfrak{K}(X) \cup \mathfrak{K}(Y), \qquad \mathfrak{K} \circ \mathfrak{K} = \mathfrak{K}$$

for all $X, Y \in \mathscr{P}(S)$.

Exercise Show that a Kuratowski operator is a Moore operator, but not conversely.

Examples of Moore operators that are not Kuratowski include $K(X) = \mathrm{pr}_1(X) \times \mathrm{pr}_2(X)$ on $S = \mathbb{R}^2$, where pr_i are the canonical projections on the factors; the logical consequence $\mathfrak{Con}$; very simply, the trivial operator $K(X) = S$.

Definition A set $C \subseteq S$ is said to be **closed** under the Moore operator K (or K-closed) if $K(C) = C$.

Exercises Prove that

i) C is K-closed $\iff K(C) \subseteq C \iff \exists Y \subseteq S$ such that $C = K(Y)$;
ii) if C is K-closed, then $X \subseteq C \iff K(X) \subseteq C$.

Lemma *Let K be a Moore operator. The intersection of K-closed sets is K-closed.*

Proof Take $\{C_i\}_{i \in I}$ all K-closed. For any $j \in I$ we have $\bigcap_{i \in I} C_i \subseteq C_j$, and by monotonicity $K\left(\bigcap_{i \in I} C_i\right) \subseteq K(C_j)$, $\forall j \in I$. Hence $K\left(\bigcap_{i \in I} C_i\right) \subseteq \bigcap_{i \in I} K(C_i) = \bigcap_{i \in I} C_i$, which means $\bigcap_{i \in I} C_i$ is K-closed. $\qquad \square$

Definition A collection of sets $\mathscr{C} \subseteq \mathscr{P}(S)$ is a **Moore family** if it is closed under intersections.

Proposition 4.3 *Let K be a Moore operator on S. The collection of K-closed subsets is a Moore family $\mathscr{C}_K$.*

Proof Exercise. $\qquad \square$

Vice versa,

Proposition 4.4 *Every Moore family $\mathscr{C} \subseteq \mathscr{P}(S)$ induces a Moore operator*

$$K_{\mathscr{C}} \colon \mathscr{P}(S) \to \mathscr{P}(S), \qquad X \mapsto K_{\mathscr{C}}(X) = \bigcap_{\substack{X \subseteq C \\ C \in \mathscr{C}}} C.$$

Proof The facts that $X \subseteq K(X)$ and K is increasing are straightforward. As $\mathscr{C}$ is Moore, $\bigcap\limits_{\substack{X \subseteq C \\ C \in \mathscr{C}}} C \in \mathscr{C}$, and $K(X) \in \mathscr{C}$ as well. Hence $K\big(K(X)\big) \subseteq K(X)$. □

The recipes we have described are canonical, in the sense that a unique Moore family is associated with every Moore operator:

Theorem *The constructions 4.3 and 4.4 are inverse to one another.*

Proof

$(K_{\mathscr{C}_K} = K)$ Call K' the operator induced by the family $\mathscr{C}_K$ of K-closed sets.

The set $K'(X) = \bigcap\limits_{\substack{X \subseteq C \\ C \in \mathscr{C}}} C$ is contained in $K(X)$, since $X \subseteq K(X) \in \mathscr{C}$.

On the other hand, as $K'(X) \in \mathscr{C}$, we have $X \subseteq K'(X) \iff K(X) \subseteq K'(X)$. Therefore $K = K'$.

$(\mathscr{C}_{K_{\mathscr{C}}} = \mathscr{C})$ Let $\mathscr{C}'$ denote the family generated by the operator K induced by $\mathscr{C}$.

If $X \in \mathscr{C}'$ then $X = K(X) = \bigcap\limits_{\substack{X \subseteq C \\ C \in \mathscr{C}}} C \in \mathscr{C}$, that is to say $\mathscr{C}' \subseteq \mathscr{C}$.

Taking $X \in \mathscr{C}$ we see $X \supseteq K(X)$, i.e. $X = K(X)$. Hence $X \in \mathscr{C}'$, and so $\mathscr{C} \subseteq \mathscr{C}'$.

□

Many times it's easier to describe Moore families $\mathscr{C}$ than Moore operators K, so the above theorem permits to find Moore operators. Classical examples arise in ☞ *universal algebra*, cf. p.61: the pair $(S, \mathscr{C})$ is said to define a **closure structure** (see Example 4.1 for the name).

Consider an algebraic structure S, say a group, a ring, a lattice etc. The collection $\mathscr{C}_S$ of sub-structures of S is a Moore family, and the corresponding operator maps a subset $X \subseteq S$ to

$$K_S(X) := \bigcap_{\substack{X \subseteq C \\ C \in \mathscr{C}}} C,$$

the set underlying the **structure generated by** X, sometimes written $\langle X \rangle$.

Exercise Show that $K_S(X)$ is the smallest sub-structure containing X.

For instance: subgroups in a group, ideals in a ring, filters in a Boolean algebra are Moore families, and in all cases K defines the subgroup generated, the ideal generated, the filter generated etc. (see also Remark 4.7 v)). It's not hard to check that all these are not Kuratowski operators.

4.2 Galois Connections

Definition Fix non-empty sets S, T. A pair of functions

$$\phi \colon \mathscr{P}(S) \to \mathscr{P}(T), \qquad \psi \colon \mathscr{P}(T) \to \mathscr{P}(S)$$

is a **Galois connection** from S to T (written $\phi \dashv \psi$) if

i) ϕ, ψ are decreasing,

ii) $X \subseteq \psi(Y) \iff Y \subseteq \phi(X)$ for all $X \subseteq S, Y \subseteq T$.   (*adjunction*)

Instead of carrying around the two maps, it's common to use the notation $X^* = \phi(X)$, $Y^* = \psi(Y)$ and $x^* = \{x\}^*$. In this way item ii) reads $X \subseteq Y^* \iff Y \subseteq X^*$, and we write X^{**} to indicate $\psi(\phi(X))$ or $\phi(\psi(X))$. Although the stars have different meaning, the common symbol is somehow justified by the fact ϕ and ψ determine each other uniquely. In fact:

$\psi(Y)$ is the largest element X in $(\mathscr{P}(S), \subseteq)$ such that $Y \subseteq \phi(X)$, and symmetrically,

$\phi(X)$ is the largest element Y in $(\mathscr{P}(T), \subseteq)$ such that $X \subseteq \psi(Y)$.

Exercise Prove this claim.

Examples

i) The notion of Galois connection originates in ☞ *Galois theory*. If $\mathbb{L}$ is an extension of a field $\mathbb{K}$, set $S = \{\mathbb{F} \text{ field} \mid \mathbb{K} \subseteq \mathbb{F} \subseteq \mathbb{L}\}$ to be the extensions inside $\mathbb{L}$ and $T = \mathrm{Aut}(\mathbb{L})^{\mathbb{F}}$ the automorphisms fixing $\mathbb{F}$. If $G \subseteq T$ is a set of automorphisms we define $\mathbb{L}^G \subseteq \mathbb{L}$ to be the subfield of G-invariant elements in $\mathbb{L}$. Then

$$\phi(\mathbb{F}) = (\mathrm{Aut}\,\mathbb{L})^{\mathbb{F}}, \qquad \psi(G) = \mathbb{L}^G$$

is a Galois connection, which defines the *Galois correspondence* between the group $\mathrm{Gal}(\mathbb{L}/\mathbb{F})$ and the field $\mathbb{L}^G$.

ii) Suppose X is a path-connected topological space. The correspondence between subgroups of $\pi_1(X)$ and covering spaces of X is a Galois connection (☞ *algebraic topology*).

iii) Let V be a vector space, $X \subseteq V$ a subspace and

$$\phi(X) := \{f \in V^\vee \mid f(X) = 0\}$$

the annihilator of X. For a subset of linear maps $Y \subseteq V^\vee$ define

$$\psi(Y) := \{\mathbf{v} \in V \mid f(\mathbf{v}) = 0 \,\forall f \in Y\},$$

the 'joint' kernel of Y, cf. Definition 15.9. This sets up a Galois connection from V to the dual $V^\vee$ (☞ *linear algebra*).

iv) Let V be an inner-product space, and for a subspace X call $\phi(X) = X^\perp$ the orthogonal subspace. Since $\left(X^\perp\right)^\perp = X$ it follows that $\phi \dashv \phi$.

Proposition 4.5 *If* $(\phi, \psi) : S \to T$ *is a Galois connection,*

a) $X \subseteq (\psi \circ \phi)X$ *and* $Y \subseteq (\phi \circ \psi)Y$ *for all* $X \subseteq S, Y \subseteq T$;
b) $\phi \circ \psi \circ \phi = \phi$, $\quad \psi \circ \phi \circ \psi = \psi$.

Proof

a) $x \in X \overset{i)}{\Longrightarrow} X^* \subseteq x^* \Longrightarrow x = x^{**} \in X^{**}$, and so $X \subseteq X^{**}$.
b) By the previous relation, $X^{***} \subseteq X^*$ is immediate if we apply $*$ one more time. On the other hand, $X^* \subseteq (X^*)^{**}$ gives the converse, and therefore $X^* = X^{***}$. The same goes for Y.

$\square$

Example In the spirit of example iii) above, the relationship between certain sets $X \subseteq S = \mathbb{C}[x_1, \ldots, x_n]$ of polynomials and their common zeroes $Y \subseteq T = \mathbb{C}^n$ is a Galois connection $\mathscr{I} \dashv \mathcal{V}$:

$$\mathcal{V}(X) := \{x \in \mathbb{C}^n \mid p(x) = 0 \text{ for every polynomial } p \in X\}$$

$$\mathscr{I}(Y) := \{p \in \mathbb{C}[x_1, \ldots, x_n] \mid p(x) = 0 \text{ for all points } x \in Y\}$$

Hilbert's *Nullstellensatz* is a core result in ☞ *algebraic geometry* implying that $X \subseteq \mathscr{I}(\mathcal{V}(X))$. The theorem allows to relate algebraic varieties to polynomial ideals, hence intertwining geometry and algebra.

Exercise Let $(X_i)_{i \in I}$ be a collection of subsets of S. Prove that $\left(\bigcup_{i \in I} X_i\right)^* = \bigcap_{i \in I} X_i^*$ and that $\left(\bigcap_{i \in I} X_i\right)^* \supseteq \bigcup_{i \in I} X_i^*$.

Theorem 4.6 *If* $(\phi, \psi) : S \to T$ *is a Galois connection, the composite maps* $\phi \circ \psi : \mathscr{P}(T) \to \mathscr{P}(S)$, $\psi \circ \phi : \mathscr{P}(S) \to \mathscr{P}(T)$ *are Moore operators.*

Proof By Proposition 4.5 we have $X \subseteq X^{**} = (\psi \circ \phi)(X)$. Since ψ, ϕ are decreasing, moreover, the composite is increasing, and $X^* = X^{***} \Longrightarrow X^{**} = (X^{**})^{**}$. Then $\psi \circ \phi$ is idempotent. The claim about $\phi \circ \psi$ is left as exercise. $\square$

Confirming—should the need arise—that no name is chosen accidentally, the following brief categorical digression [13] should explain the term 'adjunction' appearing in the initial definition. A poset $(S, \leqslant)$ can be viewed as a category S, with objects $\mathrm{Ob}(\mathsf{S}) = S$ and morphisms $\mathrm{Mor}(s, s') = \{s \to s'\}$ if $s \leqslant s'$, $\mathrm{Mor}(s, s') = \varnothing$ otherwise. A functor $\phi : \mathsf{S} \to \mathsf{T}$ is then just an increasing map.

Consider now increasing maps $\phi : S \to T$ and $\psi : T \to S$ such that $(\psi \circ \phi)s \leqslant s$, $\forall s \in S$ and $t \leqslant (\phi \circ \psi)t$, $\forall t \in T$. Then a Galois connection

$$\phi \dashv \psi : \mathsf{S} \rightleftarrows \mathsf{T}$$

is an *adjunction* of functors in the category **Poset** of partially ordered sets (☞ *category theory*).

Remark 4.7 Retaining the above conventions, a functor $\phi \colon \mathsf{S}^{\mathrm{op}} \to \mathsf{T}$ is a decreasing function. Some authors call 'monotone' Galois connection a pair of increasing maps (ϕ, ψ) satisfying $\psi(Y) \subseteq X \iff Y \subseteq \phi(X)$ (replacing ii) in the definition, which to them would be an 'antitone' connection) and $(\psi \circ \phi)X \subseteq X$ (Proposition 4.5 ii)). The difference is only a matter of taste, as shown by the following monotone connections.

i) Take $S = T$ and fix a set L. Then $\phi(X) = L \cap X$ and $\psi(Y) = Y \cup \overline{L}$ define a Galois connection. This holds in particular in Boolean algebras if we set $\phi(x) = \ell \cap x$ and $\psi(y) = y \cup \overline{\ell} = \ell \to y$ (Chap. 5).

ii) A map $f \colon S \to T$ induces an immediate Galois connection $\phi(X) = f(X)$, $\psi(Y) = f^{-1}(Y)$.
    Furthermore, $\chi(X) = \{ y \in T \mid f^{-1}(y) \subseteq X \}$ is a right adjoint to ψ, resulting in $\phi \dashv \psi \dashv \chi$ (see Exercises 3.2).

iii) Consider the inclusion $\iota \colon \mathbb{Z} \to \mathbb{R}$, and the floor / ceiling maps (3.4). Then floor $\dashv \iota \dashv$ ceiling.

iv) Take the totient function $\phi \colon \mathbb{N} \to \mathbb{N}$ (Example 10.1 v)) and let $p(n)$ be the nth prime number. Then $\phi(n) + n$ and $p(n) + n + 1$ determine a Galois connection [27].

v) Let S be a group (or an object in a small algebraic category, like **Grp, Ab, Ring, R-Mod**) and take a subgroup $X \subseteq S$. Define $\phi(X) = \langle X \rangle$ to be the free group generated by X. Its right adjoint $\psi(Y)$ is defined by the set underlying the group $Y \subseteq S$ (ψ 'forgets' the operation of Y and only retains the information of the set).

vi) In logic, the Galois connection between syntax and semantics, due to Lawvere, is described in the next section, see Theorem 4.15.

Exercise Let $(\phi, \psi) \colon S \to T$ be a Galois connection. Prove

$$X \subseteq Z^{**} \text{ for some } Z \subseteq S \iff X \text{ is } **\text{-closed} \iff X = Y^* \text{ for some } Y \subseteq T.$$

for any $X \subseteq S$. *Hint*: use Proposition 4.5.

In the section's last part we show every Galois connection is associated with a unique binary relation, explaining why Galois connections are ubiquitous.

Proposition 4.8 *Let $R \subseteq S \times T$ be a relation, and define maps $\phi \colon \mathscr{P}(S) \to \mathscr{P}(S)$, $\psi \colon \mathscr{P}(T) \to \mathscr{P}(T)$ by*

$$\phi(X) = \{ y \in T \mid xRy \ \forall x \in X \} \qquad \psi(Y) = \{ x \in S \mid xRy \ \forall y \in Y \}.$$

Then $(\phi, \psi) \colon S \to T$ is a Galois connection.

Proof Take $X \subseteq Z$. Then $xRy \ \forall x \in Z \Longrightarrow xRy \ \forall x \in X$, so that $y \in Z^* \Longrightarrow y \in X^*$.

Now suppose $X \subseteq Y^*$. Taking $y \in Y$ we claim any $x \in X$ is R-related to y. In fact, $x \in X \implies x \in Y^* \implies x R y \implies Y \subseteq X^*$. (The condition for ψ is symmetric.) □

Proposition 4.9

i) *Let $(\phi, \psi): S \to T$ be a Galois connection. The relation $R \subseteq S \times T$,*

$$x R y \iff y \in x^* (\iff x \in y^*)$$

induces the Galois connection (ϕ, ψ).

ii) *Let $R \subseteq S \times T$ be a relation, $(\phi, \psi): S \to T$ the Galois connection induced by R, and set $x R' y \iff x \in y^*$. Then $R = R'$.*

Proof

i) Let's indicate with (ϕ', ψ') the connection induced by R. Since $y \in \phi'(X) \iff y \in x^* \ \forall x \in X$, it follows $\phi'(X) = \bigcap_{x \in X} x^* = \left(\bigcup_{x \in X} \{x\} \right)^* = X^* = \phi(X)$. And similarly for $\psi' = \psi$.

ii) $x R' y \iff y \in x^* \iff y \in \phi(\{x\}) \iff z R y \ \forall z \in \{x\} \iff x R y$.

□

Observe that the Moore operator K generated by a Galois connection $(\phi, \psi): S \to T$ does not determine (ϕ, ψ) uniquely: K determines S, but we could still modify T. Suppose in fact $K = \phi \circ \psi = **$, and take an arbitrary element $\omega \notin T$. Set $T' = T \cup \{\omega\}$ then define $\phi': \mathscr{P}(S) \to \mathscr{P}(T')$, $\psi': \mathscr{P}(T') \to \mathscr{P}(S)$ by

$$\phi'(X) = \phi(X) \cup \{\omega\}, \quad \psi'(Y) = \psi(Y \setminus \{\omega\}), \quad X \subseteq S, Y \subseteq T'.$$

Clearly ϕ', ψ' are decreasing: $Y \setminus \{\omega\} \subseteq \phi(X) \iff X \subseteq \psi(Y \setminus \{\omega\})$. Finally, $\phi' \circ \psi' = \phi \circ \psi$. Therefore K spawns infinitely many Galois connections (ϕ, ψ), (ϕ', ψ'), ..., at least one for each choice of ω.

A related question regards the existence of Galois connection inducing a given Moore operator K. This is answered by the following result:

Theorem 4.10 (Representation of Moore Operators) *Let $K: S \to S$ be a Moore operator. There exist a set T and a Galois connection $(\phi, \psi): S \to T$ such that $\psi \circ \phi = K$.*

Proof Take $T = \{C \subseteq S \mid K(C) = C\}$ to be the family of K-closed sets in S. Define the relation $R \subseteq S \times T$ tautologically: $x R C \iff x \in C$, and call $(\phi, \psi): S \to T$ the connection subordinate to R:

$$C \in \phi(X) \iff x R C \ \forall x \in X \quad (X \subseteq S)$$

$$X \in \psi(\mathscr{Y}) \iff x R D \ \forall D \in \mathscr{Y} \quad (\mathscr{Y} \subseteq T).$$

Now, $C \in \phi(X) \iff \forall x \in X, \ xRC \iff x \in C \ \forall x \in X \iff X \subseteq C$ ($C \in T$ by definition of R).

We claim, moreover, that $\psi(\mathscr{Y}) = \bigcap_{C \in \mathscr{Y}} C = \bigcap \mathscr{Y}$. In fact $x \in \bigcap \mathscr{Y} \iff$ $xRC \ \forall C \in \mathscr{Y}$, which means $x \in \psi(\mathscr{Y})$.

Finally, $(\psi \circ \phi)(X) = \bigcap_{C \in \phi(X)} C = \bigcap_{\substack{X \subseteq C \\ C \in T}} C = K(X).$ □

The summary picture looks like this:

$$
\begin{array}{ccc}
\text{binary relations} & \xleftrightsquigarrow[4.9]{4.8} & \text{Galois connections} \\
 & & \Big\downarrow {\scriptstyle 4.6 \,\big|\, 4.10} \\
\text{Moore families} & \xleftrightsquigarrow[4.4]{4.3} & \text{Moore operators}
\end{array}
\tag{4.1}
$$

The vertical map is onto but highly non-injective, since a Moore operator comes from infinitely many distinct Galois connections.

4.3   Model Theory

Model theory is a large area of logic that we shall use to make the previous chapters come alive. We will only provide the opening concepts and leave the minutiae to dedicated texts.

Definition A **model** $\mathscr{M} = (M, \Phi)$ of a language $\mathscr{L}$ is a pair consisting of a set $M \neq \varnothing$ and a function Φ (the **interpretation**) that maps the extra-logical symbols of $\mathscr{L}$ to the following objects:

for every constant c		an element $\Phi(c) \in M$;
for every k-ary relation f	⇝	a function $\Phi(f) \colon M^k \to M$;
for every k-ary predicate P		a relation $\Phi(P) \subseteq M^k$.

Definition 4.11 A (relational, or first-order) **structure** is an interpretation of a language of first order. An **algebraic structure** is an interpretation of an algebraic language (that is, one without predicates).

Examples 4.12

i) An ordered space $(N, \leqslant)$ can be considered a model with one binary relation P only (no constants, nor functions).

ii) Choose as extra-logical symbols a binary function μ, a unary function h and a constant e. We interpret them as follows: $\mu(g, h) =: gh$ (multiplication), $h(g) =: g^{-1}$ (inversion) and e as neutral element 1 for μ. The three axioms

$$\forall g \quad \mu(g, e) \doteq g \doteq \mu(e, g)$$
$$\forall g \quad \exists \overline{g} \quad \mu(\overline{g}, g) \doteq e \doteq \mu(g, \overline{g})$$
$$\forall g_1, g_2, g_3 \quad \mu\big(g_1, \mu(g_2, g_3)\big) \doteq \mu\big(\mu(g_1, g_2), g_3\big)$$

define a **group structure** with unit 1.

iii) A Boolean algebra (Chap. 5) is a model for propositional calculus, as explained in Sect. 5.3.

iv) A $\mathbb{K}$-vector space V (Chap. 15) can be treated as an algebraic structure. Although the multiplication by scalars is not an operation, we may formalise it as such by collecting all constants $k \in \mathbb{K}$ and defining a family of unary operations $\{\underline{k} \colon V \to V, \ \underline{k}(x) = kx\}$ indexed by $\mathbb{K}$. In this way we obtain an enhanced structure $(V, +, (\underline{k})_{k \in \mathbb{K}}), 0, 1)$ having one binary operation, 2 constants and $|\mathbb{K}|$ functions.

Alas, the same trick won't work to axiomatise higher-order structures algebraically, such as topological spaces and the like.

Exercise Define the ingredients of the language and the interpretations necessary to model a ring, a field, or a lattice.

Remark Any model $\mathcal{M}$ of the real numbers has card $\mathcal{M} = $ card $\mathbb{R} = \aleph_1$. But the theory of real numbers has a countable axiomatisation, and the *Löwenheim–Skolem theorem* says that for countable languages of first order (such as ours) any set of formulas admits a finite or countable model. Apparently, then, the real numbers should be also described by a countable set! This is *Skolem's paradox*. Luckily there's no need to panic, despite the appearances: in order to axiomatise $\mathbb{R}$ (the issue is completeness) one needs a language of 2nd order, so the aforementioned theorem is not violated and there's no mathematical contradiction. Putnam [80] offers an interesting philosophical take on the Skolem 'paradox'.

Definition Two models $\mathcal{M}, \mathcal{M}'$ are **similar** if they are interpretations of the same language.

Example An ordered ring with unit $(A, +, \cdot, -, \leqslant, 0, 1)$ is similar to a complemented distributive lattice $(R, \cup, \cap, \bar{\ }, \leqslant, \min, \max)$. Their properties are quite different, which goes to show that the notion of similarity between models is weak.

Definition Take a set $\Gamma \subseteq \mathcal{F}$ of formulas and let

$$\text{Mod}(\Gamma) := \big\{ \mathcal{M} \colon \vDash_{\mathcal{M}} \Gamma \big\}$$

denote the collection of models in which every formula of Γ is true.

If $\mathcal{K} = \{ \mathcal{M}_1, \mathcal{M}_2, \dots \}$ is a collection of similar structures, we call

$$\text{Th}(\mathcal{K}) := \big\{ A \in \mathcal{F} \colon \vDash_{\mathcal{M}} A, \ \forall \mathcal{M} \in \mathcal{K} \big\}$$

the set of formulas that are true in every structure in $\mathcal{K}$. In case $\mathcal{K} = \{\mathcal{M}\}$ consists of one model only, $\text{Th}(\mathcal{M}) = \big\{ X \in \mathcal{F} \colon \vDash_{\mathcal{M}} X \big\}$ is called **complete theory of** $\mathcal{M}$.

Lemma $\text{Th}(\mathcal{M})$ *is a theory, and $\mathcal{M}$ a model of it.*

Proof Let $X \in \mathscr{F}$ be a formula such that $\mathrm{Th}(\mathscr{M}) \vdash X$. By soundness (Theorem 1.18) we have $\mathrm{Th}(\mathscr{M}) \vDash X$, which means $\mathscr{M}$ is a model and $X \in \mathrm{Th}(\mathscr{M})$. $\qquad\square$

Example 4.13 $\mathrm{Th}(\mathbb{N})$ is the complete theory of the natural numbers, called **elementary** (or **first-order**) **arithmetics**. Any model isomorphic to $\mathbb{N}$ is a **standard model of elementary arithmetics**. 'Standard' refers to the properties we all know.

The name 'complete theory' is justified by this:

Theorem *For a theory* $\mathbb{T}$ *the following conditions are equivalent:*

(1) $\mathbb{T}$ *is maximally non-contradictory,*
(2) $\mathbb{T}$ *is complete,*
(3) $\mathbb{T}$ *is of the form* $\mathrm{Th}(\mathscr{M})$ *for some model* $\mathscr{M}$.

Proof

(1)$\Longrightarrow$(3): $\mathbb{T}$ non-contradictory $\Longrightarrow$ coherent $\Longrightarrow$ there exists a model $\mathscr{M}$ such that $\vDash_{\mathscr{M}} \mathbb{T}$. Hence $\mathbb{T} \subseteq \mathrm{Th}(\mathscr{M})$. Since $\mathrm{Th}(\mathscr{M})$ is non-contradictory and $\mathbb{T}$ maximal, necessarily $\mathrm{Th}(\mathscr{M}) \subseteq \mathbb{T}$, and so $\mathrm{Th}(\mathscr{M}) = \mathbb{T}$.

(3)$\Longrightarrow$(2): $X \in \mathrm{Th}(\mathscr{M}) \iff \vDash_{\mathscr{M}} X \iff \nvDash_{\mathscr{M}} \neg X \iff \neg X \notin \mathrm{Th}(\mathscr{M})$. Therefore $\mathrm{Th}(\mathscr{M})$ is complete.

(2)$\Longrightarrow$(1): $\mathbb{T}$ complete $\Longrightarrow \mathbb{T} \neq \mathscr{F} \Longrightarrow \mathbb{T}$ is non-contradictory. Suppose that $\mathbb{T}$ were not maximal among non-contradictory theories. Then there would exist a non-contradictory theory $\mathbb{W} \supsetneq \mathbb{T}$. Any element $X \in \mathbb{W} \setminus \mathbb{T}$ would, by completeness, satisfy $\neg X \in \mathbb{T} \subseteq \mathbb{W}$. Eventually $\mathbb{W}$ would be contradictory ⚡. Hence $\mathbb{T}$ is maximal.

$\qquad\qquad\qquad\qquad\qquad\qquad\qquad\qquad\qquad\qquad\qquad\qquad\qquad\qquad\qquad\square$

In general,

Proposition *Let* $\mathscr{K}$ *be a collection of similar structures. Then* $\mathrm{Th}(\mathscr{K})$ *is*

– *a theory;*
– *non-contradictory if* $\mathscr{K} \neq \varnothing$ *(i.e. there are models);*
– *complete if* $\mathscr{K} = \{\mathscr{M}\}$ *only contains one model.*

Proof $A \in \overline{\mathrm{Th}(\mathscr{K})} \Longrightarrow \mathrm{Th}(\mathscr{K}) \vdash A \Longrightarrow \mathrm{Th}(\mathscr{K}) \vDash A$, that is to say $\vDash_{\mathscr{M}} A$ for any $\mathscr{M} \in \mathscr{K}$. Hence $A \in \mathrm{Th}(\mathscr{K})$, which shows $\mathrm{Th}(\mathscr{K})$ is deductively closed. Moreover $\mathrm{Th}(\varnothing) = \mathscr{F}$, and we already know (previous theorem) that $\mathrm{Th}(\mathscr{M})$ is complete. $\quad\square$

Definition Two models $\mathscr{M}, \mathscr{M}'$ are said **(elementarily) equivalent** (written $\mathscr{M} \equiv \mathscr{M}'$) if $\vDash_{\mathscr{M}} A \Longrightarrow \vDash_{\mathscr{M}'} A$ for any formula $A \in \mathscr{F}$.

Example 4.16 will give instances of elementarily equivalent models of arithmetics.

Exercise For any models $\mathscr{M}, \mathscr{M}'$, prove that

i) $\mathscr{M}, \mathscr{M}'$ similar $\Longrightarrow \mathscr{M} \equiv \mathscr{M}'$;
ii) $\mathscr{M} \equiv \mathscr{M}' \iff \mathrm{Th}(\mathscr{M}) = \mathrm{Th}(\mathscr{M}')$.

Definition 4.14 Let's indicate by **PA** (for **Peano arithmetics**) the theory axiomatised by the Peano axioms 6.1. Since the natural numbers are a model of it, another name is **intuitive arithmetics**.

Peano's axioms are more restrictive than the axioms of elementary arithmetics, though:

$$PA \subsetneq Th(\mathbb{N}),$$

because induction is a second-order property. In this sense the theory $Th(\mathbb{N})$ is a proper extension of Peano's arithmetics.

Lemma *If $\mathcal{K}$ is a collection of equivalent structures then $Th(\mathcal{K})$ is complete.*

Proof $A \in Th(\mathcal{K}) \implies \vDash_{\mathcal{M}} A, \ \forall \mathcal{M} \in \mathcal{K} \implies \nvDash_{\mathcal{M}} \neg A, \ \forall \mathcal{M} \in \mathcal{K} \implies \neg A \notin Th(\mathcal{K})$.

Conversely, $\neg A \notin Th(\mathcal{K}) \implies \exists \mathcal{M}' \in \mathcal{K}$ such that $\nvDash_{\mathcal{M}'} \neg A$. Being the models equivalent, $\nvDash_{\mathcal{M}} \neg A, \ \forall \mathcal{M} \in \mathcal{K} \implies \vDash_{\mathcal{M}} A, \ \forall \mathcal{M} \in \mathcal{K} \implies A \in Th(\mathcal{K})$. $\qquad \square$

Definition A collection of similar structures $\mathcal{K}$ is **axiomatic** if there is a set $\Gamma \subseteq \mathcal{F}$ (a presentation) such that $\mathcal{K} = Mod(\Gamma)$, and **elementary** if Γ if finite.

Exercise Suppose $\mathcal{K} = Mod(\{A_1, \ldots, A_n\})$ for some formulas $A_1, \ldots, A_n$. Show $\mathcal{K} = Mod\left(\bigwedge_{i=1}^{n} A_i\right)$. That is, an elementary structure is presented by (is model of) a single formula.

Examples It can be shown that

- the class of Abelian groups Ab is elementary;
- the class of infinite (similar) structures is axiomatic but not elementary;
- the class of finite (similar) structures, and the class of well-ordered sets, are not axiomatic;
- certain axioms of ZFC theory contain infinitely many formulas (they are called 'schemas' on p. 25). It can be proved that ZFC, albeit axiomatic, is not elementary.

The relationship between models and theories is governed by a Galois connection.

Theorem 4.15 (Mod, Th) *is a Galois connection between the collection of models and the set of formulas $\mathcal{F}$.*

Proof We leave it as exercise to show that both Mod, Th are decreasing. Regarding adjunction it's enough to find a suitable binary relation, in agreement with (4.1). Define the relation $\vDash_{\mathcal{M}} X$ between $\mathcal{M}$ and X. Then

$$Mod(\Gamma) = \left\{ \mathcal{M} : \ \vDash_{\mathcal{M}} X, \ \forall X \in \Gamma \right\} = \Gamma^*$$

$$Th(\mathcal{K}) = \left\{ X \in \mathcal{F} : \ \vDash_{\mathcal{M}} X, \ \forall \mathcal{M} \in \mathcal{K} \right\} = \mathcal{K}^*$$

in Moore's notation.   □

Therefore Mod ∘ Th, Th ∘ Mod are Moore operators, and more precisely:

Proposition $\mathfrak{Con} = \mathrm{Th} \circ Mod$.
Equivalently, $\overline{\Gamma} = \Gamma^{**}$ for every $\Gamma \subseteq \mathscr{F}$.

Proof $\Gamma^{**} = \mathrm{Th}(\mathrm{Mod}(\Gamma)) = \{X \colon \vDash_{\mathscr{M}} X, \ \forall \mathscr{M} \in \mathrm{Mod}(\Gamma)\} = \{X \mid \Gamma \vDash X\} = \overline{\Gamma}$.
                                                                                                       □

But since Γ is closed under Th ∘ Mod if and only if there exists a collection $\mathscr{K}$
such that $\mathscr{K}^* = \Gamma$, we may reformulate the result above:

Proposition $\Gamma \subseteq \mathscr{F}$ *is a theory* $\Longleftrightarrow$ Γ *is the theory of a class* $\mathscr{K}$ *of models.*
Furthermore, a collection $\mathscr{K}$ of models is closed under Mod ∘ Th *if and only if
it is axiomatic.*   □

4.4 Foundations

For roughly two millennia Euclid's main work, the Elements, provided 'the'
reference material for mathematical research and school education. Euclid proved
many theorems from five basic axioms and few primitive notions, working entirely
within his system. The set-up's coherence, i.e. the absence of contradictions, comes
from the existence of one recognised model (the Euclidean plane, in dimension two).
Over the centuries the axioms' independence has been questioned several times,
and the V Postulate's complexity prompted many, bungled, attempts to deduce it
from the other four axioms (see Chap. 18). From a methodological point of view
nothing much happened until Hilbert. He took a different approach and created a
novel system, in which he could prove all of Euclid's results. But more relevantly,
he looked at geometry from outside, meta-mathematically. His deductive system,
beside the primitive notions, consists of 21 axioms that are *independent from one
another* and *collectively complete*: none descends for the others, and together they
can decide every geometric statement, by either proving or disproving it. The key
for this is to establish the two properties, namely independence and completeness,
by exhibiting a number of models, or geometric 'worlds' (i.e., different elements
in the set Mod). For example, to show that the parallel postulate (the famous
fifth) doesn't depend on the other four axioms, it is necessary to exhibit a model
satisfying postulates I–IV but not V. Beltrami did precisely that, by devising a model
(Fig. 17.3) for the hyperbolic plane. Crucially, Beltrami's world is non-contradictory
because it lives inside the Euclidean plane, and therefore it passes the coherence
buck on to Euclidean geometry. Hilbert, instead, showed that his axioms were
complete by proving they admit a unique model. In this way there can't exist a
proposition that the axioms cannot either prove or refute, since one such statement
p would lead to the existence of two different worlds, one in which p is true and
one where it's false.

Hilbert was convinced that when we look for the foundations of mathematics, as Russell had seeked (Remark 6.44), we should require three properties: **soundness**, **completeness**, and **decidability**. As we know from Sect. 1.5, soundness expresses the impossibility of deducing false statements inside the system, while completeness entails that true statements are provable by following the rules. Decidability means there exists a method, a machine, or a test (a.k.a. an algorithm) that when applied to any statement will decide whether it can be proved or not.

Regarding the first two properties, Gödel proved that no mathematical system can be both sound and complete, and thus settled a matter that others before him, like Post and Herbrand, had only partially grasped. Specifically, with regards to intuitive arithmetics (Definition 4.14), he showed the following two groundbreaking facts

First incompleteness theorem: the theory PA is essentially incomplete (every theory extending/containing it is incomplete. In some sense, PA is incompletable).

Second incompleteness theorem: [1] PA contains a formula, let's call it $\mathfrak{Coh}$, that is true but cannot be deduced.

$\mathfrak{Coh}$ represents the meta-linguistic statement '$0 \stackrel{.}{=} 1$ *cannot be proved in theory PA*'. The layman would say that Peano's arithmetics is able to express its own consistency, but cannot prove it.

The above two results, in practice, show that it's impossible to found mathematics on an idea of proof that is finite, verifiable, trustworthy (Theorem 1.18), deterministic and that is able to prove a minimum of arithmetics (at least Theorem 6.22). In fact, $\mathfrak{Coh}$ can be proved or not. If it can be proved, then '$0 \stackrel{.}{=} 1$' is true and the system is not sound. If it cannot be proved, the system is incomplete. Hence mathematics is unsound or incomplete.

We'll give an example serving a twofold purpose. First, it gives us insight on the many-worlds issue by considering a situation we are comfortable with: the natural numbers learnt in school (spoiler alert: we won't be as confident after reading Chap. 6). Second, it helps to understand the severity of the foundational crisis raised by the incompleteness theorems. Hilbert proved his axiomatic system's completeness by transferring the problem of the coherence of geometry to the coherence of analysis, i.e. the real numbers (we'll do the same in Chap. 18). In turn, the theory of real numbers is constructed starting from $\mathbb{N}$, in particular PA (see p.156 for more on Hilbert's programme). So any issues that undermine in any way the solidity of arithmetics will, to say the least, cast a serious shadow on $\mathbb{R}$, hence on analysis, geometry and, ultimately, on the entire body of mathematics.

Example 4.16 (☞ *Non-standard Arithmetics*) The language consisting of two binary predicates (interpreted as $\leqslant$, $<$), two binary functions ($+$ and $\cdot$) and countably many constants (the numbers) has the familiar model $\mathbb{N} = (\{0, 1, 2, 3, \dots\} ; \leqslant, < ; +, \cdot)$, which we call **standard** natural numbers. Recall that in Example 4.13 we

---

[1] For *sound* systems. The incompleteness theorem for *coherent* systems is due to Rosser.

named Th($\mathbb{N}$) elementary arithmetics, and a structure isomorphic to $\mathbb{N}$ a standard model.

The reason we went to great lengths and insisted on a grandiose name is that there exists a *non-standard* model *$\mathbb{N}$ of arithmetics (originally concocted by Skolem). This model is elementarily equivalent to $\mathbb{N}$, i.e. Th($\mathbb{N}$) = Th(*$\mathbb{N}$), but it's not isomorphic to the standard numbers. It has some curious features: $\mathbb{N}$ is contained in *$\mathbb{N}$ as initial segment, and the complement set $\mathscr{Q} = $ *$\mathbb{N} \setminus \mathbb{N}$ is made of numbers that are larger than any $n \in \mathbb{N}$, called 'infinite' natural numbers. Moreover, $\mathscr{Q}$ has no minimum nor maximum, it is dense and ordered. By Theorem 8.10, then, $\mathscr{Q}$ can be immersed in $\mathbb{Q}$, so that *$\mathbb{N}$ becomes a countable non-standard model of arithmetics.

After digesting this we could move forward and define non-standard real numbers *$\mathbb{R}$, and thence do ☞ *non-standard analysis*.

Things are even worse, form a certain viewpoint. By Tarski's *theorem on the cardinality of models* there exist infinitely many non-standard models of arithmetics, all non-isomorphic. By contrast, PA is the unique theory axiomatised by the Peano axioms (a fact shown in Sect. 6.1). Also Gödel's theorems imply the existence of non-standard models. Regarding the sentence $\mathfrak{Coh}$, by the completeness theorem we have $\nvDash_{\mathcal{N}} \mathfrak{Coh}$ for some model $\mathcal{N}$ of PA. However, $\mathfrak{Coh}$ is true in the standard numbers, and therefore any model $\mathcal{N}$ in which $\neg\mathfrak{Coh}$ is true must be non-standard.

The two incompleteness theorems leave open the problem of decidability. Hilbert thought there was a unique, well-established method to decide whether something is provable. In 1928, together with his student Ackermann he came up with the aptly named *Entscheidungsproblem* (decision problem, cf. Remark 1.24). This was soon settled by *Church's undecidability theorem* (1936), which guarantees that PA is essentially undecidable, meaning it is undecidable and every extension (that has $\mathbb{N}$ as model) is undecidable, too. And almost simultaneously Turing showed, by inventing *Turing machines* and to all effects founding computer science, that there cannot be a single method to decide every statement.

What does all of this have to do with set theory? As Gödel himself recognised, a proof of coherence should follow from the axioms through the laws of logic, and in order for such a proof to be convincing we should know that the axioms and laws lead to correct results. But if we knew that, there would be no need for a proof of coherence... We'll elaborate in Sect. 7.7 on the fact that mathematics is made of two parts, which we might call 'finite' and 'transfinite'. Finite maths consists of proof methods that don't require the existence of infinite sets. Transfinite maths is all the rest, and contains almost all of existing mathematics. It's the latter that is the source of all evil. First, Gödel proved that one cannot prove the absence of contradictions in transfinite maths using deductions made of finitely many steps. Then he showed there exist statements that cannot be decided by a formal proof. This fact remains strikingly shocking even after comprehending that Gödel's theorems somehow owe their existence to the diagonal argument, and hence are directly inspired by set theory (the incompleteness to which the first theorem refers is not dissimilar to the incompleteness of $\mathbb{Q} \subset \mathbb{R}$).

On the philosophical side of the story—the most consequential and exciting one, many would argue—the incompleteness theorems demolished the three main currents in the philosophy of mathematics of the early twentieth century: Frege's *logicism* (also see p.156), Brouwer's *intuitionism* and Hilbert's *formalism*. Yet in the aftermath of this devastation Gödel's theorems also propelled new tendencies, such as neo-logicism, the ☞ *theory of types*, and various other formalisms aimed at a partial formal foundation of mathematics. A noteworthy example of neo-formalist theory is *Presburger's arithmetics*. This has $(\mathbb{N}, +, 0)$ as model and is contained in PA (think PA without multiplication). In contrast to PA, though, it is a decidable theory, and as such it has enormous applications to computer science.

Excellent texts are available on the above philosophical fallout, some of the author's favourites being [8, 22, 32, 47, 59]. A special mention goes to the lovely book [74]. A modern take on Hilbert's programme can be found in [94]. Readers interested in the mechanisms governing mathematical thought should consult [67, 77]: the former for the didactical aspects, the latter for the epistemological problems.

In 1999 Time Magazine made a list of the top achievers in twentieth century human knowledge. No wonder that they crowned Gödel mathematician of the century, Turing computer scientist of the century and Wittgenstein philosopher of the century. The thread that ties all three together is mathematical logic.

Chapter 5
Boolean Algebras ☕

The theory of Boolean algebras was founded in 1847 by Boole, who considered it a form of 'calculus' adequate for the study of logic. Apart from the crucial relationship to propositional logic, Boolean algebras enter the proofs of the completeness of first-order logic, or the independence of the axiom of choice and the continuum hypothesis in set theory (p.187).

Starting from the 1930s, after the work of Stone and Tarski, the theory freed itself of logic to become a modern independent subject, with theorems and links to countless other areas, including algebra, analysis, probability, set theory and topology. In analysis for instance, the discovery of the Stone–Čech compactification and the Stone–Weierstraß approximation theorem happened as Stone was working on Boolean algebras. Countably complete Boolean algebras, a.k.a. σ-algebras, are essential to build ☞ *measure theory*. Boolean algebras are required to study certain classes of operator algebras like *von Neumann algebras*, the cornerstone of ☞ *quantum mechanics* [7, 71, 72]. At the end of the day, Weyl [97, p. 500] famously wrote "*In these days the angel of topology and the devil of abstract algebra fight for the soul of every individual discipline of mathematics.*"

Outside the mathematical realm Boolean algebras have applications in anthropology, biology, chemistry, ecology, economics, sociology and especially in philosophy and computer science (to model electronic circuits, programming languages, databases and for ☞ *complexity theory*). Wonderful references to learn the subject are [45, 87].

As for us, this chapter on one side exemplifies an important type of axiomatisation called 'algebraic theory'. On the other it contextualises the existing bridge between logic and topology (Sect. 5.3).

Pre-requisites: topology for Example 5.19 and Remark 5.18.

© The Author(s), under exclusive license to Springer Nature Switzerland AG 2021
S. G. Chiossi, *Essential Mathematics for Undergraduates*,
https://doi.org/10.1007/978-3-030-87174-1_5

5.1 Basics

Definitions A **Boolean algebra** is a structure

$$\mathbb{B} := (B, \cap, \cup, {}^-, 0, 1)$$

where B is a non-empty set, $\cap, \cup$ are binary operations, $^-$ a function and $0 \neq 1$ elements of B such that

(*commutativity*) $a \cup b = b \cup a,$ $a \cap b = b \cap a$
(*associativity*) $(a \cup b) \cup c = a \cup (b \cup c)$ $(a \cap b) \cap c = a \cap (b \cap c)$
(*distributivity*) $a \cup (b \cap c) = (a \cup b) \cap (a \cup c)$ $a \cap (b \cup c) = (a \cap b) \cup (a \cap c)$
(*neutral elements*) $a \cup 0 = a,$ $a \cap 1 = a$
(*complementation*) $a \cup \bar{a} = 1,$ $a \cap \bar{a} = 0$

for every $a, b, c \in B$.

We call **unit** the element 1, **zero** the element 0, **union/intersection** the operations $\cup/\cap$, while $\bar{a}$ is the **complement** of a.

Examples 5.1

i) The simplest instance is $B = \mathbb{Z}_2 = \{0, 1\}$, with complementation $\bar{1} = 0, \bar{0} = 1$ and operations

$\cup$	0	1
0	0	1
1	1	0

$\cap$	0	1
0	0	0
1	0	1

For reasons that will become clear later, we shall indicate the Boolean algebra by

$$2 := (\mathbb{Z}_2, \cap, \cup, {}^-, 0, 1).$$

ii) $(\mathscr{F}, \wedge, \vee, \neg, \bot, \top)$, using $\models\!\mid$ as equality predicate (Proposition 1.7).
iii) $(\mathscr{P}(X), \cap, \cup, {}^-, \varnothing, X)$ for any set $X \neq \varnothing$ (Theorem 2.3).
iv) $B = \{x \in \mathbb{N} \mid x \text{ divides } 30\}$, where $\cup = \text{lcm}, \cap = \text{lcm}, \bar{a} = 30/a$. The number 30 is the zero, 1 the unit.

Exercises Prove that

i) in a Boolean algebra the elements 1, 0, and $\bar{a}$ are unique, for each $a \in B$;
ii) associativity is a consequence of the other axioms;
iii) example iv) above can be generalised as follows. The set of divisors of a square-free integer $n > 1$ (square-free means it's not divisible by any perfect square k^2) form a Boolean algebra where $\cup = \text{lcm}, \cap = \text{lcm}, \bar{a} = n/a$.
Convince yourself that square-freeness is an essential hypothesis.

iv) Recall that a subset $Z \subseteq Y$ is *cofinite* if its complement $Y \setminus Z$ is finite. Prove that $\mathbb{B} = \{ X \subseteq \mathbb{N} \mid X \text{ is finite or cofinite} \} \subseteq \mathscr{P}(\mathbb{N})$ is a Boolean algebra with the set-theoretical operations of $\mathscr{P}(\mathbb{N})$.

Remark 5.2 If we swap $\cap$ and $\cup$, and 1 and 0, the axioms do not change. This phenomenon is known as **duality principle** for Boolean algebras. Formally, there exists a duality map $\eth : \mathbb{B} \to \mathbb{B}$, such that

$$\eth(a \cap b) = a \cup b, \quad \eth(a \cup b) = a \cap b, \quad \eth(0) = 1, \quad \eth(1) = 0.$$

Now, complementation is the map $\mathfrak{n} : \mathbb{B} \to \mathbb{B}$, $\mathfrak{n}(a) = \overline{a}$. Let $\eth^{\mathfrak{n}} : \mathbb{B} \to \mathbb{B}$ be the composite $\eth^{\mathfrak{n}} := \mathfrak{n} \circ \eth$. For instance, $\eth^{\mathfrak{n}}\left(a \cap \left(b \cup (\overline{a} \cap 0) \right) \right) = \overline{a} \cap \left(\overline{b} \cup (a \cap 0) \right)$.

$\circ$	$\mathbb{1}$	$\eth$	$\mathfrak{n}$	$\eth^{\mathfrak{n}}$
$\mathbb{1}$	$\mathbb{1}$	$\eth$	$\mathfrak{n}$	$\eth^{\mathfrak{n}}$
$\eth$	$\eth$	$\mathbb{1}$	$\eth^{\mathfrak{n}}$	$\mathfrak{n}$
$\mathfrak{n}$	$\mathfrak{n}$	$\eth^{\mathfrak{n}}$	$\mathbb{1}$	$\eth$
$\eth^{\mathfrak{n}}$	$\eth^{\mathfrak{n}}$	$\mathfrak{n}$	$\eth$	$\mathbb{1}$

The four symmetries $\mathbb{1}, \eth, \mathfrak{n}, \eth^{\mathfrak{n}}$ form the Klein group D_2, see Example 14.3.

Exercise Prove

i) $\mathfrak{n} \circ \eth = \eth \circ \mathfrak{n}$
ii) the maps $\mathfrak{n}, \eth, \eth^{\mathfrak{n}}$ are involutions.

In the next result every item consists of dual statements ($\eth$-invariant).

Theorem *For any* $a, b, c \in B$

i) $a \cap (a \cup b) = a = a \cup (a \cap b)$ *(absorption)*
ii) $a \cap a = a, \quad a = a \cup a, \quad a = \overline{\overline{a}}$ *(idempotence)*
iii) $a \cup 1 = 1, \quad a \cap 0 = 0$
iv) $\overline{0} = 1, \quad \overline{1} = 0$
vi) $\overline{a \cup b} = \overline{a} \cap \overline{b}, \quad \overline{a \cap b} = \overline{a} \cup \overline{b}$ *(De Morgan laws)*
vii) $\left((a \cap c = b \cap c) \wedge (a \cup c = b \cup c) \right) \Longrightarrow a = b$ *(cancellation)*

Proof Exercise. $\square$

Exercise

i) Prove that $a \cap b = a \iff a \cup b = b \iff a \cap \overline{b} = 0 \iff b \leqslant \overline{a}$ for all $a, b \in B$. In the algebras $\mathscr{F}$ and $\mathscr{P}(X)$, what does $\leqslant$ correspond to?
ii) Relative to the previous theorem, item vii), find an example for which $a \cap c = b \cap c$ is not enough to cancel c and obtain $a = b$.

Definition A Boolean algebra $\mathbb{B}$ is partially ordered by the relation

$$a \leqslant b \iff a \cap b = a.$$

Exercise Check that $\leqslant$ is indeed a partial order on the base set B.

The converse is false. Consider $B = \{0, a, b, c, 1\}$ with ordering defined by the 'tri-rectangular' Hasse diagram.

tri-rectangular lattice

The reason is that this order is not total, and reflects the fact that it's impossible to define a Boolean structure: the complement $\overline{a}$ cannot be defined uniquely (there are two elements $x = b, c$ satisfying $a \cup x = 1$, $a \cap x = 0$).

This also hints at the fact that ☞ *lattice theory* is more general and far-fetching than the theory of Boolean algebras. The next result stands at the base of the equivalence between the category **BAlg** of Boolean algebras and the category **BLat** of Boolean lattices.

Theorem *In* $(\mathbb{B}, \leqslant)$ *the element* 1 *is the maximum,* 0 *the minimum. Furthermore, all sets of two elements* $\{a, b\} \subseteq B$ *have supremum and infimum*

$$a \cap b = \inf\{a, b\}, \qquad a \cup b = \sup\{a, b\}.$$

(One says $(\mathbb{B}, \leqslant)$ *is a Boolean lattice.)*

Proof Exercise.   □

Examples In Example 5.1 iv) we have $a \leqslant b \iff b \mid a$, whence 30 is the minimum and 1 the maximum.

The tri-rectangular lattice, on the other hand, is not Boolean.

Definition A **Boolean ring** is a ring $\mathscr{B} = (B, +, \cdot)$ with unit 1 in which every element b is idempotent: $b^2 = b$, $\forall b \in B$.

Corollary *A Boolean ring* $\mathscr{B}$ *has characteristic* 2 *and is commutative.*

Proof For any $a, b \in B$

$$a + b = (a + b)^2 = a^2 + b^2 + ab + ba = a + b + ab + ba \implies 0 = ab + ba.$$

Then, on the one hand, by taking $a = b$ we deduce $0 = a^2 + a^2 = a + a = 2a$. On the other, $ab = -ba = ba$.   □

Examples

i) The simplest instance is the ring of mod 2 integers, cf. (2.1): $(\mathbb{Z}_2, +, \cdot)$ with operations

+	0	1
0	0	1
1	1	0

$\cdot$	0	1
0	0	0
1	0	1

ii) If X is any set, $(\mathscr{P}(X), \triangle, \cap)$ is a Boolean ring.

iii) If $\mathscr{B} = (B, +, \cdot)$ is a Boolean ring, the set of maps $\{f \colon B \to \mathbb{Z}_2\}$ is a Boolean ring with pointwise operations:

$$(f + g)(b) := f(b) + g(b) \quad \text{and} \quad (f \cdot g)(b) := f(b)g(b), \text{ for all } b \in B.$$

Definition 5.3 The operation on $\mathbb{B}$

$$a \triangle b := (a \cap \overline{b}) \cup (b \cap \overline{a})$$

is called **symmetric difference** of a and b.

Exercises

i) Prove $\triangle$ is commutative, associative, and that $a \triangle a = 0$ for any $a \in \mathbb{B}$.

ii) Prove, for all $a, b, c \in \mathbb{B}$

$$a \triangle 0 = a, \qquad (a \triangle b) \cap c = (a \cap c) \triangle (b \cap c), \qquad \overline{a} \triangle \overline{b} = a \triangle b.$$

iii) The abstract operation $\triangle$ corresponds to what in the concrete algebra $\mathscr{P}(X)$?

iv) Show that the Boolean sum of 2 is

$\triangle$	0	1
0	0	1
1	1	0

Theorem *Let* $\mathbb{B} = (B, \cap, \cup, \overline{}, 0, 1)$ *be a Boolean algebra and define*

$$a \cdot b := a \cap b, \qquad a + b := a \triangle b.$$

Then $R(\mathbb{B}) := (B, +, \cdot)$ *is a Boolean ring with zero* 0 *and unit* 1.

This justifies calling $\triangle$ **Boolean sum** as well, since it truly is an addition in a ring.

Proof Exercise. □

Theorem 5.4 *Let* $\mathscr{B} = (B, +, \cdot)$ *be a Boolean ring with zero* 0 *and unit* 1. *Set*

$$a \cap b := a \cdot b, \qquad a \cup b := a + b + a \cdot b, \qquad \overline{a} := 1 + a.$$

Then $B(\mathscr{B}) := (B, \cap, \cup, \overline{}, 0, 1)$ *is a Boolean algebra.*

Proof Exercise.   □

Exercise Prove that $R(B(\mathcal{B})) = \mathcal{B}$ and $B(R(\mathbb{B})) = \mathbb{B}$. This is an equivalence of categories, between **BAlg** and Boolean rings **BRing**.

Definition Let $\mathbb{B} = (B, \cap, \cup, {}^{-}, 0, 1)$ and $\mathbb{B}' = (B', \cap', \cup', {}^{-\prime}, 0', 1')$ be Boolean algebras. Borrowing names from universal algebra, we call **Boolean homomorphism** a map $h \colon B \to B'$ that preserves the whole structure:

$$h(a \cap b) = h(a) \cap' h(b) \qquad h(a \cup b) = h(a) \cup' h(b)$$

$$h(\overline{a}) = \overline{h(a)}' \qquad h(0) = 0' \qquad h(1) = 1'$$

for all $a, b \in B$. We'll write $h \in \mathrm{Hom}\,(\mathbb{B}, \mathbb{B}')$.

Exercise 5.5 Prove that

i) $h \in \mathrm{Hom}\,(\mathbb{B}, \mathbb{B}') \iff h \colon B \to B'$ preserves union and complementation;
ii) the functions $\mathfrak{d}, \mathfrak{n}, \mathfrak{d}^n$ of Remark 5.2 are Boolean automorphisms.

Example Let X be a non-empty set. The structure $2 = (\mathbb{Z}_2, \cup, \cap, \triangle, \leqslant)$ transfers term-by-term to the set $2^X := \{f \colon X \to 2\}$ of maps $X \to \mathbb{Z}_2$, provided we define:

$$f \leqslant g \iff f(x) \leqslant g(x), \qquad (f * g)(x) = f(x) * g(x) \quad \text{where } * = \cup, \cap, \triangle$$

for all $x \in X$. (See Proposition 3.4 as well.) Hence 2^X turns into a Boolean algebra.

The choice of indicating truth values of propositional logic (Definition 1.1) by 0, 1 comes from the following consideration.

Proposition 5.6 *Every truth function of propositional calculus is a Boolean homomorphism* $\upsilon \in \mathrm{Hom}\,(\mathscr{F}, 2)$.

Proof Due to Proposition 1.3 it is straightforward to verify that $\upsilon \colon (\mathscr{F}, \vee, \wedge, \neg, \bot, \top) \to 2$ satisfies

$$\upsilon(\neg A) = 1 + \upsilon(A) = \overline{\upsilon(A)} \qquad \upsilon(\bot) = 0 \qquad \upsilon(\top) = 1$$
$$\upsilon(A \wedge B) = \upsilon(A)\upsilon(B) \qquad \upsilon(A \vee B) = \upsilon(A) + \upsilon(B) + \upsilon(A)\upsilon(B).$$

By Theorem 5.4, therefore, we can think of υ as a homomorphism mapping $\mathscr{F}$ to the algebra 2.   □

For example, the formula $A \wedge \neg A$ is a contradiction, corresponding to

$$\upsilon(A \wedge \neg A) = \upsilon(A)(1 + \upsilon(A)) = \upsilon(A) + \upsilon(A)^2 = \upsilon(A) + \upsilon(A) \equiv_2 0,$$

for any υ. Moreover, it's easy to see (exercise)

$$\upsilon(A \to B) = 1 + \upsilon(A)(1 + \upsilon(B))$$

$$\upsilon(A \leftrightarrow B) = 1 + \upsilon(A) + \upsilon(B).$$

Definition A Boolean algebra $\mathbb{B}$ is said **complete** if every non-empty subset $B' \subseteq B$ admits supremum sup B'.

Exercise Prove that $\mathbb{B}$ is complete $\Longleftrightarrow$ every non-empty subset $B' \subseteq B$ admits inf B'.

Examples We consider algebras of sets with the standard structure.

i) An infinite subset in a Boolean algebra might not have supremum. The algebra

$$\mathbb{B} = \big\{ X \subseteq \mathbb{N} \ \big| \ \text{finite or cofinite} \big\} \subseteq \mathscr{P}(\mathbb{N})$$

is incomplete, since $P = \big\{ \{2n\} \ \big| \ n \in \mathbb{N} \big\}$ doesn't have supremum. If that were not the case, say $S = \sup P$, we would have $\{2n\} \subseteq S$ for any n, and any set Y containing the even numbers would contain S. Hence $S \subseteq \mathbb{B}$ would be infinite, and so cofinite, implying S would contain (all even numbers and) infinitely many odd numbers. Let b be an arbitrary odd number. Then $Y = S \setminus \{b\}$ would contain the even numbers, at the same time being a proper subset and an upper bound for P. But $Y \subseteq S$, forcing S to not be the supremum $\notmid\notmid$.

(The analogue phenomenon in topology occurs with the *Alexandrov compactification*.)

ii) Lebesgue-measurable subsets in $\mathbb{R}$ form an incomplete algebra (without proof).

Definition An element a is called an **atom** of the Boolean algebra $\mathbb{B}$ if it is a non-null minimal element:

$$a \neq 0 \ \text{ and } \ b \leq a \ \implies \ b = a, \quad \forall b \neq 0.$$

The algebra $\mathbb{B}$ is said **atomic** if every $x \neq 0$ has at least one atom preceding it:

$$\mathfrak{A}(x) := \big\{ a \in \mathbb{B} \ \big| \ a \text{ atom} \leqslant x \big\} \neq \varnothing,$$

and we set $\mathfrak{A}(\mathbb{B}) := \mathfrak{A}(1)$ to denote the set of atoms.

Exercises 5.7 Prove

i) in Example 5.1 iv) the atoms are 6, 10, 15.

ii) $a \in \mathfrak{A}(\mathbb{B}) \iff \forall x \in \mathbb{B} \ a \leqslant x \ aut \ a \leqslant \overline{x}$.

iii) $a \in \mathfrak{A}(\mathbb{B}) \iff \forall x, y \in \mathbb{B} \ a \leqslant x \cup y \implies a \leqslant x \ \vee a \leqslant y$.

iv) A finite Boolean algebra is atomic.

v) $\mathscr{P}(X)$ is atomic and its atoms are the singletons $\{x\}$, $x \in X$.

The converse of the last item is that complete, atomic Boolean algebras are power sets:

Theorem 5.8 (Lindenbaum–Tarski) *Every atomic and complete Boolean algebra $\mathbb{B}$ is isomorphic to the algebra $\mathscr{P}(\mathfrak{A}(\mathbb{B}))$.*

Proof The map

$$
\begin{aligned}
h: \mathbb{B} &\longrightarrow \mathscr{P}(\mathfrak{A}(\mathbb{B})) \\
0 \neq b &\longmapsto \mathfrak{A}(b) \\
0 &\longmapsto \varnothing
\end{aligned}
$$

is compatible with $\cup$, $^{-}$ by Exercises 5.7 ii-iii), hence it's a Boolean homomorphism.

It is 1-1 since $\mathbb{B}$ is atomic: suppose $h(a) = h(b)$ for some $a, b \in B$. Then $h(a \cap \overline{b}) = \varnothing$, and so $a \cap \overline{b} = 0$. Similarly $h(a \cup \overline{b}) = \mathbb{B}$, so $a \cup \overline{b} = 1$. In conclusion $a = b$.

It is onto because $\mathbb{B}$ is complete: we claim $\sup Y \in h^{-1}(Y)$ for any non-empty $Y \subseteq B$. For any atom $a \leqslant \sup Y$ we have $a = a \cap \sup Y = \sup_{y \in Y} \{a \cap y\}$. Take y such that $a \cap y \neq 0$ (y exists since $a \neq 0$). As y is an atom, $a = a \cap y = y \implies a \in Y \implies h(\sup Y) \subseteq Y$. From $Y \subseteq h(\sup Y)$ the claim follows.

An alternative argument is to exhibit the inverse homomorphism, which in the present case is the map sending $\varnothing \mapsto 0$, $Y \mapsto \sup Y$.   □

A careful inspection of the above proof detects its essence:

Corollary *Let $\mathbb{B}$ be complete. Then $\mathbb{B}$ is atomic if and only if $1 = \sup \mathfrak{A}(\mathbb{B})$.*

Example Returning to Example 5.1 iv), where $\mathfrak{A}(\mathbb{B}) = \{6, 10, 15\}$, the Lindenbaum–Tarski isomorphism reads:

$\mathbb{B}$	1	2	3	5	6	10	15	30
$\updownarrow$	$\updownarrow$	$\updownarrow$	$\updownarrow$	$\updownarrow$	$\updownarrow$	$\updownarrow$	$\updownarrow$	$\updownarrow$
$\mathscr{P}(\mathfrak{A}(\mathbb{B}))$	$\mathfrak{A}(\mathbb{B})$	$\{6, 10\}$	$\{6, 15\}$	$\{10, 15\}$	$\{6\}$	$\{10\}$	$\{15\}$	$\varnothing$

Having said that, bear in mind that atomless algebras do exist:

Examples

i)   the Lindenbaum–Tarski algebra (5.1);
ii) the algebra of half-open real intervals $[x, y) \subseteq \mathbb{R}$ (with the usual set-theoretical operations);
iii) any infinite and free Boolean algebra contains no atoms: i) and ii) are examples.

5.2   Filters and Representations

Filters go back to George Birkhoff's work in topology, whose aim was formalising the abstract notion of 'bigness' in set theory. They were defined by Henri Cartan only in 1937. They possess deep applications in almost all areas,[1] typically relative to the use of the *forcing* technique pioneered by Cohen regarding the axiom of choice and the continuum hypothesis in ☞ *topology, logic and measure theory*. The convergence of maximal filters, for instance, characterises compactness in topological spaces, a task that sequences cannot do in general (in spaces that aren't first countable).

Definition 5.9 Let $\mathbb{B} = (B, \cap, \cup, ^-, 0, 1)$ be a Boolean algebra. A **filter** is a non-empty subset $F \subseteq B$ such that

$$F \cap F \subseteq F \quad \text{and} \quad \forall b \in B \quad b \geqslant F \Longrightarrow b \in F.$$

It's not hard to see the notion is dual to that of ideal,[2] since complementation converts filters into ideals and vice versa:

$$F \text{ is a filter} \iff \overline{F} = B \setminus F \text{ is an ideal.}$$

Examples

  i) $\{1\}$ and B are filters of $\mathbb{B}$ (the obvious ones).
 ii) Consider $h \in \text{Hom}\,(\mathbb{B}, \mathbb{B}')$ and a filter $F \subseteq B'$. Then $h^{-1}(F) \subseteq B$ is a filter.
      If $G \subseteq B$ is a filter, is it true that $h(G) \subseteq B'$ is a filter?
iii) Fix $X \subseteq B$. The intersection $(X) := \bigcap_{X \subseteq F_i} F_i$ of all filters containing X is the
      filter generated by X.
 iv) The filter $F_x := (\{x\})$ generated by one element $x \in B$ is called the **principal filter** generated by x. A non-principal filter is said **free**.

Proposition 5.10 *In a Boolean algebra $\mathbb{B}$ the following statements are equivalent:*

  i) *F is a filter*
 ii) *$F \cap F \subseteq F$, $\quad F \cup B \subseteq F$,*
iii) *$a, \overline{a} \cup b \in F \Longrightarrow b \in F$ for all $a, b \in B$.*

Proof Exercise. □

Exercises 5.11 Let $F \subseteq \mathbb{B}$ be a filter. Show

  i) F is proper if and only if $0 \notin F$.
 ii) $1 \in F$ (i.e. $F = \varnothing \iff 1 \notin F$).

---

[1] Samuel Gomes da Silva [42] explained to me why the importance of filters cannot be overestimated.

[2] Ideals were introduced by Dedekind in [25].

iii) $F_x = \{b \in B \mid x \leqslant b\}$ for any $x \in B$.
iv) (X) is the smallest filter containing a subset $X \subseteq B$.
v) $F \cup \{x \in \mathbb{B} \mid \overline{x} \in F\}$ is a Boolean (sub)algebra.

Remark A filter $F \subseteq \mathbb{B}$ is said **complete** if it is closed under supremum: $\sup G \in F$ for any $G \subseteq F$. It follows that a filter in a complete Boolean algebra is complete. So, complete filters may be considered the Boolean analogue of upper sets R, which are characterised by being 'closed from above': $r \in R \ \wedge \ s \geqslant r \implies s \in R$. See Definition 8.2.

A broader definition holds in general ordered structures. A **filter** in a poset $(X, \leqslant)$ is a non-empty upper set $F \subseteq X$ such that

$$\forall x, y \in F \ \exists z \in F : z \leqslant x \wedge z \leqslant y.$$

Example: consider $S \neq \varnothing$ and the power set $(\mathscr{P}(S), \subseteq)$ as poset. A filter F on S is a collection of subsets of S (hence $F \subseteq \mathscr{P}(S)$) that is (1) closed under finite intersections and (2) an upper set in $\mathscr{P}(S)$ (entailing $S \in F$).

Pressing on with the filter-ideal correspondence, the dual notion to a maximal (or prime) ideal is that of an ultrafilter. The concept, again introduced by H. Cartan in 1937, was (in)famously invoked by Gödel in the 'proof' of the existence of God (!).

Definition A filter $U \subsetneq \mathbb{B}$ is called an **ultrafilter** if it is a maximal (and proper) element in $(\mathscr{P}(\mathbb{B}), \subseteq)$. The collection of ultrafilters is denoted by

$$\mathsf{S}(\mathbb{B}) := \{ U \text{ ultrafilter } \subset \mathbb{B} \}.$$

Proposition 5.12 *The following statements are equivalent:*

i) $U \subset \mathbb{B}$ *is an ultrafilter;*
ii) *for any* $b \in \mathbb{B}, \ \ b \in U$ *aut* $\overline{b} \in U$;
iii) $0 \notin U$ *and for all* $a, b \in \mathbb{B}$ $a \cup b \in U \iff b \in U \vee a \in U$ *(U is a 'prime' filter).*

Proof Exercise.   □

Example The power set algebra $\mathbb{B} = \mathscr{P}(\{1, 2, 3, 4\})$ is finite, hence atomic, and its singletons are the atoms: $\mathfrak{A}(\mathbb{B}) = \{\{1\}, \{2\}, \{3\}, \{4\}\}$. The lattice structure is represented in the Hasse diagram of Fig. 5.1.

The set $F_{\{1,4\}} = \{\{1, 4\}, \{1, 2, 4\}, \{1, 3, 4\}, \{1, 2, 3, 4\}\}$ is the principal filter generated by $\{1, 4\}$: it's contained in $F_{\{1\}} = \{\{1\}, \{1, 2\}, \{1, 3\}, \{1, 4\}, \{1, 2, 3\},$ $\{1, 2, 4\}, \{1, 3, 4\}, \{1, 2, 3, 4\}\}$, so it's not maximal. $F_{\{1\}}$, on the other hand, is maximal: by Proposition 5.12 ii), every element belongs either to $F_{\{1\}}$ or to the complement $\mathsf{C}F_{\{1\}} = \{\{2, 3, 4\}, \{3, 4\}, \{2, 4\}, \{2, 3\}, \{4\}, \{3\}, \{2\}, \varnothing\}$.

There are no free ultrafilters because the algebra is finite (Exercise 5.7 v)).

Exercise Let $F \subseteq \mathbb{B}$ be a filter. Prove that F is an ultrafilter in $F \cup \{x \in \mathbb{B} \mid \overline{x} \in F\}$.

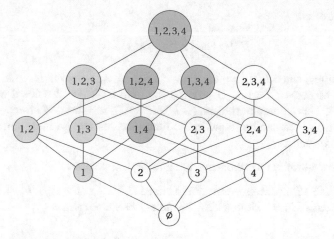

Fig. 5.1 The lattice of $\mathscr{P}(\{1, 2, 3, 4\})$

Proposition *The principal filter generated by $a \in B$ is an ultrafilter if and only if a is an atom:*

$$F_a \in S(\mathbb{B}) \iff a \in \mathfrak{A}(\mathbb{B}).$$

Therefore there's a 1-1 correspondence between atoms and principal ultrafilters.

Proof Consequence of Exercises 5.7 ii)-iii) and 5.11 iii). □

Still, infinite Boolean algebras always contain free ultrafilters [45].

The next results depends on the axiom of choice in such an inseparable way that it is not possible to prove it in ZF theory (without AC).

Lemma 5.13 (Ultrafilter Lemma) *Every proper filter $F \subset \mathbb{B}$ can be extended to an ultrafilter $U \supseteq F$.*

Proof Consider the collection $\{F_i\}_{i \in I} \subseteq \mathscr{P}(B)$ of proper filters containing F (the family is non-empty as it contains F itself). Every bounded chain in $\{F_i\}_{i \in I}$ has a maximal element by Zorn's lemma, namely the union $\bigcup_{i \in I} F_i$. This is a filter (Proposition 5.10) and proper (every F_i is proper), hence it's an ultrafilter. □

The ultrafilter containing F won't almost ever be unique; in many cases there is an exorbitant number of ultrafilters extending a given filter.

Recall $\mathrm{Hom}(\mathbb{B}, 2)$ is the set of Boolean homomorphisms. Sometimes they are called 'maximal' Boolean homomorphisms because of a relationship to ultrafilters: it can be proved that ultrafilters are kernels (to be defined with care) of morphisms h, as the usual recipe of universal algebra prescribes that.

Proposition 5.14 *The map $u\colon \mathrm{Hom}(\mathbb{B}, 2) \to S(\mathbb{B})$, $h \mapsto u(h) = \{x \in \mathbb{B} \mid h(x) = 1\}$, is a 1-1 correspondence.*

Proof First of all

$$h^{-1}(1) = \left\{ x \in \mathbb{B} \mid h(x) = 1 \right\}$$

is an ultrafilter, since $h(x) = 1 \iff h(\overline{x}) \neq 1$, meaning $x \in u(h) \iff \overline{x} \notin u(h)$. Moreover, for any $U \in \mathsf{S}(\mathbb{B})$ the map $h \colon \mathbb{B} \to 2$ defined by $h(x) = 1 \iff x \in U$ is a homomorphism. As $h \in u^{-1}(U)$, u is onto. We forego the proof of injectivity, which can be found in the literature, lest we go into minutiae about *Boolean congruences*.   □

If we view ultrafilters as elements in $\mathrm{Hom}\,(\mathbb{B}, 2)$, we have

Theorem 5.15 (Stone Representation Theorem, 1936) *Every Boolean algebra* $\mathbb{B}$ *can be embedded in* $\mathscr{P}\big(\mathrm{Hom}\,(\mathbb{B}, 2)\big)$ *by the 1-1 homomorphism*

$$\begin{aligned} \sigma \colon \mathbb{B} &\longrightarrow \mathscr{P}\big(\mathrm{Hom}\,(\mathbb{B}, 2)\big) \\ b &\longmapsto \sigma(b) \colon h \mapsto h(b) \end{aligned},$$

called **Stone homomorphism**.

Proof Take arbitrary $a, b \in \mathbb{B}$, $h \in \mathrm{Hom}\,(\mathbb{B}, 2)$. Since

$$\begin{aligned} \sigma(a \cup b)(h) &= h(a \cup b) = h(a) \cup h(b) = \sigma(a)(h) \cup \sigma(b)(h) \\ \sigma(\overline{a})(h) &= h(\overline{a}) = \overline{h(a)} = \overline{\sigma(a)}(h), \end{aligned}$$

σ is a homomorphism. Injectivity is a consequence of the following observation (not proved here): for any $a \neq 0$ there exists an $h \in \mathrm{Hom}\,(\mathbb{B}, 2)$ such that $h(a) = 1$. In turn, this follows from the next exercise.   □

Exercise Let $F \subseteq \mathbb{B}$ be a filter and $a \neq 0$ an element. Then there exists an ultrafilter $U \supseteq F$ such that $a \in U$.

We have seen in Definition 2.4 that $2^X = \mathscr{P}(X)$ seen as sets. In truth that identification is structural:

Exercise If X is a non-empty set, Exercise 5.5 guarantees 2^X is a Boolean algebra. Prove that 2^X is isomorphic to $\mathscr{P}(X)$.

Hint: the inspiration comes from Proposition 5.14. Define $u \colon 2^X \to \mathscr{P}(X)$ by $u(f) = f^{-1}(1)$, with inverse $u^{-1}(Y) = f$ where $f(x) = 1 \iff x \in Y$.

Theorem 5.16 *Every Boolean algebra* $\mathbb{B}$ *is isomorphic to the power set* $\mathscr{P}(X)$ *of some set* X.

Proof Using the previous exercise, where we took $X = \mathrm{Hom}\,(\mathbb{B}, 2)$, we can consider the map $u \colon 2^{\mathrm{Hom}\,(\mathbb{B}, 2)} \to \mathscr{P}(\mathrm{Hom}\,(\mathbb{B}, 2))$. From Theorem 5.15 the map $u \circ \sigma \colon \mathbb{B} \to \mathscr{P}(2^{\mathbb{B}})$ is an immersion, so $\mathbb{B} \cong u \circ \sigma(\mathbb{B}) \subseteq \mathscr{P}(\mathrm{Hom}\,(\mathbb{B}, 2))$.   □

Let's summarise the morphisms of **BAlg** handled so far:

- $\sigma : \mathbb{B} \hookrightarrow \mathscr{P}\big(\mathrm{Hom}\,(\mathbb{B}, 2)\big)$ (Stone),
- $u : 2^{\mathrm{Hom}\,(\mathbb{B}, 2)} \cong \mathscr{P}(\mathrm{Hom}\,(\mathbb{B}, 2))$ (previous theorem),
- $\tilde{u} : \mathscr{P}(2^{\mathbb{B}}) \cong \mathscr{P}(S(\mathbb{B}))$ (induced by the identification $2^{\mathbb{B}} \cong S(\mathbb{B})$ of Proposition 5.14).

If we compose them,

$$\sigma' = \tilde{u} \circ u^{-1} \circ \sigma : \mathbb{B} \to \mathscr{P}(S(\mathbb{B}))$$

becomes an immersion that maps $a \in \mathbb{B}$ to the collection of ultrafilters that contain a. Therefore Theorem 5.15 can be reformulated:

Theorem 5.17 (Stone Representation Theorem, v2) *Every Boolean algebra $\mathbb{B}$ is isomorphic to a subalgebra of $\mathscr{P}\big(S(\mathbb{B})\big)$ under the 1-1 homomorphism $\sigma' = \tilde{u} \circ u^{-1} \circ \sigma$ (**Stone homomorphism**), explicitly defined by*

$$\sigma' : \mathbb{B} \longrightarrow \mathscr{P}\big(S(\mathbb{B})\big)$$
$$b \longmapsto \big\{ \mathrm{U} \in S(\mathbb{B}) \mid b \in \mathrm{U} \big\}.$$

This result is a special case of *Stone duality*, a broad framework linking topological spaces[3] and ☞ *lattice theory*. Stone's representation theorem has been generalised and extended in several directions and its applications have reached into almost every area of modern mathematics. As it lies at the frontier of algebra, geometry, topology and ☞ *functional analysis*, the corpus of mathematics which has arisen is seldom seen as a whole. The textbook [54] collects many of its consequences.

Remark 5.18 The following construction puts on the set $\mathrm{Hom}\,(\mathbb{B}, 2)$ a topological structure.

Take a Boolean algebra $\mathbb{B}$. The **Stone space** or **dual space** is the set of its ultrafilters $S(\mathbb{B}) = 2^{\mathbb{B}}$, equipped with the Stone topology whereby $\mathscr{O}(2^{\mathbb{B}}) := \big\{ u \circ \sigma(x) \mid x \in \mathbb{B} \big\}$ is a basis of clopen sets, generated by neighbourhoods $\big\{ \mathrm{U} \in S(\mathbb{B}) \mid \mathrm{U} \ni x \big\}$. Thus $S(\mathbb{B})$ is compact, Hausdorff and 0-dimensional (hence totally disconnected). Starting from the other end, the space of clopen sets $S(X) := \mathscr{O}(X)$ of an arbitrary topological space X inherits a Boolean structure, in which filters correspond to closed sets and atoms to isolated points. The procedure effectively sets up a *Stone functor* $S : \mathbf{Top} \to \mathbf{BAlg}$, by which one proves the equivalence between $\mathbf{BAlg}$ and compact, Hausdorff and 0-dimensional spaces $\mathbf{CHaus}_0$.

If we weaken compactness to local compactness the majority of the theory remains valid. The dual of a *Boolean space* (zero-dimensional locally compact Hausdorff space) is a Boolean ring without unit, and conversely. The example to bear in mind is a compact Boolean space with one point removed: reinstating

---

[3] We shouldn't shy away from topology: as Stone said, "*A cardinal principle of modern mathematical research may be stated as a maxim: One must always topologize.*" [93].

the point corresponds to the Alexandrov compactification in topology, and to the creation of unity in the Boolean ring [45].

Example 5.19 Let X be a discrete infinite space. The dual $\beta X := \mathsf{S}(X)$ (with the Stone topology) is called *Stone–Čech compactification* of X. Every continuous map $f : X \to K$, for any compact Hausdorff space K, factors through βX to a unique continuous map $\beta f : \beta X \to K$:

$$
\begin{array}{ccc}
X & \xrightarrow{\ \ f\ \ } & K \\
& \searrow_{\text{1-1}} \quad \nearrow_{\beta f} & \\
& \beta X &
\end{array}
$$

The inclusion $X \subseteq \beta X$ is dense, and the universal property characterises βX up to homeomorphisms in **CHaus**. The elements of X are mapped to principal ultrafilters in βX.

We may equivalently construct $\beta X := \mathsf{S}(\mathscr{P}(X))$ as dual to the complete algebra $\mathscr{P}(X)$, since $\mathsf{S}(\mathscr{P}(X)) \overset{\text{1-1}}{\longleftrightarrow} \mathsf{S}(X)$.

$$
\begin{array}{ccc}
\mathsf{BAlg} & \overset{\mathsf{S}}{\longleftrightarrow} & \mathsf{CHaus}_0 \\
\iota \downarrow & & \uparrow \beta \\
2^{\mathsf{BAlg}} & \overset{\mathscr{P}}{\longleftrightarrow} & \mathsf{Set}
\end{array}
$$

A softer first contact with the Stone–Čech compactification goes through a typical construction of ☞ *functional analysis*: consider the space $\mathscr{C} := \mathscr{C}^0(X, I)$ of continuous maps from X to the real interval $I = [0, 1]$. Define $g : X \to I^{\mathscr{C}}$, $x \mapsto \big(f(x)\big)_{f \in \mathscr{C}}$. Then $\beta X = \overline{g(X)} \subseteq I^{\mathscr{C}}$.

In ☞ *operator theory* the Gelfand dual $\widehat{B(X)}$ of the C^*-algebra of bounded real maps $B(X)$ is isomorphic to βX.

5.3 Boolean Logic

As a final demonstration of the reach of filters we present an application to logic.

Definition Fix an (infinite) language of first order with set of formulas $\mathscr{F}$. The algebraic structure

$$
\mathbb{LT} = \big(\mathscr{F}/_{\models\mid} , \wedge, \vee, \neg, \bot, \top, \leqslant \big) \tag{5.1}
$$

is called **Lindenbaum–Tarski algebra** of propositional calculus. The canonical quotient map $\pi : \mathscr{F} \to \mathbb{LT}$, $\pi(P) = [P]$ is defined by logically equivalent formulas, and the order relation is

$$[P] \leqslant [Q] \text{ if and only if } \models P \to Q.$$

Exercise Show that $\mathbb{LT}$ is a Boolean algebra. (*Hint*: Proposition 5.6.)
    Besides, why do we insist on the language being infinite?

With the understanding that in $\mathbb{LT}$ tautologically equivalent formulas are equal,
we can avoid using cosets $[P]$ and just use P.

Lemma $\mathbb{LT}$ *has no atoms.*

Proof Take an arbitrary formula $A \in \mathscr{F}$ that is not a contradiction (the Boolean
0), and let $p \in V$ be a variable that doesn't show up in A. Consider truth functions
υ, ω such that $\upsilon(A) = 1$, $\omega(p) = 0$ and $\omega\big|_{V \setminus \{p\}} = \upsilon\big|_{V \setminus \{p\}}$. Then $\omega(A) = 0$, so
$A \not\leqslant p$. In a similar way, we may choose ω so that $\omega(p) = 1$, and obtain $A \not\leqslant \overline{p}$.
The consequence is that A is not an atom in $\mathbb{LT}$.
    An alternative argument would be to show $\mathbb{LT}$ is free. Since it's infinite, that
would force it to be atomless.   □

Under Stone's representation we then have $\mathbb{LT} \hookrightarrow \mathscr{P}(\mathsf{S}(\mathbb{LT}))$. But we can do
better:

Theorem *Let $\mathscr{L}$ be an infinite language. The dual space to the Lindenbaum–Tarski
algebra is homeomorphic to the space of models*

$$\mathsf{S}(\mathbb{LT}) \cong 2^{\mathscr{F}}.$$

Therefore models form a compact, Hausdorff and totally disconnected space.

Proof Introduce the homomorphism $\tilde{\upsilon} \in \mathrm{Hom}\,(\mathbb{LT}, 2)$, $\tilde{\upsilon}[A] := \upsilon(A)$ that makes

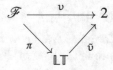

commute. Now set

$$\psi : 2^{\mathscr{F}} \longrightarrow \mathsf{S}(\mathbb{LT})$$
$$\upsilon \longmapsto \mathrm{U}_{\upsilon}$$

where $\mathrm{U}_{\upsilon} := \big\{ [A] \in \mathbb{LT} \mid \tilde{\upsilon}[A] = 1 \big\} = \big\{ [A] \mid \upsilon(A) = 1 \big\}$ is the ultrafilter
associated with $\upsilon \in \mathrm{Hom}\,(\mathscr{F}, 2)$. We claim is a homeomorphism.
    First, ψ is onto: if $\mathrm{U} \subseteq \mathbb{LT}$ is an ultrafilter, take the associated homomorphism
$\tilde{\upsilon} : \mathbb{LT} \to 2$ and define $\upsilon = \tilde{\upsilon} \circ \pi$, so that $\psi(\upsilon) = \mathrm{U}$.
    Second, ψ is 1-1: take truth functions $\upsilon \neq \omega$, so there exists $A \in \mathscr{F}$ such
that $\upsilon(A) \neq \omega(A)$. This means $\upsilon(A) = 1 - \omega(A)$. Suppose $\upsilon(A) = 1$ (the other
possibility is completely similar). Then $[A] \in \mathrm{U}_{\upsilon} = \psi(\upsilon)$ and $[A] \notin \mathrm{U}_{\omega} = \psi(\omega)$,
whence $\mathrm{U}_{\upsilon} \neq \mathrm{U}_{\omega}$.

Third, ψ is bi-continuous: because $U \in \sigma'[A] \overset{5.17}{\iff} [A] \in U \iff$
$\psi^{-1}(U)(A) = 1$, it follows that

$$\{\psi^{-1}(U) \mid U \in \sigma'[A]\} = \{\psi^{-1}(U) \mid [A] \in U\} = \{\upsilon \mid \upsilon(A) = 1\}$$

is an open neighbourhood of A in the space of models. And conversely.   □

Lemma 5.20 *In the algebra* $\mathbb{LT}$

i) *a filter F is an MP-closed set of formulas;*
ii) *(the equivalence classes of) the axioms Λ form an ultrafilter.*

Proof

i) For this we'll make use of the notation $a \to b$ to indicate $\bar{a} \cup b$ in $\mathbb{LT}$. This is
    no problem since we know that $a \to b$ and $\bar{a} \cup b$ are tautologically equivalent
    formulas. But now the characterisation of a filter F given in Proposition 5.10 iii)
    becomes very suggestive, logically speaking:

$$a \in F \text{ and } (a \to b) \in F \implies b \in F.$$

Wearing the spectacles of logic, this is precisely saying F is closed under modus
ponens.
ii) Exercise.

                                                                                        □

Now, because ultrafilters are maximal,

Proposition *The pair $(\mathbb{LT}, \Lambda)$ defines a consistent and complete theory.*

Proof Straightforward consequence of Theorem 1.21, Proposition 5.12.   □

Let $\Gamma \subseteq \mathbb{LT}$ be a subset of formulas. We know $t \in \mathscr{F}$ can be deduced from Γ
($\Gamma \vdash t$) if there is a finite sequence in $\mathbb{LT}$ (a proof) $b_1, b_2, \ldots, b_n = t$ such that

$$b_i \in \Gamma \text{ for all } i = 1, \ldots, n \quad \text{or} \quad b_k \doteq b_j \to b_i \text{ for some } j, k < i,$$

see Definition 1.12. Hence Γ satisfies the axioms of propositional calculus, and
so the deduction theorem as well. Its closure under operator 4.2 are the theorems
$\mathfrak{Con}(\Gamma) = \mathfrak{Thm}_\Gamma$ of Γ.

Proposition $\mathfrak{Con}(\Gamma) \subseteq \mathbb{LT}$ *is the filter generated by* $\Gamma \cup \Lambda$.

Proof Exercise. (*Hint*: look at Lemma 5.20.)   □

After we have fixed the ultrafilter of axioms, Definition 1.15 reads as follows:

Definition A deductively closed set Γ in theory $(\mathbb{LT}, \Lambda)$, i.e. $\mathfrak{Con}(\Gamma) \subseteq \Gamma$, is called
axiomatic system.

Therefore:

Theorem *An axiomatic system in* $(\mathbb{LT}, \Lambda)$ *is a filter that contains the ultrafilter* Λ *of axioms.*

Proof An axiomatic system Γ contains Λ and is MP-closed. Conversely, if $\Gamma \supseteq \Lambda$ is a filter, then $\mathfrak{Con}(\Gamma) = \Gamma$. $\qquad\qquad\square$

It won't have gone amiss that the entire discussion is completely general, and doesn't depend on the specific properties of the algebra $\mathbb{LT}$. That's why one could define axiomatic systems in an arbitrary Boolean algebra $\mathbb{B}$ with a prescribed ultrafilter U.

Remark 5.21 (Intuitionistic Logic) In the same way classical logic can be reduced to Boolean algebras, as we have done above, the algebraic skeleton of intuitionistic logic is a *Heyting algebra*. The latter is a bounded lattice $(H, \vee, \wedge, \leqslant, 0, 1)$ equipped with an 'implication' $\rightarrow$ defined by:

$$c \wedge a \rightarrow b \; \dashv\vdash \; c \leqslant a \rightarrow b.$$

In a logic of this kind $a \rightarrow b$ is too weak a formula for the MP to be valid. The consequence of this weakening is the loss of the law of excluded middle. In H we have:

$$a \leqslant \overline{\overline{a}} \quad \text{but} \quad \overline{\overline{a}} \nleqslant a \qquad \forall a \in H.$$

From that, a formula a is not always equivalent to $\neg\neg a$. This fact alters the face of the entire body of mathematics that we are familiar with.

Exercise Show $B = \{\overline{a} \mid a \in H\}$, even if not a Heyting subalgebra, is a Boolean algebra.

Heyting algebras therefore extend the notion of Boolean algebras.

Part II
Numbers and Structures

Chapter 6
Intuitive Arithmetics

This chapter intends to discuss the simplest properties of the numerical sets used everyday: natural numbers, integers and rational numbers. The (logical and formal, but practical as well) complexity of real numbers is deferred to Chap. 7. A detailed study of each of these sets should deserve a whole semester.

6.1 Natural Numbers

Every one of us knows from early on how to handle the set

$$\mathbb{N} = \{0,\ 1,\ 2,\ 3,\ 4,\ 5,\ 6,\ 7,\ \dots\}.$$

It's one of those notions considered to be glaringly obvious. Yet there is something deeply unsatisfactory with such an incontrovertible description, namely the suspicious use of the ellipsis '...', which attempts to convince us that the counting goes on indefinitely. Our first hunch proves right, for the above is not rigorous, even if very intuitive. Besides, what does 'counting' mean? And 'indefinitely'? The naive notion of natural number crumbles under a technical analysis, and it turns out that the simplest way to define natural numbers is to characterise them by the following axioms.

Definition 6.1 (Dedekind–Peano) Take an arbitrary set N. The four statements that follow are called **Peano axioms** (PA) for N:

I. $\exists s \colon N \to N,\ n \mapsto s(n)$ (called **successor** function)
II. $\forall n, m \in N \quad s(n) = s(m) \Longrightarrow n = m$ (s is 1-1)
III. $\exists! \, 0_N \in N$ (called **zero**) such that $\forall n \in N \quad 0_N \neq s(n)$ (0 is not a successor)
IV. Take a subset $X \subseteq N$. (**principle of finite induction**)

© The Author(s), under exclusive license to Springer Nature Switzerland AG 2021
S. G. Chiossi, *Essential Mathematics for Undergraduates*,
https://doi.org/10.1007/978-3-030-87174-1_6

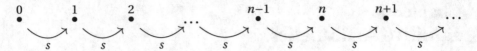

Fig. 6.1 Visual Peano axioms

$$\text{If} \quad \begin{cases} 0 \in X \\ \forall n \in N \quad n \in X \implies s(n) \in X \end{cases} \quad \text{then} \quad X = N.$$

Let's start to examine the PA and their fallout in a bit more detail. The first property we show is that, with the exception of 0, all other elements are successors of something. Put differently, 0 is the 'first' element, and so it plays a special role (Fig. 6.1).

Corollary *If a set N satisfies Peano's axioms then for any $n \neq 0$ in N there exists an $m \in N$ such that $n = s(m)$.*

Proof Define $X = \{0\} \cup s(N) \subseteq N$. Then clearly $0 \in X$, and $s(n) \in X$ for any n (note: this is true even if $n \notin X$!). By induction, then, $X = N$. □

Exercise Prove that if N satisfies the PA, then $n \neq s(n)$ for any $n \in N$.
Hint: define $X = \{n \in N \mid n \neq s(n)\}$ and apply induction.

The point is that the Peano axioms completely characterise the set N:

Theorem 6.2 *Suppose the two structures $(N, 0_N, s_N)$ and $(M, 0_M, s_M)$ satisfy Peano's axioms. Then N is isomorphic to M, that is: there exists a bijection $f : N \to M$ such that $f(0_N) = 0_M$ and $f(s_N(n)) = s_M(f(n))$ for any $n \in N$.*

$$
\begin{array}{ccc}
N & \xrightarrow{\ f\ } & M \\
\downarrow{\scriptstyle s_N} & & \downarrow{\scriptstyle s_M} \\
N & \xrightarrow{\ f\ } & M
\end{array}
$$

Proof We won't prove this here, but just mention it is a non-trivial consequence of Theorem 6.6, see [41, p35]. □

In 1964 Lawvere, extending the work of Dedekind (1888), proved that a model of the PA is a so-called initial object in the category of triples $(N, 0_N, s_N)$.

More interesting to us is Peano's observation that axioms II, III, IV are mutually independent: there exist models satisfying any two of the three only. For example, in the model $\mathcal{N} = (\mathbb{N} \cup \{\infty\}, 0, s)$ where $s(n) = n + 1$ and $s(\infty) = \infty$,

we have $\mathcal{N} \models \text{II} \land \text{III} \land \neg\text{IV}$.

Exercise Find *finite* models for the other cases: $\{\neg\text{II}, \text{III}, \text{IV}\}$, or $\{\neg\text{II}, \neg\text{III}, \text{IV}\}$, and so on.

Here is a suggestive and explicit construction that will serve as blueprint for the set $\mathbb{N}$:

Definition (von Neumann 1923) Take a set N and define recursively the following map on sets

$$N \mapsto N^+ := N \cup \{N\}.$$

In this way taking $N = \varnothing$ produces

$$\varnothing^+ = \{\varnothing\}, \quad \{\varnothing\}^+ = \{\varnothing, \{\varnothing\}\}, \quad \{\varnothing, \{\varnothing\}\}^+ = \Big\{\varnothing, \{\varnothing\}, \{\varnothing, \{\varnothing\}\}\Big\}, \quad \dots$$

There is a subtlety to observe here, namely: $N \in N^+$ and $N \subseteq N^+$ (N is both an element and a subset of N^+), but $N \notin N$ (it's not a member of itself). This will be crucial in view of Remark 6.44.

Definition A set N is said **inductive** if $\varnothing \in N$ and for any $X \in N$ we have $X^+ \in N$.

The existence of inductive sets is warranted by the axioms of ZF theory.

Exercise Show that the intersection of two inductive sets is inductive.

Definition Let N be an arbitrary inductive set. The set of **natural numbers** $\mathbb{N}$ is the intersection of all subsets $X \subseteq N$:

$$\mathbb{N} := \bigcap \mathscr{P}(N) = \bigcap_{X \subseteq N} X.$$

Theorem 6.2 proves the uniqueness of $\mathbb{N}$ as the smallest subset in N satisfying the axiomatic properties. Moreover, it guarantees the above definition doesn't depend on the choice of the initial inductive set N.

Exercise Take any set X and define on it a map $s(x) = x^+ := x \cup \{x\}$. Suppose they satisfy the PA, and show X is inductive.

So now we can view $\mathbb{N}$ as made of

$$\{\varnothing, \ \varnothing^+, \ \varnothing^{++}, \ \varnothing^{+++}, \ \varnothing^{++++}, \ \dots\},$$

which led someone to say half-seriously that arithmetics, and hence mathematics, is about the empty set! In an even more suggestive way we may define the natural numbers as

$$0 := \varnothing$$

$$1 := 0^+ = 0 \cup \{0\} = \{0\}$$

$$2 := 1^+ = 1 \cup \{1\} = \{0\} \cup \{1\} = \{0, 1\}$$ (6.1)

$$3 := 2^+ = 2 \cup \{2\} = \{0, 1\} \cup \{2\} = \{0, 1, 2\}$$

and so forth. This definition lends itself to many generalisations, first and foremost ☞ *ordinal arithmetics*, see Remark 8.16.

We'll denote by $\mathbb{N}^*$ the set $\mathbb{N} \setminus \{0\} = \{1, 2, 3, \dots\}$.

Proposition *Up to isomorphism $\mathbb{N}$ is the smallest set satisfying the Peano axioms.*

Proof Exercise. By isomorphism we mean a map that behaves as in Theorem 6.2 (it preserves the zero and commutes with the successor map).   □

Now let's take a look at operations. If we write the successor map as $n + 1 := s(n)$, we can define on $\mathbb{N}$ two operations recursively:

Definition 6.3 For any $n, m \in \mathbb{N}$

[addition] $n + 0 = n$ $n + s(m) = s(n + m)$
[multiplication] $n \cdot 0 = 0$ $n \cdot s(m) = (n \cdot m) + n.$

In terms of the successor map, every $k \in \mathbb{N}$ can be recast in the form

$$k = \underbrace{(s \circ \dots \circ s)}_{k \text{ times}}(0) = s^k(0),$$

and the above operations become

$$n + m = s^m(n), \quad \forall m \neq 0 \quad n \cdot m := \underbrace{n + n + \dots + n}_{m \text{ times}} = s^{m-1}\left(s^n(0)\right) = s^{n(m-1)}(0).$$

Exercise Convince yourself of these two properties. Then prove that addition and multiplication as defined above satisfy: commutativity, associativity, uniqueness of the zero (implying the algebraic structures $(\mathbb{N}, +)$ and $(\mathbb{N}, \cdot)$ are commutative monoids), and distributivity of $\cdot$ over $+$.

Definition 6.4 $\mathbb{N}$ possesses an order relation $\leqslant$ characterised by four properties:

i) $\nexists m < 0$ ii) $0 \leqslant n$

iii) $m < s(n) \implies m \leqslant n$ iv) $s(m) \leqslant n \implies m < n$

for all $m, n \in \mathbb{N}$.

One writes $n < m$ to indicate $n \leqslant m$ but $n \neq m$, whilst $m \geqslant n$ means $n \leqslant m$, and $m > n$ means $n < m$. In this way $\mathbb{N}^* = \{n \in \mathbb{N} \mid n > 0\}$.

Exercises Prove that

i) the relation $\leqslant$ can be defined equivalently by requesting:

$$n \leqslant m \overset{\text{def}^n}{\iff} \exists p \in \mathbb{N} \text{ such that } n + p = m.$$

ii) $(\mathbb{N}, \leqslant)$ is totally ordered.

Further properties of the (well) ordering of $\mathbb{N}$ will be examined in Sect. 8.2. Let's remind, in conclusion, that the entire theory of numerical sequences and series depends essentially on the ordered structure of $\mathbb{N}$, perhaps explaining why arithmetics lies at the heart of analysis.

6.2 Principle of Induction

The axiom of induction is probably the most difficult to grasp, being the least intuitive of the lot. In this section we aim to explain its power, in relationship to its use in certain arguments, called proofs by induction.

Consider a property $P(n)$ that depends on a natural number $n \in \mathbb{N}$. We may think of the statement '3 *divides* $n^3 - n$' for something concrete. We want to prove that $P(n)$ is true for every $n \in \mathbb{N}$. To do that let's define the subset

$$X = \{n \in \mathbb{N} \mid P(n)\} \subseteq \mathbb{N}$$

and call

- **inductive base** the proposition $P(0)$,
- **inductive hypothesis** the proposition $P(k)$ where $k \in \mathbb{N}$ is a fixed—but arbitrary—number.

The principle of induction's recipe consists of two steps:

(base step) prove the base $P(0)$, that is: $0 \in X$;
(inductive step) prove $P(k + 1)$ under assumption $P(k)$, with $k \in \mathbb{N}$ arbitrary but given. In other words: $k \in X \implies k + 1 \in X$, $\forall k \in \mathbb{N}$.

The Peano axiom will then force $X = \mathbb{N}$, and $P(n)$ will be true for every $n \in \mathbb{N}$.

The domino effect is the ultimate incarnation of induction and a perfect way to visualise it. In standing dominoes, tile nr. $k + 1$ falls when knocked down by tile number k, which is pushed by number $k - 1$ et cetera. Going back along the line, tile 1 is knocked down by the initial tile (number '0', by convention). To trigger the chain reaction it's enough to push tile 0 and place tiles at a small distance, so to guarantee that each tile will fall when hit by tile immediately preceding it.

Example Consider the proposition $P(n)$: '3 *divides* $n^3 - n$', i.e. $\exists q \in \mathbb{N}$ such that $3q = n^3 - n$. Set $X = \{n \in \mathbb{N} \mid P(n)\}$.

Base step: $P(0)$ is true since $3 \cdot 0 = 0^3 - 0$. Hence $0 \in X$.

Inductive hypothesis: suppose $P(k)$ holds for some k (arbitrary). The goal is proving $P(k + 1)$. Since

$$(k + 1)^3 - (k + 1) = \underbrace{k^3 - k}_{3q} + 3k^2 + 3k = 3(q + k^2 + 3k),$$

then 3 divides $(k + 1)^3 - (k + 1)$, and the induction principle guarantees $X = \mathbb{N}$.

Exercises

i) Prove that $\displaystyle\sum_{k=0}^{n} k := 0 + 1 + 2 + 3 + \ldots + (n - 1) + n = \frac{n(n + 1)}{2}$ for any $n \in \mathbb{N}$ (the symbol $\sum$ is explained on p. 123).

ii) Find the flaw in the following reasoning, which allegedly proves that all the students in the classroom are fans of the same football team. Let $P(n)$ be the proposition '*in every set of n students, everybody is a Fluminense[1] fan*'. Clearly $P(0)$ and $P(1)$ are true (the first by contradiction, the second tautologically). Then suppose $P(k)$ holds, and consider a set with $k + 1$ students $\{s_1, s_2, \ldots, s_k, s_{k+1}\}$. By induction hypothesis every student in $\{s_1, s_2, \ldots, s_k\}$ (k members) is a Fluminense fan, and so are students in $\{s_2, \ldots, s_k, s_{k+1}\}$ (k members). Therefore all students in $\{s_1, s_2, \ldots, s_k\} \cup \{s_2, \ldots, s_k, s_{k+1}\} = \{s_1, s_2, \ldots, s_k, s_{k+1}\}$ support the same team, meaning $P(k + 1)$ is true.

Hint: consider $k = 2$ and go over this argument.

The induction principle is also employed to prove properties that hold starting from a certain number $n_0 > 0$. In a statement like '$P(n)$ *holds for every $n \geqslant n_0$*' we must take $P(n_0)$ as base, because the Peano axioms do not distinguish $\mathbb{N}$ from a tail $\{n_0, n_0 + 1, n_0 + 2, \ldots\}$. As a matter of fact, there is a 1-1 correspondence $k \mapsto k + n_0$ between the two sets. This implies $\{n_0, n_0 + 1, n_0 + 2, \ldots\}$ and $\mathbb{N}$ have the same cardinality (roughly, the same quantity of elements) for any n_0, and the consequence is that induction works if we pull from $\mathbb{N}$ any subset of the type $\{0, 1, \ldots, n_0 - 1\}$ (later on called initial segment, or lower set). This will be put into fuller context with Theorems 8.10 and 8.13.

---

[1] Given the sensitivity of the matter, the recommendation is for teachers to choose the team carefully.

Example Let's demonstrate the *Bernoulli inequality*

$$(1+x)^n \geqslant 1 + nx, \qquad n \geqslant 1, \ x \in (-1, +\infty).$$

Base (here, $n = 1$): $1 + x \geqslant 1 + x$ for any $x > -1$ ✓.

Induction step: suppose the inequality holds for a certain $k > 1$. Hence $(1 + x)^{k+1} = (1+x)^k(1+x) \geqslant (1+kx)(1+x) = 1 + (k+1)x + kx^2 \geqslant 1 + (k+1)x$, proving the inequality for $k + 1$ ✓, and we're done.

Exercise Prove that $n^2 - n - 6 \geqslant 0$, $\forall n \geqslant 3$.

Take a predicate $P(n)$ and let's recast the induction principle in this way:

$$(\mathfrak{I}) \quad \begin{cases} P(0) \\ \forall k \in \mathbb{N} \quad P(k) \Longrightarrow P(k+1) \end{cases} \quad \Longrightarrow \quad P(n) \ \forall n \in \mathbb{N}.$$

There is another version of induction that goes as follows:

$$(\mathfrak{I}^*) \quad \begin{cases} P(0) \\ \forall k \in \mathbb{N} \quad P(0) \wedge P(1) \wedge \ldots \wedge P(k) \Longrightarrow P(k+1) \end{cases} \quad \Longrightarrow \quad P(n) \ \forall n \in \mathbb{N}.$$

At first sight the latter seems stronger that $(\mathfrak{I})$. Sometimes $(\mathfrak{I}^*)$ is referred to as complete induction. Before we sort out the relationship between the two versions, it is useful to show how to use $(\mathfrak{I}^*)$ for proving, for example, a divisibility property of the integers (cf. Definition 6.18):

Proposition 6.5 (Prime Factorisation) *Every natural number $n > 1$ can be written as product of primes (if n itself is prime, a product with only one factor).*

Proof Take a natural number $n \geqslant 2$ and let $P(n)$ be the proposition 'n *can be written as product of primes*' (with the understanding that if n itself is prime, n is its own factorisation). The sentence $P(2)$ is true. So suppose $P(2), \ldots, P(k)$ all hold, for some arbitrary $k \in \mathbb{N}$. Observe that any number n between 2 and k is a product of primes. Now consider $k + 1$. If it's prime, we're done. If it's not prime, $k + 1 = ab$ for some $a, b \in \mathbb{N}$ smaller than $k + 1$. By the inductive hypothesis of $(\mathfrak{I}^*)$ we have $a = p_{i_1} \cdots p_{i_a}$ and $b = q_{j_1} \cdots p_{j_b}$, where the p_i, q_j are all prime. Therefore $k + 1 = p_{i_1} \cdots p_{i_a} \cdot q_{j_1} \cdots p_{j_b}$ is a product of primes. □

Along these lines, to prove $P(15)$ using $(\mathfrak{I}^*)$ we will only need $P(3)$ and $P(5)$ ($15 = 3 \cdot 5$), whereas principle $(\mathfrak{I})$ only allows us to use $P(14)$.

Theorem *Finite and complete induction are equivalent:* $(\mathfrak{I}) \iff (\mathfrak{I}^*)$.

Proof $(\mathfrak{I}) \implies (\mathfrak{I}^*)$: let $P(n)$ be a predicate such that $P(0)$ is true, and $\{P(0), P(1), \ldots, P(k)\} \implies P(k+1)$ for every k. Define $Q(k) := P(0) \wedge P(1) \wedge \ldots \wedge P(k)$. First, $Q(0)$ is true. Moreover, $Q(k) \iff P(0) \wedge P(1) \wedge \ldots \wedge P(k) \overset{\text{ind. hyp.}}{\implies} P(k+1)$. Hence $P(0) \wedge P(1) \wedge \ldots \wedge P(k) \wedge P(k+1) = Q(k+1)$, and by $(\mathfrak{I})$ we know $Q(n)$ holds for every n. In particular $P(n)$ holds for every n.

$(\mathfrak{I}^*) \Longrightarrow (\mathfrak{I})$: let $P(n)$ be a predicate such that $P(0)$ holds, and $P(k) \Longrightarrow P(k+1)$, $\forall k$. If $P(0) \wedge P(1) \wedge \ldots \wedge P(k)$ is true also $P(k)$ is true, and by induction hypothesis $P(k+1)$ holds, too. Therefore $(\mathfrak{I}^*)$ implies $P(n)$ holds for every n. $\square$

Example In order to show that 12 divides $n^4 - n^2$ for every $n \in \mathbb{N}$ we shall use $(\mathfrak{I}^*)$. More precisely, taking the predicate $P(n)$: '12 *divides* $n^4 - n^2$' we'll prove $P(1)$, ..., $P(6)$ are true. For the inductive step we set $m = n - 5$, so that $n + 1 = m + 6$. Then we use that 12 divides $m^4 - m^2$ to show that 12 divides $(n+1)^4 - (n+1)^2$.

Exercises

(1) Show that using only \$3 and \$5 bills it's possible to produce any sum ($\geqslant$ \$8). In other words, $\forall n \geqslant 8 \; \exists r, s \in \mathbb{N}$ such that $3r + 5s = n$.

(2) Prove that $\displaystyle\sum_{k=1}^{n} k^2 = \frac{n(n+1)(2n+1)}{6}$ and $\displaystyle\sum_{k=1}^{n} k^3 = \left(\sum_{k=1}^{n} k\right)^2 = \frac{n^2(n+1)^2}{4}$, $\forall n \in \mathbb{N}$.

The first one can be proved also by observing

$$3k^2 + 3k + 1 = (k+1)^3 - k^3 \Longrightarrow 3\sum_{k=1}^{n} k^2 + 3\sum_{k=1}^{n} k + 1 = \sum_{k=1}^{n}(k+1)^3 - \sum_{k=1}^{n} k^3 \Longrightarrow$$

$$3\sum_{k=1}^{n} k^2 + 3\frac{n(n+1)}{2} + 1 = (n+1)^3 - 1 \Longrightarrow \sum_{k=1}^{n} k^2 = -\frac{n(n+1)}{2} - 1 + \frac{(n+1)^3 - 1}{3}.$$

In general one can compute $\displaystyle\sum_{k=1}^{n} k^p$ by knowing $\displaystyle\sum_{k=1}^{n} k^{p-1}$, and it can be proved that

$$\sum_{k=1}^{n} k^p = \frac{1}{p+1} n^{p+1} + \frac{1}{2} n^p + \sum_{k=2}^{p+1} \frac{\beta_k}{k} \binom{p}{k-1} n^{p-k+1}.$$

The coefficients β_k are the *Bernoulli numbers* (Example 14.11c)) and have to do with ☞ *umbral calculus*.

(3) Show $\displaystyle\sum_{p=1}^{n} \frac{1}{p(p+1)} = \frac{n}{n+1}$, $\forall n \in \mathbb{N}$.

(4) Prove, for any $n \in \mathbb{N}$,

$$7 | 2^{2+n} + 3^{2n+1}, \qquad 9 | n^3 + (n+1)^3 + (n+2)^3, \qquad 3 | (2^n)^2 - 1, \qquad 8 | 3^{2n} - 1.$$

(5) Let a, b be natural numbers and $P(n)$ the statement 'if $\max\{a, b\} = n$ then $a = b$'. Clearly, $P(0)$ holds, since $\max\{a, b\} = 0$ forces both numbers to be zero, so $a = b$. Now suppose $P(k)$ true, and take a, b such that $\max\{a, b\} = n + 1$. Then $\max\{a - 1, b - 1\} = n$, and by inductive hypothesis $a - 1 = b - 1$,

i.e. $a = b$. This would prove $P(n)$ for any n, which is obviously ridiculous since it would say $a < b \implies a = b$ (e.g., $\max\{2, 3\} = 3 \implies 2 = 3$) ⚡⚡.

So where's the mistake in the argument then?

(6) Prove $a^{n+m} = a^n a^m$, $(a^n)^m = a^{nm}$ for every $a \neq 0$ and $n, m \in \mathbb{N}$.

Convince yourselves using examples that $(a^n)^m \neq a^{(n^m)}$, i.e. that it's important to pay attention to brackets. For this reason the convention for iterated powers is that $a^{n^m} := a^{(n^m)}$.

(7) Fix $a \in \mathbb{N}^+$. Find the mistake in the following reasoning: first of all $a^0 = 1$. Assuming $a^k = 1$ for every $k \leqslant n$, we would have

$$a^{n+1} = a^n \cdot a = a^n \cdot \frac{a^n}{a^{n-1}} \overset{\text{ind. hyp.}}{=} 1 \cdot \frac{1}{1} = 1.$$

Apparent conclusion: $a^n = 1, \forall n \neq 0$. But this cannot be, since $3^4 \neq 1$.

(8) For any $a \neq 1$ prove $\displaystyle\sum_{j=0}^{n} a^j = \frac{a^{n+1} - 1}{a - 1}$; see also Example 9.2.

(9) Prove that $n! > 2^n$ for any $n \geqslant 4$, and $2^n > n^2$ for every $n \geqslant 5$.

(10) Prove that

- the product of two consecutive natural numbers is even;
- if the roots of $x^2 + ax + b = 0$ are consecutive natural numbers, then a is odd and b even;
- if $x^2 + ax + b = 0$ has two even solutions, then a, b must be even.

(11) Show $\displaystyle\left(\sum_{i=1}^{n} x_i\right)^2 \leqslant n \sum_{i=1}^{n} x_i^2$ for every $n \geqslant 1$ and any $x_1, \ldots, x_n \in \mathbb{R}$.

Then prove there is equality in the above formula if and only if $x_1 = \ldots = x_n$.

(12) Take real numbers $x, y \geqslant 0$, $\lambda \in [0, 1]$. Prove

$$\left(\lambda x + (1 - \lambda)y\right)^n \leqslant \lambda x^n + (1 - \lambda)y^n, \qquad \text{for any } n \geqslant 1.$$

This property translates the geometrical convexity of the function x^n, cf. Definition 10.6.

6.3 Recursion ☕

Perhaps the most relevant fallout of Dedekind's method for constructing $\mathbb{N}$ is the celebrated recursion theorem. It gives rise to new functions using primitive recursion, and generates every *partially recursive* function (which, by the *Church–Turing thesis*, means *computable*). ☞ *Recursion theory*, created by Church, Gödel, Herbrand, Kleene, Turing and Post, picks up from this point. Besides, we should not forget that Dedekind was the first to highlight the importance of recursive reasoning [24].

The principle of induction is useful for both proving certain properties and defining objects as well. For example consider $Y \subseteq \mathbb{N}$, an element $y_0 \in Y$ and a map $f : Y \to Y$. We would like to construct a sequence in Y using f, in a recursive way:

$$u_0 = y_0, \quad u_1 = f(u_0), \quad u_2 = f(u_1), \quad \ldots$$

Setting $u(s(n)) = f(u(n))$ won't do, because induction implies there exists at most one such map u, but doesn't guarantee its existence. What we need is

Theorem 6.6 (Primitive Recursion, Dedekind [24]) *Let N be a set satisfying the Peano axioms, and $y_0 \in Y$ an element in an arbitrary non-empty set. For any map $f : N \times Y \to Y$ there exists a unique map $u : N \to Y$ such that*

$$u(0) = y_0$$

$$u(s(n)) = f(n, u(n))$$

for every $n \in N$.

Proof For simplicity we'll assume f does not depend on n, i.e. $f : Y \to Y$ (the general case is treated in [41, p.97]). Consider

$$S = \left\{ n \in \mathbb{N} \mid \exists g : \{0, 1, \ldots, n\} \to Y, g(0) = y_0, \ u(s(k)) = f(g(k)) \ \text{for } k < n \right\}.$$

Then $0 \in S$, and if $n \in S$ there exists $g_n : \{0, 1, \ldots, n\} \to Y$ such that $g_n(0) = y_0$, $g_n(s(k)) = f(g_n(k))$ for $k < n$. Set

$$g(k) = \begin{cases} g_n(k) & 0 \leqslant k \leqslant n \\ f(g_n(n)) = f(g(n)) & k = s(n). \end{cases}$$

Then $s(n) \in S$, and so $S = \mathbb{N}$. Notice that for every chosen n the map g_n is unique. To finish it suffices to set $u(n) = g_n(n)$. $\square$

One very practical version of the theorem is the following:

Theorem 6.7 (Primitive Recursion in $\mathbb{N}$) *Consider maps $g : \mathbb{N}^k \to \mathbb{N}$, $\phi : \mathbb{N}^{k+2} \to \mathbb{N}$. There exists a unique map $f : \mathbb{N}^{k+1} \to \mathbb{N}$ such that*

$$f(x, 0) = g(x)$$

$$f(x, y + 1) = \phi\big(x, y, f(x, y)\big)$$

for every $x \in \mathbb{N}^k$, $y \in \mathbb{N}$. *One says* f *is obtained by primitive recursion (PR) from* g, ϕ.

This result is crucial in that ϕ inherits from f, g the property of being computable. It allows, moreover, to use recursive definitions. In particular it permits to prove the existence of maps $\mathbb{N}^2 \rightarrow \mathbb{N}$ such as the addition $+$ and the multiplication $\cdot$. In first-order arithmetics this wouldn't be a problem anyway, since those operations are primitive notions. But in set theory, where they are introduced through definitions, it's necessary to justify them.

The proof of the PR theorem requires general facts of set theory and cannot be proved simply by tapping into Peano's axioms. Actually, once we have sufficiently strong axioms to deduce PR, we can develop arithmetics to the point of proving the Peano axioms. Thus PA tend to loose their role of ultimate axiomatic set, because they end up being theorems in farther-fetching theories. The latter, by the way, are necessary to make a proper and concrete use of the Peano axioms.

Examples 6.8 The following operations are obtained by PR.

i) **Addition**, as per Definition 6.3. Iterating (!), we define the sum of $x_1, \ldots, x_n$ by

$$\sum_{i=1}^{n} x_i := \begin{cases} x_1 & \text{if } n = 1 \\ \sum_{i=1}^{n-1} x_i + x_n & \text{if } n > 1 \end{cases}$$

ii) **Multiplication**, as per Definition 6.3. The product of $x_1, \ldots, x_n$ is

$$\prod_{i=1}^{n} x_i := \begin{cases} x_1 & \text{if } n = 1 \\ \left(\prod_{i=1}^{n-1} x_i \right) \cdot x_n & \text{if } n > 1 \end{cases}$$

iii) **Exponentiation** in base $a \neq 0$ with exponent n is

$$a^0 := 1, \qquad a^n := a^{n-1} a \text{ if } n > 0.$$

We set $0^n = 0$ for every $n \neq 0$ (and note the symbol 0^0 hasn't been defined).

iv) The **factorial** $n!$ of a natural number $n \in \mathbb{N}$:

$$0! := 1$$

$$(n+1)! := n!(n+1) \quad \forall n \geqslant 0. \tag{6.2}$$

Exercise Prove the relation $n!\big((n+2)! - (n+1)!\big) = \big((n+1)!\big)^2$.

A couple of handy consequences of Theorem 6.6 are next:

Corollary 6.9 *Consider a set $X \neq \varnothing$ and maps $r_m \colon X^m \to X$, $m \in \mathbb{N}^*$. Then for every $x \in X$ there exists a unique map $f \colon \mathbb{N} \to X$ such that*

$$f(1) = x, \qquad f(n+1) = r_n\big(f(1), f(2), \ldots, f(n)\big) \quad \forall n \geqslant 1.$$

Corollary *If $X \subseteq \mathbb{N}$ is an infinite set there exist an increasing bijective map $f \colon \mathbb{N} \to X$.*

Proof If $Y \subseteq X$ is finite the complement $X \setminus Y$ isn't empty and so it has minimum by Theorem 8.12. To conclude it suffices to define $f(1) = \min X$, $f(n+1) = \min \big(X \setminus \{f(1), f(2), \ldots, f(n)\}\big)$ recursively. $\qquad\qquad\square$

Example 6.10 ([70]) Corollary 6.9 allows to show that every real number is the limit of a sequence of rational numbers. Fix $x \in \mathbb{R}$ and let's construct a bijective map $g \colon \mathbb{N} \to \mathbb{N}$ such that

$$\lim_{n \to \infty} \sum_{i=1}^{n} \frac{(-1)^{g(i)}}{g(i)} = x.$$

Set $L(n) := \displaystyle\sum_{i=1}^{n} \frac{(-1)^{g(i)}}{g(i)}$. Call $P, I \subseteq \mathbb{N}$ the sets of even and odd natural numbers, then define $g \colon \mathbb{N} \to \mathbb{N}$ by $g(1) = 1$ and

$$g(n+1) = \begin{cases} \min \big(P \setminus \{g(1), g(2), \ldots, g(n)\}\big) & \text{if } L(n) \leqslant x, \\[2mm] \min \big(I \setminus \{g(1), g(2), \ldots, g(n)\}\big) & \text{if } L(n) > x. \end{cases}$$

Since $\displaystyle\sum_{n \in P} \frac{1}{n} = \sum_{n \in I} \frac{1}{n} = +\infty$ the map g is bijective. Therefore $\displaystyle\lim_{n \to \infty} g(n) = +\infty$ and hence $\displaystyle\lim_{n \to \infty} \big|L(n) - x\big| = 0$.

Exercises 6.11 Show that the following maps are obtained by primitive recursion:

i) predecessor: $\mathfrak{a}(0) := 0$, $\mathfrak{a}(n) := n - 1$ for $n > 0$;

ii) truncated difference: $n \dot{-} m := \max \{n - m, 0\}$.
   Hint: $n \dot{-} 0 = n$, $n \dot{-} (m+1) = \mathfrak{a}(n \dot{-} m)$;

iii) **absolute value**: $|n - m| := (n \dot{-} m) + (m \dot{-} n)$;

iv) **sign**: $\mathrm{sgn}(0) := 0$, $\mathrm{sgn}(n) := 1$ if $n > 0$.
   Hint: $\mathrm{sgn}(n) = n \dot{-} \mathfrak{a}(n)$;

v) **Kronecker delta**: $(n, m) \mapsto \delta_{mn} := \begin{cases} 1 & \text{if } n = m \\ 0 & \text{if } n \neq m \end{cases}$.
   Hint: $\delta(n, m) = 1 \dot{-} \mathrm{sgn}\big(|n - m|\big)$.

vi) $H(n) = \frac{1}{2}n(n+1)$.

     Hint: $H(0) = 0$, $H(n+1) = H(n) + n + 1$.

    Aside, note that H is increasing, and $n \leqslant H(n)$ for any $n \in \mathbb{N}$.

vii) $\sigma(n, m) = \frac{1}{2}(n+m)(n+m+1) + m$.

     Hint: $\sigma(n, m) = H(n+m) + m$.

It's not hard to believe that every time we repeatedly apply the primitive recursion theorem (in $\mathbb{N}$) we generate maps that grow faster than the previous ones. Imagine we applied the theorem infinitely many times: eventually we would obtain a (computable) map that is not recursive. The sequence of operations $f_0 = s$ (successor), $f_1 = +$ (addition), $f_2 = \cdot$ (multiplication), $f_3 = $ exponentiation, ... (Examples 6.8) can be defined in one go as follows:

$$\begin{cases} f_{n+1}(x, 0) = 1 \\ f_{n+1}(x, y+1) = f_n\big(x, f_{n+1}(x, y)\big) \end{cases}$$

The sequence of *hyperoperations* $\{f_k\}$ is related to Lambert's W function (Exercise 14.10v)).

Note that f_k grows faster than f_i for any $k > i$. Put otherwise, for any y after a certain x we have $f_i(x, y) < f_k(x, y)$:

$$1 + y < x + y < xy < x^y < \dots \quad \text{for suitable numbers } x.$$

The function of 3 variables $F(n, x, y) := f_n(x, y)$ grows faster than any of the f_n. To make the comparison more effective take a special case: even $G(x, y) := F(x, x, y)$ grows faster than every f_n.

Example 6.12 The Ackermann function $\text{Ack} : \mathbb{N}^2 \to \mathbb{N}$

$$\text{Ack}(m, n) = \begin{cases} n + 1 & \text{if } m = 0 \\ \text{Ack}(m - 1, 1) & \text{if } m > 0 \text{ and } n = 0 \\ \text{Ack}\big(m - 1, \text{Ack}(m, n - 1)\big) & \text{if } m > 0 \text{ and } n > 0 \end{cases}$$

cannot be obtained by PR. We can't prove this fact here, but we can at least try to make a vague sense of its behaviour. For instance, to see how fast its growth is, it helps to compute a simple expression, such as

$$\text{Ack}(4, 3) = \dots = \text{Ack}(3, \text{Ack}(3, 65533)) = 2^{2^{2^{2^{2^{2}}}}} - 3 = 2^{2^{65536}} - 3.$$

This number is so massive that the majority of computers will return ∞ already if we feed them the exponent 2^{65536} (☞ *computability/recursion theory*).

6.4   Integer Numbers

Integers, usually denoted by $\dots,\ -2,\ -1,\ 0,\ 1,\ 2,\ \dots,\ n-1,\ n,\ n+1,\ \dots$
are in reality cosets in a quotient.

Definition Consider on $\mathbb{N}$ the equivalence relation

$$[(m,n)] \sim [(r,s)] \iff m+s = n+r.$$

The quotient set $\mathbb{Z} := \dfrac{\mathbb{N} \times \mathbb{N}}{\sim}$ is called set of **integer numbers**.

The choice of letter comes from the German word for number (*Zahl*). Intuitively,

the equivalence class $[(m,n)] \in \mathbb{Z}$ represents $m-n \in \{0, \pm 1, \pm 2, \dots\}$.

It's necessary to go through an equivalence relation in order to identify distinct representations of the same number, like $2-3 = -1 = 7-8$, i.e. $[(2,3)] = [(7,8)]$.

Definition We call sum and product on $\mathbb{Z}$ the operations

$$[(m,n)] + [(r,s)] := [(m+r, n+s)] \quad [(m,n)] \cdot [(r,s)] := [(mr+ns, ms+nr)]$$

We'll denote by $\mathbb{Z}^* := \mathbb{Z} \setminus \{0\}$ non-zero integers.

Whenever we introduce objects by a quotient procedure, we should make sure the definitions don't depend on the representatives chosen in equivalence classes. We'll show this for $+$ and leave the product to the reader to sort out. Take $(m', n') \in [(m,n)]$ and $(r', s') \in [(r,s)]$. Then $m'+n = n'+m, r'+s = s'+r$, adding which gives $(m'+r') + (n+s) = (n'+s') + (m+r)$. But now

$$[(m',n')] + [(r',s')] = [(m'+r', n'+s')] = [(m+r, n+s)] = [(m,n)] + [(r,s)],$$

so the sum is well defined on $\mathbb{Z}$.

Theorem $(\mathbb{Z}, +, \cdot)$ *is a commutative ring with unit, i.e. the operations satisfy*

(1) (associativity) $\quad (a+b) + c = a + (b+c), \qquad a(bc) = (ab)c$
(2) (distributivity) $\quad c(a+b) = ca + cb, \qquad (a+b)c = ac + bc$
(3) (additive neutral element) $\quad \exists\, 0 \in \mathbb{Z}$ *such that* $a + 0 = a = 0 + a$ *(called* zero*)*
(4) (multiplicative neutral element) $\quad \exists\, 1 \in \mathbb{Z}^*$ *such that* $a \cdot 1 = a = 1 \cdot a \qquad$ *(one, unit)*
(5) (additive inverse) $\quad \forall a\ \exists x$ *such that* $a + x = 0 = x + a \qquad$ *(opposite* $=: -a$*)*
(6) (commutativity) $\quad a+b = b+a, \qquad ab = ba$

for every $a, b, c \in \mathbb{Z}$.

Proof Exercise.   □

Property (5) permits to introduce the difference of two integers $a - b := a + (-b)$, and thus define the operation of **subtraction** on $\mathbb{Z}$.

Exercise 6.13 Prove that

i) $0 := [(0, 0)]$ is the zero, and $[(-m, -n)]$ the opposite of $[(m, n)]$;
ii) $1 := [(1, 0)]$ is the multiplicative unit;
iii) for any $a, b \in \mathbb{Z}$ $a \cdot 0 = 0$, $(-a)(-b) = ab$.

Definition The set $\mathbb{Z}$ carries an order relation:

$$[(m, n)] \geqslant [(0, 0)] \iff \exists p \in \mathbb{N} : m = n + p.$$

(More informally, $\forall a, b \in \mathbb{Z}$ $a \leqslant b \iff b - a \in \mathbb{N}$.)

Proposition *The structure* $(\mathbb{Z}, \leqslant)$ *is totally ordered.*

Proof Exercise.   □

So every number $a \in \mathbb{Z}^*$ is either positive ($0 \leqslant a$ and $0 \neq a$, written $0 < a$ or $a \in \mathbb{Z}^+$), or negative ($a \leqslant 0$ and $0 \neq a$, written $a < 0$ or $a \in \mathbb{Z}^-$). Furthermore, a is said non-negative if $0 \leqslant a$, non-positive if $a \leqslant 0$.

Exercise Prove that $a^2 := a \cdot a \geqslant 0$ for any $a \in \mathbb{Z}$.

Proposition 6.14 *The integers form an integral domain:* $ab \neq 0 \iff a, b \neq 0$.
In particular, $ac = bc \implies c = 0$ *or* $a = b$ *(cancellation property).*

Proof ($\implies$) See Exercise 6.13 iii).
($\impliedby$) Take representatives (x, y), (z, w) of a, b fulfilling $[(x, y)] \cdot [(z, w)] = [(0, 0)]$, so that $xz + yw = xw + zy$. If $z = w$ then $[(z, w)] = [(0, 0)]$ and we are done. Supposing $z > w$, there is a $p \in \mathbb{N}^+$ such that $w + p = z$. Hence (in $\mathbb{N}$) $x(w + p) + yw = xw + y(w + p) \implies xp = yp \implies x = y$, which forces $[(x, y)] = [(0, 0)]$.   □

Definition 6.15 An **ordered domain** $(D, +, \cdot, \leqslant)$ is an integral domain equipped with a total-order relation $\leqslant$ such that

$$a \leqslant b \implies a + c \leqslant b + c, \qquad 0 \leqslant a, b \implies 0 \leqslant ab$$

for all $a, b, c \in D$.

Exercises 6.16 Prove the following statements.

i) If D is an ordered domain, the subset $D^+ := \{x \in D \mid 0 < x\}$ is closed under sums and products.
ii) Let $(D, +, \cdot)$ be an integral domain, and define a subset D^+ with the properties:

   • $x \in D^+$ *aut* $-x \in D^+$ for any $x \in D \setminus \{0\}$
   • $D^+ + D^+ \subseteq D^+$, $D^+ \cdot D^+ \subseteq D^+$.   (closure under sums and products)

Then the relation $x \leqslant y \iff y - x \in D^+ \cup \{0\}$ turns D into an ordered domain.

iii) In an ordered domain D, squares are non-negative: $x^2 > 0$ for any $x \in D \setminus \{0\}$.

Above we proved $D = \mathbb{Z}$ is an ordered domain, with 'positive' subset $D^+ = \mathbb{N}^*$. As a matter of fact it can be shown $\mathbb{Z}$ is the smallest ordered domain; put equivalently, any ordered domain contains a subset isomorphic (as ordered space) to $\mathbb{Z}$. See also Theorem 7.2.

6.4.1 Divisibility

Proposition 6.14 ensures that there are no **zero-divisors** in $\mathbb{Z}$. This will be automatic once we learn that the integers are a subring of a field (the rationals). But this section wishes to emphasise the importance of the notion of integral domain, and therefore we'll discuss the essential aspects of divisibility in $\mathbb{Z}$. For the record, there exist spaces with zero divisors:

- the ring $\mathbb{Q}[x, y]/(xy)$, quotient by the prime ideal generated by xy: x^3 and y^3 are non-zero, yet their product is null in the quotient ring;
- every $\mathbb{Z}_n$ with n not prime; for example, in $\mathbb{Z}_{12}$ one has $3 \neq 0 \neq 4$ and $3 \cdot 4 = 0$;
- the algebra of matrices: $\begin{pmatrix} 0 & 1 \\ 0 & 0 \end{pmatrix}\begin{pmatrix} 1 & 0 \\ 0 & 0 \end{pmatrix} = \begin{pmatrix} 0 & 0 \\ 0 & 0 \end{pmatrix}$.
- trivially, any space containing nilpotent elements.

Theorem 6.17 (Division Algorithm) *For every $a \in \mathbb{Z}, b \in \mathbb{Z}^+$ there exist unique numbers $r \in \mathbb{Z}$ (called **remainder**) and $q \in \mathbb{Z}$ such that*

$$a = bq + r, \quad with \quad 0 \leqslant r < b.$$

Proof Define $W := \{a - bt \mid t \in \mathbb{Z}\}$ and $W^+ := W \cap \mathbb{N}$. By Zorn's lemma there exists $r = \min W^+ \geqslant 0$. Hence there is a $q \in \mathbb{Z}$ such that $r = a - bq$. We claim $r < b$. By contradiction, if $r = a - bq$ were larger than or equal to b we would have $r > a - (q + 1)b \in W^+$ ⚡⚡. $\square$

The above property of $\mathbb{Z}$ is called being a 'Euclidean domain' in algebra. It depends on the absolute value function, which will be introduced in Definition 7.3 when we'll discuss real numbers. The 'real' version of the previous theorem will be proved in Corollary 7.10, but from another point of view.

Definition 6.18 An integer m is a **multiple** of $n \in \mathbb{Z}$ when there exists a $k \in \mathbb{Z}$ such that $m = nk$. The set of integer multiples of n is written $n\mathbb{Z}$.

If $n \neq 0$, we say n **divides** m (or that it is a **divisor** or a **factor** of m): $n \mid m$.

Using Definition 2.1 we may recast this as

$$n\mathbb{Z} = [0]_{\equiv_n} \qquad n \mid m \iff m \equiv_n 0.$$

An integer m is **even** if $2|m$, and **odd** otherwise, i.e. $m \equiv_2 1$.

Exercises 6.19 Take $m, n \in \mathbb{Z}$ and show

i) $1|n; \qquad \forall m \neq 0 \ m|0; \qquad m|1 \Longrightarrow m = \pm 1$

ii) $m|n, \ n|q \Longrightarrow m|q$

iii) $m|n, \ n|m \Longrightarrow m = \pm n$

iv) $m|n, \ m|q \Longrightarrow m|(sn + tq) \ \forall s, t \in \mathbb{Z}$.

v) If $m \geqslant 2$, the set of integers not divisible by m is $[1]_{\equiv_m} \cup [2]_{\equiv_m} \cup \cdots \cup [m-1]_{\equiv_m}$.

Definition 6.20 Let $m, n \in \mathbb{Z}$ be not both zero. The **greatest common divisor** of them, indicated by $\gcd(m, n)$, is

$$\gcd(m, n) = \max \left\{ M \in \mathbb{Z}^+ \text{ such that } M|m, \ M|n \right\} \in \mathbb{Z}^+$$

Proposition 6.21 *For every $m, n \in \mathbb{Z}$ not both zero, $\gcd(m, n)$ exists and is unique. Moreover, there exist integers $\rho, \sigma \in \mathbb{Z}$ such that $\gcd(m, n) = \rho m + \sigma n$.*

Proof First we show the existence of the greatest common divisor. The set $A = \{rm + sn \mid r, s \in \mathbb{N}\}$ contains 0. If $x \in A$ is a negative number, then $-x \in A$. Hence A contains positive elements, and by the well-ordering principle it has a minimum $a := \min A$. Therefore $a = \rho m + \sigma n$ for some ρ, σ. We claim $a = \gcd(m, n)$.

By Exercise 6.19 iv) it's enough to show $a|m, n$. Theorem 6.17 guarantees that $m = aq + r, 0 \leqslant r < a$. Then $m = (\rho m + \sigma n)q + r$, i.e. $r = m = (1 - \rho q) - \sigma nq$. But this means $r \in A$. Now since r is smaller than a, necessarily r cannot be positive. Hence $r = 0$, i.e. $a|m$. In the same way one shows $a|n$.

Uniqueness: if $a' > 0$ divides m, n, by definition and by Exercise 6.19 iii) we have $a = \pm a'$. But both a, a' are positive, so $a' = a$. $\qquad \qquad \square$

The integers $\rho, \sigma \in \mathbb{Z}$ in the linear combination are far from unique: $\gcd(24, 9) = 3 = 3 \cdot 9 + (-1) \cdot 24 = (-5) \cdot 9 + 2 \cdot 24$.

Example The division algorithm provides a recursive method to compute $\gcd(a, b)$ for any $a > b > 0$. Assuming $a = bq_0 + r_1$ with $0 \leqslant r_1 < b$, first of all observe $\gcd(a, b) = \gcd(b, r_1)$. Now the problem has become simpler and we can iterate, dividing b by r_1: $b = q_1 r_1 + r_2, \ 0 \leqslant r_2 < q_1$. From that $r_1 = q_2 r_2 + r_3, \ldots$, until we arrive at $r_{n-1} = q_n r_n$, or in other words $r_{n+1} = 0$. The process terminates because remainders form a decreasing sequence $r_1 > r_2 > \cdots r_n \geqslant 0$, and since $\mathbb{N}$ is well ordered the sequence is finite, see Sect. 8.2). By definition

$$\gcd(a, b) = \gcd(b, r_1) = \gcd(r_1, r_2) = \ldots = \gcd(r_{n-1}, r_n) = r_n.$$

A concrete example:

$$
\begin{array}{llll}
\gcd(100, 28) = \gcd(28, 16) & \text{because} & 100 = 3 \cdot 28 + 16 & (r_1 = 16) \\
\gcd(28, 16) = \gcd(16, 12) & & 28 = 1 \cdot 16 + 12 & (r_2 = 12) \\
\gcd(16, 12) = \gcd(12, 4) & & 16 = 1 \cdot 12 + 4 & (r_3 = 4) \\
\gcd(12, 4) = 4 & & 12 = 3 \cdot 4 + 0 & (r_4 = 0)
\end{array}
$$

To find $\rho, \sigma \in \mathbb{Z}$ such that $\gcd(m, n) = \rho m + \sigma n$ it suffices to substitute back in the above relations:

$$
\begin{aligned}
\gcd(100, 28) = 4 &= 16 + (-1)12 \\
&= 16 + (-1)\big(28 + (-1)16\big) \\
&= [100 - 3 \cdot 28] + (-1)\big(28 + (-1)[100 - 3 \cdot 28]\big) \\
&= 2 \cdot 100 - 7 \cdot 28 \quad \Longrightarrow \quad \rho = 2, \ \sigma = -7.
\end{aligned}
$$

Definition Two integers p_1, p_2 are said **coprime** if they don't have common divisors except for 1

$$
\gcd(p_1, p_2) = 1.
$$

Exercises Prove the following.

i) p_1, p_2 are coprime if and only if there exist $\rho, \sigma \in \mathbb{Z}$ such that $1 = \rho p_1 + \sigma p_2$.
ii) If p_1, p_2 are coprime, $p_1 \mid q p_2 \Longrightarrow p_1 \mid q$ for any $q \in \mathbb{Z}$.
iii) Additionally, if $\gcd(p_1, p) = 1$ and $\gcd(p_2, p) = 1$ then $\gcd(p_1 p_2, p) = 1$ for any $p \in \mathbb{Z}$.

Another consequence of Lemma 6.21 is the well-known

Theorem 6.22 (Chinese Remainder Theorem) *Let* $n_1, \ldots n_k > 1$ *be pairwise coprime integers, and* $N = \prod_{i=1}^{k} n_i$. *For any* $a_1, \ldots, a_k \in \mathbb{Z}$ *there exists a unique* $x \in \mathbb{Z}$ (mod N) *such that* $x \equiv_{n_i} a_i$, $i = 1, \ldots, k$.

Proof Exercise: start by proving the theorem for $k = 2$ using Lemma 6.21, then extend by induction. $\qquad\qquad\square$

More abstractly, the Chinese remainder theorem says that

$$
x \ \mathrm{mod} \ N \mapsto \big(x \ \mathrm{mod} \ n_1, \ldots, x \ \mathrm{mod} \ n_k\big)
$$

is a ring isomorphism $\mathbb{Z}_N \cong \mathbb{Z}_{n_1} \times \cdots \times \mathbb{Z}_{n_k}$. Among the countless applications we mention the Gödel numbering (p. 46, important for the incompleteness theorem in logic) and ☞ cryptography (signatures of digital certificates within the *https* protocol, cf. Remark 6.27).

Prime numbers have fascinated mathematicians since the dawn of time.

Definition An integer $p > 1$ is called **prime** when its only divisors are 1 and p itself.

Proposition p *is prime* $\iff$ *for any* $a \in \mathbb{Z}$: $p|a$ *or* $\gcd(p, a) = 1$.

Proof ($\implies$) If p is prime, and $\gcd(a, p) \neq 1$, then $\gcd(a, p)|a, p$. But by hypothesis $p = \gcd(a, p)|a$.
($\impliedby$) Suppose $a \neq 1$ is a divisor of p, i.e. $p = ab$. By assumption $p|a$ or $\gcd(p, a) = 1$. In the former case $a = p$ by Exercise 6.19 iii). In the latter, a being a divisor, necessarily $a = 1$. $\qquad\square$

Corollary *Distinct primes are coprime.*

Proof Exercise. $\qquad\square$

The next result shows that any integer can be factorised into primes.

Theorem 6.23 (Fundamental Theorem of Arithmetics) *Every integer* $a \neq 0, 1$ *can be written as product:*

$$a = \pm p_1^{a_1} p_2^{a_2} \cdots p_k^{a_k}$$

of ± 1 *and primes* $p_1 < p_2 < \ldots < p_k$, *with exponents* $a_1, \ldots, a_k > 0$.

The ordering of the factors is required only to get unique decompositions, up to permutations.

Proof We may assume $a > 1$ (for if $a < -1$ we can just multiply by -1).
    Existence: proved in Proposition 6.5.
    Uniqueness: suppose $a > 1$ is written as $a = p_1 p_2 \cdots p_m = q_1 q_2 \cdots q_n$ (with possible repeated factors, in presence of multiplicity). Let's show that $m = n$ and the q_j are the p_i, in some order. The prime p_1 divides one of the q_j. Relabelling the q_j if, if necessary, we can assume $p_1|q_1$. But q_1 is prime, so $p_1 = q_1$ and

$$\frac{a}{p_1} = p_2 \cdots p_m = q_2 \cdots q_n.$$

Similarly, the prime p_2 equals one of the remaining q_j, and again we can assume $p_2 = q_2$, whence

$$\frac{a}{p_1 p_2} = p_3 \cdots p_m = q_3 \cdots q_n.$$

We can do this for every prime p_i (the product is finite), so $m \leqslant n$ and every p_i is one of the q_j. Now swap the roles of p_i and q_j to obtain $n \leqslant m$ and that every q_j is a p_i. All in all, $m = n$ and $q_i = p_i$ for all $i = 1, \ldots, n$ up to rearranging factors (i.e. a bijection $i \mapsto j$). $\qquad\square$

Note: the main reason for 1 not being considered prime is that if it were, we would have distinct decompositions for the same number, e.g. $2 = 2 \cdot 1 = 2 \cdot 1 \cdot 1 = 2 \cdot 1 \cdot 1 \cdot 1 \cdot 1 \cdot 1 \cdot 1 \ldots$

In abstract algebra, the above theorem is phrased by saying $\mathbb{Z}$ is a 'unique factorisation domain'.

Remark 6.24 The problem of deciding whether a certain number N is prime is complicated, but since ancient times (more or less practical) techniques are available to find primes not exceeding a certain threshold, starting from the famous *Eratosthenes sieve*. More interesting is the prime-factorisation question, that is, the computational problem of determining the explicit prime factorisation of a specific integer N. Formulated as a decision problem, it asks to decide whether N has or not a factor less than a given p. The fact that no efficient factorisation algorithm is known lies at the heart of various cryptography systems in use.

The **P versus NP** question is an unsolved problem in ☞ *computer science*. It asks whether all problems whose solution can be verified rapidly can be solved rapidly as well. The vague word 'rapidly' means there exists an algorithm doing the job, which is executed in polynomial time, so that the time to complete the job varies as a polynomial function of the input's size, as opposed to—say—exponential time. The class of questions which an algorithm can answer in polynomial time is called 'P'. For certain questions there's no known way to find a solution quickly, but it's still possible to verify the answer (if known) quickly. The questions whose answer can be verified in polynomial time form class 'NP' (non-deterministic polynomial time).

In Sūdoku, for example, it's easy to check whether you have won once the game is over. But it's usually extremely tricky to determine in advance (calculate) whether or not you will succeed. If we have a partially filled grid (Fig. 6.2 left), of any size, does there exist a solution? Every proposed solution (blue) is easily verifiable, and the time needed to check grows slowly (polynomially) with the grid's size. But all the known algorithms for solving the puzzle take exponential time. Hence Sūdoku is NP (quickly verifiable) but doesn't seem to be P (quickly solvable). Integer factorisation is another NP problem.

8	9			4			5	6
1				9				4
			8		7			
		8				2		
5	7						3	9
		2				7		
			2		9			
7				5				3
2	4			7			1	8

8	9	7	1	4	2	3	5	6
1	2	6	5	9	3	7	8	4
4	5	3	8	6	7	1	9	2
6	1	8	9	3	4	2	7	5
5	7	4	6	2	1	8	3	9
9	3	2	7	8	5	7	4	1
3	8	5	2	1	9	4	6	7
7	6	1	4	5	8	9	2	3
2	4	9	3	7	6	5	1	8

Fig. 6.2 An NP game

Since $P \subseteq NP$, the point is to establish or disprove the converse. Although related issues have been tackled in the 1950s by Nash, Gödel and von Neumann, the problem was properly formulated only in 1971, and is considered the most important open problem in the area [31]. It is in fact one of the Clay Mathematics Institute's *Millennium Problems* [101]. The answer to $P \stackrel{?}{=} NP$ would determine whether polynomially verifiable problems, like Sūdoku, can be solved in polynomial time or not. If we could prove $P \neq NP$, there would exist NP problems more difficult to solve than to verify: they could not be solved in polynomial time, but any solution could be checked in polynomial time. Theoretical computer science apart, any proof would have a tremendous impact in mathematics (game theory, cryptography—it would seal the fate of public encryption keys, hence of our credit cards!), artificial intelligence, multimedia processing, philosophy, economics and scores of other domains.

Corollary 6.25 *There exist infinitely many prime numbers.*

Proof This classical argument is attributed to Euclid. Suppose by contradiction that the list of primes is finite, so that the product of all primes $P = \prod_{p \text{ prime}} p$ is finite. The integer $1 + P$ can't be prime, as it is bigger than any p in the list. Moreover none of the p divides $1 + P$ (the division's remainder is 1), so $1 + P$ cannot admit a prime factorisation. But this violates the fundamental theorem of arithmetics $\frac{1}{2}\frac{1}{2}$.

Kummer's variation on the same theme goes as follows: with the above notation, $P - 1 \in \mathbb{N}$ is a product of primes, so it has a prime divisor p in common with P, forcing p to divide $(P - (P - 1)) = 1$ $\frac{1}{2}\frac{1}{2}$.   □

We shall present two more arguments.

Second Proof of Corollary 6.25 The first is topological in nature (although this won't transpire in the wording) and is due to Furstenberg [35]. Recall $[a]_{\equiv_m} = a + m\mathbb{Z}$ denotes an infinite *arithmetic progression*, so finite sets cannot be unions of arithmetic progressions. Otherwise said, a finite intersection of arithmetic progressions is either $\varnothing$ or infinite. But $[a]_{\equiv_m} = \mathbb{Z} \setminus \bigcup_{j=1}^{m-1} [a + j]_{\equiv_m}$. From this we deduce $\mathbb{Z} \setminus \{\pm 1\} = \bigcup_{p \text{ prime}} [0]_{\equiv_p}$. If there were only finitely many prime numbers, by Exercise 6.19 v) the right-hand-side set would be a finite union $\frac{1}{2}\frac{1}{2}$.   □

There's also a sophisticated proof based on the following famous fact:

Theorem 6.26 (Euler 1748) *The zeta function* $\zeta(s) := \sum_{n=1}^{\infty} \dfrac{1}{n^s}$ *is equal to the infinite product* $\prod_{p \text{ prime}} (1 - p^{-s})^{-1}$.

Proof Observe preliminarily that $\dfrac{\zeta(s)}{2^s} = \dfrac{1}{2^s} + \dfrac{1}{4^s} + \dfrac{1}{6^s} + \dfrac{1}{8^s} + \cdots$. A little manipulation shows $\zeta(s)\left(1 - \dfrac{1}{2^s}\right) = 1 + \dfrac{1}{3^s} + \dfrac{1}{5^s} + \dfrac{1}{7^s} + \dfrac{1}{9^s} + \cdots$, where we have eliminated from the series the term containing 2^{-s}. Similarly, $\zeta(s)\left(1 - \dfrac{1}{2^s}\right)\left(1 - \dfrac{1}{3^s}\right) = 1 + \dfrac{1}{5^s} + \dfrac{1}{7^s} + \dfrac{1}{9^s} + \cdots$, where now the 3^{-s} term has disappeared. Continuing in the elimination process, by Theorem 6.23 we inductively obtain $\zeta(s) \prod\limits_{p} \left(1 - \dfrac{1}{p^s}\right) = 1$, which is our claim.   □

Third Proof of Corollary 6.25 Take the limit as $s \longrightarrow 1$ in Euler's formula: $\sum\limits_{n=1}^{\infty} \dfrac{1}{n} = \prod\limits_{p} \left(1 - p^{-1}\right)^{-1}$. If the number of primes were finite, the product on the right would be finite, but the harmonic series still diverges ⚡.

Many other proofs of 6.25 can be found in [4, 84].

Remark Another very active line of research studies the distribution of primes. The survey [62] places emphasis on explaining the main ideas rather than the most general results available.

Dirichlet (1837) showed that if a, b are coprime, there exist infinitely many primes of the form $p = a \mod b$. The *Green-Tao theorem* (2004, 2008) generalises this fact, proving that the set of prime numbers contains arithmetic progressions $\{a + bm \mid m = 1, \ldots, k\}$ of any length (any k). Ulam (1963) observed that plotting primes along a spiral seems to reveal alignments along certain diagonals. This depends on certain quadratic polynomials producing more primes than others.

Even if the problem of the distribution of primes has been satisfactorily solved, see Eq. (10.2), there are plenty of conjectures still lacking a formal proof, albeit confirmed by experiments. One such is the known *Goldbach conjecture* (1742), whereby any integer larger than 4 is the sum of two odd primes. The source of its arduousness is that primes are defined by means of the multiplication, whereas the above property involves a sum. The 'ternary' Goldbach conjecture, for which any integer larger than 7 is the sum of three odd primes, was proved by Helfgott in 2012.

A second popular conjecture, that of *twin primes*, speculates that there exist infinitely many primes p such that $p+2$ or $p-2$ is prime. Examples of twin primes include $(3, 5)$, $(5, 7)$, $(11, 13)$, and the largest pair known is $2, 996, 863, 034, 895 \cdot 21, 290, 000 \pm 1$. The larger is the interval we examine, the scarcer twin primes become, in agreement with the fact the gap between adjacent primes tends to become wider as the numbers increase. Let $\phi_2(x)$ denote the number of primes $p \leqslant x$ such that $p + 2$ is prime, and set $C = \prod\limits_{p \geqslant 3} \left(1 - \dfrac{1}{(p-1)^2}\right)$. The *Hardy–*

Littlewood conjecture, similar in spirit to the asymptotic formula (10.2), alleges that

$$\lim_{x \to \infty} \phi_2(x) = \lim_{x \to \infty} 2C \int_2^x \frac{dt}{(\log t)^2}. \text{ See [43] for a panoramic view.}$$

All this was about general primes. Along those lines, ☞ *number theory* is repleted with open problems concerning Mersenne numbers, Fermat numbers, Catalan numbers, Fibonacci numbers, or the more amusingly named: pernicious, apocalyptic, untouchable, deficient, astonishing, abundant, sociable, evil, hoax, lucky, odious, practical, admirable, economical, wasteful numbers et al.

`Dually to Definition 6.20, the **least common multiple** of $m, n \in \mathbb{Z}^+$ is the number

$$\text{lcm}(m, n) := \min\left\{s \in \mathbb{Z}^+ \text{ such that } m|s \text{ and } n|s\right\}.$$

It can be proved that $\text{lcm}(m, n)$ enjoys analogous properties to the gcd.

Exercises

i) Let $a = p_1^{a_1} p_2^{a_2} \cdots p_k^{a_k}$ and $b = p_1^{b_1} p_2^{b_2} \cdots p_k^{b_k} \cdots p_h^{b_h}$ be prime decompositions. Prove that

$$\gcd(a, b) = p_1^{c_1} \cdots p_k^{c_k} \quad \text{where } c_i = \min\{a_i, b_i\}$$

for all $i = 1, \ldots, k$. Find a similar expression for $\text{lcm}(a, b)$.

ii) Prove that $ab \neq 0 \implies \text{lcm}(a, b) \gcd(a, b) = ab$.

iii) Explain why, if we wish to show that a number $n > 0$ is prime, it suffices to prove n is not divisible by any prime $p \leqslant \sqrt{n}$.

iv) If p is prime, show there cannot exist integers a, b such that $a^2 = pb^2$. Equivalently: p prime $\implies \sqrt{p}$ is irrational.

Regarding the last exercise,

Proposition *If $s \in \mathbb{Z}^+$ is square-free (i.e. not divisible by any perfect square k^2) then $\sqrt{s} \notin \mathbb{Q}$.*

Proof We shall replicate a simple proof due to Estermann. Since $s \in \mathbb{Z}^+$ is square-free there exists n such that $n^2 < s < (n+1)^2$, and so $0 < \sqrt{s} - n < 1$. By contradiction, suppose $\sqrt{s} = a/b$ were rational. We may assume b is the smallest positive denominator representing the fraction $\sqrt{s}$. Hence b is also the smallest positive integer such that $b\sqrt{s} \in \mathbb{Z}$. Consequently $(\sqrt{s} - n)b\sqrt{s} = b\sqrt{s} - nb\sqrt{s}$ is an integer. But $0 < \sqrt{s} - n < 1$, and so $(\sqrt{s} - n)b < b$. Therefore $N = (\sqrt{s} - n)b$ is an integer smaller than b that satisfies $N\sqrt{s} \in \mathbb{Z}$ ⚡.   □

Exercise What does 'square-free' mean in terms of the prime-factorisation theorem?

In conclusion,

Proposition *The subset of integers* $\{[(m,0)] \in \mathbb{Z} : m \in \mathbb{N}\}$ *satisfies the Peano axioms.*

Proof Exercise.   □

In view of this result one says that $\mathbb{N} \subset \mathbb{Z}$, in the sense that the subset $\mathbb{Z}^+ \cup \{0\} \subseteq \mathbb{Z}$ can be identified with $\mathbb{N}$, as follows:

$$[(m,n)] \rightsquigarrow m - n \in \mathbb{Z}$$

$$[(m,0)] \rightsquigarrow m \in \mathbb{N}$$

Said better: the restrictions of the operations and the order relation on $\mathbb{Z}$ coincide with the operations and ordering of $\mathbb{N}$: we have, indeed, isomorphisms of commutative monoids

$$(\{0\} \cup \mathbb{Z}^+, +, \leqslant) \cong (\mathbb{N}, +, \leqslant) \cong (\mathbb{Z}^- \cup \{0\}, +, \geqslant)$$

$$(\mathbb{Z}^+, \cdot) \cong (\mathbb{N}^*, \cdot).$$

The key difference between $\mathbb{Z}$ and $\mathbb{N}$ is that the former is unbounded below, so on $\mathbb{Z}$ the successor and predecessor maps are inverse to one another: $\mathfrak{a}(s(n)) = s(\mathfrak{a}(n)) = n$, $\forall n \in \mathbb{Z}$.

Remark The construction of $\mathbb{Z}$ starting from $\mathbb{N}$ is categorical. One can in fact define a functor $\mathsf{Mon} \to \mathsf{Ab}$ mapping a commutative monoid M to its (Abelian) Grothendieck group $K_0(M)$ (☞ *K-theory*). In our case $K_0(\mathbb{N}) = \mathbb{Z}$.

Remark 6.27 (☞ Cryptography) The *RSA cryptosystem* (after Rivest, Shamir and Adleman) is one the first public-key encryption systems (1977) and is widely used today for secure data transmission. It involves a public key, known by everyone and used for coding messages, and a private key (kept secret, for decryption). The whole idea is that a message encrypted with the public key can only be deciphered in a reasonable amount of time by knowing the private key, because of the incredible difficulty of factoring the product of two unknown large primes p, q. For instance, can you factor $n = 3233$, and how long does it take you? (This is considered tiny by modern supercomputers, which would take less than a billionth of a second to do it.)

Here is how the algorithm works in an example:

Key generation: choose distinct primes $p = 61, q = 53$ and compute $\rho = \mathrm{lcm}(p - 1, q - 1) = 780$ (all these numbers are kept secret).
Pick an integer $1 < e < \rho$ such that $\gcd(e, \rho) = 1$. We'll take the amenable $e = 2^4 + 1$. The public key consists of e and $n = pq = 3233$.

Encryption: to encrypt the message $m = 65$ use the function $c(m) = m^e$ (mod n) = 2790.

Decryption: to decrypt the encrypted message $c = 2790$ we need to find d satisfying $de \equiv_\rho 1$. In our case $d = 413$. The private key consists of n and d. The decryption function is $m(c) = c^d$ (mod n) = 65.

The simplicity of the example doesn't do justice to the problem's complexity. The principle behind RSA is that finding three very large numbers e, d, n solving

$$(m^e)^d \equiv_n m, \quad \text{for all integers } 0 \leqslant m < n$$

is viable. But even if we do know e, n and m, determining d can be fiendishly involved.

6.5 Rational Numbers

Also the familiar rational numbers are, like the integers, defined as equivalence classes.

Definition Consider the Cartesian product $\mathbb{Z} \times \mathbb{Z}^*$ and the equivalence relation

$$(p, q) \sim (r, s) \iff ps = qr.$$

The **rational numbers** are the elements of the quotient $\mathbb{Q} := \dfrac{\mathbb{Z} \times \mathbb{Z}^*}{\sim}$.

The symbol, introduced by Peano in 1895, comes from the Latin *quotiente*. In the practice the number $[(p, q)] \in \mathbb{Q}$ is written as a fraction $\dfrac{p}{q} = p/q$, because the definition implies $\dfrac{p}{q} = \dfrac{kp}{kq}$ for any $k \in \mathbb{Z}^*$, making the cosets well defined. The non-unique representation motivates the need of the relation $\sim$. For instance, the fraction $\dfrac{28}{12}$ (reducible, since $\gcd(28, 12) = 4$) represents the same rational number as $\dfrac{7}{3}$ (irreducible, $\gcd(7, 3) = 1$). Cross-multiplying, in fact, we see $28 \cdot 3 = 7 \cdot 12$. The issue of the (ir)reducibility of fractions probably kick-started the development of Hellenic geometry, culminating in Thales' theorem 18.5.

Let's define $\mathbb{Q}^* := \mathbb{Q} \setminus \{[(0, 1)]\}$.

Definition The operations on $\mathbb{Q}$:

$$[(p, q)] + [(r, s)] := [(ps + qr, qs)] \qquad [(p, q)] \cdot [(r, s)] := [(pr, qs)]$$

are called addition and multiplication.

Proposition *With the above operations $(\mathbb{Q}, +, \cdot)$ is a field, where*

- *the zero is* $[(0, 1)]$,
- *the (additive) opposite of* $[(p, q)]$ *is* $[(-p, q)]$,
- *the (multiplicative) unit is* $[(1, 1)]$,
- *the (multiplicative) inverse of* $[(p, q)] \in \mathbb{Q}^*$ *is* $[(p, q)]^{-1} = [(q, p)]$.

Proof Direct application of the operations' definition.   □

All the issues on divisibility seen in the previous section lose their meaning in $\mathbb{Q}$, since every rational is divisible by any non-zero rational.

Exercise Explain rapidly why

i) $\dfrac{0}{1} = \dfrac{0}{s} = 0$ for every $s \in \mathbb{Z}^*$

ii) $\dfrac{-p}{q} = \dfrac{p}{-q} = -\dfrac{p}{q}$ for any $p/q \in \mathbb{Q}$

iii) $\left(\dfrac{p}{q}\right)^{-1} = \dfrac{q}{p}$ for any $p/q \in \mathbb{Q}^*$

iv) $q^{-1} = \dfrac{1}{q}$ for every $q \in \mathbb{Z}^*$.

The subset $\{[(p, 1)] \in \mathbb{Q}\}$ can be identified with $\mathbb{Z}$, in the sense that the map

$$\mathbb{Q} \ni [(p, 1)] \longleftrightarrow p \in \mathbb{Z}$$

is a ring isomorphism. That's why we usually write $\mathbb{Z} \subset \mathbb{Q}$.

Remark 6.28 The construction of the rationals is functorial: any integral domain D possesses a *field of fractions* $K(D)$, which is the smallest field containing D. Examples include:

- $\mathbb{Q} = K(\mathbb{Z})$ is the field of fractions of the integers,
- $\mathbb{R}(x) := K(\mathbb{R}[x])$ are the fractions of polynomial maps $\mathbb{R}[x]$, called **rational functions**.

Definition Define the subset $\mathbb{Q}^+ \subseteq \mathbb{Q}$ of 'positive' rationals by

$$\frac{p}{q} \in \mathbb{Q}^+ \iff pq >_{\mathbb{Z}} 0$$

and then set $\mathbb{Q}^- := -\mathbb{Q}^+ = \{-q \mid q \in \mathbb{Q}^+\}$, the 'negative' rationals.

Exercise Verify that

i) $\mathbb{Q} = \mathbb{Q}^+ \sqcup \{0\} \sqcup \mathbb{Q}^-$

ii) $\mathbb{Q}^+$ is closed under addition and multiplication: $\mathbb{Q}^+ + \mathbb{Q}^+ \subseteq \mathbb{Q}^+$, $\mathbb{Q}^+ \cdot \mathbb{Q}^+ \subseteq \mathbb{Q}^+$.

iii) $\mathbb{Q}^- + \mathbb{Q}^- \subseteq \mathbb{Q}^-$, $\mathbb{Q}^- \cdot \mathbb{Q}^- \subseteq \mathbb{Q}^+$, $\mathbb{Q}^- \cdot \mathbb{Q}^+ \subseteq \mathbb{Q}^-$.

It follows from Exercise 6.16, then, that $\mathbb{Q}^+$ induces on $\mathbb{Q}$ a total order $\leqslant$. We may also forget that and prove it directly:

Proof Reflexivity is obvious. Skew-symmetry descends from the observation that if $q \geqslant q' \implies q = q' + a$ for some $a \geqslant 0$, and $q' \geqslant q \implies q' = q + a'$ for a $a' \geqslant 0$, then $q = q + a + a'$. Hence $a + a' = 0$, meaning $a = a' = 0$.

Transitivity: $q \geqslant q' \implies q = q' + a'$ for some $a' \geqslant 0$, and $q' \geqslant q'' \implies q' = q'' + a''$ for some $a'' \geqslant 0$. Then $q = q'' + (a' + a'')$, i.e. $q \geqslant q''$.

Totality: suppose $p/q \neq r/s$. Then $ps \neq rq$ in $\mathbb{Z}$, which means $ps > rq$ *aut* $ps < rq$, or equivalently $ps - rq > 0$ *aut* $ps - rq < 0$. The latter implies $p/q - r/s = (ps - rq)(qs)$ is larger *aut* smaller than 0. $\qquad\square$

From the easy consequence

Corollary *The relation $\leqslant$ is compatible with addition:* $q \leqslant q' \iff q + c \leqslant q' + c$ $\forall q, q', c \in \mathbb{Q}$.

Proof Exercise. $\qquad\square$

we deduce

Proposition

i) $(\mathbb{Q}, +, \cdot, \leqslant)$ *is an ordered field.*
ii) *The canonical inclusion* $(\mathbb{Z}, +_{\mathbb{Z}}, \cdot_{\mathbb{Z}}, \leqslant_{\mathbb{Z}}) \hookrightarrow (\mathbb{Q}, +_{\mathbb{Q}}, \cdot_{\mathbb{Q}}, \leqslant_{\mathbb{Q}})$ *is a 1-1 homomorphism of ordered domains.*

Proof Exercise. $\qquad\square$

Just to set the record straight, $q < q'$ stands for $q \leqslant q' \land q \neq q'$, while $q > q'$ means $-q < -q'$.

Exercises Prove, for any $q, q' \in \mathbb{Q}^*$,

i) $1/q > 0 \iff q > 0$
ii) $q^2 > 0$
iii) if $qq' > 0$ then: $q < q' \iff 1/q' < 1/q$
iv) if $qq' < 0$ then: $q < q' \iff 1/q < 1/q'$.

Note the order relation defined above is not a well ordering. The set $\{q \in \mathbb{Q} \mid q^2 > 5\}$ doesn't have minimum. On the contrary,

Lemma 6.29 *The set $\mathbb{Q}$ is **dense in itself**:*

$$\forall q \neq q' \in \mathbb{Q} \quad \exists t \in \mathbb{Q} \quad \text{such that } q < t < q'.$$

Proof It's enough to take the average $t = (q + q')/2 \in \mathbb{Q}$. $\qquad\square$

From this we infer that, in reality, there exist infinitely many rationals t_i lying between $q \neq q'$: the sequence $t_1 = (q + q')/2, t_j = (q + t_{j-1})/2, j > 1$, for example. See (8.1).

The density of $\mathbb{Q}$ lies at the foundations of the topological difference between $\mathbb{Q}$ and $\mathbb{Z}$: the integers are discrete (every singleton $\{n\} \subseteq \mathbb{Z}$ is open), whereas rational points $\{q\} \subseteq \mathbb{Q}$ are not open, cf. Theorem 8.10.

6.5.1 Decimal Representation

In this section we describe an application of the algorithm of integer division. The discussion will pick up in Sect. 7.5 with the extension to real numbers.

Definition A **decimal number** is a rational number of the form $\dfrac{m}{10^{k_0}}$, with $m \in \mathbb{Z}, k_0 \in \mathbb{N}$.

As any integer $m \in \mathbb{Z}^+$ is determined by a (finite) number of digits $d_k = 0, \ldots, 9$:

$$m = d_0 d_1 d_2 \cdots d_{k_0} = 10^{k_0} d_0 + 10^{k_0 - 1} d_1 + 10^{k_0 - 2} d_2 + \cdots + 10 d_{k_0 - 1} + d_{k_0},$$

we have $\dfrac{m}{10^{k_0}} = d_0 + \dfrac{d_1}{10} + \dfrac{d_2}{10^2} + \ldots + \dfrac{d_{k_0}}{10^{k_0}}$. By this the d_k are actually the remainders of the successive divisions by powers 10^j, $j = 0, 1, \ldots, k_0$, called decimal digits. Already from school we're used to writing, symbolically,

$$\frac{m}{10^{k_0}} = d_0 . d_1 \, d_2 \cdots d_{k_0}. \tag{6.3}$$

The set of decimal numbers contains $\mathbb{Z}$, and forms a proper subring of $\mathbb{Q}$, because there are rationals than can't be put in the above form, as explained by the following

Exercise Let $\dfrac{p}{q} \in \mathbb{Q}$ be irreducible, with $q > 0$. Prove $\dfrac{p}{q}$ is decimal if and only if the prime decomposition of q is of the form $q = 2^a 5^b$ for certain $a, b \in \mathbb{N}$.

Theorem 6.17 ensures a non-decimal rational $d_0.d_1 d_2 d_3 \cdots$ will have infinitely many decimal digits, since the Euclidean algorithm computing p/q doesn't stop. Its decimal expression must be periodic: the sequence of digits $\{d_k\}_{k>0}$ is periodic starting from a certain index $k_0 > 0$. For any $k > k_0$, in other words, there exists an $N \geqslant 1$ such that

$$d_k = d_{sN+k}, \qquad \forall s \in \mathbb{N}.$$

The periodic expansion is shortened as follows

$$d_0 . d_1 \cdots d_{k_0} \overline{d_{k_0+1} \cdots d_{k_0+N}} \tag{6.4}$$

where $d_{k_0+1} \cdots d_{k_0+N}$ are the digits repeating themselves indefinitely.

Lemma *The number in* (6.4) *is rational, but not decimal.*

Proof Call a the number in question. As $10^{k_0+N} a$ and $10^{k_0} a$ have the same decimal digits $\overline{d_{k_0+1} \cdots d_{k_0+N}}$, their difference $10^{k_0+N} a - 10^{k_0} a = 10^{k_0}(10^N - 1)a$ is an integer, say $m \in \mathbb{Z}$. Hence

$$d_0.d_1 \cdots d_{k_0}\overline{d_{k_0+1} \cdots d_{k_0+N}} = a = \frac{m}{10^{k_0}(10^N - 1)} \in \mathbb{Q}.$$

But as $9 \,|\, 10^N - 1$, the denominator contains the prime factor 3, and a cannot be decimal by the exercise above.   □

Example Let's compute the rational number represented by the periodic decimal expression $a = 234.51\overline{749} = 234.51749749749749749 \cdots$

We have $k_0 = 2$ (nr. of non-periodic digits), $N = 3$ (nr. of periodic digits) and $k_0 + N = 5$. Computing

$$10^5 a - 10^2 a = 23451749.\overline{749} - 23451.\overline{749} = 32428298 \in \mathbb{Z}$$

we get

$$a = \frac{32428298}{10^5 - 10^2} = \frac{32428298}{10^2(10^3 - 1)} = \frac{32428298}{99900}.$$

To formalise a little,

Definition Let $(d_k)_{k \in \mathbb{N}}$ be a sequence of digits $d_k \in \{0, 1, \ldots, 8, 9\}$. A series of the form

$$\sum_{i=0}^{\infty} 10^{-i} d_i = d_0 + \frac{d_1}{10} + \frac{d_2}{10^2} + \ldots + \frac{d_n}{10^n} + \ldots$$

is called **decimal expansion**.

The convention is to indicate a decimal expansion by $d_0 . d_1 d_2 d_3 \cdots$. If there exists an $N \in \mathbb{N}$ such that $d_k = 0$ for all $k > N$, one speaks of a finite expansion. It is trivially periodic, with 'period' $\overline{0}$. To sum up, we have proved

Proposition 6.30 *There exists a 1-1 correspondence between $\mathbb{Q}$ and periodic decimal expansions. More precisely,*

i) *decimal numbers correspond to finite expansions* (6.3);
ii) *non-decimal rationals correspond to infinite expansions* (6.4).

We'll address in Sect. 7.5 the meaning of non-periodic expansions.

Beware that the decimal representation is not unique: the rational number 1 can be written in finite form $1.\overline{0}$ or in periodic form $0, \overline{9}$. The same then happens to any decimal number

$$d_0.d_1 \cdots d_{k_0} = d_0.d_1 \cdots (d_{k_0} - 1)\overline{9}.$$

For example: $38.4 = \dfrac{384}{10} = \dfrac{383 + 1}{10} = \dfrac{383 + 0, \overline{9}}{10} = 38.3\overline{9}.$

6.6 A Tasting of Cardinal Numbers

This section's purpose is to get acquainted with the concept of cardinality of a set. At least naively, it should be (made) accessible to first-year students, too.

Definition 6.31 Two sets X, Y are said to have the **same cardinality** if there exists a 1-1 correspondence $f : X \to Y$. If so, we write

$$\text{card } X = \text{card } Y \quad \text{or} \quad |X| = |Y| \quad \text{or} \quad X \approx Y.$$

We may think of a **cardinal number** $|X|$ as the collection of all sets with the same cardinality of X. Very roughly, $|X|$ should convey the idea of the number of members of the set, and having the same cardinality should make us think of being 'equi-numerous', in some sense. But equi-numerosity is not an equivalence relation (as such it would have to be defined on the set of all sets, see Remark 6.44). The formula $|X| = |Y|$ is a binary predicate rather than a relation. Although here we cannot define the notion of cardinal number rigorously, which would require the *theory of ordinal numbers*,[2] the above idea provides us with the essential: a 'quantity' $|X|$ associated to every set in 1-1 correspondence to X, and only those. One could argue that it's immaterial to be able to determine explicitly what a cardinal number actually *is* (as Frege, Russell and von Neumann, among others, did). In fact the characterising property allows to reduce any issue about equality of cardinals to a question on bijections between sets.

Example The function of Example 3.5 is bijective, so $\mathbb{N} \approx \{n \in \mathbb{N} \mid n = 2p\}$. The first shock emerging from the theory of cardinal numbers is that there are as many even numbers and natural numbers, despite the fact that $\{2p \in \mathbb{N} \mid p \in \mathbb{N}\}$ is strictly contained in $\mathbb{N}$.

Definition 6.32 A set X is said **finite** if $X \approx \{1, 2, 3, \dots, n\}$ for some $n \in \mathbb{N}$. We write $|X| = n$ to indicate that the cardinality of X equals the number of its elements. Another frequent symbol is $\#X = n$.

By convention $|\varnothing| = 0$, and $\varnothing$ is finite.

Exercises Show that

i) for any finite sets A, B,

$$|A \cup B| = |A| + |B| - |A \cap B|, \qquad |A \times B| = |A| \cdot |B|.$$

This implies, in particular, that $A \cup B$ and $A \times B$ are finite.

ii) For n finite sets $A_1, \dots, A_n$ the *inclusion-exclusion principle* holds:

---

[2] In a way, in ZF cardinals are just a notation, whereas in ZFC they are ordinals, see Remark 8.16.

$$\left| \bigcup_{i=1}^{n} A_i \right| = \sum_{i=1}^{n} |A_i| - \sum_{i<j} |A_i \cap A_j| + \sum_{i<j<k} |A_i \cap A_j \cap A_k| - \cdots + (-1)^{n-1} \left| \bigcap_{i=1}^{n} A_i \right|$$

$$(6.5)$$

(☞ *probability theory*). *Hint*: exercise 3.17. Another proof comes from 14.5 iii).

iii) Using ii), prove that if $n = p_1^{a_1} p_2^{a_2} \cdots p_r^{a_r}$ is a prime factorisation, Euler's totient function (10.2) equals

$$\phi(n) = n - \sum_{i=1}^{r} \frac{n}{p_i} + \sum_{1 \le i < j \le r} \frac{n}{p_i p_j} - \cdots = n \prod_{i=1}^{r} \left(1 - \frac{1}{p_i} \right). \qquad (6.6)$$

A set that isn't finite is said to be **infinite**. For instance $\mathbb{N}$, prime numbers, even integers, odd integers, $\mathbb{Z}, \mathbb{Q}, \mathbb{R}$ are all infinite.

Definition The cardinality of natural numbers is denoted by

$$\text{card } \mathbb{N} =: \aleph_0$$

(read 'aleph nought', from the first letter of the Hebrew alphabet). Any set A of cardinality $\aleph_0$ is called **countable**.

Countable sets form the foundation of the branch called ☞ *discrete mathematics*.

The existence of a bijection $a : \mathbb{N} \to A$, $n \mapsto a(n) =: a_n$ justifies the intuitive idea of counting the infinitely many elements of A: $a_1, a_2, a_3, \ldots,$ $a_{n-1}, a_n, a_{n+1}, \ldots$

Note that a countable set is infinite. (Warning: in certain books the cardinality $\aleph_0$ is called countably infinite or enumerable, and countable means either finite or countably infinite.)

Examples

i) Odd natural numbers are countable.

ii) $\mathbb{N}^*$ is countable since $s \colon \mathbb{N}^* \to \mathbb{N}$, $s(n) = n + 1$ is bijective.

iii) The Cartesian product $\mathbb{N} \times \mathbb{N}$ is countable. To see this consider

$$C = \{ (x, y) \in \mathbb{Z}^2 \mid 0 \le x \le y \}.$$

First, for any integer $n \ge 0$ there exists a unique pair $(x, y) \in C$ such that $n = x + \sum_{i=0}^{y} i$ (see Example 6.41). This makes C countable. Moreover,

$$\mathbb{N}^* \times \mathbb{N}^* \to C, \qquad (a, b) \mapsto (a, a + b)$$

is bijective, so $\mathbb{N}^* \times \mathbb{N}^*$ is countable. Now use item ii).

Moving on to integers, we have

Proposition 6.33 card $\mathbb{Z} = \aleph_0$.

Proof We claim the map $h : \mathbb{N} \to \mathbb{Z}$, $h(n) = (-1)^{n+1}\dfrac{n}{2} + \dfrac{1 - (-1)^n}{4}$, is bijective. Spelt out,

$$h(n) = \begin{cases} -\dfrac{n}{2} & \text{if } n \text{ is even} \\ \dfrac{n+1}{2} & \text{if } n \text{ is odd} \end{cases}.$$

Now it's not hard to check $h^{-1} : \mathbb{Z} \to \mathbb{N}$, given by $h^{-1}(n) = \begin{cases} -2n & \text{if } n > 0, \\ 2n - 1 & \text{if } n \leqslant 0 \end{cases}$ is the inverse.   □

Exercise Show that the multiples $m\mathbb{Z} := \{mn \mid n \in \mathbb{Z}\}$ of an integer $m \in \mathbb{Z} \setminus \{0\}$ are countable.

Definition Let A, B be arbitrary disjoint sets with cardinality $|A| = \alpha$, $|B| = \beta$. The sum, product and exponentiation of cardinals is defined as follows:

$$\alpha + \beta := |A \sqcup B| \qquad \alpha \cdot \beta := |A \times B| \qquad \alpha^\beta := \left|A^B\right|.$$

A comment: card $\{\clubsuit, \spadesuit\} = 2$, card $\{\spadesuit\} = 1$ but card $\left(\{\clubsuit, \spadesuit\} \cup \{\spadesuit\}\right) \neq 2 + 1$, so it's important to remember the sum formula holds for a disjoint union, for otherwise we should expect something along the lines of the exercise on p.143.

The other two formulas seem easier to digest. The Cartesian product consists of pairs (a, b), so counting its elements boils down to counting the choices of b for each choice of a. Similarly, counting maps $f : B \to A$ reduces to counting the possible images $f(b) \in A$ for each choice of $b \in B$. This is straightforward if the sets in question are finite, but once again in the infinite case the spectre of the axiom of choice is lurking in the background.

Proposition 6.34 *There are as many subsets of a set $A \neq \varnothing$ as maps $A \to \{0, 1\}$:*

$$\mathscr{P}(A) \approx \{0, 1\}^A.$$

Consequently, $\left|\mathscr{P}(A)\right| = 2^{|A|}$.

Proof By Definition 3.3 any $B \subseteq A$ has a characteristic function $\chi_B \in \{0, 1\}^A$. Conversely, from any map $f : A \to \{0, 1\}$ we can define a subset of A, namely the pre-image $f^{-1}(1)$. The two constructions are one the inverse of the other, because the map $\chi : \mathscr{P}(A) \to \{0, 1\}^A$, $B \mapsto \chi_B$ is

    one-to-one: $\chi_B = \chi_{B'} \implies B = \chi_B^{-1}(1) = \chi_{B'}^{-1}(1) = B'$,

    onto: $\chi_{f^{-1}(1)} = f$.

    To finish, $\left|\mathscr{P}(A)\right| = \left|\{0, 1\}^A\right| = 2^{|A|}$.   □

In case A is finite, this fact has a nice proof, based on induction, that expounds the role of the characteristic function. Fix $n \in \mathbb{N}$. If $n = 0$ ($A = \varnothing$), then $2^{\varnothing} = \{\varnothing\}$ and $|2^{\varnothing}| = 1 = 2^0$.

Now suppose $|A| = k$, so $|2^A| = 2^k$. Consider a set $\{a_1, a_2, \ldots, a_{k+1}\}$ of $k + 1$ elements. Its subsets B can be of two types: those that contain a_{k+1} and those not containing a_{k+1}. The number of subsets of the first type equals the number of subsets of the second type, since

> if $a_{k+1} \in B$ (type 1) then $B \setminus \{a_{k+1}\}$ is of type 2;
> if $a_{k+1} \notin B$ (type 2) then $B \cup \{a_{k+1}\}$ is of type 1.

Type 1 are subsets in $\{a_1, a_2, \ldots, a_k\}$, and by inductive hypothesis there are 2^k of them. The same holds for type-2 subsets. Overall, therefore, we have $2^k + 2^k = 2^{k+1}$ subsets of $\{a_1, a_2, \ldots, a_{k+1}\}$.   □

Example (Cantor's Diagonal Argument) Here we'll show that there are infinite sets with different cardinality. We shall compare $\mathbb{N}^+$ and the interval $(0, 1)$, consisting of decimal expressions $0.i_1i_1i_3i_4i_5 \cdots$ The claim is that their cardinal numbers are distinct. Assume, by contradiction, $\mathbb{N}^+ \to (0, 1)$, $n \to r_n$ were a 1-1 correspondence, and represent each r_n in decimal form:

$$r_1 = 0.a_{11}a_{12}a_{13} \cdots a_{1n} \cdots$$
$$r_2 = 0.a_{21}a_{22}a_{23} \cdots a_{2n} \cdots$$
$$r_3 = 0.a_{31}a_{32}a_{33} \cdots a_{3n} \cdots$$
$$\vdots \qquad \vdots \qquad \ddots \qquad \vdots$$
$$r_n = 0, a_{n1}a_{n2}a_{n3} \cdots a_{nn} \cdots$$
$$\vdots \qquad \vdots \qquad \qquad \vdots \quad \ddots$$

Define the number $x = 0.x_1x_2x_3 \ldots \in (0, 1)$ as follows:

> if $a_{11} = 3$ set $x_1 = 2$ (or any digit $\neq 3$); otherwise $x_1 = 3$;
> if $a_{22} = 3$, set $x_2 = 2$ (or any digit bar 3); otherwise $x_2 = 3$.

In general, set x_k equal to any digit different from a_{kk}, for every k. The consequence is that $x \neq r_n$ for every n, since x and r_n differ by the nth digit: $x_n \neq a_{nn}$. Therefore the map $n \to r_n$ isn't onto ⚡.

Definition 6.35 We write $|A| \leqslant |B|$ whenever there exists an injective map $f : A \hookrightarrow B$.

As a consequence, there exists a subset of B with the same cardinality of A (the image $f(A)$ springs to mind).

In the case of finite sets A, B with $|A| > |B|$, the non-existence of 1-1 maps $f : A \to B$ is a well-known fact, inspired by the following situation: suppose we have to house m pigeons in a dovecote with $n < m$ holes. Necessarily one hole will host more than one, in fact at least $\left\lceil \dfrac{m}{n} \right\rceil > 1$ pigeons. This fact is absolutely patent, and certainly doesn't lack intuitive justifications. At the same time it has a host of

non-trivial applications [69, 82], one example among all being ☞ *Ramsey theory.*
The precise statement, seemingly formalised by Dirichlet,[3] goes as follows:

Proposition 6.36 (Pigeonhole Principle) *Let P, $H \neq \varnothing$ be finite sets (pigeons &*
holes) and consider a map $f : P \to H$ (placing the pigeons in the holes). Then
there exists $h \in H$ such that $\left| f^{-1}(h) \right| \geqslant \dfrac{|P|}{|H|}$.

Proof To fix ideas take $P = \{p_1, \ldots, p_m\}$ and $H = \{h_1, \ldots, h_n\}$, of cardinality
$\#P = m$, $\#H = n$. By contradiction, suppose $\left| f^{-1}(h) \right| < m/n$ for every $h \in H$.
Call $f(P) = \{d_1, \ldots, d_k\} \subseteq H$ the image, so $k \leqslant n$. Then $P = \displaystyle\bigsqcup_{i=1}^{k} f^{-1}(d_i)$ is a
partition, and therefore $m = \displaystyle\sum_{i=1}^{k} \left| f^{-1}(d_i) \right| < k \cdot \dfrac{m}{n} \leqslant n \cdot \dfrac{m}{n} = m$ ↯.   □

When we apply the principle to infinite sets the proposition becomes sen-
sationally false—as Hilbert's hotel 7.16 shows—and explains why the need for
Definition 6.35.

Examples 6.37 Here are two situations where the apparently harmless pigeonhole
principle guarantees the existence of configurations which, at least at first sight, are
not straightforward.

i) Is it possible to cover a chessboard with domino tiles and leave uncovered just
   two squares in opposite corners?
      The answer is no. Squares in opposite corners have the same colour, so if we
   disregard them the number of squares of one colour would exceed by 2 the other
   colour's number. But a domino tile always occupies a black and a white square,
   so any tiling defines a 1-1- correspondence between white and black squares.
   If these had different cardinality, the pigeonhole principle would preclude the
   existence of a bijection.
ii) Among the people going to a dinner party, some know each other, others don't.
   ('Knowing each other' is a symmetric, non-reflexive relation.) We claim there
   exists two people that know the same number of others. Call N the number of
   participants. If everyone knew a different number of people there would be a
   person knowing $N - 1$ others (all, except themselves), and a person knowing
   nobody ↯.

Exercises Problems of the ensuing kind can be tackled by means of Proposition 6.36. A large selection, including the ones here, can be found in [12].

---

[3] The word pigeonhole is nowadays exclusively used for a small space in a cabinet or wall for
sorting things (like letters and papers in a post office, or room keys in a hotel), which resembles the
structures that used to house pigeons. Add this to the fact Dirichlet's father was a postman, back
when the birds were employed to deliver messages.

i) Choose five nodes in a square grid. Why can you be certain that there is a segment joining two chosen nodes whose midpoint falls on one of the grid's nodes?

ii) Show that any polyhedron always has two faces with the same number of edges.

iii) Let $p(x)$ be a polynomial with integer coefficients. Suppose there are numbers $a, b, c \in \mathbb{Z}$ different from each other such that $p(a) = p(b) = p(c) = 2$. Prove $p^{-1}(3)$ contains no integers.

iv) There are N light bulbs arranged in a circle, numbered 1 to N, and all turned on. At time t we look at light bulb t: if it's on, we flick the switch of light bulb $t + 1$ (mod N): we turn it off if it was on, and turn it on if it was off. If light bulb t is off, instead, we don't do anything. Show that there is a moment when all light bulbs will be back on again.

Now let's return to the general picture.

Proposition 6.38 *Take sets $A, B \neq \varnothing$. The existence of a 1-1 map $f : A \hookrightarrow B$ is equivalent to the existence of an onto map $g : B \twoheadrightarrow A$.*

Proof If $f : A \to B$ is 1-1, define $g : B \to A$ as follows: fix $a_0 \in A$ and set $g(b) = f^{-1}(b)$ if $b \in f(A)$, $g(b) = a_0$ otherwise. Vice versa, if $g : B \to A$ is onto, by the axiom of choice there is a section $f : A \to B$. The latter is 1-1 since $f(a_1) = f(a_2) \implies a_1 = g(f(a_1)) = g(f(a_2)) = a_2$.   □

Here's another result with a complicated history: it was published by Schröder and Bernstein in 1897 (Cantor only stated it, in 1887). But actually the first proof is due to Dedekind (1887).

Theorem 6.39 (Cantor–Schröder–Bernstein) *If $|A| \leqslant |B|$ and $|B| \leqslant |A|$ then $|A| = |B|$.*

Proof [70, ch.2.3] We'll prove first   □

Lemma *For any maps $f : A \to B$, $g : B \to A$, there exists a subset $A' \subseteq A$ such that*

$$A' \cap g(B \setminus f(A')) = \varnothing, \qquad A' \cup g(B \setminus f(A')) = A.$$

Proof Consider the family of subsets

$$\mathscr{A} = \{ B' \subseteq A \mid B' \cap g(B \setminus f(B')) = \varnothing \}.$$

As $\varnothing \in \mathscr{A}$, it is not empty. We claim the union $A' = \bigcup \mathscr{A}$ is what we're looking for. Take $x \in A'$, so there is a $B' \in \mathscr{A}$ such that $x \in B'$. Then $x \notin g(B \setminus f(B'))$ and so $x \notin g(B \setminus f(A'))$. If there existed $x \notin A' \cup g(B \setminus f(A'))$, setting $C = A' \cup \{x\}$ we would obtain $g(B' \setminus f(C)) \subseteq g(B \setminus f(A'))$, in other words $C \cap g(B \setminus f(C)) = \varnothing$ ↯.   □

Resuming the proof of Theorem 6.39, by the lemma there is an $A' \subseteq A$ such that, setting $B' = B \setminus f(A')$,

$$A' \cap g(B') = \varnothing \quad \text{and} \quad A' \cup g(B') = X.$$

The definition of B also says $B' \cap f(A') = \varnothing$ and $B' \cup f(A') = Y$, whilst the injectivity of f, g forces the restriction maps $f: A' \to f(A')$ and $g: B' \to g(B')$ to become bijective. To conclude it's enough to observe that

$$h: A \to B, \qquad h(x) = \begin{cases} f(x) & \text{if } x \in A' \\ g^{-1}(x) & \text{if } x \in g(B') \end{cases}$$

is a bijection. $\qquad\qquad\qquad\qquad\qquad\qquad\qquad\qquad\qquad\qquad\qquad\qquad\qquad\qquad\square$

A more articulated version of the previous lemma serves as good training:

Exercise Let $f: A \to B, g: B \to A$ be 1-1, but not onto, maps. Define subset sequences

$$A_0 = A \setminus \text{Im}(g), \quad A_{n+1} = g\big(f(A_n)\big)$$
$$B_n = f(A_n).$$

for any $n \in \mathbb{N}$. If $a \notin A_0$ indicate by $b = g^{-1}(a) \in B$ the unique element such that $a = g(b)$. Prove that $h : A \to B$,

$$h(a) = \begin{cases} f(a) & \text{if } a \in A_k \text{ for some } k \in \mathbb{N} \\ g^{-1}(a) & \text{otherwise} \end{cases}$$

is bijective.

We put Theorem 6.39 to good use by showing rational numbers are countable, see Fig. 6.3:

Proposition 6.40 card $\mathbb{Q} = \aleph_0$.

Proof To start with, $\aleph_0 \leqslant |\mathbb{Q}|$ since $n \mapsto \dfrac{n}{1}$ is 1-1. Conversely, pick a non-zero rational, and write it as irreducible fraction $(-1)^p \dfrac{m}{n}$, with $m, n > 0$, $\gcd(m, n) = 1$, and $p = 0, 1$ depending on the sign. The function $f: \mathbb{Q} \to \mathbb{N}$

$$f(0) = 0, \qquad f\left((-1)^p \frac{m}{n}\right) = 2^m 3^n 5^p$$

is 1-1 by the fundamental theorem of arithmetics 6.23, and so $|\mathbb{Q}| \leqslant \aleph_0$. The Cantor–Schröder–Bernstein theorem allows to conclude. $\qquad\qquad\qquad\qquad\qquad\qquad\qquad\square$

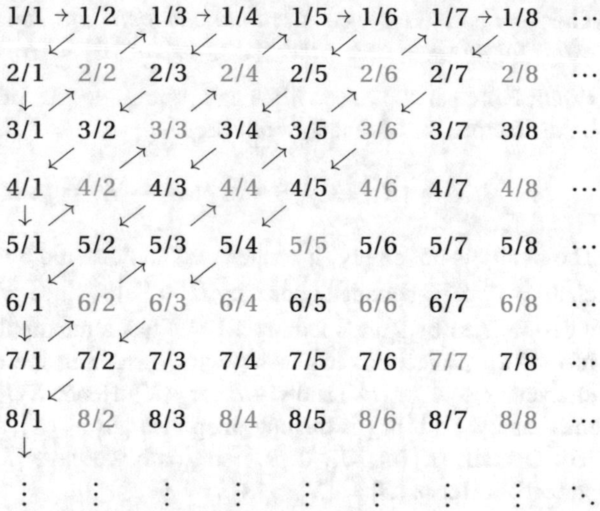

Fig. 6.3 Enumeration of $\mathbb{Q}^+$.
(Grey numbers could be
skipped as they have already
appeared in the snake)

$$1/1 \to 1/2 \qquad 1/3 \to 1/4 \qquad 1/5 \to 1/6 \qquad 1/7 \to 1/8 \quad \cdots$$

$$2/1 \quad 2/2 \quad 2/3 \quad 2/4 \quad 2/5 \quad 2/6 \quad 2/7 \quad 2/8 \quad \cdots$$

$$3/1 \quad 3/2 \quad 3/3 \quad 3/4 \quad 3/5 \quad 3/6 \quad 3/7 \quad 3/8 \quad \cdots$$

$$4/1 \quad 4/2 \quad 4/3 \quad 4/4 \quad 4/5 \quad 4/6 \quad 4/7 \quad 4/8 \quad \cdots$$

$$5/1 \quad 5/2 \quad 5/3 \quad 5/4 \quad 5/5 \quad 5/6 \quad 5/7 \quad 5/8 \quad \cdots$$

$$6/1 \quad 6/2 \quad 6/3 \quad 6/4 \quad 6/5 \quad 6/6 \quad 6/7 \quad 6/8 \quad \cdots$$

$$7/1 \quad 7/2 \quad 7/3 \quad 7/4 \quad 7/5 \quad 7/6 \quad 7/7 \quad 7/8 \quad \cdots$$

$$8/1 \quad 8/2 \quad 8/3 \quad 8/4 \quad 8/5 \quad 8/6 \quad 8/7 \quad 8/8 \quad \cdots$$

$$\vdots \qquad \vdots \qquad \vdots \qquad \vdots \qquad \vdots \qquad \vdots \qquad \vdots \qquad \vdots \qquad \ddots$$

Example 6.41 There are many more ways to define a 1-1 map $\mathbb{Q} \to \mathbb{N}$, all more or less based on Theorem 6.23. For instance the composite map

$$
\begin{array}{ccccc}
\mathbb{Q} \to & \mathbb{Z} \times \mathbb{Z} & \to & \mathbb{N} \\
0 \leqslant \frac{m}{n} \mapsto & (2m, 2n) & \mapsto & 2^{2m} 3^{2n} \\
0 > \frac{m}{n} \mapsto & (-2m - 1, 2n) & \mapsto & 2^{-2m-1} 3^{2n}
\end{array}.
$$

Alternatives include the use of Proposition 6.30 (an incarnation of Theorem 7.22), or the following numbering. Recall the function $H(n) = \dfrac{n(n+1)}{2}$ (from Exercise 6.11) and consider $\sigma : \mathbb{N}^2 \to \mathbb{N}$, $\sigma(a, b) = a + H(a + b)$. Relying on Exercises 6.11 vi)-vii) we'll show σ is bijective.

(Injectivity) Take $(x, y) \neq (u, v)$ and suppose $u + v \leqslant x + y$ without loss of generality.

If $u + v < x + y$ define $z = u + v - x - y - 1$, so

$$\sigma(u, v) = H(x + y + z + 1) + v = H(x + y + z) + x + y + z + 1 + v$$

$$\geqslant H(x + y) + y + x + z + v + 1 > \sigma(x, y);$$

If $u + v = x + y$, and supposing $y \neq v$, we have $\sigma(x, y) = H(x + y) + y \neq H(u + v) + v = \sigma(u, v)$.

(Surjectivity) We'll use induction on $z = \sigma(x, y)$. For $z = 0$ we have $\sigma(0, 0) = 0$. For $z > 0$ there are two cases: if $x = 0$ then $\sigma(0, y + 1) = H(y) + y + 1 = \sigma(0, y) + 1 = z + 1$; and if $x > 0$ then $\sigma(x - 1, y + 1) = H(x + y) + y + 1 = \sigma(x, y) + 1 = z + 1$.

The next result is, as it happens, equivalent to the axiom of choice.

Theorem 6.42 (Hartogs) *Cardinal numbers are always comparable: in other words for any sets A, B either $|A| \leqslant |B|$, or $|B| < |A|$, exclusively.*

Proof For a set $X \subseteq A \times B$ we call $\mathrm{pr}_A \colon X \to A$, $\mathrm{pr}_B \colon X \to B$ the projections on the factors, and define the collection

$$\mathscr{X} = \{X \subseteq A \times B \mid \mathrm{pr}_A \colon X \to A, \ \mathrm{pr}_B \colon X \to B \text{ are 1-1}\}.$$

The family is not empty (it contains $\varnothing$) and ordered by inclusion. Moreover, every chain $\mathscr{C} \subseteq \mathscr{X}$ is bounded above, by $C = \bigcup \mathscr{C}$, since $\mathrm{pr}_A \colon C \to A$, $\mathrm{pr}_B \colon C \to B$ are 1-1. Then by Zorn's lemma 3.12 $\mathscr{X}$ has a maximal element X. We claim one of the two projections has to be onto. Assuming the contrary, there would exist elements $x \in A \backslash \mathrm{pr}_A(X)$ and $y \in B \backslash \mathrm{pr}_B(X)$. Hence $X \cup \{(x, y)\} \in \mathscr{X}$, contradicting maximality. So if pr_A is the onto map then $|A| = |X|$, and since pr_B is 1-1, $|X| \leqslant |B|$. Overall, $|A| \leqslant |B|$. If pr_A isn't onto we have $|X| < |A|$, but pr_B must be surjective. Hence $|B| = |X| < |A|$. $\square$

Proposition *The relation $\leqslant$ between cardinal numbers is a total order.*

Proof Reflexivity is due to the fact the identity $\mathbb{1}$ is 1-1. As for skew-symmetry, see Theorem 6.39. Transitivity: if $f \colon A \hookrightarrow B$, $f' \colon B \hookrightarrow C$ are 1-1, so is $f' \circ f$ (without loss of generality we may assume the composite map is defined, possibly by considering different sets with the cardinality of B).

  Totality: see Theorem 6.42, or 8.14. $\square$

Theorem 6.43 (Cantor) card $\mathscr{P}(A) >$ card A, *for any set A.*

Proof The map $A \hookrightarrow 2^A, a \mapsto \{a\}$ is 1-1, so card $\mathscr{P}(A) \geqslant$ card A. There remains to show card $\mathscr{P}(A) \neq$ card A, for which we'll argue by contradiction. Suppose equality holds, then, so there is bijection $h \colon A \to \mathscr{P}(A), a \mapsto f(a) \subseteq A$. Consider

$$\mathbb{Y} = \{a \in A \mid a \notin f(a)\}.$$

This set belongs to $\mathscr{P}(A)$, so it must have a unique pre-image $r \in A$ under h, i.e. $f(r) = \mathbb{Y}$. Now, by definition of $\mathbb{Y}$, it follows that:
  if $r \in \mathbb{Y} = f(r)$, then $r \notin \mathbb{Y}$ ⨳.
  if $r \notin \mathbb{Y} = f(r)$, then $r \in \mathbb{Y}$ ⨳. $\square$

  Iterating this idea we can generate a sequence of power sets

$$\mathbb{N} \subset \mathscr{P}(\mathbb{N}) \subset \mathscr{P}(\mathscr{P}(\mathbb{N})) \subset \mathscr{P}(\mathscr{P}(\mathscr{P}(\mathbb{N}))) \subset \dots$$

whose (infinite) cardinalities balloon: put differently, infinite sets come in infinitely many 'sizes'

$$\aleph_0 < 2^{\aleph_0} < 2^{2^{\aleph_0}} < 2^{2^{2^{\aleph_0}}} < \cdots$$

Remark 6.44 The theory of sets we have presented is the one on which (most) mathematicians rely on a daily basis, and many of the theorems accepted by the community are founded on it. But it's no secret that Cantor's theory is not problem-free. Notwithstanding the insurmountable contradictions it engenders, we shouldn't abandon it. The main point in favour of keeping it is its axiomatisation. Cantor's theory is based on the conclusion that the known contradictions have to do with very large sets.[4] Its cornerstone is that we can subconsciously and effectively decide which objects are sets and which aren't. The naivety of set theory might be responsible for the theory's contradictions that are commonly called paradoxes. As the inherent connotation of a paradox is that of a fake or cosmetic contradiction, whereas the propositions we are talking about are genuine contradictions, we should rather call them antinomies.

The *abstraction principle* of naive set theory—a.k.a. postulate V in Frege's intensional logic—says that any property (predicate) determines a set (a list of elements). It can be considered the Achilles's heel of the theory itself. When invoked carelessly, the principle gives rise to at least three logical contradictions that go back to Russell, Cantor and Burali-Forti.

Russell's antinomy, described below, emerges naturally from the proof of theorem 6.43, more specifically out of the non-existence of surjections $A \twoheadrightarrow \mathscr{P}(A)$. According to the naive definition we gave at the very start of the chapter, a set is determined by its elements (with no further condition). It would appear the same would hold for the set X of all sets (the one whose elements are all possible sets). This alleged X would then have the curious property that $X \in X$. If we accept that a set can belong to itself, we may as well fathom the opposite situation of a set whose members are not elements of themselves:

$$\mathrm{Russ} = \{X \mid X \notin X\}.$$

Now what about the set Russ? Does it or doesn't it belong to itself? It's easy to see that

$$\mathrm{Russ} \notin \mathrm{Russ} \iff \mathrm{Russ} \in \mathrm{Russ},$$

a contradiction showing that the predicate $\in$ should at the same time be autological and heterological[5].

By the way, speaking of Russell's demolition of naive set theory (and of Frege's logic), it's riveting to note Frege never conceived the possibility of removing Postulate V from his system in order to salvage it. Postulate V is only used to prove *Hume's law*, which is the extra-logical theorem codified by Definition 6.31. It is a

---

[4] Who thinks to eschew the issue will be disappointed: *inaccessible cardinals* are necessary in *category theory*.

[5] A variation on the same theme: is 'heterological' a heterological word?

fact that Frege used Hume's law to prove all of the axioms of 2nd order arithmetics, thus 'arithmetising' analysis.

Russell rejected $\texttt{Russ}$ because its definition is impredicative (read: it can't be a set). On a semi-philosophical note, observe that $\texttt{Russ}$ (assuming such a thing does exist) must be a very large object. Its defining property is satisfied by all objects that are not sets (not being sets, they don't have elements and so they can neither belong to themselves). At the same time the property is satisfied by most sets: a number set is not a number, so it doesn't belong in itself.

When facing Russell's antinomy we have three possible escape routes: use intuitionistic logics (which violate the law of excluded middle, see Remark 5.21), waive the predicate $\in$, or change the definition of what a set is. Zermelo–Fraenkel theory chooses the third way: it denies the existence of the set of all sets, stripping $\texttt{Russ}$ of any ontological meaning. Let's take a closer look at the *extension axiom*

$$\forall A, B \ (x \in A \leftrightarrow x \in B) \to A = B.$$

It says sets with the same members are equal (the conditional $\to$ can be replaced by $\leftrightarrow$ actually, because the converse is obvious). As a consequence a set is uniquely determined by its elements ('extensional' logic). The axiom can be used in formulas of the type '*there exists a set A such that for all x, $x \in A \iff \vDash p(x)$*' where p is a unary predicate that doesn't mention A, in order to define a unique set A whose members are exactly the elements satisfying p.

Parallel to this, the *specification schema* stipulates that any property p determines a set ('intensional' logic) as a list of elements:

$$\forall A \ \exists B : \forall x \ x \in B \leftrightarrow (x \in A \wedge \vDash p(x)).$$

In this way Russell's antinomy is sorted, since now $B = \left\{ x \mid \vDash p(x) \right\}$ is a set, defined unambiguously by the extension axiom.

Later generalisations and variations of ZF theory amend the above defect by introducing a hierarchy among set-like objects. The von Neumann–Bernays–Gödel theory and the Morse–Kelley theory [57] introduce *classes* (collections of sets, like $\texttt{Russ}$), conglomerates (collections of classes) etc. The Tarski–Grothendieck theory adds one axiom to ZF to define universes. An intuitive example, which was sneaked in at the beginning of Chap. 2), would be $\mathscr{U} := \left\{ x \mid x \doteq x \right\}$.

Chapter 7
Real Numbers

We know what it means to count to an arbitrarily large number n. But ever since the pre-Socratic philosopher Zeno came up with his famous paradoxes we understand less clearly how to move between two points drawn on a line, the reason being that the notion of infinity is interwoven in the idea of moving. The Greeks (most notably, Eudoxus) improved the picture, but not enough for any substantial progress to be made. We need to jump ahead two millennia to reconcile the 'continuous' with the 'discrete'.

Having mentioned drawing, every one of us from an early age deals with the real number line in one form or another. This chapter wishes to clarify what on earth this real line is. The intuition of a straight line on a plane or in space derives from the practice of handling segments, but it doesn't explain what it means that this line extends indefinitely in space. Nor what the single real numbers are. The subconscious model for a real number is that of a dot drawn with a pen on paper. But the dot we plot actually corresponds to an infinity of points ... and that's a load of ink.

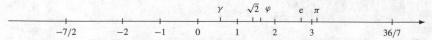

An attractive construction of rational numbers is founded in Euclidean geometry. In that model the real line, freed of its arithmetics, is taken as a primitive concept and subjected to the Euclidean axioms (Euclid himself was convinced his constructions didn't hit any holes). Then there's the issue of existence. The standard model for Euclidean geometry is the Cartesian plane that consists of pairs of real numbers, and verifying the axioms depends on the properties of the real line. Defining real numbers starting from the Euclidean line creates a vicious circle. (That's why in Sect. 18.1 we'll do exactly the opposite.) Speaking of incoherencies, the circular reasoning of certain school textbooks is not that subtle: after presenting the rationals, they define irrational numbers as those real numbers that aren't rational, and then go on to say that real numbers aggregate rationals and irrationals!

© The Author(s), under exclusive license to Springer Nature Switzerland AG 2021
S. G. Chiossi, *Essential Mathematics for Undergraduates*,
https://doi.org/10.1007/978-3-030-87174-1_7

It is possible to define axiomatically and abstractly real numbers (along the lines of Theorem 7.11 essentially): the axiomatic system's primitive concepts are points (real numbers), the operations of addition and multiplication and the order relation. The list of axioms is long-ish but, bar one, they aren't so difficult to grasp because they are familiar from $\mathbb{Q}$. The exception is the completeness axiom, which says the real line has no holes. Thus the real line will be a set whose arithmetics and order satisfy the axioms of the real numbers.

More interesting is to specify a concrete construction and only afterwards prove the properties we wish real numbers to fulfil. There are three equivalent methods commonly used to construct real numbers. Each one has its own merits and drawbacks, and each generates a model of the real numbers. We'll call them Weierstraß–Stolz model (via decimal expansions—the most intuitive model), Dedekind model (through Dedekind cuts—the most efficient model), Méray–Cantor model (metric-space completion—the model with biggest reach). Spivak amusingly refers to the latter two as 'real numbers of the algebraists' and 'real numbers of the analysts' respectively [88]. At present we'll discuss the Dedekind model, leaving the others to Sects. 7.5 and 19.6.

The common heuristics of all these approaches is that we seek a numerical set $\mathbb{R}$ that contains measures, and just taking $\mathbb{Q}$ is not enough. For instance $\sqrt{2}$ is the length of the hypothenuse of a right triangle with equal legs of length 1. But it's easy to show $\sqrt{2} \notin \mathbb{Q}$:

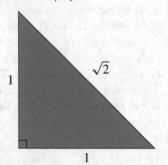

Suppose, by contradiction, there existed $p \in \mathbb{N}, q \in \mathbb{N}^+$ with $\mathrm{lcm}(q, p) = 1$ and such that $p/q = \sqrt{2}$. Then $2q^2 = p^2$, and so $2|p$, i.e. $p = 2r$ for some r. Substituting would give $q^2 = 2r^2$, now implying q is even. But p, q were taken coprime ⨇.

Even so, it's well known there are approximations (sequences of rationals) that 'converge' to $\sqrt{2}$

$$1.4 \quad 1.41 \quad 1.414 \quad 1.4142 \quad 1,41421 \ldots \longrightarrow \sqrt{2} \text{ (from below), and}$$
$$1.5 \quad 1.42 \quad 1.415 \quad 1.4143 \quad 1,41422 \ldots \longrightarrow \sqrt{2} \text{ (from above).}$$

Hence we could identify $\sqrt{2}$ with the pair of sets of rationals

$$\alpha = \left\{ q \in \mathbb{Q} \mid q < 0 \vee q^2 < 2 \right\} \quad \text{and} \quad \mathbb{Q} \setminus \alpha = \left\{ q \in \mathbb{Q} \mid q^2 > 2 \right\}. \tag{7.1}$$

Note how these sets are, in some still unclear way, 'separated' by $\sqrt{2}$

$$a < \sqrt{2} < b, \quad \forall a \in \alpha, \ b \in \mathbb{Q} \setminus \alpha.$$

Notation: if α, β denote subsets in a poset $(X, \leqslant)$, we'll write $\alpha \leqslant \beta$ to mean $a \leqslant b, \forall a \in \alpha, b \in \beta$.

Definition 7.1 Let $(X, \leqslant)$ be a partially ordered space. A subset $\alpha \neq \varnothing$ is called **(open) lower set** if

i) for any $x \in \alpha, y \in X, \ y \leqslant x \Longrightarrow y \in \alpha$
ii) for any $x \in \alpha$ there exists $y \in \alpha$ such that $x < y$.

The significance of these properties is the following:

i) says anything preceding an element ('lower' than it) is still an element;
ii) ensures α doesn't have a maximum (topologically, it translates openness).

Taking the complement $\complement \alpha = X \setminus \alpha$ of an open lower set (an upper set) gives a partition $\alpha \sqcup \complement \alpha = X$, where moreover $\alpha < \complement \alpha$. Put equivalently, $(\alpha, \complement \alpha)$ is a Dedekind cut in $\mathbb{Q}$, in agreement with Definition 8.2. For this reason we'll sometimes interchange the names (open lower set and Dedekind cut) and treat them as synonyms. Beware that the terminology is not standard: we have chosen—mainly for didactical purposes—lower set, others use left half-line, or left ray, or other locutions still.

Definition (Dedekind, 1872[1]) A **real number** is an open lower set in $\mathbb{Q}$ (or equivalently, a Dedekind cut of $\mathbb{Q}$ as per Definition 8.2). The set of real numbers, called **real line**, is denoted by

$$\mathbb{R} := \{\alpha \subset \mathbb{Q} \mid \alpha \text{ is an open lower set}\}.$$

Historically the adjective 'real' was coined to distinguish the reals from imaginary numbers (Chap. 13). Dedekind's construction of $\mathbb{R}$ is as simple as it is profound. It is simple, because (see Theorem 7.2) the rationals can be identified with a subset of the real line

$$\mathbb{N} \subset \mathbb{Z} \subset \mathbb{Q} \subset \mathbb{R}.$$

Thus $\mathbb{R}$ is essentially made of $\mathbb{Q}$ and its gaps. At the same time the process that led to the reals starting from the natural numbers is quite involved. Indeed, from the definitions we see that

$$\mathbb{R} \subseteq \mathscr{P}\Big(\mathscr{P}\big(\mathscr{P}(\mathbb{N} \times \mathbb{N}) \times \mathscr{P}(\mathbb{N} \times \mathbb{N})\big)\Big).$$

This is probably the reason why the construction, which appeared at the beginning in the first edition of Rudin's classic textbook 'Principles of Mathematical Analysis'

---

[1] This year was particularly momentous: Dedekind and Cantor defined the continuum $\mathbb{R}$, Felix Klein gave the world the Erlangen Programme, and Bertrand Russell was born. (This book came close to being entitled *1872*.) Connoisseurs will know it as the year when Monet started *Impression, Sunrise* that gave the name to the Impressionist movement, and when Dostoevsky's novel *Demons* and Nietzsche's first book *The Birth of Tragedy* came out.

(winner of the Steele Prize for Mathematical Exposition), was relegated to an appendix in the third edition!

We should also remember that the idea outlined above is simple in retrospective, but historically it was very hard for mathematicians to accept this point of view (and one we would never demand of school kids). This approach, to all effects, released the notion of real numbers from its geometrical origin. As mathematicians like Weierstraß and Dedekind were preparing their calculus classes, they became increasingly aware, over the years, that the concept of real line didn't have a solid foundation. Although various ideas had been in the air for some time, the critical year in the development of the notion of real line was 1872, when Dedekind's pamphlet [23] and papers by Méray, Cantor and Heine (Weierstraß's student) saw the light. The 'de-geometrisation' of real numbers did not happen without scepticism, since many mathematicians thought it was artificial and abstruse to treat irrational numbers formally, without the notion of geometric magnitude. Nor is it clear that Dedekind himself had fully understood the importance of what he had done. The separation of the algebraic and geometric constructions of $\mathbb{R}$ might be compared to the divorce of the foundations of geometry from their origins in Euclidean space. These proceedings lasted until the nineteenth century, with the axiomatic approaches ensuing the dawn of non-Euclidean geometry, and beyond.

Besides, [23] might have been the single act of creation of logicism, whose champions were Frege, Russell and Whitehead. A construction of $\mathbb{R}$ based on $\mathbb{Q}$ and set theory befits the logicist programme aiming at reducing mathematics to logic. Dedekind cuts, alongside Peano's reduction of $\mathbb{Q}$ to $\mathbb{N}$, propelled a movement seeking to 'arithmetise' analysis. In this context it's important to remark Dedekind didn't actually construct the natural numbers; he characterised them, through Definition 6.1.

7.1   Algebraic Properties and Ordering

We shall endow open lower subsets of $\mathbb{Q}$ with the familiar structure.

Definition (**Addition on** $\mathbb{R}$) The sum of real numbers $\alpha, \beta \in \mathbb{R}$ is given by

$$\alpha + \beta := \{x + y \mid x \in \alpha, y \in \beta\}.$$

Lemma *If $\alpha, \beta \in \mathbb{R}$ then $\alpha + \beta \in \mathbb{R}$.*

Proof The set $\alpha + \beta$ is not empty, and is a proper subset because there exist $x' \in \mathbb{Q} \setminus \alpha$, $y \in \mathbb{Q} \setminus \beta$ with $\alpha < x'$, $\beta < y'$, and so $\alpha + \beta < x' + y'$, implying $x' + y' \notin \alpha + \beta$.

To prove i): take $x + y \in \alpha + \beta$ and $z \leqslant x + y$. Then $z - y \leqslant x \in \alpha$ implies $z - y \in \alpha$, and so $z = (z - y) + y \in \alpha + \beta$.

For ii): suppose $x + y \in \alpha + \beta$. We know there exists $x' \in \alpha$ such that $x < x'$ and similarly there is a $y' \in \beta$ with $y < y'$. Hence $x + y < x' + y'$.   □

To define the opposite of a real number α the obvious idea of symmetrising the complement set and taking $\{-x \mid x \in \mathbb{Q} \setminus \alpha\}$ won't do, since the latter is not an open lower set. We need a subtler construction:

Definition (Additive Opposite) The opposite of $\alpha \in \mathbb{R}$ is

$$-\alpha := \{ -x \mid \exists y > 0 \text{ such that } x - y \in \mathbb{Q} \setminus \alpha \}.$$

Exercise Prove that

i) for any $\alpha \in \mathbb{R}$, $-\alpha$ is an open lower set in $\mathbb{Q}$.
ii) $0 := \{ q \in \mathbb{Q} \mid q \leqslant 0 \}$ is the additive neutral element (the zero).

Definition (Ordering of $\mathbb{R}$) A real number $\alpha \neq 0$ is called **positive**, written $\alpha > 0$, when $0 \in \alpha$. It is called **negative**, written $\alpha < 0$, when $0 \in \complement\alpha$. The corresponding sets are indicated by

$$\mathbb{R}^+ := \{ \alpha \in \mathbb{R} \mid \alpha > 0 \}, \qquad \mathbb{R}^- := \{ \alpha \in \mathbb{R} \mid \alpha < 0 \}.$$

Moreover, we write $\alpha < \beta$ whenever $\alpha - \beta < 0$. All this induces a partial order relation

$$\alpha \leqslant \beta \overset{\text{def}^n}{\Longleftrightarrow} \alpha \subseteq \beta$$

$$\text{(as numbers)} \qquad\qquad \text{(as sets)}$$

Accounting for the fact lower sets are nested, we have

Proposition *The ordering $\leqslant$ is total: every $\alpha \in \mathbb{R} \setminus \{0\}$ is either positive or negative.*

Proof Exercise. $\qquad\qquad\qquad\qquad\qquad\qquad\qquad\qquad\qquad\qquad\qquad\qquad\qquad$ $\square$

Exercise For any $a, b, c \in \mathbb{R}$, show that $a < b \iff a + c < b + c$.

The consequence is that any summand $c \in \mathbb{R}$ can be cancelled (simplified) in equations and inequalities.

Let's tackle the product of real numbers now. If we take $\alpha = \beta = 1$ then $-2 \in \alpha$ and $-2 \in \beta$, but the product $(-2)(-2) = 4$ doesn't belong in the lower set we'd like $\alpha\beta$ to be. Hence we can't just define multiplication by the naive product $\{xy \in \mathbb{Q} \mid x \in \alpha, y \in \beta\}$, i.e. term by term.

Definition (Multiplication on $\mathbb{R}$) The product of positive real numbers $\alpha, \beta > 0$ is

$$\alpha \cdot \beta := \{ z \in \mathbb{Q} \mid \exists x \in \alpha, y \in \beta \text{ with } x > 0, y > 0 \text{ and } z \leqslant xy \}.$$

Having define the product of positive reals we can now generalise to arbitrary α, β:

$$\alpha \cdot \beta := \begin{cases} 0 & \text{if } \alpha = 0 \text{ or } \beta = 0 \\ (-\alpha) \cdot \beta & \text{if } \alpha < 0, \beta > 0 \\ \alpha \cdot (-\beta) & \text{if } \alpha > 0, \beta < 0 \\ (-\alpha) \cdot (-\beta) & \text{if } \alpha, \beta < 0 \end{cases}$$

And the last ingredient to complete the algebraic structure:

Proposition *The multiplicative inverse of* $\alpha \in \mathbb{R} \setminus \{0\}$ *is*

$$\alpha^{-1} = \begin{cases} \mathbb{Q}^- \cup \{q \in \mathbb{Q}^+ \mid q^{-1} \in \mathbb{Q} \setminus \alpha\} & \text{if } \alpha > 0 \\ -((-\alpha)^{-1}) & \text{if } \alpha < 0 \end{cases}$$

Proof Exercise.   □

The crowning point of the previous definitions/properties is the following

Theorem $(\mathbb{R}, +, \cdot, \leqslant)$ *is an ordered field.*

Proof Exercise.   □

To tell the truth $\mathbb{R}$ is a very special ordered field, as Theorem 7.11 elucidates.

Corollary *There are no zero-divisors in* $\mathbb{R}$:

$$ab \neq 0 \iff a, b \neq 0 \qquad \forall a, b \in \mathbb{R}.$$

Proof By contrapositive, suppose $ab = 0$. There are two cases: if $a = 0$ the proof ends; if $a \neq 0$ the inverse a^{-1} exists, so $b = 1 \cdot b = (a^{-1}a)b = a^{-1}(ab) = a^{-1} \cdot 0 = 0$.

The implication $a = 0 \lor b = 0 \implies ab = 0$ is immediate by the definition.
                                                                                    □

So, we are allowed to cancel factors in equations whenever they are non-zero, and only in this case.

Exercise Prove that the inverse is unique, and $(a^{-1})^{-1} = a$ for every $a \in \mathbb{R} \setminus \{0\}$.

Definition Take a real number $a \neq 0$ and $b \in \mathbb{Z}$. We set

$$a^b := \begin{cases} 1 & \text{if } b = 0 \\ a^{b-1}a & \text{if } b > 0 \\ (a^{-1})^{-b} & \text{if } b < 0 \end{cases}.$$

For $b \in \mathbb{Z}^*$ we put $0^b = 0$ (the expression 0^0 is undefined).

Exercises Show, for every $a, b \in \mathbb{R}$,

$$a < b \iff -b < -a \iff ca < cb \ (\forall c \in \mathbb{R}^+) \iff da > db \ (\forall d \in \mathbb{R}^-),$$

$a \neq 0 \iff a^2 > 0, \qquad a > 0 \iff a^{-1} > 0, \qquad ab > 0 \iff a, b > 0 \lor a, b < 0.$

Borrowing notation 8.1 we write

$$(-\infty, +\infty) := \mathbb{R},$$

although $\pm\infty$ are nothing but symbols, and not numbers: $2 + \infty$, $0 \cdot \infty$ do not make sense in $\mathbb{R}$. (Yet, see Sect. 7.6.)

Exercise Show that any real interval can be obtained from intervals of type $(a, +\infty)$ and $(-\infty, b)$, $a, b \in \mathbb{R}$, through set-theoretical operations.

On one hand this justifies the name half-line or ray for real numbers, since at an intuitive level we're identifying a number $\alpha \in \mathbb{R}$ with our idea of an interval $(-\infty, \alpha)$. On the other hand it's the standard way to endow $\mathbb{R}$ with a topology (Sect. 8.1).

Example Knowing that $b \in (-\infty, 2]$ we wish to find the smallest interval containing the number $1 - 2b$.

Since $b \in (-\infty, 2] \iff b \leqslant 2 \iff -2b \geqslant -4 \iff 1 - 2b \geqslant -3$, the required interval is $[-3, +\infty)$.

Now we can explain why we may regard $\mathbb{Q}$ as a subset of $\mathbb{R}$ (note the subtlety).

Theorem 7.2 *Let* $(\mathbb{K}, +, \cdot, 0_\mathbb{K}, 1_\mathbb{K}, \leqslant)$ *be an ordered field. There exists a unique 1-1 map* $\psi : \mathbb{Q} \to \mathbb{K}$ *preserving the ordered-field structure.*

Proof Because

$$(*) \qquad 0_\mathbb{K} < 1_\mathbb{K} < 1_\mathbb{K} + 1_\mathbb{K} < 1_\mathbb{K} + 1_\mathbb{K} + 1_\mathbb{K} < \dots$$

are all distinct elements, we set

$$(**) \qquad m_\mathbb{K} = \begin{cases} \underbrace{1_\mathbb{K} + \dots + 1_\mathbb{K}}_{m \text{ times}} \geqslant 0_\mathbb{K}, & m \in \mathbb{Z}^+ \\ 0_\mathbb{K}, & m = 0 \\ -_\mathbb{K}(-m)_\mathbb{K}, & m \in \mathbb{Z}^- \end{cases}.$$

Now define a map ψ by $\psi(0) = 0_\mathbb{K}$, $\psi(1) = 1_\mathbb{K}$ and

$$\psi\left(\frac{m}{n}\right) = \psi(m)\psi(n^{-1}) = \frac{m_\mathbb{K}}{n_\mathbb{K}}, \qquad m \in \mathbb{Z}, n \in \mathbb{N}^*.$$

The reader might show that the above doesn't depend on the representatives chosen, and hence ψ is unique. We'll prove the properties of ψ over $\mathbb{N}$ and leave it as exercise to extend them to $\mathbb{Z}$ and then $\mathbb{Q}$, using $(**)$:

- ψ is 1-1: $m \neq n$ implies $\psi(m) \neq \psi(n)$, due to $(*)$;
- it preserves addition: $\psi(m + n) = (m + n)_{\mathbb{K}} = \underbrace{1_{\mathbb{K}} + \ldots + 1_{\mathbb{K}}}_{m+n} = m_{\mathbb{K}} + n_{\mathbb{K}} = \psi(m) + \psi(n)$;
- it preserves multiplication:

$$\psi(mn) = \psi(\underbrace{n + \ldots + n}_{m}) = \underbrace{\psi(n) + \ldots + \psi(n)}_{m} = \underbrace{n_{\mathbb{K}} + \ldots + n_{\mathbb{K}}}_{m}$$

$$= \underbrace{1_{\mathbb{K}}n_{\mathbb{K}} + 1_{\mathbb{K}}n_{\mathbb{K}} \ldots + 1_{\mathbb{K}}n_{\mathbb{K}}}_{m} = (\underbrace{1_{\mathbb{K}} + \ldots + 1_{\mathbb{K}}}_{m})n_{\mathbb{K}} = m_{\mathbb{K}}n_{\mathbb{K}} = \psi(m)\psi(n);$$

- it preserves the order relation: if $m \leqslant n$ then $n = m + k$, $k \geqslant 0$. Hence $\psi(n) = \psi(m) + \psi(k) = \psi(m) + k_{\mathbb{K}}$, and since $\mathbb{K}$ is ordered $k_{\mathbb{K}} \geqslant 0_{\mathbb{K}}$. Therefore $\psi(n) \geqslant \psi(m)$, thus ending the proof that ψ is a 1-1 homomorphism of ordered fields. $\square$

Corollary *The set $\mathbb{R}$ contains a copy of $\mathbb{Q}$.*

Proof The image $\psi(\mathbb{Q}) \subseteq \mathbb{R}$ under the theorem's 1-1 homomorphism is isomorphic to the ordered field $\mathbb{Q}$. $\square$

Exercise Prove the above embedding $\psi : \mathbb{Q} \hookrightarrow \mathbb{R}$ is explicitly given by

$$q \mapsto \{y \in \mathbb{Q} \mid y \leqslant q\}.$$

A second consequence of the proof of Theorem 7.2 is that an ordered field must be infinite, and therefore its *characteristic* is 0. An immediate example is the field $\mathbb{C}$, see Remark 13.2.

Then no finite field can be ordered. As an exercise the reader should try to define an order on $\mathbb{Z}_p = \{0, 1, 2, \ldots, p - 1\}$ to understand where the problem lies.

Definition A real number $r \in \mathbb{R}$ is said to be **irrational** if $r \notin \mathbb{Q}$.

Exercises Consider any $q \in \mathbb{Q}^*$, $r \notin \mathbb{Q}$ and prove

i) $q + r \notin \mathbb{Q}$, $qr \notin \mathbb{Q}$
ii) $r^{-1} \notin \mathbb{Q}$
iii) $p \in \mathbb{N}$ prime $\implies \sqrt{p} \notin \mathbb{Q}$.
   Find a counterexample to invalidate the converse.

Exercises (challenging) Show that

i) for any $r < s \in \mathbb{Q}$, there exist infinitely many irrational numbers in the interval (r, s);
ii) there exists no open interval $\varnothing \neq (a, b) \subseteq \mathbb{R}$ such that $(a, b) \cap \mathbb{Q} = \varnothing$;

iii) there exists no open interval $\varnothing \neq (a, b) \subseteq \mathbb{R}$ such that $(a, b) \cap (\mathbb{R} \setminus \mathbb{Q}) = \varnothing$.

In a nutshell, any non-trivial real interval contains an infinity of rationals and irrationals.

7.2 Absolute Value

The concept of absolute value represents a routine obstacle in the learning process. Despite many technical conversations with specialists in ☞ *didactics of mathematics*, the true root of the problem remains mysterious to me, so I ended up including a whole section on the topic.

Definition 7.3 One calls **absolute value** (or *modulus*) of the number $x \in \mathbb{R}$ the quantity

$$|x| := \max \left\{ x, -x \right\}.$$

Examples: $|3/5| = 3/5$, $|-3| = 3$, $|0| = 0$. Regrettably, but effectively, one says the absolute value forgets the sign of its argument.

With the next corollary we'll provide an alternative definition, but before getting to that

Definition 7.4 One calls *positive part* and *negative part* of a real number x

$$x^+ := \max \left\{ x, 0 \right\} \geqslant 0 \qquad \text{and} \qquad x^- := \max \left\{ -x, 0 \right\} = -\min \left\{ x, 0 \right\} \geqslant 0.$$

Exercises Prove

i) $|x| = x^+ + x^-$, $x = x^+ - x^-$, or alternatively: $x^+ = \dfrac{|x| + x}{2}$, $x^- = \dfrac{|x| - x}{2}$

ii) $\max\{x, y\} = \dfrac{|x-y| + x - y}{2} + y$, $\min\{x, y\} = \dfrac{x - y - |x-y|}{2} + y$

for any $x, y \in \mathbb{R}$.

Corollary *For any* $x \in \mathbb{R}$

$$|x| = \begin{cases} x, & x \geqslant 0 \\ -x, & x < 0 \end{cases}.$$

Proof Exercise.   □

Note the absolute value of x is a non-negative number: $|x| \geqslant 0$ (actually, it is positive-definite: $|x| > 0$ for all $x \in \mathbb{R}^*$, see Sect. 15.5).

Exercises 7.5 For any $a, b \in \mathbb{R}$ prove

  i) $a \leqslant |a|$

  ii) $|a| = |b| \iff a = b \vee a = -b$ (written $a = \pm b$ for short)

  iii) $|a| < b$ (with $b > 0$) $\iff a \in (-b, b)$

  iv) $|ab| = |a| \cdot |b|$ (whence $|a^2| = |a|^2 = a^2$).

Example Solve the equation $|x - 3| = 2x$ for $x \in \mathbb{R}$.

If $x - 3 \geqslant 0$, i.e. $x \geqslant 3$, then $|x - 3| = x - 3$: $x - 3 = 2x \implies x = -3$ ✗

If $x - 3 < 0$, i.e. $x < 3$, we have $|x - 3| = 3 - x$: $3 - x = 2x \implies x = 1$ ✓

Hence the solution is $x = 1$.

Equally (or more) important than the definition itself is the geometric meaning of the absolute value. By definition

$$\text{dist}(a, b) := |a - b| \tag{7.2}$$

is the **distance** between a and b, that is to say the length of the segment joining a, b.

Exercises Show

  i) $\text{dist}(a, b) = \text{dist}(b, a)$

  ii) $\text{dist}(a, b) = \text{dist}(-a, -b)$

  iii) $\text{dist}(a, 0) = |a| = |-a| = \text{dist}(-a, 0)$;

      therefore $|a|$ is the distance of both $\pm a$ to the origin 0 of the real line.

Examples

  i) Solving $|x - 3| = 5$ means we have to find the points $x \in \mathbb{R}$ whose distance from point 3 equals 5:

    (when $x \geqslant 3$) $x = 3 + 5 = 8$;   (when $x < 3$) $x = 3 - 5 = -2$.

    In one go, $|x - 3| = 5 \implies x - 3 = \pm 5 \implies x = -2, 8$.

  ii) $|x - 3| \geqslant 5$ is satisfied by points $x \in \mathbb{R}$ whose distance from 3 is larger than or equal to 5:

    (if $x \geqslant 3$) $x \geqslant 8$;   (if $x < 3$) $x \leqslant -2$,   so $x \in (-\infty, -2] \cup [8, +\infty)$.

  iii) $|x - 3| < 5$ is satisfied by the $x \in \mathbb{R}$ that lie less than 5 units away from point 3, hence $x \in (-2, 8)$.

  iv) Solve $|2x - 3| < 4|x - 1|$.

    There are $2^2 = 4$ cases to consider due to the four possible combinations of signs inside the absolute values:

$$\left. \begin{array}{l} \text{if} \quad 2x - 3 \geqslant 0, x - 1 \geqslant 0, \quad \text{i.e.} \quad x \geqslant \tfrac{3}{2} \\ \text{the inequality reads } 2x - 3 < 4(x - 1) \iff x > \tfrac{1}{2} \end{array} \right\} \quad \text{then} \quad x \geqslant \frac{3}{2};$$

$$\left. \begin{array}{l} \text{if} \quad 2x - 3 \geqslant 0, x - 1 < 0, \quad \text{i.e.} \quad x \in \varnothing \\ \qquad \text{(the inequality is immaterial)} \end{array} \right\} \quad \text{has no solution;}$$

$$\left. \begin{array}{l} \text{if} \quad 2x - 3 < 0, x - 1 \geqslant 0, \quad \text{i.e.} \quad x \in [1, \tfrac{3}{2}) \\ \qquad -(2x - 3) < 4(x - 1) \iff x > \tfrac{7}{6} \end{array} \right\} \quad \text{then} \quad x \in \left(\tfrac{7}{6}, \tfrac{3}{2} \right);$$

$$\left. \begin{array}{l} \text{if} \quad 2x - 3 < 0, x - 1 < 0, \quad \text{i.e.} \quad x < 1 \\ \qquad -(2x - 3) < -4(x - 1) \iff x < \tfrac{1}{2} \end{array} \right\} \quad \text{then} \quad x < \tfrac{1}{2}.$$

Solution: $x \in \left(-\infty, \dfrac{1}{2} \right) \cup \left(\dfrac{7}{6}, +\infty \right)$.

v) The **midpoint** of distinct points $a \neq b \in \mathbb{R}$ on the real line is by definition the solution m to equation $|m - a| = |m - b|$, i.e. a point equidistant from a and b:

$$|m - a| = |m - b| \implies m - a = \pm(m - b) \implies m = \frac{a + b}{2}.$$

The geometric midpoint coincides with the **arithmetic mean** of two numbers.

Proposition 7.6 (Triangle Inequality) *For any $a, b \in \mathbb{R}$*

$$|a + b| \leqslant |a| + |b|.$$

Corollary 18.3 will have more to say about the name, see also Fig. 15.1.

Proof As $|a + b|$ and $|a| + |b|$ are non-negative, the claim is equivalent to proving $\big(|a + b|\big)^2 \leqslant \big(|a| + |b|\big)^2$. Remembering exercise 7.5 we have

$$|a + b| \leqslant |a| + |b| \iff \big(|a + b|\big)^2 \leqslant \big(|a| + |b|\big)^2$$
$$\iff (a + b)^2 \leqslant \big(|a| + |b|\big)^2$$
$$\iff a^2 + b^2 + 2ab \leqslant |a|^2 + |b|^2 + 2|a||b|$$
$$\iff ab \leqslant |a||b| = |ab|.$$

$\square$

Exercise Prove $|a - b| \leqslant |a| + |b|$ and $\big||a| - |b|\big| \leqslant |a - b|$, for any $a, b \in \mathbb{R}$.

We might ask what is the absolute value's theoretical importance, apart from the practical use. There are two basic reasons: first of all the absolute value represents a norm on $\mathbb{R}$ seen as a metric space, as explained in (7.2). Secondly, it is the function that ensures $\mathbb{R}$ is a Euclidean domain, see Corollary 7.10.

Examples The aforementioned algebraic properties are used over and over to solve equations and inequalities, as we set out to exemplify.

i) Find all $x \in \mathbb{R}$ such that

$$\frac{2x^2 - 5x}{x - x^3} = 0.$$

We are allowed to divide by $x - x^3$ (= multiply by $(x - x^3)^{-1}$) only if that quantity is different from zero. Hence we must find out when

$$x - x^3 \neq 0 \iff x(1 - x^2) = x(1 - x)(1 + x) \neq 0$$
$$\iff x \neq 0 \ \lor \ x \neq 1 \ \lor \ x \neq -1.$$

We then view $x \in \mathbb{R} \setminus \{0, 1, -1\}$ as the necessary condition for the existence of solutions.

Since now the denominator doesn't vanish we can multiply both sides by it

$$(x - x^3)\frac{2x^2 - 5x}{x - x^3} = (x - x^3) \cdot 0 \iff \cancel{(x - x^3)}\frac{2x^2 - 5x}{\cancel{x - x^3}} = 0$$
$$\iff 2x^2 - 5x = 0$$
$$\iff x(2x - 5) = 0$$
$$\iff x = 0 \ \lor \ 2x - 5 = 0$$
$$\iff x = 0 \ \lor \ x = \frac{5}{2}.$$

Of these putative solutions only those fulfilling the existence condition are true solutions. Therefore the initial equation has a unique solution $x = \frac{5}{2}$.

ii) Find the mistake in the following argument (which would prove $1 = 0$):

$$x = 1 \overset{(\star)}{\iff} x \cdot x = x \cdot 1 \iff x^2 = x \iff x^2 - 1 = x - 1$$
$$\iff (x + 1)(x - 1) = x - 1 \overset{(\star\star)}{\iff} (x + 1)\cancel{(x - 1)} = \cancel{(x - 1)}$$
$$\iff x + 1 = 1 \iff x = 0.$$

There are two errors. The first takes place in $(\star)$, since $x = 1 \implies x^2 = x$ but $x^2 = x \not\implies x = 1$. It is true, though, that $(x \neq 0 \ \land \ x^2 = x) \implies x = 1$.

The second occurs in $(\star\star)$: the logical equivalence holds only for $x - 1 \neq 0$.

iii) Find the numbers $x \in \mathbb{R}$ such that

$$\frac{x^2 - 6x + 5}{x(x^2 + 1)} > 0.$$

Among the solutions we'll find by algebraic manipulations we must exclude those x such that $x(x^2 + 1) \doteq 0$, i.e. $x = 0$ (as $x^2 + 1 \geqslant 1 > 0$ for any x). The displayed inequality is therefore equivalent to

$$0 < \frac{x^2 - 6x + 5}{x} = \frac{(x-1)(x+5)}{x},$$

which can be solved quickly by comparing the signs of $x - 1, x + 5, x$.

Interval:	$(-\infty, -5)$	$(-5, 0)$	$(0, 1)$	$(1, +\infty)$
Sign of $x - 1$	$-$	$-$	$-$	$+$
Sign of $x + 5$	$-$	$+$	$+$	$+$
Sign of x	$-$	$-$	$+$	$+$
Sign of $\dfrac{(x-1)(x+5)}{x}$	$-$	$+$	$-$	$+$

Hence the solutions are all the numbers in $(-5, 0) \cup (1, +\infty)$.

7.3 Completeness

At least two elementary completeness' notions for $\mathbb{R}$ are known. The first is usually part of the content of analysis lectures, where one proves $\mathbb{R}$ is **Cauchy complete** (Theorem 19.19). The rationals, instead, are not Cauchy complete, for there exist sequences of rationals $(q_n)_{n \in \mathbb{N}}$ that converge outside the set $\mathbb{Q}$:

$$3, \frac{31}{10}, \frac{314}{100}, \frac{3141}{1000}, \frac{31415}{10000}, \frac{314159}{100000}, \ldots \longrightarrow \pi.$$

Here we shall discuss a second notion, due to Dedekind, that originates from the ordered structure of real numbers.

Definition 7.7 An ordered space $(X, \leqslant)$ is **Dedekind complete** if any non-empty, upper bounded subset admits least upper bound:

$$\varnothing \neq Y \subseteq X, \ \mathfrak{B}^+(Y) \neq \varnothing \implies \exists \sup Y \in X.$$

Two examples of the utmost importance are treated next.

Theorem 7.8 $(\mathbb{Q}, +, \cdot, \leqslant)$ *is Dedekind incomplete.*

Proof The set $A = \{q \in \mathbb{Q} \mid q > 0, \ q^2 < 2\}$ is non-empty ($1 \in A$) and upper bounded by $2 \in \mathfrak{B}^+(A) \subset \mathbb{Q}$. If $\mathbb{Q}$ were complete we could put $c = \sup A$, and so $c^2 < 2$ or $c^2 > 2$. In the former case we define $d = \dfrac{1}{2} \min \left(\dfrac{2 - c^2}{(c+1)^2}, 1 \right) > 0$. Then $c + d$ would be rational, larger than c and its square would be less than 2:

$$(c + d)^2 < c^2 + d(c+1)^2 < 2.$$

Consequently $c + d \in A$ ⨹.

Hence necessarily $c^2 > 2$, in which case we take $d = \dfrac{c^2 - 2}{2(c+1)^2}$, so that $c - d \in \mathbb{Q}^+, c - d < c$ and eventually

$$(c - d)^2 < c^2 - d(c+1)^2 > 2,$$

another contradiction. □

Theorem $(\mathbb{R}, +, \cdot, \leqslant)$ *is Dedekind complete.*

Proof Suppose $A \subseteq \mathbb{R}$ is upper bounded by $M \in \mathbb{R}$ (seen as lower set), and not empty. Define

$$\lambda := \bigcup_{\alpha \in A} \alpha \subseteq \mathbb{Q},$$

the union of all open lower sets of A. First, we claim λ is an open lower set. In fact, $\lambda \neq \varnothing$ as union of non-empty sets. Moreover, λ is a proper subset of $\mathbb{Q}$: for any $\alpha \in A$ we have $\alpha \subseteq M$, so an element $x \notin M$ cannot belong in any $\alpha \in A$ and *a fortiori* it won't belong in their union λ. Pick $z \in \lambda$ and $w \leqslant z$, so that $z \in \alpha$ for some α. Then $w \in \alpha$ and hence $w \in \lambda$. Finally, any $x \in \lambda$ belongs to an $\alpha \subseteq \lambda$, and so there exists $\alpha \ni y > x$. Which implies $y \in \lambda$.

Now it's easy to see that $\lambda = \sup A$. First, $\alpha \leqslant \lambda$ for every $\alpha \in \lambda$. Second, if there is an upper bound $\mu \geqslant \alpha$ for every $\alpha \in \lambda$, then $\lambda \subseteq \mu$, which means $\lambda \leqslant \mu$. □

It is no accident that the proof mimics that of Proposition 2.12.

Corollary 7.9 *The field $\mathbb{R}$ is **Archimedean**:*

$$\forall a < b \in \mathbb{R}^+ \ \exists n \in \mathbb{N} \ \ \text{such that } na > b.$$

Proof (By contradiction) suppose there existed α, β such that $n\alpha \leqslant \beta$ for any $n \in \mathbb{N}$. The set $S = \mathbb{N}\alpha$ would be upper bounded by β. Set $\beta^* = \sup S$, so $\beta^* \geqslant n\alpha$ for any $n \in \mathbb{N}$, and then $\beta \geqslant (m+1)\alpha = m\alpha + \alpha$. Therefore $\beta - \alpha \geqslant m\alpha$, making $\beta - \alpha$ an upper bound of S that is smaller than β ⨹. □

Corollary 7.10 (Euclidean Division) *For any $a, b > 0$ there exist unique numbers $q \in \mathbb{N}, r \in [0, b)$ such that*

$$a = bq + r.$$

Proof When $a < b$ we take $q = 0$ and $r = a$.

When $a \geqslant b$, by Corollary 7.9 there is an $n \in \mathbb{N}$ such that $nb > a$. Now it suffices to choose $q = \min \{n \in \mathbb{N} \mid nb > a\} - 1$ ($\mathbb{N}$ is well ordered) and $r = bq - n$. Uniqueness is evident. □

Corollary $\mathbb{Q}$ *is dense in* $\mathbb{R}$: $\forall c < d$ *in* $\mathbb{R}$ *there exists* $\dfrac{m}{n} \in \mathbb{Q}$ *such that* $c < \dfrac{m}{n} < d$.

Proof Since $d - c > 0$ the Archimedean property forces $n(d - c) > 0$ for some $n \in \mathbb{N}$. The set of such n is well ordered, so let m denote the minimum integer with that property. We claim m/n is the number we're seeking. Namely, $m > nc \implies$

$$\frac{m-1}{n} \leqslant c \implies c < \frac{m}{n} = \frac{m-1}{n} + \frac{1}{n} < c + (d - c) = d. \qquad \square$$

To better grasp the idea of density, consider two examples at the opposite end of the spectrum: between two consecutive integers $m, m + 1 \in \mathbb{Z}$ there is no integer (but infinitely many rationals and irrationals). Analogously for $\left\{\dfrac{1}{n}, \, n \in \mathbb{N}^*\right\}$: no number of the form $\dfrac{1}{k}$ lies between a pair $\dfrac{1}{n}, \dfrac{1}{n+1}$. The latter set and $\mathbb{Z}$ are very far from being dense in anything (for which reason they are called 'nowhere dense'). Following a known staple in ☞ *measure theory*, the most famous instance of this kin might be the Cantor set (Example 3.20).

Corollary *For any $r \in \mathbb{R}$ there exists a set $S \subset \mathbb{R}$ such that $r = \sup S$.*

Proof Define $S := \left\{\dfrac{m}{n} \in \mathbb{Q} \, \middle| \, \dfrac{m}{n} \leqslant r\right\}$. The previous corollary tells $r = \sup S$. □

The relationship between Dedekind completeness and Cauchy completeness is articulated by the following fact (without proof):

Theorem *An ordered field $\mathbb{K}$ (Definition 6.15) is Cauchy complete and Archimedean if and only if it is Dedekind complete.* □

It descends from this that $\mathbb{R}$ is Cauchy complete, too.

Theorem 7.11 (Characterisation of $\mathbb{R}$ as Ordered Field) *Up to isomorphisms, $(\mathbb{R}, +, \cdot, \leqslant)$ is the unique Dedekind complete, ordered field.*

Proof The statement is about uniqueness (the rest was proven earlier). We'll invoke Theorem 7.2, whereby for any ordered field $\mathbb{K}$ there is a unique embedding $\psi \colon \mathbb{Q} \hookrightarrow \mathbb{K}$ of ordered fields. Take two complete ordered fields $\mathbb{K}_1, \mathbb{K}_2$, so we have embeddings $\psi_1 \colon \mathbb{Q} \to \mathbb{K}_1, \psi_2 \colon \mathbb{Q} \to \mathbb{K}_2$. Setting $Q_1 = \psi_1(\mathbb{Q}), Q_2 = \psi_2(\mathbb{Q})$, we have that $\psi = \psi_2 \circ \psi_1^{-1} \colon Q_1 \to Q_2$ is an isomorphism. By completeness define

$$\phi(x) = \sup \{\psi(y) \mid y \in Q_1, \ y < x\}, \quad x \in \mathbb{K}_1.$$

The map $\phi \colon \mathbb{K}_1 \to \mathbb{K}_2$ is an isomorphism: for $r \in \mathbb{K}_2, x \in \mathbb{K}_1$ such that $\phi(x) = r$, we then have $x = \sup \{\psi^{-1}(q) \mid q \in Q_2, q < r\}$.   □

Exercise Consider positive reals a, b such that $a/b \notin \mathbb{Q}$. Prove the set of integer linear combinations $\{ma + nb \mid m, n \in \mathbb{Z}\}$ (a.k.a. the integer span of a, b) is dense in $\mathbb{R}$.

Hint: fix $N > 0$ consider linear combinations $ma + nb$ with $|m|, |n| \leqslant N$. Show these numbers are all distinct and contained in the interval $[-N(|a| + |b|), N(|a| + |b|)]$. Conclude there are two combinations whose distance doesn't exceed $\frac{2N(|a|+|b|)}{(2N+1)^2-1} = \frac{|a|+|b|}{2(N+1)}$. Finally subtract one from the other...

Why have we requested $a/b \notin \mathbb{Q}$? What happens if the quotient is rational? (Come up with an example.)

7.4   Roots

One major hurdle we face when dealing with roots of real numbers for the first time is understanding that the following is a definition.

Definition For $n \in \mathbb{N}$ odd, one calls n^{th} **root of** $b \in \mathbb{R}$ the unique number $\sqrt[n]{b} := \xi \in \mathbb{R}$ such that $\xi^n = b$.

For $n \in \mathbb{N}$ even, one calls n^{th} **root of** $b \geqslant 0$ the unique non-negative number $\sqrt[n]{b} := \zeta \geqslant 0$ such that $\zeta^n = b$.

In either case the number b under the root sign is called **radicand**, and n is the **index**.

First of all we must realise a $2m^{\text{th}}$ root is well defined only for non-negative numbers:

$$\zeta^{2m} = b \geqslant 0 \iff \sqrt[2m]{b} := \zeta,$$

and that the $2m^{\text{th}}$ root is, by utter definition, either zero or positive.

For odd roots, instead,

$$\xi^{2m+1} = b \iff \sqrt[2m+1]{b} := \xi.$$

The origin of the $\sqrt{}$ symbol seems to be the work of Stevin (1634).

Examples The **square** root ($n = 2$) of the positive number 9 is $\sqrt{9} := \sqrt[2]{9} = 3 > 0$.

There is no square root of -16 (there is no real number ξ such that $\xi^2 = -16$, since $\xi^2 \geqslant 0$).

The **cubic** root ($n = 3$) of -27 equals $\sqrt[3]{-27} = -3$.

Lemma *For any real $\delta > 1$ there exists a real $\sigma > 1$ such that $\sigma^2 \leqslant \delta$.*

Proof Pick a $\gamma \in \mathbb{R}$ such that $1 < \gamma < \delta$. We claim that either γ or δ/γ fulfil the requirement. If $\gamma^2 \leqslant \delta$ we can take $\sigma = \gamma$. Otherwise, $\gamma^2 \leqslant \delta$ implies $\delta/(\gamma^2) < 1$, so $(\delta/\gamma)^2 < \delta$ and we may take $\sigma = \delta/\gamma$. $\qquad\square$

Theorem (Existence of Square Roots) *For each $b \in \mathbb{R}^+$ the equation $\xi^2 = b$ has a unique solution in $\mathbb{R}^+$.*

Proof Since $r < s \Longrightarrow r^2 < s^2$ in $\mathbb{R}^+$, the equation has one solution at most. The cases $b = 0, 1$ are trivial, so we consider $b > 1$ (if $0 < b < 1$ it's enough to solve $z^2 = 1/b$ and take $x = 1/z$).

Define $L = \{l \in \mathbb{R}^+ \mid l^2 \leqslant b\}$, $R = \{r \in \mathbb{R}^+ \mid r^2 \geqslant b\}$. Because of $1 \in L$, $b \in R$ and $L \leqslant b \leqslant R$ we have a Dedekind cut, and hence a separating element ξ. We claim $\xi^2 = b$. By contradiction, suppose $\xi^2 < b$ or $\xi^2 > b$. In the first situation $1 < b/\xi^2$, and the previous lemma gives us a $\sigma > 1$ such that $\sigma^2 \leqslant b/\xi^2$, i.e. $\sigma\xi \leqslant b$. Therefore $\sigma\xi \in L$. But $\sigma\xi < \xi$ and ξ is separating $\not\!\not$.

In the second case $\xi^2/b > 1$. Take $\sigma > 1$ such that $\xi^2/b \geqslant \sigma^2$, i.e. $(\xi/\sigma)^2 \geqslant b$. Then $\xi/\sigma \in R$, so $\xi/\sigma < \xi$ but ξ is separating $\not\!\not$. $\qquad\square$

The punch line is that it's wrong to say $\sqrt{9}$ equals ± 3, albeit $(-3)^2 = 9$.

Exercises Take any $a, b \geqslant 0$ real and prove

i) $\dfrac{a+b}{2} \geqslant \sqrt{ab}$ (the **arithmetic mean** is greater than or equal to the **geometric mean**);

ii) $\sqrt{a^2 + b^2} \leqslant |a| + |b|$ (for all $a, b \in \mathbb{R}$);

iii) $\sqrt{a + b} \leqslant \sqrt{a} + \sqrt{b}$;

iv) $\sqrt{a^2} = |a|$ (for all $a \in \mathbb{R}$);

v) $\sqrt[3]{a^3 + b^3} \leqslant \sqrt[3]{a} + \sqrt[3]{b} \leqslant a + b$;

vi) $\sqrt{1 + n^2} \notin \mathbb{Q}$ for every $n \in \mathbb{Z}^*$;

vii) $r > 0$ irrational $\Longrightarrow \sqrt[n]{r} \notin \mathbb{Q}$.

Examples

i) Solve $x^2 = 25$.

By definition of root, if x is positive then $x = \sqrt{x^2} = \sqrt{25} = 5$.

But when x is negative we have $-x = \sqrt{x^2} = \sqrt{25} = 5$.

By the previous exercise, in fact, $|x| = \sqrt{x^2} = \sqrt{25} = 5$, i.e. $x = \pm 5$.

ii) Solve $x - 4\sqrt{x} + 2 = 0$.

We may view the equation as if it were quadratic in the variable $\sqrt{x}$:

$$(\sqrt{x})^2 - 4\sqrt{x} + 2 = 0 \Longrightarrow \sqrt{x} = \frac{4 \pm \sqrt{16 - 8}}{2} = 2 \pm \sqrt{2} \geqslant 0 \quad (\text{both } \checkmark)$$

so that $x = (2 \pm \sqrt{2})^2 = 6 \pm 4\sqrt{2}$.

The next notation turns out particularly useful when manipulating powers and roots:

Definition Let $\zeta \neq 0$ be real number and $p/q \in \mathbb{Q}$. One sets

$$\zeta^{p/q} := \begin{cases} 1 & \text{if } p = 0 \\ \left(\sqrt[q]{\zeta}\right)^p & \text{if } p \neq 0 \end{cases}.$$

with the understanding that $\zeta > 0$ in case q is even.

Exercise Consider arbitrary numbers $x, y \in \mathbb{R}$. Prove

$$\zeta^x > \zeta^y \iff x > y \quad \text{for } \zeta > 1,$$
$$\zeta^x > \zeta^y \iff x < y, \quad \text{for } 0 < \zeta < 1.$$

Proposition 7.12 *For any real number $\zeta > 0$ and any $x, y \in \mathbb{R}$,*

$$\zeta^{-1} = \frac{1}{\zeta}, \qquad \zeta^{x+y} = \zeta^x \zeta^y, \qquad (\zeta^x)^y = \zeta^{xy}.$$

Proof Exercise (challenge). *Hint*: start with $\zeta > 1$. Take $\alpha \in \mathbb{R}$ (an open lower set $\alpha \subset \mathbb{Q}$) and show $(\zeta^\alpha, \zeta^{\complement\alpha})$ is a Dedekind cut (that defines the number $\zeta^\alpha \in \mathbb{R}$). $\square$

7.5 Decimal Approximation

We saw in Sect. 6.5.1 every rational number admits a decimal expansion, which can be finite (for decimal numbers) or infinite and periodic (for non-decimal numbers). Now we wish to address non-periodic infinite expressions, which correspond to irrational numbers.

If $\gamma > 0$ is a real number, Corollary 7.9 ensures we can find n such that $10^{-n} < \gamma$. Mimicking that argument we could prove that decimal numbers are dense in $\mathbb{R}$.

Let's discuss a method for approximating real numbers by means of decimal numbers. With a fixed n, we seek $m/10^n$ that approximates in the best possible way the real number $x \in \mathbb{R}$. Said equivalently, we want to find $m \in \mathbb{Z}$ such that

$$\frac{m}{10^n} \leqslant x < \frac{m+1}{10^n}.$$

When $n = 0$ we have $m = \lfloor x \rfloor$ and we set $d_0 = \lfloor x \rfloor$.

When $n = 1$, the largest number $m/10 \leqslant x$ will be larger than d_0, so $r_0 = x - d_0$ belongs in $[0, 1)$. So call d_1 the largest integer such that $d_1/10 \leqslant r_0$, so $1 \leqslant d_1 \leqslant 9$. We may write $r_0 = d_1/10 + r_1$ for some $0 \leqslant r_1 < 1/10$. Iterating the argument by induction we obtain the recursive formula

$$r_{k+1} = d_k 10^{-k} + r_k, \qquad 0 \leqslant r_k < \frac{1}{10^k}.$$

In case $x = \dfrac{\tilde{m}}{10^{\tilde{n}}}$ is already decimal, we'll have $r_{\tilde{n}} = 0$ and hence $x = \displaystyle\sum_{k=0}^{\tilde{n}} \frac{d_k}{10^k} =$ $d_0.d_1 d_2 \cdots d_{\tilde{n}}$ gives the right answer, cf. (6.3).

We can do the same thing if if none of the r_j vanishes, as in (6.4) (i.e., when x is not decimal). Notice that

$$d_0.d_1 d_2 \cdots d_n \leqslant x < d_0.d_1 d_2 \cdots d_n + \frac{1}{10^n}. \tag{7.3}$$

In all cases, we will always obtain a **proper decimal sequence** $d_0.d_1 d_2 \cdots d_n \cdots$ in this way. By this locution we mean a sequence of decimal digits that doesn't stabilise to 9. Here is why:

Lemma *If there exists k such that $d_j = 9$ for every $j \geqslant k$, then $d_0.d_1 d_2 \cdots d_n \cdots$ is not the best possible approximation (from below) of x by decimal numbers $m/10^n$.*

Proof By contradiction, suppose $9 = d_k = d_{k+1} = d_{k+2} = \ldots$. Then

$$d_0.d_1 d_2 \cdots d_n = d_0.d_1 d_2 \cdots d_k + \frac{9}{10^{k+1}} + \frac{9}{10^{k+2}} + \ldots + \frac{9}{10^n}$$

$$= d_0.d_1 d_2 \cdots d_k + \frac{9}{10^{k+1}} \frac{1 - 1/10^{n-k}}{1 - 10}$$

$$= d_0.d_1 d_2 \cdots d_k + \frac{1}{10^k} \left(1 - \frac{1}{10^{n-k}} \right)$$

$$= d_0.d_1 d_2 \cdots d_k + \frac{1}{10^k} - \frac{1}{10^n}.$$

On the other hand $x \geqslant d_0.d_1 d_2 \cdots d_n = d_0.d_1 d_2 \cdots d_k + \dfrac{1}{10^k} - \dfrac{1}{10^n}$. Taking in account that the latter must hold for every $n \geqslant k$ we obtain $d_0.d_1 d_2 \cdots d_k + \dfrac{1}{10^k} \leqslant x$ ⨏⨏. $\qquad\square$

Stolz remarked in 1886 that the set D of proper decimal sequences could be used to define $\mathbb{R}$. His model, where arithmetics and ordering are introduced on D, establishes the most intuitive method for constructing the real numbers.

Theorem 7.13 (Weierstraß–Stolz) *Decimal approximation defines a 1-1 correspondence* $\mathbb{R} \cong D$.

Proof Let's call

$$d \colon \mathbb{R} \to D, \ d(x) = d_0.d_1 d_2 \cdots$$

the map associating with x its approximation. We begin showing d has an inverse map $s \colon D \to \mathbb{R}$. There's no doubt that $x = \sup\{d_0.d_1d_2\cdots d_n \mid n \in \mathbb{N}\}$, because (7.3) implies $d_0.d_1d_2\cdots d_n \leqslant x$ and we know $x - 10^{-n} < d_0.d_1d_2\cdots d_n$. Furthermore, for any $x' < x$ there exists an n large enough so that $x - 10^{-n} > x'$, and then $x' < x - 10^{-n} < d_0.d_1d_2\cdots d_n$.

Defining

$$s \colon d_0.d_1d_2\cdots d_n \mapsto \sup\{d_0.d_1d_2\cdots d_n \mid n \in \mathbb{N}\}$$

immediately shows $s \circ d = \mathbb{1}_{\mathbb{R}}$. So now we aim for $d \circ s = \mathbb{1}_D$. It's clear that the supremum of $\{d_0.d_1d_2\cdots d_n \mid n \in \mathbb{N}\}$ is finite, because $d_0.d_1d_2\cdots d_n < d_0 + 1$. Calling it $x = \sup\{d_0.d_1d_2\cdots d_n \mid n \in \mathbb{N}\}$, we claim that, for any n, $d_0.d_1d_2\cdots d_n$ approximates x from below with an error less than $1/10^n$. Clearly $d_0.d_1d_2\cdots d_n \leqslant x$. Take $k > n$ such that $d_k < 9$ (here we use the hypothesis the sequence is proper). The fact that

$$\frac{d_{n+1}}{10^{n+1}} + \frac{d_{n+2}}{10^{n+2}} + \ldots + \frac{d_k + 1}{10^k} < \frac{1}{10^n}$$

implies

$$x = \sup\{d_0.d_1d_2\cdots d_m \mid m \in \mathbb{N}\} \leqslant d_0.d_1d_2\cdots d_n \cdots (d_k + 1)$$

$$< d_0.d_1d_2\cdots d_n + \frac{1}{10^n},$$

so that the approximating error is smaller than $1/10^n$. $\qquad\qquad\square$

The Weierstraß–Stolz model of $\mathbb{R}$ is the most appealing to children, and the most comforting one for instructors, who can explain real numbers using decimal expansions and experiments. The method's disadvantage is that verifying the arithmetic axioms is rather exacting.

Proposition 19.21 relies on Cauchy sequences as an alternative, but is similar in spirit.

Remark Algorithm 6.17 for finding the greatest common divisor of two integers suggests a method for representing the quotient as iterated fraction. Dividing 840 by 611 for instance, we have

$$\frac{840}{611} = 1 + \frac{229}{611} \qquad \frac{611}{229} = 2 + \frac{153}{229} \qquad \frac{229}{153} = 1 + \frac{76}{153} \qquad \frac{153}{76} = 2 + \frac{1}{76}$$

which we might subsume in the expression

$$\frac{840}{611} = 1 + \cfrac{1}{2 + \cfrac{1}{1 + \cfrac{1}{2 + \cfrac{1}{76}}}},$$

called *continuous fraction*. Every rational number can be put in this form. For irrational numbers, on the contrary, Euclidean division never stops and each generates an 'infinite' fraction. For example the positive solution to $0 = x^2 - 2x - 1 = x(x - 2) + 1$

$$x = \cfrac{1}{2 - x} = \cfrac{1}{2 - \cfrac{1}{2 - \cfrac{1}{2 - \cfrac{1}{2 - \ddots}}}} = 1 + \sqrt{2}$$

may be regarded as more significant an expression than the expansion $2.41421356237\cdots$ (which in turn doesn't reveal any regularity in the digits). Truth be told, every quadratic irrational (Sect. 13.3) possesses a periodic continuous fraction, exactly as rational numbers have periodic decimal expansions.

The number $\varphi = \dfrac{1 + \sqrt{5}}{2}$, called the **golden ratio**, is the positive solution to $\varphi = 1 + \dfrac{1}{\varphi}$. The symbol is meant to commemorate the greatest sculptor of ancient times, Phidias (Φειδίας, ~480–430 aC). It has the simplest and most regular non-trivial continuous fraction (Fig. 7.1):

$$\varphi = 1 + \cfrac{1}{1 + \cfrac{1}{1 + \cfrac{1}{1 + \ddots}}}$$

Fun fact: the golden ratio and $1 + \sqrt{2}$ are the first entries in a sequence. The positive root of $x^2 - nx - 1 = 0$, $n \in \mathbb{N}^*$, is called n^{th} metallic mean. The first metallic mean is φ, and the case $n = 2$ is the cheaper 'silver ratio' $1 + \sqrt{2}$.

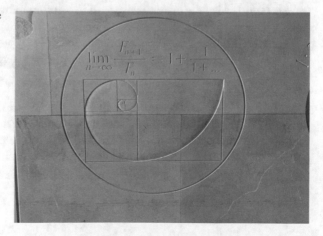

Fig. 7.1 The bas-relief at the Simons Centre for Geometry and Physics, Stony Brook University shows the asymptotic relationship between the Fibonacci numbers (10.1) and the golden ratio (Photo courtesy of the author)

There exist aperiodic continuous fractions, too. Two remarkable instances, whose pattern can't go amiss, are due to Euler:

$$\frac{\pi}{4} = 1 + \cfrac{1}{1 + \cfrac{1^2}{2 + \cfrac{3^2}{2 + \cfrac{5^2}{2 + \cfrac{7^2}{2 + \ddots}}}}}, \qquad e = 2 + \cfrac{1}{1 + \cfrac{1}{2 + \cfrac{2}{3 + \cfrac{3}{4 + \cfrac{4}{5 + \ddots}}}}}.$$

For the theory of continuous fractions of real numbers consult [48, 56].

7.6 Extending the Real Line ☕

Sometimes it's convenient to view $\sup \mathbb{R} = +\infty$ and $\inf \mathbb{R} = -\infty$ as if they were numbers-of-sorts, or 'infinitely large' and 'infinitely small' concrete quantities. This might be dictated by reasons of formal uniformity, but becomes particularly valuable in infinitesimal calculus and more generally in ☞ *topology* and *measure theory*.

One defined the **extended real numbers** by extending the real line with two additional elements

$$\overline{\mathbb{R}} := \mathbb{R} \cup \{-\infty, +\infty\}.$$

The extended line inherits a total order where

$$-\infty = \min \overline{\mathbb{R}}, \qquad +\infty = \max \overline{\mathbb{R}},$$

and we write $\overline{\mathbb{R}} = [-\infty, +\infty]$ to remind ourselves it's a form of compactification (☞ *topology*). Also the operations extend in a commutative, yet partial, way:

$$s + (+\infty) = +\infty \ (s \neq -\infty), \qquad r + (-\infty) = -\infty \ (r \neq +\infty),$$
$$p(\pm\infty) = \pm\infty \quad (p \in \mathbb{R}^+), \qquad q(\pm\infty) = \mp\infty \quad (q \in \mathbb{R}^-).$$

Moreover, one sets $|+\infty| = |-\infty| = +\infty$.

That said, $\overline{\mathbb{R}}$ is not a semigroup because the sum $+\infty + (-\infty)$ and the products $0 \cdot \pm\infty$ are not defined (and, hence, neither are $\pm\infty/\pm\infty$, $1^{\pm\infty}$, $\pm\infty^0$). Traditionally these expressions are referred to as 'indeterminacies', and learning how to deal with them is the job of infinitesimal calculus. In this framework we may extend the tangent function by setting $\tan\left(\pm\frac{\pi}{2}\right) = \pm\infty$. Thus the extension $\tan\colon \left[-\frac{\pi}{2}, \frac{\pi}{2}\right] \to \overline{\mathbb{R}}$ becomes bijective.

There is another extension, way more appealing and of different nature, obtained by adding to $\mathbb{R}$ just one extra point: ∞. As before, the latter is merely a symbol denoting an element that doesn't belong to the initial set $\mathbb{R}$. The new object, indicated by

$$\mathbb{RP}^1 := \mathbb{R} \cup \{\infty\},$$

is called **real projective line**, and is a type of *one-point compactification* in ☞ *topology*.

Proposition 7.14 *There exists a 1-1 correspondence $S^1 \cong \mathbb{RP}^1$.*

Proof Let

$$S^1 := \{(x, y) \in \mathbb{R}^2 \mid x^2 + y^2 = 1\}$$

be the unit circle centred at the origin of the Cartesian plane $\mathbb{R}^2$ and $r \cong \mathbb{R}$ the x-axis. If $N = (0, 1) \in S^1$ is the circle's north pole, we call $\pi_N(P)$ the central projection of $P = (x, y) \in S^1 \setminus \{N\}$ on r (Fig. 7.2). The system of equations $\begin{cases} \pi_N(P) = N + t(P - N) \\ y = 0 \end{cases}$ is compatible for $t = \frac{x}{1-y}$, and has one, and only one, solution $\left(\frac{x}{1-y}, 0\right)$. Therefore it defines a bijection $P \mapsto \frac{x}{1-y}$ that identifies $S^1 \setminus \{N\} \cong \mathbb{R}$.

Let's extend it by adding one extra point:

Definition The map $\pi_N : S^1 \longrightarrow r \cup \{\infty\}$

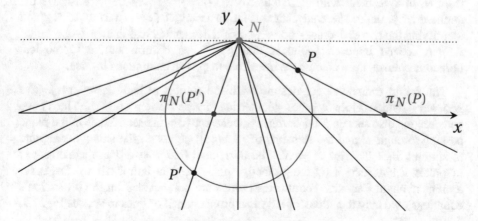

Fig. 7.2 1-dimensional stereographic projection

$$(x, y) \longmapsto \begin{cases} \dfrac{x}{1 - y} & \text{if } (x, y) \neq (0, 1) \\ \infty & \text{if } (x, y) = (0, 1) \end{cases}$$

is called **1-dimensional stereographic projection**.

We claim the map

$$\xi \longmapsto \begin{cases} \left(\dfrac{2\xi}{\xi^2 + 1}, \dfrac{\xi^2 - 1}{\xi^2 + 1} \right) & \text{if } \xi \in \mathbb{R} \\ (0, 1) & \text{if } \xi = \infty \end{cases} .$$

is the inverse projection $\pi_N^{-1} : \mathbb{RP}^1 \to S^1$, since

$$\pi_N \left(\pi_N^{-1}(\xi) \right) = \frac{\frac{2\xi}{\xi^2 + 1}}{1 - \frac{\xi^2 - 1}{\xi^2 + 1}} = \frac{2\xi}{\xi^2 + 1 - (\xi^2 - 1)} = \xi, \qquad \pi_N \left(\pi_N^{-1}(\infty) \right) = \pi_N(0, 1) = \infty$$

$$\pi_N^{-1} \left(\pi_N(x, y) \right) = \left(\frac{\frac{2x}{1-y}}{\left(\frac{x}{1-y} \right)^2 + 1}, \frac{\left(\frac{x}{1-y} \right)^2 - 1}{\left(\frac{x}{1-y} \right)^2 + 1} \right) = (x, y), \qquad \pi_N^{-1} \left(\pi_N(0, 1) \right) = \pi_N^{-1}(\infty) = (0, 1).$$

Therefore π_N is a 1-1 correspondence between the real projective line and the unit circle. $\qquad\qquad\qquad\qquad\qquad\qquad\qquad\qquad\qquad\qquad\qquad\qquad\qquad\qquad\qquad\qquad\qquad\qquad$ □

Exercise Determine the projection's fixed points. Show that π_N maps the upper half-circle (minus N) to $(-\infty, -1) \cup (1, +\infty)$ and the lower half-circle to $(-1, 1)$.

Remarks The stereographic construction destroys any notion of order inherited from $\mathbb{R}$, in agreement with the circularity (!) of $\mathbb{RP}^1$. That's because $\mathbb{RP}^1$ is the quotient of $\overline{\mathbb{R}}$ under the equivalence relation fixing $\mathbb{R}$ ($x \sim x$ if $x \in \mathbb{R}$), and identifying the two endpoints $-\infty \sim +\infty$ of the extended real line.

A second important point is that π_N is, as it turns out, a topological *homeomorphism*, i.e. a continuous invertible map with continuous inverse.

The fitting framework for dealing with the subject at hand is ☞ *projective geometry*, which gives a more geometrical interpretation of $\mathbb{RP}^1$. The above construction also sets up a 1-1 correspondence between the set of lines in the plane passing through N and the elements of $r \cup \{\infty\}$: points on the axis r correspond to straight lines PN, for $P \neq N$. The horizontal line $y = 1$ doesn't intersect r: or rather, it intersects it (in the projective plane $\mathbb{RP}^2$) at infinity. If we forget for a moment about Cartesian coordinates (which are only needed for the projection's equations), we may then think of $\mathbb{RP}^1$ as a proper pencil of lines in $\mathbb{R}^2$, see Fig. 7.2.

By the way, as $\pi_N^{-1}(\xi) \in S^1$ we can write $\dfrac{2\xi}{\xi^2 + 1} = \cos\theta$, $\dfrac{\xi^2 - 1}{\xi^2 + 1} = \sin\theta$ for

some θ. Hence the parameter $\xi = \tan\left(\dfrac{\theta}{2} \right)$ defines the angle $\theta/2 \in [-\pi/2, \pi/2]$ formed by the y-axis (oriented downwards) and the line PN:

$$|\theta| < \pi \quad \leftrightsquigarrow \quad P \neq N \quad \leftrightsquigarrow \quad PN \cap r = \{P\} \quad \leftrightsquigarrow \quad \xi \in \mathbb{R}$$

$$\theta = \pm\pi \quad \leftrightsquigarrow \quad P \equiv N \quad \leftrightsquigarrow \quad \{y = 1\} \cap r = \varnothing \quad \leftrightsquigarrow \quad \xi = \infty$$

Half-jokingly, π_N assigns a value to $\tan\left(\dfrac{\pi}{2}\right)$ and rubber-stamps the forbidden division that calculus students secretly covet (or openly dread): '$1/0 = \infty$'.

7.7 Transfinite Arithmetics ☕

This is the second serving of cardinal numbers and the ideal continuation of Sect. 6.6. As we shan't go into the theory of ordinals (the soul of set theory) we will only be able to address certain aspects.

The addition, multiplication and exponentiation of cardinal numbers extend some of the algebraic properties of natural numbers (commutativity, associativity). At the same time there are a few surprises and unexpected facts. First, cardinal numbers do not form a set but a class (*Burali-Forti antinomy*). Second, 0, 1 are not the only neutral elements for $+, \cdot$ respectively (Propositions 7.20, 7.26).

Definition A set A is **Dedekind infinite** if there exists a proper subset $B \subsetneq A$ with $|A| = |B|$.

It's not uncommon to see this as the definition of infinite sets in ZFC theory, replacing Definition 6.32. This is because

Proposition *A set A is infinite if and only if it is Dedekind infinite.*

Proof

($\Longleftarrow$) Let $B \subsetneq A$ be a subset with $|A| = |B|$. If, by contradiction, A were finite, $|A| \in \mathbb{N}$, then $B \subset A$ would be finite too, and of cardinality strictly less as proper ⚡.

($\Longrightarrow$) Define $f : \mathbb{N} \to \mathscr{P}(A)$ so that $f(n) = \{B \subseteq A \mid \mathrm{card}\, B = n\}$, for any n. Note $f(n)$ is never empty, for otherwise A would be finite (prove this by induction on n). The image of f is the countable set $\{f(n)\}_{n \in \mathbb{N}}$, whose members are infinite sets (and possibly uncountable). Using the axiom of choice we pick one element in each member set, which has to be a finite subset in A. More precisely, there exists a (countable) set $\{\psi_n\}_{n \in \mathbb{N}}$ such that $\psi_n \in f(n)$ for any n, and so $\psi_n \subseteq A$ is finite, of cardinality n.

Now set $\Psi := \bigcup_{n \in \mathbb{N}} \psi_n \subseteq A$. This is a countable set, so there exists a 1-1 correspondence $h : \mathbb{N} \approx \Psi$. Consider the map

$$\Psi \sqcup \complement\Psi = A \longrightarrow A \setminus \{h(0)\}$$
$$\complement\Psi \ni a \longmapsto \qquad a \qquad .$$
$$\Psi \ni h(n) \longmapsto \quad h(n+1)$$

As the above map is bijective, it follows A is Dedekind infinite.

□

So, in ZFC theory the two notions are indistinguishable. We should mention, though, there exist models of ZF (the crux is the absence of the AC) containing infinite sets that are not Dedekind infinite.

Exercise Let X be an infinite set, $x_0 \in X$ any point, and suppose there is a 1-1 (non-bijective) map $X \to X$. Show that $X \setminus \{x_0\}$ is infinite.

Let's begin to examine the *capax infiniti* then.

Proposition 7.15 *If A is infinite then $|A| \geqslant \aleph_0$. Rephrasing: every infinite set contains a countable subset.*

Proof A is infinite and hence not empty, so let $x_0 \in A$. By the exercise $A \setminus \{x_0\}$ is infinite, hence non-empty, and the AC allows us to pick an element x_1, necessarily different from x_0. And so forth, each time choosing $x_i \in A \setminus \{x_0, x_1, \ldots, x_{i-1}\} \neq \varnothing$. The set of choices $\{x_i\}_{i \in \mathbb{N}} \subseteq A$ is countable by construction.   □

Example 7.16 The Hilbert hotel is a fictional establishment with $\aleph_0$ rooms, all booked. Even so, there always are spare rooms: if a new guest arrives they are given room 1, the occupant of room 1 moves to room 2, the occupant of 2 moves to 3, and so on: the guest in room n goes to room $n + 1$, $\forall n$. (This works out because $1 + \aleph_0 = \aleph_0$.) The hotel manager could also decide to shift guests starting from the mth one, thus leaving the first $m - 1$ where they are. Whatever the manager does, infinitely many guests must always change rooms:

$$\operatorname{card} \mathbb{N}^* = \operatorname{card} \big(\mathbb{N}^* \setminus \underbrace{\{1, 2, \ldots, m\}}_{\text{initial segment}} \big),$$

or in fancier terms: $\forall n \in \mathbb{N} \quad n + \aleph_0 = \aleph_0$, see Proposition 7.20.

Proposition 7.17 *For any pairwise disjoint sets X, Y, Z there exist 1-1 correspondences*

  i) $X^{Y \cup Z} \approx X^Y \times X^Z$
  ii) $(X \times Y)^Z \approx X^Z \times Y^Z$
  iii) $\left(X^Y\right)^Z \approx X^{Y \times Z}$ **(*exponential law*)**

Proof

i) Associate with $f: Y \cup Z \to X$ its restrictions $f|_Y : Y \to X$, $f|_Z : Z \to X$. Then we have a bijection $f \mapsto (f|_Y, f|_Z)$.

ii) Take a map $f: Z \to X \times Y$ and call π_X, π_Y the canonical projections (Example 3.5 iii)). The map $f \mapsto (\pi_X \circ f, \pi_Y \circ f)$ is bijective:

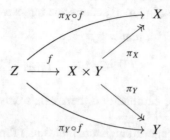

iii) For $f: Y \times Z \to X$ we define the variable-raising operator $f \mapsto \hat{f}$ as follows:

$$\hat{f}: Z \to X^Y \text{ acts by } z \mapsto \hat{f}(z): Y \to X, \text{ where } \hat{f}(z)(y) = f(y, z).$$

Reciprocally, for $g: Z \to X^Y$ define the variable-lowering operator $g \mapsto \check{g}$ by letting

$$\check{g}: Y \times Z \to X, \quad \check{g}(y, z) = g(z)(y).$$

The claim (exercise) is that the raising and lowering operators are inverse to each other: $\hat{\check{g}} = g$, $\check{\hat{f}} = f$.

□

Remark 7.18 The exponential law allows to view $h: Y \times Z \to X$ as a collection of maps $h_z : Y \to X$ parametrised by the set Z and, vice versa, to encode a family $\{h_z : Y \to X\}$ in a single function:

$$h: Y \times Z \to X \quad \overset{1-1}{\longleftrightarrow} \quad \{h_z : Y \to X\}_{z \in Z}.$$

In ☞ *computer science* the operator $\check{}: h(y, z) \mapsto h_z(y)$ is called *currying* (after the logician Haskell Curry, not the spice mixture).

More abstractly, Set is Cartesian-closed (☞ *category theory*), so $\mathrm{Set}(Y \times Z, X) \cong \mathrm{Set}(Z, X^Y)$. It's difficult to overstate the impact of this observation in the whole of mathematics. Unbeknownst to us, we've already seen situations that can be explained in terms of the exponential law:

– in logic, the tautological equivalence of formulas $(Y \wedge Z \to X)$ and $(Z \to (Y \to X))$, see Exercises 1.6;
– in set theory, the characteristic functions $\{\chi_B : A \to \mathbb{Z}_2\}_{B \in \mathscr{P}(A)}$ can be moulded into a single map $\chi: A \times \mathscr{P}(A) \to \mathbb{Z}_2$ (which was used in Proposition 6.34);

– the evaluation map between a finite-dimensional vector space V and its bidual $V^{\vee\vee}$, described in (15.3).

Manifestations of the same phenomenon are legion, e.g.:

– the relationship between the functors $\otimes$ and *Hom* crops up concretely in many constructions of ☞ *algebraic topology* and is the reason for the sheer existence of ☞ *homological algebra*;
– the identification between group actions and representations (Chap. 17).

Corollary *For any cardinal numbers a, b, c*

$$a^0 = 1, \quad a^1 = a, \quad a^{b+c} = a^b a^c, \quad (ab)^c = a^c b^c, \quad (a^b)^c = a^{bc}.$$

Proof For each set A of cardinality a there exists only one function $\varnothing \to A$ (the empty function), so $a^0 = |A^\varnothing| = 1$.

If $\{\star\}$ denotes the generic singleton, there are a functions (constants) $\{\star\} \to A$, one for each element of A (here we use the AC), and therefore $a^1 = |A^{\{\star\}}| = |A| = a$.

The remaining properties are immediate consequences of Proposition 7.17. □

Corollary 7.19 *For any cardinal numbers $a \leqslant b$ and c*

$$a + b \leqslant b + c, \quad ac \leqslant bc, \quad a^c \leqslant b^c, \quad c^a \leqslant c^b.$$

Proof Call A, B, C pairwise-disjoint sets of cardinality a, b, c, and let $i : A \to B$ be a 1-1 map. Then

$\quad j : A \cup C \to B \cup C$, given by $j\big|_A = i$, $j\big|_C = \mathbb{1}_C$, is 1-1;

$\quad i \times \mathbb{1}_C : A \times C \to B \times C$, $(i \times \mathbb{1}_C)(x, y) := (i(x), \mathbb{1}_C(y))$ is 1-1;

$\quad$ taking $f \in A^C$, we may assume $i \circ f \in B^C$ is well defined (taking equinumerous sets). We claim the function $A^C \to B^C$, $f \mapsto i \circ f$ is injective: $i \circ f = i \circ g \implies i(f(c)) = i(g(c))$ for all $c \in C$, and since i 1-1 we have $f(c) = g(c)$, that is $f = g$;

$\quad$ the mapping $C^A \to C^B$, $\phi \mapsto \phi \circ i$ is one-to-one. □

With these general preliminaries in place we can examine the arithmetics of countable sets.

Proposition 7.20 *For any $n \in \mathbb{N}^+$ and $a \geqslant \aleph_0$*

i) $n + \aleph_0 = \aleph_0$, $\quad\quad \aleph_0 + \aleph_0 = \aleph_0$, $\quad\quad \aleph_0 + a = a$,

ii) $\aleph_0 \cdot n = \aleph_0$, $\quad\quad \aleph_0 \cdot \aleph_0 = \aleph_0$, $\quad\quad (\aleph_0)^n = \aleph_0$.

Proof The map $\{a_0, \ldots, a_{n-1}\} \cup \mathbb{N} \to \mathbb{N} : a_j \mapsto j, k \mapsto n + k$ is 1-1 and onto, as seen with Example 7.16. Actually, $\{a_k\}_{k \in \mathbb{N}} \cup \{b_k\}_{k \in \mathbb{N}} \to \mathbb{N} : a_k \mapsto 2k, b_k \mapsto 2k+1$ is a bijection (Proposition 6.33), proving the first two properties.

Suppose $|A| = a$ and a countable subset $B \subseteq A$. Since $A \setminus B, B$ are disjoint $|A| = |A \setminus B| + |B| = \kappa + \aleph_0$. Therefore $a + \aleph_0 = (\kappa + \aleph_0) + \aleph_0 = \kappa + (\aleph_0 + \aleph_0) = \kappa + \aleph_0 = a$.

As for products: Proposition 6.40 says $\aleph_0^2 = \aleph_0$, but as $n \leqslant \aleph_0$ the previous corollary implies $\aleph_0 = \aleph_0 \cdot 1 \leqslant \aleph_0 \cdot n \leqslant \aleph_0^2 = \aleph_0$.   □

Proposition 7.21 *For any $a \leqslant \aleph_0$ we have $a \cdot \aleph_0 = \aleph_0$.*

Proof Let A be a set of cardinality a. Let's order the family

$$\mathscr{E} = \left\{(E, f) \mid E \subseteq A, \ f \colon E \times \mathbb{N} \to E \text{ is one-to-one}\right\}$$

by declaring

$$(E, f) \leqslant (H, g) \iff E \subseteq H \text{ and } g \colon H \times \mathbb{N} \to H \text{ extends } f \colon E \times \mathbb{N} \to E.$$

Any chain $\left\{(E_i, f_i)\right\}_{i \in I}$ in $\mathscr{E}$ is bounded by the union $\left(\bigcup_i E_i, f = \bigcup_i f_i\right)$, where $f\big|_{E_i \times \mathbb{N}} = f_i$ for any $i \in I$. By Zorn's lemma there is a maximal element $(M, f^M) \in \mathscr{E}$, and we purport $|M| = |A|$. Suppose $|M| < |A|$ by contradiction. Then $A \setminus M$ is infinite and contains a countable set B. Choose a 1-1 map $g \colon B \times \mathbb{N} \to B$, and define $h \colon (M \cup B) \times \mathbb{N} \to M \cup B$ by

$$h(x, n) = \begin{cases} f^M(x, n) & x \in M \\ g(x, n) & x \in B \end{cases}.$$

This map h is injective and restricts to f^M, so $(M \cup B, h) \geqslant (M, f^M)$. But (M, f^M) is maximal ⨽⨽.   □

Lemma *If A, B are countable set then $A \cup B$ is countable.*

Proof If $A \approx \mathbb{N} \approx B$ there exist onto maps $f_0 \colon \mathbb{N} \times \{0\} \twoheadrightarrow A$, $f_1 \colon \mathbb{N} \times \{1\} \twoheadrightarrow B$. Hence $f_0 \cup f_1 \colon \mathbb{N} \times \{0, 1\} \twoheadrightarrow A \cup B$ is onto, and so $\aleph_0 = |A| \leqslant |A \cup B| \leqslant |\mathbb{N} \times \{0, 1\}| = \aleph_0$.   □

Theorem 7.22 *If $\mathscr{A} = \{A_i\}_{i \in I}$ is a countable collection of countable sets A_i, their union $\bigcup \mathscr{A}$ is countable.*

Proof ([92, p. 91]) As $\mathscr{A}$ is countable, we have an onto map $g \colon \mathbb{N} \twoheadrightarrow \mathscr{A}$. But for any $n \in \mathbb{N}$ the set $g(n)$ is countable, so there exists a map $f_n \colon \mathbb{N} \times \{n\} \twoheadrightarrow g(n)$. Now $\bigcup_{n \in \mathbb{N}} f_n \colon \mathbb{N} \times \mathbb{N} \twoheadrightarrow \bigcup \mathscr{A}$ is surjective, which proves $\bigcup \mathscr{A}$ is countable.

Another short argument can be found in [70].   □

Remark 7.23 The proof above requires the axiom of choice in an essential manner. The sets of functions $F_n := g(n)^{\mathbb{N} \times \{n\}}$ are non-empty and pairwise disjoint. And $f_n \in F_n$ for every n. But it's precisely the AC that allows us to select the function f_n among all others.

Also observe the half-hidden use of currying (Remark 7.18).

At this juncture we are ready to leap into the uncountable world.

Definition The cardinality of $\mathbb{R}$ is called **continuum**: $\mathfrak{c} := |\mathbb{R}|$.

The first thing to observe, as $\mathfrak{c} = |\mathbb{R}| \geqslant |(0, 1)| > |\mathbb{N}| = \aleph_0$ by Cantor's diagonal argument, is that $\mathbb{R}$ cannot be countable.

Proposition *Any non-trivial interval* $(a, b) \subseteq \mathbb{R}$ $(a \neq b)$ *has cardinality* $\mathfrak{c}$.

Proof Pick any $c \in (a, b)$ and consider $f : (a, b) \to \mathbb{R}$,

$$
f(x) = \begin{cases} \dfrac{x - c}{x - a} & \text{if } a < x \leqslant c \\[2mm] \dfrac{x - c}{b - x} & \text{if } c \leqslant x < b \end{cases}.
$$

It stretches $(a, c]$ to $(-\infty, 0]$ and $[c, b)$ to $[0, +\infty)$. If we consider the two expressions separately, $f\big|_{(a,c]}$ has inverse $f^{-1}\big|_{(-\infty,0]}(x) = \dfrac{ax - c}{x - 1}$ and $f\big|_{[c,b)}$ has inverse $f^{-1}\big|_{[0,+\infty)}(x) = \dfrac{bx + c}{x + 1}$. Hence $(a, b) = (a, c] \cup [c, b) \approx (-\infty, 0] \cup [0, +\infty) = \mathbb{R}$. $\qquad\square$

By the same argument used in Hilbert's hotel $\mathfrak{c} \leqslant 1 + \mathfrak{c} \leqslant \aleph_0 + \mathfrak{c} = \mathfrak{c}$. In this way card $(0, 1) = \mathfrak{c} = 1 + \mathfrak{c} = \text{card}\,[0, 1)$, and then $\mathbb{R}$ has the same cardinality of an arbitrary non-empty interval $I \subseteq \mathbb{R}$.

Exercise Here are a few explicit bijections that identify the real line with a suitable finite interval:

rational maps: $x \mapsto \dfrac{2x - 1}{x(1 - x)}, \quad x \mapsto \dfrac{x}{1 + |x|}$

exponential-like: $x \mapsto e^{-x^2}$ (*normal distribution*, Example 10.7 iii)), $x \mapsto \tanh x$

trigonometric: $x \mapsto \arctan x, \quad x \mapsto \tan x$.

Determine the interval in each case. *Hint*: Figs. 12.6, 12.16, 12.17.

Theorem 7.24 $2^{\aleph_0} = \mathfrak{c}$.

Proof We'll construct an injective map $f : [0, 1) \hookrightarrow \{0, 1\}^{\mathbb{N}}$ using the **bisection method**. The latter is an important tool in a host of other contexts, for instance for proving Theorem 19.19.

We shall define $f(x) = 0.a_1 a_2 a_3 \cdots$, with digits $a_i = 0, 1$ for every $i \in \mathbb{N}$. First, identify $f(x)$ with the sequence $(a_i)_{i \in \mathbb{N}}$ in $\{0, 1\}^{\mathbb{N}}$. It's instructive to read the recipe over the tree below:

(O) if $x \in [0, 1/2)$, set $a_1 = 0$; (I) if $x \in [1/2, 1)$, set $a_1 = 1$.

In case (O) : (OO) if $x \in [0, 1/4)$ we set $a_2 = 0$; (OI) if $x \in [1/4, 1/2)$, then $a_2 = 1$.

In case (I) :   (IO) if $x \in [1/2, 3/4)$ we set $a_2 = 0$;   (II) if $x \in [3/4, 1)$, then $a_2 = 1$.

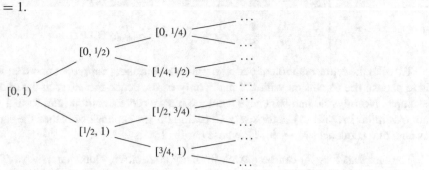

Continue like this, halving the interval $[a, b) = \left[a, \frac{a+b}{2}\right) \cup \left[\frac{a+b}{2}, b\right)$ chosen in the previous step.

The claim is we cannot go down path (IIIII···) indefinitely. Put otherwise x cannot always belong to the right-most interval:

$$\Upsilon := \bigcap_{n>0}[1 - 2^{-n}, 1) = \varnothing.$$

By contradiction, if there were an $x \in \Upsilon$ we'd have $x \geqslant 1 - \dfrac{1}{2^n}$ for every $n > 0$. But $\lim_{n \to \infty} 2^{-n} = 0$ would force $x \geqslant 1$ ✂✂. (In fancier terms, $[0, 1)$ doesn't have the finite-intersection property (3.10).) Now, $\Upsilon = \varnothing$ means we will never construct $0.1111111\dots$ ($a_j = 1 \; \forall j$) in the binary search. Therefore f must be 1-1, i.e. $2^{\aleph_0} \geqslant \mathfrak{c}$.

The difference $\{0, 1\}^{\mathbb{N}} \setminus f([0, 1))$ is made by numbers with infinite strings of consecutive ones, whose cardinality is $\aleph_0$ (the number of digits preceding the sequence of 1s is finite). Consequently $\mathfrak{c} \geqslant 2^{\aleph_0}$, and the Cantor–Schröder–Bernstein theorem delivers what we want. □

Much in the same spirit

Example 7.25 Write numbers of $x \in [0, 1]$ in binary form

$$x = \sum_{i=1}^{\infty} \frac{b_i}{2^i} = 0, b_1 b_2 \cdots, \qquad b_i = 0, 1.$$

One can show that a rational number $x \in [0, 1]$ of the form $m/2^n$ (these are called dyadic) has two binary expressions. You might want to refer to (8.1) and take the following as guidance

$$\frac{3}{4} = \begin{cases} \dfrac{0}{2^0} + \dfrac{1}{2^1} + \dfrac{0}{2^2} + \dfrac{1}{2^3} + \dfrac{1}{2^4} + \cdots =_{\text{base } 2} 0.10\overline{1} \\[2ex] \dfrac{0}{2^0} + \dfrac{1}{2^1} + \dfrac{1}{2^2} + \dfrac{0}{2^3} + \dfrac{0}{2^4} + \cdots =_{\text{base } 2} 0.11\overline{0} \end{cases}.$$

To make the representation of periodic (rational) numbers unique, our convention is to choose the expansion with tail made only of 0s, hence the latter in the above example. Non-dyadic numbers have a unique binary representation (and form a set not dissimilar in spirit to Cantor's set). Counting binary expansions is then the same as counting sequences $\mathbb{N} \to \{0, 1\}$, whose cardinality is $2^{\aleph_0}$.

The fact that $\mathfrak{c} > \aleph_0$ can be proved in many alternative, elementary ways. The shortest argument is possibly this: suppose the real numbers in $(0, 1)$ could be listed as $(a_i)_{i \in \mathbb{N}}$. Enclose each a_j by an interval of width 10^{-j}. Then $(0, 1)$ would be covered by a union of intervals of total length $\displaystyle\sum_{j=1}^{\infty} 10^{-j} = \frac{1}{9} \notin$.

Another, lengthier, argument is based on completeness and the Archimedean property:

Theorem *The set $\mathbb{R}$ is uncountable.*

Proof Suppose by contradiction $f : \mathbb{N} \approx \mathbb{R}$ is an enumeration of the reals. Consider the first number $f(0)$ in the list. As all other real numbers are larger or smaller than it, we pick the fist number $f(k_1) > f(0)$, i.e. $k_1 = \min\{k \mid f(k) > f(0)\}$ (possible since $\mathbb{R}$ is lower bounded).

For every $k < k_1$ we have $f(k) \notin (f(0), f(k_1))$, so we define $k_2 = \min\{k \mid f(0) < f(k) < f(k_1)\}$.

Then $f(k) \notin (f(0), f(k_1))$ $\forall k < k_2$, and we set $k_3 = \min\{k \mid f(k_2) < f(k) < f(k_1)\}$.

As $f(k) \notin (f(k_2), f(k_1))$ for all $k < k_3$, define $k_4 = \min\{k \mid f(k_2) < f(k) < f(k_3)\}$, and so on. At each iteration the interval considered narrows, because the left end increases and the right end decreases:

$$k_{2n+1} = \min\{k \mid f(k_{2n}) < f(k) < f(k_{2n-1})\},$$

$$k_{2n} = \min\{k \mid f(k_{2n-2}) < f(k) < f(k_{2n-1})\},$$

so that in general $f(k_{2n-2}) < f(k_{2n}) < f(k_{2n+1}) < f(k_{2n-1})$.

The set of left ends $\{f(k_{2n}) \mid n \in \mathbb{N}\}$ is non-empty so it admits least upper bound

$$r = \sup\{f(k_{2n}) \mid n \in \mathbb{N}\}.$$

This number is finite, less than $f(k_1)$, and an upper bound: $r \geqslant f(k_{2n+2}) > f(k_{2n})$. On the other hand each right end ('even') is larger than every left end ('odd'), because $f(k_{2n}) < f(k_{2n+2m}) < f(k_{2n+2m+1}) < f(k_{2m+1})$ for all m, n. Hence

r is the supremum: $r \leqslant f(k_{2n+3}) < f(k_{2n+1})$. Summarising:

$$(\dagger) \qquad f(k_{2n}) < r < f(k_{2n+1}) \quad \forall n.$$

It's not hard to show (exercise) that $k_{2n} \geqslant 2n$. Hence property 7.9 furnishes, for any K, an N such that $2N > K$, and so $k_{2N} \geqslant 2N > K \implies K < k_{2N} \implies f(K) = r \notin (f(k_{2N-2}), f(k_{2N-1}))$ ⚡. The contradiction emerges from assuming $f(K) = r$ (f onto): in that case for $n \gg 0$ we'd have $k_{2n} > K$ and K shouldn't belong in the interval $(f(k_{2n}), f(k_{2n+1}))$, violating $(\dagger)$. □

Further arithmetic properties of the continuum:

Proposition 7.26

i) $c + c = c$, $\quad c^n = c \quad \forall n > 0$, $\quad c \cdot \aleph_0 = c$;
ii) the space of real sequences $\mathbb{R}^{\mathbb{N}}$ has cardinality c;
iii) the space of real functions $\mathbb{R}^{\mathbb{R}}$ has cardinality 2^c.

Proof Below, passages marked with $(*)$ make use of the exponential law (Proposition 7.17).

i) $c + c = 2^{\aleph_0} + 2^{\aleph_0} = 2^{\aleph_0+1} = c$, and using Proposition 7.20 we deduce $c^n = (2^{\aleph_0})^n \overset{(*)}{=} 2^{n\aleph_0} = 2^{\aleph_0} = c$. Moreover, $c \leqslant \aleph_0 \cdot c \leqslant c \cdot c = c$, and so $c \cdot \aleph_0 = c$ by skew-symmetry.

ii) $\text{card}\left(\mathbb{R}^{\mathbb{N}}\right) = c^{\aleph_0} = (2^{\aleph_0})^{\aleph_0} \overset{(*)}{=} 2^{\aleph_0 \cdot \aleph_0} = 2^{\aleph_0} = c$.

iii) $\text{card}\left(\mathbb{R}^{\mathbb{R}}\right) = (2^{\aleph_0})^c \overset{(*)}{=} 2^{\aleph_0 \cdot c} = 2^c$.

□

Regarding property ii) in Proposition 7.26,

Exercise (guided proof that $\left|\mathbb{R}^{\mathbb{N}}\right| = |\mathbb{R}|$) Call S the set of increasing sequences $(a_k) \in \mathbb{N}^{\mathbb{N}}$ of natural numbers, and define the partial sums $\tilde{a}_k = \sum_{i=1}^{k} a_i$. Convince yourselves that

$$\mathbb{N}^{\mathbb{N}} \to S, \qquad (a_n)_{n \in \mathbb{N}} \mapsto (\tilde{a}_k)_{k \in \mathbb{N}}$$

is a bijection (it's the not-so-secret reason why the theory of series is part of the theory of sequences). Then $\text{card}\, S = \aleph_0^{\aleph_0} = 2^{\aleph_0}$.

Now take the map $(1, +\infty) \to S$ that sends x to the integer part $\lfloor 10^n x \rfloor$. As this is manifestly 1-1, $|\mathbb{R}| \leqslant |S|$. But $S \to \mathbb{R}$: $(a_n) \mapsto \sum_{n=1}^{+\infty} \frac{1}{10^{a_n}}$ is also injective, so $|\mathbb{R}| \geqslant |S|$. Eventually $|\mathbb{R}| = |S|$.

Example 7.27 The Cantor set C, described in Example 3.20, has cardinality c. For this we'll exhibit a 1-1 correspondence between C and the unit interval. Numbers in C have ternary expansion

$$c = \sum_{i=1}^{\infty} \frac{c_i}{3^i} = 0.c_1 c_2 \cdots \quad \text{where } c_i = 0, 2 \quad \text{(note: } c_i \neq 1\text{)}.$$

From Example 7.25 we know any $x \in [0, 1]$ has a binary representation $0.b_1 b_2 b_3 \cdots$, with digits $b_i = 0, 1$. The mapping $[0, 1] \to C$ that doubles each digit $b_i \mapsto c_i := 2b_i$ (mod 2) is easily 1-1 and onto.

Aside, this shows C contains irrational numbers, too. For example the non-periodic ternary number $0.0200200020002000002 \cdots$ belongs in C.

With the exception of sets with 0 or 1 element only ($\varnothing$ and singletons), no finite set X has the same cardinality as it Cartesian square $X \times X$, because $n < n^2$ for all $1 < n < \aleph_0$. For infinite sets the picture is the exact opposite, as the following characterisation (equivalent to the AC) explains.

Theorem (Tarski) *If X is infinite then* card $X =$ card X^2.

Proof We'll argue in analogy to Proposition 7.21. Consider the collection $\mathscr{A} = \{(E, f)\}$ where $\varnothing \neq E \subseteq X$ and $f : E \times E \to E$ is 1-1. By Proposition 7.15 X contains countable sets, so $\mathscr{A} \neq \varnothing$. Let's order $\mathscr{A}$ by extension, i.e.. $(E, f) \leqslant (H, g) \iff E \subseteq H$ and $g\big|_{E \times E} = f$. Since every chain is bounded (exercise), Zorn's lemma gives us a maximal element (A, f). From the injectivity of f we deduce $|A^2| = |A|$. Hence it suffices to show $|A| = |X|$. Suppose $|A| < |X|$ by contradiction. Then $X \setminus A$ contains some set B with the same cardinality of A, and in particular

$$|B| = |B \times A| = |A \times B| = |B^2|.$$

But $(A \cup B) \times (A \cup B) = A^2 \cup (A \times B) \cup (B \times A) \cup B^2$, and by Remark 7.23 we have $B \approx (A \times B) \cup (B \times A) \cup B^2$. Hence we can extend f to $A \cup B$, against the maximality ⚡⚡.   □

Corollary *If $|X| \geqslant \aleph_0$ then $|X^n| = |X|$ for any $n > 0$.*
The union $\bigcup_{i \in \mathbb{N}} X_i$ of $\aleph_0$ copies of $X \approx X_i$ has cardinality $|X|$.

Proof Use induction.   □

Exercises 7.28

i) Let $\mathbb{K}$ be an infinite field (say, $\mathbb{R}$). Prove card $\mathbb{K}[x] =$ card $\mathbb{K}$.

ii) (challenge) Let X be infinite, $\mathbb{K}$ a field and B a basis of the vector space of maps $\mathbb{K}^X = \{f : X \to \mathbb{K}\}$. Show that $|B| > |X|$. (In this respect, see also Exercise 15.4 ii)).
   Deduce that an infinite-dimensional vector space V cannot be isomorphic to its algebraic dual $V^{\vee} := \{f : V \to \mathbb{K} \text{ linear map}\}$.

Coming full circle, Propositions 7.20, 7.21, 7.26 can be incorporated into one result, again equivalent to the axiom of choice:

Proposition (Tarski) *Let $a \geqslant \aleph_0$ and $b > 0$ be cardinal numbers. Then*

$$ab = \max\{a, b\} = a + b.$$

Proof Suppose $a \geqslant b$. Corollary 7.19 and Theorem 7.7 tell us $a = a \cdot 1 \leqslant ab \leqslant a^2 = a$.

As for the second equality: identify $A \cup B \approx (A \times \{b_0\}) \cup (\{a_0\} \times B)$, for some $a_0 \in A, b_0 \in B$. Thus we embed $A \cup B \hookrightarrow A \times B$, and so $a + b \leqslant ab$. By the previous part, then, $a \leqslant a + b \leqslant ab = a$. □

To close the chapter on a high, let's speculate a bit.

Continuum Hypothesis (Cantor 1878) There is no intermediate cardinal between $\aleph_0$ and $2^{\aleph_0} = \mathfrak{c}$.

Suppose we believe in the existence of a smallest uncountable cardinal (without the theory of ordinals this is no more than an act of faith), and let's call it $\aleph_1 = \min\{c \mid c > \aleph_0\}$.

What the continuum hypothesis postulates is that

$$2^{\aleph_0} = \aleph_1.$$

It a well-known fact that this statement is undecidable in ZF. From a certain viewpoint the motive is that defining a well ordering on $\mathbb{R}$ is an axiomatic fact (cf. p.63).

For even bigger cardinals the template is similar:

Generalised Continuum Hypothesis (GCH) $\nexists c$ such that $a < c < 2^a$, for any $a \geqslant \aleph_0$.

Theorem 7.29 (Cohen 1963, Gödel 1940) *The generalised continuum hypothesis is independent of ZF theory.*

Crucially, in showing the result Cohen introduced a powerful new method he named *forcing*, which takes a small model of ZF and adds elements in such a way that the axioms remain satisfied, but other specific sentences are violated. In particular, he showed that it is possible to simultaneously satisfy the ZFC axioms and admit almost any value for $\aleph_1$.

Although the GCH and the axiom of choice are both independent of ZF theory, Sierpiński showed that the axiom of choice is a theorem of ZF if we assume the generalised continuum hypothesis: GCH $\vdash_{ZF}$ AC.

Chapter 8
Totally Ordered Spaces ☕

This shorter chapter explores properties with topological nuance, induced by total-order relations. The attention is placed on numerical sets. A set's attributes of being discrete or complete/connected mirror the concrete ideas of counting and measurement, and in ancient times might have prodded the need for the conceptual development of the notion of number.

For Sect. 8.1 a tiny background on topology could be beneficial, but is not strictly necessary.

Notation: we write $A \leqslant B$ to mean $a \leqslant b$ for all $a \in A, b \in B$. Similarly, $a \leqslant B$ stands for $a \leqslant b \; \forall b \in B$.

Definition 8.1 A subset $I \neq \varnothing$ in a totally ordered space $(X, \leqslant)$ is an **interval** if it is **convex**: $\forall a, b \in I, \forall x \in X \; a < x < b \Longrightarrow x \in I$.

The standard notation for intervals is:

$$
\text{\textbf{bounded}}
\begin{cases}
[a, b] := \{x \in X \mid a \leqslant x \leqslant b\} \\
(a, b) := \{x \in X \mid a < x \leqslant b\} \\
[a, b) := \{x \in X \mid a \leqslant x < b\} \\
(a, b] := \{x \in X \mid a < x < b\}
\end{cases}
\left.
\begin{array}{l}
(-\infty, b) := \{x \in X \mid x < b\} \\
(a, +\infty) := \{x \in X \mid a < x\} \\
[a, +\infty) := \{x \in X \mid a \leqslant x\} \\
(-\infty, b] := \{x \in X \mid x \leqslant b\}
\end{array}
\right\}
\text{\textbf{unbounded}}
$$

We must stress that $\pm\infty$ are just symbols. (At the same time, see Sect. 7.6.)

Definition 8.2 A **Dedekind cut** of a totally ordered space $(X, \leqslant)$ is a partition in two sets $\varnothing \neq L, R \subset X$ such that $L < R$.

Since $R = X \setminus L$ is the complement of L and is thus determined by L, a Dedekind cut of X in practice can be identified with a non-empty **lower set** L:

$$
\forall l \in L, x \in X \quad x \leqslant l \Longrightarrow x \in L.
$$

That's why in Definition 7.1 we chose to concentrate on lower sets to define real numbers, rather than abiding by the traditional definition. Symmetrical to a lower set is an **upper set** R:

© The Author(s), under exclusive license to Springer Nature Switzerland AG 2021
S. G. Chiossi, *Essential Mathematics for Undergraduates*,
https://doi.org/10.1007/978-3-030-87174-1_8

$$\forall r \in R, x \in X \quad x \geqslant r \Longrightarrow x \in R.$$

Exercises

i) Show that (L, R) is a Dedekind cut if and only if $\{L, R\}$ is a covering such that $L < R$.
ii) Take $X = \mathbb{R}$ and identify which among the above eight interval types are lower sets, which are upper sets.
iii) The intersection and union of lower sets is again a lower set.

Definition Dedekind cuts (L, R) come in three flavours:

(1) **jump**: $\exists \max L$ and $\exists \min R$
(2) **gap**: $\nexists \max L$ and $\nexists \min R$
(3) **normal cut**: $\exists \max L$ but $\nexists \min R$ (or the other way around).

These names are by no means standard. A synonym of gap is lacunary cut, and normal cut is a personal choice of the author.

Examples 8.3

i) The pair $(\{m \in \mathbb{Z} \mid m < 5/2\}, \{m \in \mathbb{Z} \mid m > 13/4\})$ is a jump in $\mathbb{Z}$. Actually, any cut of $\mathbb{Z}$ must be a jump (Theorem 8.10).
ii) The real number $\sqrt{2}$, cf. (7.1), is a gap in $\mathbb{Q}$. We will show that every gap in $\mathbb{Q}$ defines an irrational number.
iii) Every Dedekind cut of $\mathbb{Q}$ is either normal or a gap, due to the density (see Lemma 6.29). Theorem 8.6 will explain that any real number ζ is the least upper bound of a lower set L, and as such ζ is the separating element of the cut $(L, \mathbb{Q} \setminus L)$.

8.1   Order Topology

One-sided unbounded intervals of type $(a, +\infty)$ and $(-\infty, b)$ generate a topology on an ordered space $(X, \leqslant)$ called **order topology**. Technically, these intervals are a neighbourhood system, because any open set is obtainable as union of a finite intersection of such. We shall not be concerned with this. What matters is that in this topology $[a, b]$ is **closed**, whereas $(-\infty, b)$, (a, b), $(a, +\infty)$ are **open**, which justifies the common terminology whereby we call $[a, b]$ closed interval (endpoints included) and (a, b) open interval (endpoints excluded).

Definition A totally ordered set X is **connected** if X does not have topological partitions. That's to say, there do not exist non-empty, open and disjoint subsets $L, R \subseteq X$ such that $L \cup R = X$.

Since a normal Dedekind cut is not a topological partition (one of the sets is not open), immediately

Corollary 8.4 *If X has jumps or gaps, X is not connected.*

Proof Exercise. □

Lemma 8.5 *An ordered space $(X, \leqslant)$ is dense if and only if it has no jumps.*

Proof By definition the existence of a jump (L, R) precludes elements from lying between $\max L$ and $\min R$. And vice versa, too. □

Definition A function $f\colon (X, \leqslant_X) \to (Y, \leqslant_Y)$ between totally ordered spaces is **monotone** if it preserves the orderings:

$$x_1 \leqslant_X x_2 \Longrightarrow f(x_1) \leqslant_Y f(x_2), \qquad x_1, x_2 \in X.$$

In case f is onto and strictly monotone:

$$x_1 <_X x_2 \Longrightarrow f(x_1) <_Y f(x_2)$$

(and hence 1-1), we call it *order isomorphism*. We will use the symbol $(X, \leqslant_X) \cong (Y, \leqslant_Y)$, as universal algebra teaches us to do, to indicate order-isomorphic spaces.

Exercise Show that monotonicity may be equivalently defined by asking $x_1 \leqslant_X x_2 \iff f(x_1) \leqslant_Y f(x_2)$.

Definition Let $(X, \leqslant)$ be a totally ordered space and $A, B \subseteq X$ non-empty subsets such that $A \leqslant B$. An element $\xi \in X$ such that $A \leqslant \xi \leqslant B$ is called **separating element** of the pair (A, B).

There are three mutually exclusive possible scenarios:

$$\xi \in A, \ \xi \notin B \ (\text{or } \xi \notin A, \ \xi \in B), \qquad A \cap B = \{\xi\}, \qquad A < \xi < B.$$

The existence of a separating element is not guaranteed in general, nor is its uniqueness:

in $\mathbb{Q}$ the pair $\{q \mid q^2 < 5\} < \{q \mid q^2 > 5\}$ is a partition, but admits no separating element;

in $\mathbb{Z}$ the sets $\mathbb{Z}^-$ and $\{z \in \mathbb{Z} \mid z > 3\}$ are separated by $0, 1, 2, 3$.

In the special case where (A, B) is a Dedekind cut, the above middle option cannot occur.

Definition When every Dedekind cut $A < B$ admits a separating element one says the space $(X, \leqslant)$ satisfies **Dedekind's axiom**.

If we consider Dedekind cuts $(\alpha, \complement\alpha)$ in $X = \mathbb{Q}$, we have

Theorem 8.6 *The Dedekind axiom on $\mathbb{Q}$ is equivalent to the completeness of $\mathbb{R}$.*

Proof $(\Longleftarrow)$ Take $\varnothing \neq A, B \subseteq \mathbb{R}$ such that $A \leqslant B$ and let's show $\lambda = \bigcup_{a \in A} a$ is a separating element. Clearly $A \leqslant \lambda$, and since $a \subseteq b$ for every $a \in A, b \in B$, on one hand any b is an upper bound, on the other the union λ of the $a \in A$ is contained in (is smaller than) every b. Therefore λ is the smallest upper bound, i.e. $A \leqslant \lambda = \sup A \leqslant B$.

($\Longrightarrow$) Let $\varnothing \neq A \subseteq \mathbb{R}$ be upper bounded. Then $B = \{k \in \mathbb{R} \mid k \geqslant A\}$ is not empty, and obviously $A \leqslant B$. By assumption there exist an element $\xi \in \mathbb{R}$ separating A and B. Now it's easy to see $\xi = \sup A$. In fact ξ bounds A, so by definition $B \ni \xi$. But as $\xi \leqslant B$, ξ is the least of all upper bounds. □

We remind that $(X, \leqslant)$ is called **complete** (Definition 7.7) if any non-empty, upper bounded subset admits supremum. In that case, the upper set R of any Dedekind cut (L, R) in X must possess a minimal element ξ, making it either equal to $[\xi, +\infty)$ or to $(\xi, +\infty)$. Consequently L must be either $(-\infty, \xi)$ or $(-\infty, \xi]$, respectively. This means that ξ is a separating element:

$$(-\infty, \xi) < \xi \leqslant [\xi, +\infty) \qquad \text{or} \qquad (-\infty, \xi] \leqslant \xi < (\xi, +\infty).$$

In both cases ξ represents (read: completely determines) the Dedekind cut.

Proposition 8.7 *An ordered space* $(X, \leqslant)$ *is complete if and only if it doesn't have gaps.*

Proof ($\Longrightarrow$) We prove the contrapositive statement. If X has a gap (L, R) the set of upper bounds is $\mathfrak{C}^+(L) = R$ and $\sup L = \min R$. But R doesn't have minimum, and so X is incomplete.

($\Longleftarrow$) Suppose X incomplete (by contradiction), i.e. assume there exists an upper bounded set $A \subseteq X$ for which $\nexists \sup A$. By definition $R = \mathfrak{C}^+(A)$ is an upper set, so its complement $L = X \setminus R$ is a lower set and together they form a Dedekind cut $(L < R$ since $\exists \min R)$. Moreover, $A \cap R = \varnothing \Longrightarrow A \subseteq L$. Consequently L has maximum $z = \max L$ (otherwise there would be a gap). This element $z \neq \max A$ cannot belong in A, implying $z \in R$ ⚡. □

In a complete space X, any lower bounded subset A admits $\inf A$, hence the existence of the sup for upper bounded sets is equivalent to the existence of the inf for lower bounded sets.

Proposition 8.8 *In an ordered space* $(X, \leqslant)$ *the following are equivalent:*

  i) X is connected
 ii) X has no jumps nor gaps
iii) X is dense and complete.

Proof i) $\Longrightarrow$ ii): this is Corollary 8.4.

iii) $\Longleftrightarrow$ ii): consequence of Lemma 8.5 and Proposition 8.7.

There remains iii) $\Longrightarrow$ i). By contradiction, suppose $X = U \sqcup V$, with $U, V \neq \varnothing$ open (X disconnected). Take $U \ni u < v \in V$ and consider the set

$$H = \{x \in X \mid [u, x] \subseteq X\}.$$

It isn't empty ($u \in H$), and lies in $[u, v)$. By completeness set $z = \sup H$, so either $H = [u, z)$ or $H = [u, z]$. The density guarantees that every neighbourhood of z also contains points of U. Hence $z \in \overline{U} = U$, i.e. $z \in H$. But this implies $z < v$ and also that any neighbourhood of z contains points of V. Consequently $z \in \overline{V} = V$ ⨲. $\qquad\square$

Proposition 8.9 *Let $(X, \leqslant)$ be ordered and $I \subseteq X$ a subset. Then*

$$I \text{ is connected} \iff I \text{ is an interval.}$$

Proof ($\Longrightarrow$) Let $a, b \in I$ and $x \in X$ be elements such that $a < x < b$. If, by contradiction, $x \notin I$, we would have a partition $(-\infty, x) \sqcup (x, +\infty)$ of I ⨲.

($\Longleftarrow$) We'll show I has no jumps or gaps (using Proposition 8.8). If there were a jump between $a, b \in I$, the pair $\big((-\infty, a], [b, +\infty)\big)$ would disconnect X ⨲. If (H, K) were a gap, the lack of max H, min K would produce a gap: if $h \in H, k \in K$ then $\big((-\infty, h) \cup H, K \cup (k, +\infty)\big)$ would force X to be disconnected ⨲. $\qquad\square$

Definition An ordered space $(X, \leqslant)$ is called

– **separable**, if there exists a dense and countable subset;
– **discrete**, when all of its Dedekind cuts are jumps.

Example $\mathbb{R}$ is separable because of $\mathbb{Q}$, whilst any $X \subseteq \mathbb{Z}$ is discrete.

Let's gather up results Definition 6.4, Examples 8.3, Lemma 6.29, Proposition 6.40 and Theorem 7.11:

Theorem 8.10 (Characterisation of Number Sets as Ordered Spaces) *Up to (order) isomorphisms the following are the unique sets satisfying the listed properties:*

$\mathbb{N}$:	$\exists \min$,	$\nexists \max$,	*discrete*	
$\mathbb{Z}$:	$\nexists \min$,	$\nexists \max$,	*discrete*	
$\mathbb{Q}$:	$\nexists \min$,	$\nexists \max$,	*dense,*	*countable*
$\mathbb{R}$:	$\nexists \min$,	$\nexists \max$,	*connected,*	*separable*

Proof $\mathbb{N}$) Take X discrete, with no maximum, and let $a_0 := \min X$. Since there is no max, there exists an $x > a_0$, and $(\{a_0\}, \{x > a_0\})$ is a jump. Define $a_1 := \min\{x > a_0\}$, so that a_1 is the successor of a_0. By induction $(\{a_0, a_1, \ldots, a_k\}, \{x > a_k\})$ is a jump, so set

$$a_{k+1} := \min \big\{x \in X \mid x > a_k > a_{k-1} > \cdots > a_0 \big\}.$$

$\aleph_0$ iterations produce $A := \{a_i\}_{i \in \mathbb{N}} \subseteq X$, which is countable by construction: $A \approx \mathbb{N}$.

If $X \setminus A \neq \varnothing$ then $(A, X \setminus A)$ would be a jump, so A would have maximum ⨲. Therefore $X = A \approx \mathbb{N}$. $\qquad\square$

$\mathbb{Z}$) Every cut of $\mathbb{Z}$ is of type $\big(\mathbb{Z} \cap (-\infty, k], [k+1, +\infty) \cap \mathbb{Z}\big)$, i.e. a jump. $\qquad\square$

	$\frac{1}{16}$	$\frac{1}{8}$	$\frac{3}{16}$	$\frac{1}{4}$	$\frac{5}{16}$	$\frac{3}{8}$	$\frac{7}{16}$	$\frac{1}{2}$	$\frac{9}{16}$	$\frac{5}{8}$	$\frac{11}{16}$	$\frac{3}{4}$	$\frac{13}{16}$	$\frac{7}{8}$	$\frac{15}{16}$
$\mathfrak{D}_1=$								•							
$\mathfrak{D}_2=$				•						•			•		
$\mathfrak{D}_3=$		•		•		•		•		•		•		•	
$\mathfrak{D}_4=$	•	•	•	•	•	•	•	•	•	•	•	•	•	•	•

Fig. 8.1 The dyadic pyramid

Ⓠ) We claim that any X with those properties is in 1-1 correspondence to a model set $\mathfrak{D}$.

Fix $n \in \mathbb{N}$ and let

$$\mathfrak{D}_n := \left\{ \frac{p}{2^n} \mid p \in \mathbb{N} \cap (0, 2^n) \right\}.$$

These form an increasing chain of nested sets (Fig. 8.1):

$$\varnothing = \mathfrak{D}_0 \subset \mathfrak{D}_1 \subset \mathfrak{D}_2 \subset \cdots \subset \mathfrak{D}_n \subset \mathfrak{D}_{n+1} \subset \cdots$$

The union

$$\mathfrak{D} := \bigcup_{n \in \mathbb{N}} \mathfrak{D}_n \tag{8.1}$$

is the set of **dyadic numbers**. By construction these are the rational numbers in $(0, 1)$ with finite binary expansion. Any interval $\left(\frac{p}{2^n}, \frac{q}{2^n} \right)$ contains $\frac{p+q}{2^{n+1}}$, so $\mathfrak{D}$ is dense in itself. It has no maximum ($\sup \mathfrak{D} = 1 \notin \mathfrak{D}$) nor minimum ($\inf \mathfrak{D} = 0 \notin \mathfrak{D}$). It is also countable, as countable union of finite sets.

We claim that any $X = \{x_1, x_2, \dots \}$ satisfying the four demands is order-isomorphic to $\mathfrak{D}$. Define $f : \mathfrak{D} \to X$ step by step:

on $f(\mathfrak{D}_1)$: $\quad f(1/2) := x_1$

on $f(\mathfrak{D}_2)$: $\quad f(1/4) := x_h, \; h = \min_{i \in \mathbb{N}} \{x_i < x_1\} \qquad f(3/4) := x_k, \; k = \min_{i \in \mathbb{N}} \{x_i > x_1\}$

Then $\{x_1, x_2\} \subseteq f(\mathfrak{D}_2)$ where $x_2 = x_h \vee x_2 = x_k$.

Next, to attain $f(1/8) < f(1/4) < f(3/8) < f(1/2)$ we set

on $f(\mathfrak{D}_3)$: $\quad f(1/8) := x_\ell$, where $\ell = \min_{i \in \mathbb{N}} \left\{ x_i < f(1/4) \right\}$

$\qquad\qquad f(3/8) := x_m$, where $m = \min_{i \in \mathbb{N}} \left\{ f(1/4) < x_i < f(1/2) \right\}$

and so forth. Now suppose we have defined $f(\mathfrak{D}_n)$, and that $\{x_1, \dots, x_n\} \subseteq f(\mathfrak{D}_n)$. Set

$$S_n := \mathfrak{D}_n \setminus \mathfrak{D}_{n-1} = \left\{ 2^{-n} p \mid p \in \mathbb{N} \cap (0, 2^n) \text{ odd} \right\}.$$

Since $f(\mathfrak{D}_{n+1}) = f(\mathfrak{D}_n \cup S_{n+1})$, if the element x_{n+1} hasn't already shown up it has to appear somewhere, so $\{x_1, \ldots, x_{n+1}\} \subseteq f(\mathfrak{D}_{n+1})$. By construction f is increasing, and onto because $x_n \in f(\mathfrak{D}_n)$. Consequently $X \approx \mathfrak{D}$. $\qquad\square$

$\mathbb{R}$) Let's suppose X fulfils the four requirements. A countable dense subset $A \subseteq X$ has the same cardinality as $\mathbb{Q}$, so there's a bijection $f : \mathbb{Q} \to A$. The idea is to extend it to a map $f : \mathbb{R} \to X$.

Pick $x \in \mathbb{R} \setminus \mathbb{Q}$ and set

$$Q^{<x} = \{q \in \mathbb{Q} \mid q < x\}, \quad Q^{>x} = \{q \in \mathbb{Q} \mid q > x\}.$$

We claim $x = \sup Q^{<x}$. Suppose $\sup f(Q^{<x}) = m \in A$ by contradiction, so that there exists a $q \in \mathbb{Q}$ such that $f(q) = m = f(x)$. Because $f(Q^{<x}) \cup f(Q^{>x}) = A$, we deduce that either $m \in f(Q^{<x})$ or $m \in f(Q^{>x})$. In the former case $\exists q < x$ such that $f(q) = m$. But as x is the least upper bound, there exists a $q \leqslant q' \leqslant x$. Hence $m = f(q) \leqslant f(q')$ ↯. The other case is completely similar.

Now we have $f(Q^{<x}) < f(Q^{>x})$, and then $f(x) = f(\sup Q^{<x}) \notin A$. We consider three cases $\mathbb{Q} \ni q < x$, $\mathbb{Q} \ni p > x$ and $x < y \notin \mathbb{Q}$ (note $\exists x < q < y$, by density) separately: this shows f is monotone, and therefore 1-1.

Eventually, for any $z \in X \setminus A$ we have $\sup \{a \in A \mid a < z\} = z$. Hence there exists an x such that $f^{-1}(\{a \in A \mid a < z\}) = Q^{<x}$. But then $f^{-1}(x) = z$, meaning f is onto. $\qquad\square$

Note that appending to $\mathbb{R}$ a maximum or a minimum, as we did in Sect. 7.6, doesn't alter the intimate order structure. In other words the characteristic features of real intervals (inherited from $\mathbb{R}$) are: density, completeness and separability.

Corollary 8.11 *A countable ordered set* $(X, \leqslant)$ *is isomorphic to* $(\mathbb{Q}, \leqslant)$.

Proof This is Theorem 8.10 $\mathbb{Q}$). $\qquad\square$

Corollary *A connected ordered space* $(X, \leqslant)$ *is uncountable:* $|X| \geqslant \aleph_1$.

Proof Corollary 8.11 forces a countable ordered set to be (isomorphic to) a subset of $\mathbb{Q}$. But the only intervals (connected subsets) in $\mathbb{Q}$ are the singletons $\{q\}$, so by Proposition 8.9 the rational numbers are totally disconnected. $\qquad\square$

Other topological properties of ordered spaces $(X, \leqslant)$, on which we won't elaborate much, include:

- X is **Hausdorff**, i.e.: distinct points $x < y$ belong in disjoint open sets.

  There are two possibilities: either there is an element a between x, y or not. In the former situation the open sets $(-\infty, a) \ni x$, $(a, +\infty) \ni y$ separate the points. In the latter, we have a hole between x, y (jump), so we take open sets $(-\infty, x)$, $(y, +\infty)$.

- X is compact $\iff$ complete and bounded.

  ($\Longrightarrow$) If X were unbounded, say $\nexists \max X$, the open covering $\{(-\infty, x) \mid x \in X\}$ would not admit a finite subcovering ↯. We claim the space must be complete too. If there were a gap $L < R$, $\nexists \max L$, $\nexists \min R$ and $X = L \cup R$, then from the nested open covering $\{(-\infty, l) \mid l \in L\} \cup \{(r, +\infty) \mid r \in R\}$ we wouldn't be able to extract a finite subcovering ⚡⚡.

($\Longleftarrow$) Since X is bounded call $m = \min X$, $M = \max X$. Take an open covering $\{U_i\}_{i \in I}$ an open covering. The set

$$H := \{x \in X \mid [m, x] \text{ can be covered by finitely many } U_i\}$$

is not empty since $m \in H$, and it has supremum $z = \sup H$ by completeness. We claim $z \in H$. First, z belongs to one U_{i_0}, and actually $z \in (a, b) \subseteq U_{i_0}$. Hence $(a, b) \cap H \neq \varnothing$. Take $y \in (a, b) \cap H$, so $a < y \leqslant z$. As $y \in H$, the interval $[m, y]$ is covered by, say, $U_{i_1} \cup \cdots \cup U_{i_n}$. But $(y, z] \subseteq U_{i_0}$, therefore $[m, y] \subseteq U_{i_1} \cup \cdots \cup U_{i_n} \cup U_{i_0}$. This implies $z \in H$. Now we'll show $z = \max H$. If $z = M$ we are done. If $z < M$, then $z \in U_{i_0}$ cannot have an immediate successor (otherwise this element would live in H and we could cover). The upper bounds of z don't have minimum, and so $U_{i_0} \cap (z, M] \neq \varnothing$. But then we can cover up to points $w > z$, $w \in (a, b) \subseteq U_{i_0}$, implying $[m, w]$ can be finitely covered. This contradicts the fact that z is the supremum of H.

- if X is connected, subsets $Y \subseteq X$ are compact and connected if and only if they are closed and bounded intervals $Y = [a, b]$ (without proof).

8.2 Well-Ordered Spaces

Recall (Definition 2.13) that a **chain** C in a poset $(X, \leqslant)$ is a totally ordered subset, and a (strictly) **descending chain** is a chain C admitting maximum:

$$\max C = x_1 > x_2 > x_3 > \cdots$$

Definition An (infinite) poset $(X, \leqslant)$ is said to satisfy the **descending chain condition (DCC)** if there are no infinite descending chains of elements. That is, any descending chain is finite.

Exercise Prove that the DCC is the same as asking that every weakly descending (i.e. non-ascending) sequence $x_1 \geqslant x_2 \geqslant x_3 \geqslant \cdots$ eventually stabilises: $\exists m > 0 \colon x_k = x_m$ for every $k \geqslant m$.

Remark Chain conditions are important in ☞ *ring and module theory*. For instance, an *Artin ring* is a ring that satisfies the DCC on ideals. A *Noetherian ring* satisfies the **ascending chain condition (ACC)** (no infinite ascending chains) on ideals. Although the DCC appears to be dual to the ACC, in rings it is the stronger condition (an Artin ring is Noetherian). Examples: the fields $\mathbb{Q}$, $\mathbb{R}$ are Artin. The integers $\mathbb{Z}$ are a Noetherian ring but not an Artin ring. Ideals in $\mathbb{Z}$ are the sets of multiples $\mathbb{Z}m$ of natural numbers $m \in \mathbb{N}$, and

$$\mathbb{Z}p \subseteq \mathbb{Z}q \iff q \mid p.$$

Hence any chain of proper ideals containing the ideal $\mathbb{Z}m$ is allowed to decrease arbitrarily ($\mathbb{N}m$ is unbounded above)

$$\cdots \subset \mathbb{Z}m' \subset \cdots \subset \mathbb{Z}m \subset \cdots \subset \mathbb{Z}m_0,$$

but cannot go up indefinitely, since m has a finite number of natural divisors: $m_0 = \min_{n>1} \{n \mid m\}$.

Definition A poset $(X, \leqslant)$ is **well ordered** (and $\leqslant$ is a **well ordering**) if it satisfies the DCC.

Exercises Prove that

i) $(X, \leqslant)$ is well ordered $\iff$ any $Y \subseteq X$ admits minimum $y = \min Y$.
   In particular, X well ordered $\implies \exists \min X$.
ii) Let $(X, \leqslant)$ be well ordered. Then any subset $Y \subseteq X$ is well ordered (by $\leqslant \mid_Y$).
iii) A well ordering is total. *Hint*: consider a two-element subset $\{z, y\} \subseteq X$.

Peano's axiom III (see Definition 6.1) ensures that $\min \mathbb{N} = 0$. As a consequence of Theorem 3.13, $(\mathbb{N}, \leqslant)$ is well ordered. This fact usually goes by

Theorem 8.12 (Principle of Least Integer) *Every non-empty subset $X \subseteq \mathbb{N}$ has minimum.*

Exercise Prove the least-integer principle by showing that it is equivalent to induction principle ($\mathfrak{I}^*$), see p. 119. *Hint*: use $Y = \mathbb{N} \setminus X$.

Corollary

a) *There is no natural number n such that $0 < n < 1$.*
b) *For any $m \in \mathbb{N}$, there's no $n \in \mathbb{N}$ such that $m < n < m + 1$.*
c) *If $m, n \in \mathbb{N}$, then $m < n \iff m + 1 \leqslant n$.*

Proof a) If there were an n between 0 and 1, the set $\{n \in \mathbb{N} \mid 0 < n < 1\} \neq \varnothing$
   would have a minimum $n_0 > 0$. Therefore $0 < n_0^2 < n_0 < 1$ ↯.
b) If $m < n < m + 1$ then $m - m < n - m < m + 1 - m$, falling back in case a).
c) Exercise.

□

At this juncture it's no surprise that the least-integer principle emerges through the discrete topology of $\mathbb{N}$.

Proposition 8.13 (Characterisation of $\mathbb{N}$ as Ordered Set) $(\mathbb{N}, \leqslant)$ *is the unique ordered set fulfilling the three properties:*

i) $\mathbb{N}$ *is well ordered.*
ii) *Any n has a successor* $(\forall n \; \exists m : s(n) = m)$.
iii) *Any $n > \min \mathbb{N}$ is a successor* $(\forall n \neq \min \mathbb{N} \; \exists m : s(m) = n)$.

Proof Exercise.   □

What about $\mathbb{Z}$ then? The standard order relation of the integers is not a well ordering, since $\mathbb{Z}$ itself has no minimum. But if we accept Zermelo's theorem 3.13 (i.e., the axiom of choice), in principle there must exist another exotic ordering $\preccurlyeq$ where there is a minimum integer. For example $0 \preccurlyeq 1 \preccurlyeq -1 \preccurlyeq 2 \preccurlyeq -2 \preccurlyeq \cdots$

Examples

i) The set of positive rationals of the form $\dfrac{1}{n}$, $n \in \mathbb{N}^*$, doesn't have smallest element. In fact $0 < \dfrac{1}{n+1} < \dfrac{1}{n}$ shows it's impossible to find a 'first' rational $\dfrac{1}{n}$, at least with respect to the standard order.

ii) The set $\left\{1 - \frac{1}{n} \mid n \geqslant 1\right\} \cup \left\{2 - \frac{1}{n} \mid n \geqslant 1\right\} \subseteq \mathbb{Q}$ is well ordered. Note that $1 = 2 - \frac{1}{1}$ has an infinite number of predecessors.

iii) Consider $\left\{1 - \frac{1}{n} \mid n > 0\right\} \cup \left\{1 + \frac{1}{m} \mid m > 0\right\} \cup \{l \in \mathbb{R} \mid l > 2\}$. It goes to show that condition iii) in Proposition 8.13 is necessary.

Definition Let $(X, \leqslant)$ be partially ordered and $x \in X$ a point. We'll indicate by

$$I_x := \left\{\xi \in X \mid \xi < x\right\}, \qquad \overline{I_x} := \left\{\xi \in X \mid \xi \leqslant x\right\}$$

the (open and closed) lower sets determined by x. These are sometimes called **initial segments**.

Examples

i) An initial segment in $\mathbb{N}$ has the form $\{0, 1, \ldots, m-1, m\}$ for some $m \in \mathbb{N}$.

ii) A finite set $\{a_0, \ldots, a_k\}$ is in 1-1 correspondence with the initial segment $\{0, \ldots, k\}$.

iii) Initial segments in a well-ordered set with minimum x_0 are bounded intervals: $I_x = [x_0, x)$, $\overline{I_x} = [x_0, x]$.

The relationship between initial segments and upper sets is clarified by the next result.

Proposition *A map $f \colon (X, \leqslant_X) \to (Y, \leqslant_Y)$ between posets is decreasing if and only if the pre-image of any closed initial segment in Y is an upper set in X.*

Proof ($\Longrightarrow$) Take an initial segment $I_y \subseteq Y$ and $x \in f^{-1}(I_y)$. We then have $f(x) \in I_y$ and so $f(x) \leqslant y$. On the other hand any $z \geqslant x$ satisfies $f(z) \leqslant f(x) \leqslant y \Longrightarrow f(z) \in I_y \Longrightarrow z \in f^{-1}(I_y)$. Hence $f^{-1}(I_y)$ is an upper set.

($\Longleftarrow$) For any x we know that $f(x) \in I_{f(x)}$, and so $x \in f^{-1}(I_{f(x)})$. But $f^{-1}(I_{f(x)})$ is an upper set, so $x \leqslant y \Longrightarrow y \in f^{-1}I_{f(x)}$, meaning $f(y) \in I_{f(x)} \Longrightarrow f(y) \leqslant f(x)$.   □

The next result in practice extends the principle of induction to well-ordered sets:

Theorem (Transfinite Induction) *Let $(X, \leqslant)$ be well ordered, $x \in X$ an element and $A \subseteq X$ a subset.*

$$If \quad \begin{cases} x \in A \\ \forall a \in X \quad [x, a) \in A \Longrightarrow a \in A \end{cases} \quad then \quad A = X.$$

Proof By contradiction, if $X \setminus A$ were not $\varnothing$, due to the well ordering there would exist $h = \min(X \setminus A)$. Then $[x, h) \subseteq A$, and by induction hypothesis $h \in A$ ⚡. □

So, it makes sense to speak of successor of an element a in a well-ordered space, it being

$$a^+ := \min \left(X \setminus I_a \right).$$

If $x_0 = \min X$, successor elements are determined by initial segments

$$a^+ \overset{1-1}{\longleftrightarrow} [x_0, a],$$

which means by sets of predecessors. The phenomenon is particularly evident in $\mathbb{N}$, where we saw that $n^+ = \{n, \{n\}\}$ is indeed the set $\{0, 1, \dots, n-1\}$, cf (6.1). For that reason transfinite induction is an honest generalisation of the finite induction of natural numbers.

Lemma *Let $(X, \leqslant)$ be well ordered, $f \colon X \to X$ an increasing map. Then $x \leqslant f(x)$ for any $x \in X$.*

Proof Define $A := \{x \in X \mid x > f(x)\}$ and suppose by contradiction $A \neq \varnothing$. Since $a := \min A > f(a)$ it follows $f(a) > f(f(a))$ by monotonicity, hence $f(a) \in A$. But a is the minimum ⚡. □

Proposition *A well-ordered set $(X, \leqslant)$ cannot be order-isomorphic to any proper initial segment of itself.*

Proof Call $x_0 = \min X$, so that an initial segment must look like $I_k = [x_0, k)$. Suppose $f \colon X \to I_k$ is the alleged isomorphism. The composite of f with the inclusion $I_k \hookrightarrow X$ would be monotone. Therefore $f(k) \in I_k$ and $f(k) < k$, violating the previous lemma ⚡. □

Proposition *The only order-automorphism of a well-ordered set $(X, \leqslant)$ is $\mathbb{1}_X$.*

Proof If $f \in \mathrm{Aut}\,(X)$ by the previous lemma $x \leqslant f(x)$ for any $x \in X$. Suppose by contradiction there existed an element y strictly smaller that its image $f(y)$. Then we could define $m = \min\{x \mid x < f(x)\}$. Since f is bijective, the element $a = f^{-1}(m)$ would exist, and satisfy $f(a) = m < f(m)$, implying $a < m$ ⚡. □

In a sense we won't explore, the above proposition guarantees that the 'type' of a well ordering is unique, a fact that opens the door to the theory of ☞ *ordinal numbers.* The next result can be easily adapted to show that the relation $\leqslant$ between ordinals is total.

Theorem 8.14 *The cardinalities of two well-ordered spaces are always comparable.*

Proof Let A, B be well ordered with $a = \min A, b = \min B$, and assume by contradiction that $|A| \not\leqslant |B|$ and $|B| \not\leqslant |A|$. Rephrasing, suppose A isn't order-isomorphic to any initial segment $[b, b') \subset B$ and, conversely, $B \not\cong [a, a')$ for no $a' \in A$. Define

$$H = \big\{x \in A \mid [a, x] \text{ is order isomorphic to a proper initial segment of } B\big\}.$$

Since $\{a\} = [a, a] \cong [b, b] = \{b\}$, $H \ni a$ is not empty. For any $x \in H$ we have an order isomorphism $[a, x] \cong [b, y]$ for some $y \in B$, and we define a map $f \colon H \to B$ by $f(x) := y$. Consequently f is monotone, and its image $f(H)$ is an interval of B. By hypothesis $H \neq A$ and $f(H) \neq B$, so H is a proper initial segment, say $H = [a, h)$, where $h := \min\big\{x \mid x \notin H\big\}$. Therefore $f(H) = [b, k)$. Now construct

$$F(x) := \begin{cases} f(x) & \text{if } x < h \\ k & \text{if } x = k \end{cases}.$$

This map is an order isomorphism from $[a, h + 1)$ to $[b, k + 1)$. But $h \notin H$, so in the end either $H = A$ or $f(H) = B$ ⚡.   □

The theorem is saying that between any two well-ordered sets A, B there always exists a 1-1 map (from the 'smaller' one to the 'larger' one). Theorem 6.42 is clearly stronger that this, but it works less well with ordinals.

Exercises 8.15

i) Let I be an ordered set of indices and $\mathscr{X} = \{X_i\}_{i \in I}$ a family of pairwise-disjoint ordered sets. Show $\sum \mathscr{X} := \left(\bigcup \mathscr{X}, \leqslant_{\text{lex}}\right)$ is totally ordered under **lexicographic order**:

$$\forall x \in X_i, y \in X_j \quad x \leqslant_{\text{lex}} y \iff (i <_I j) \vee (i = j \wedge x \leqslant_{X_i} y).$$

ii) Take A, B well ordered and show $A + B$ is well ordered.
   (*Hint*: take $Y \subseteq A \cup B$ and consider the cases $Y \cap A = \varnothing$ and $Y \cap A \neq \varnothing$.)

Remark 8.16 (The Idea Behind ☞ *Ordinal Numbers)* A finite set $X = \{0, 1, 2, \dots, n - 1\}$ of cardinality $|X| = n$ can be totally ordered in $n!$ ways, which reduce to 1 by applying an order isomorphism (a permutation). In other words to any finite cardinal number n there corresponds one 'order type' $[0, n)^{\mathscr{O}} := n$, its ordinal number. (Compare to (6.1).) Since $[0, m) \cong [n + 1, n + m + 1)$, finite ordinals behave like natural numbers:

$$[0, n)^{\mathcal{O}} + [0, m)^{\mathcal{O}} = [0, n)^{\mathcal{O}} + [n+1, n+m+1)^{\mathcal{O}} = \big([0, n) + [n+1, n+m+1)\big)^{\mathcal{O}}$$

$$= [0, n+m+1)^{\mathcal{O}} = [0, m+n+1)^{\mathcal{O}}$$

$$= \big([0, m) + [m+1, m+n+1)\big)^{\mathcal{O}}$$

$$= [0, m)^{\mathcal{O}} + [0, n)^{\mathcal{O}}.$$

Things get interesting for infinite sets. The type of order of $\mathbb{N}$ is indicated by ω_0. It follows from Proposition 8.13 that appending to $\mathbb{N}$ an additional element $\star$ can change the ordering:

$\{\star, 0, 1, \ldots\}$ is order-isomorphic to $\mathbb{N}$: $\star$ becomes the new minimum but not much else changes: $1 + \omega_0 = \{\star, 0, 1, \ldots\}^{\mathcal{O}} = \omega_0$. Nor does placing $\star$ somewhere in the middle, $\{0, 1, \ldots, n, \star, n+1, \ldots\}$, make much of a difference. The Peano axioms guarantee we still have $\mathbb{N}$ essentially.

$\{0, 1, \ldots, \star\}$, on the other hand, has a completely different structure: now $\star$ is the maximum, and has no immediate predecessor: $\omega_0 + 1 = \{0, 1, \ldots, \star\}^{\mathcal{O}} \neq \omega_0$.

That goes to show that infinite ordinals are not like infinite cardinals. Moreover, an infinite set can be ordered in very different ways. Theorem 8.10 says $\mathbb{Z}$, $\mathbb{Q}$ have a fundamentally different ordinal from ω_0, and Corollary 8.11 suggests that all countable ordinals can be realised by subsets of $\mathbb{Q}$.

Exercise Is $\mathbb{N} + \mathbb{N}$ order-isomorphic to $\mathbb{N}$?

Part III
Elementary Real Functions

Chapter 9
Real Polynomials

In this chapter we'll undertake the study of polynomials with real coefficients in one variable. They are important because they are a major source of functions. A comprehensive reference is [52].

Pre-requisites: the proof of Cardano's formula (9.5) requires the knowledge of complex numbers, and the proof of Descartes' sign rule needs a minimum of infinitesimal calculus.

9.1 The Algebra of Polynomials

Definition A **polynomial** with real coefficients is a function $a : \mathbb{N} \to \mathbb{R}$ that stabilises at 0, i.e. a sequence $a = (a_i)_{i \in \mathbb{N}}$ of real numbers called **coefficients**, for which there exists an $n \in \mathbb{N}$ such that $a_i = 0$ for every $i > n$

$$a_0, \; a_1, \; a_2, \; \ldots, \; a_{n-1}, \; a_n, \; 0, \; 0, \; 0, \; \cdots$$

Concretely, if we ignore the infinite tail of zeroes we have finitely many non-null terms $a_0, a_1, \ldots, a_n$, and for operational reasons one thinks of a polynomial as a formal expression of the type

$$a(x) = \sum_{j=0}^{n} a_j x^j = a_n x^n + a_{n-1} x^{n-1} + \ldots + a_1 x + a_0$$

where x is a symbol, called **variable**. Each morsel $a_j x^j$ in the sum is a **monomial**, and a_0 is the **constant term**. The biggest index $j \geqslant 0$ such that $a_j \neq 0$ is called **degree** $\deg(a)$. In a polynomial of degree n the coefficient a_n is called leading coefficient, and a polynomial with $a_n = 1$ is **monic**.

Immediate consequence of the definition is

© The Author(s), under exclusive license to Springer Nature Switzerland AG 2021
S. G. Chiossi, *Essential Mathematics for Undergraduates*,
https://doi.org/10.1007/978-3-030-87174-1_9

Corollary *Two polynomials* $a(x) = \sum_{i=0}^{n} a_i x^i$ *and* $b(x) = \sum_{j=0}^{m} b_j x^j$ *are equal when*

i) $\deg(a) = \deg(b)$ $(n = m)$
ii) $a_i = b_i$ *for every* $i = 0, \ldots, \deg(a)$.

The space of polynomials with real coefficients is denoted by $\mathbb{R}[x]$.

Exercise Explain why, as a set, card $\mathbb{R}[x] = \aleph_1$.

The space $\mathbb{R}[x]$ also comes equipped with a sum, a product and a multiplication by scalars, all defined term by term.

Theorem 9.1 *Let* $a(x) = \sum_{i=0}^{n} a_i x^i$ *and* $b(x) = \sum_{j=0}^{m} b_j x^j$ *be in* $\mathbb{R}[x]$. *The operations*

$$a(x) + b(x) := (a_0 + b_0) + (a_1 + b_1)x + \ldots + (a_s + b_s)x^s, \qquad s = \max\{m, n\}$$

$$a(x) \cdot b(x) = c_0 + c_1 x + \ldots + c_{n+m} x^{n+m}, \text{ where } c_k := \sum_{j=0}^{k} a_j b_{k-j},$$

$$\lambda\, a(x) := (\lambda a_0) + (\lambda a_1)x + \ldots + (\lambda a_n)x^n, \quad \lambda \in \mathbb{R}$$

turn $\mathbb{R}[x]$ *in a commutative* $\mathbb{R}$-*algebra with unit* $1 := 1 + \sum_{j>0} 0 \cdot x^j$ *of infinite dimension.*

Proof Regarding the structure, it's all fault of $\mathbb{R}$ being a field. As for the dimension, polynomials $p(x) = \sum_{i=0}^{n} a_i x^i$ are (finite) linear combinations of monomials $x^k \in \mathbb{R}[x]$. The latter form a countable basis (see Exercise 15.1 vi)), whence $\dim_{\mathbb{R}} \mathbb{R}[x] = \infty$. $\qquad \square$

We'll indicate by $0 := \sum_j 0 \cdot x^j$ the **zero** polynomial (neutral element of $+$), for which $\deg(0) := -\infty$ by decree.

Example

$$(x^2 + 2x - 5) + (x^3 - x + 2) = (0x^3 + x^2 + 2x - 5) + (x^3 + 0x^2 - x + 2)$$

$$= (0 + 1)x^3 + (1 + 0)x^2 + (2 - 1)x + (-5 + 2)$$

$$= x^3 + x^2 + x - 3$$

$$(4x^3 + x - 1)(x^2 + x + 1) = x^2(4x^3 + x - 1) + x(4x^3 + x - 1) + (4x^3 + x - 1)$$

$$= 4x^5 + x^3 - x^2 + 4x^4 + x^2 - x + 4x^3 + x - 1$$

$$= 4x^5 + 4x^4 + 5x^3 - 1$$

Exercises

i) Prove, for any $a(x), b(x) \in \mathbb{R}[x]$,

$$\deg(a + b) \leqslant \max\{\deg a, \deg b\} \qquad \deg(ab) = \deg a + \deg b,$$

with the convention that $-\infty + \mathbb{N} = -\infty = -\infty + (-\infty)$, and $-\infty < 0$, cf. Sect. 7.6.

ii) Show that the polynomials of degree 0 can be identified with the real numbers.

Immediately then,

Corollary $\mathbb{R}[x]$ *is an integral domain.*

Proof If $p(x), q(x) \in \mathbb{R}[x]$ have zero product $p(x)q(x) = 0$, then $\deg(p) + \deg(q) = -\infty$. Hence $\deg(p) = -\infty$ or $\deg(q) = -\infty$, and one of p, q is the zero polynomial. $\qquad\square$

Theorem (Euclidean Division) $\mathbb{R}[x]$ *is a Euclidean domain: for any* $a(x), b(x) \in \mathbb{R}[x]$ *with* $b \neq 0$ *and* $\deg(a) \geqslant \deg(b)$*, there exist unique polynomials* $q(x), r(x) \in \mathbb{R}[x]$ *such that*

$$a(x) = b(x)q(x) + r(x) \quad \text{with } \deg(r) < \deg(b).$$

Proof The existence of quotient and remainder is automatically warranted in case $\deg(a) < \deg(b)$, because $a(x) = b(x) \cdot 0 + a(x)$.

Suppose $a(x) = \sum_{i=0}^{n} a_i x^i$ and $b(x) = \sum_{j=0}^{m} b_j x^j$, $b_m \neq 0$, have degrees $\deg(a) \geqslant \deg(b) = m$, and let's use induction on $\deg(a)$.

For $\deg(a) = k + 1 > m$ the polynomial $r_1(x) = a(x) - \dfrac{a_{k+1}}{b_m} x^{k+1-m} b(x)$ has degree less than $k+1$. By induction hypothesis $r_1(x) = q_1(x)b(x) + r_2(x)$, $\deg r_2 < \deg b$. Hence

$$a(x) = r_1(x) + \frac{a_{k+1}}{b_m} x^{k+1-m} b(x) = \left(q_1(x) + r_2(x) + \frac{a_{k+1}}{b_m} x^{k+1-m} \right) b(x) + r_2(x).$$

Uniqueness: assuming there exist $q_1(x), q_2(x), r_1(x), r_2(x)$ satisfying $b(x)q_1(x) + r_1(x) = a(x) = b(x)q_2(x) + r_2(x)$ and $\deg r_1, \deg r_2 < \deg b$, we obtain

$$b(x)(q_1(x) - q_2(x)) = r_2(x) - r_2(x).$$

If $q_1(x) - q_2(x) \neq 0$ we'd have $\deg(r_1 - r_2) = \deg b + \deg(q_1 - q_2) \geqslant \deg b$, and then $\deg(r_1 - r_2) \leqslant \max\{\deg r_1, \deg r_2\} < \deg b \; \maltese$. Consequently $q_1(x) = q_2(x)$, which in turn implies $r_2(x) - r_2(x) = 0$ because b is non-null. $\qquad\square$

Example The division $\dfrac{2x^4 + x^3 - x + 2}{x^2 + 3} = 2x^2 + x - 6 + \dfrac{-4x + 20}{x^2 + 3}$ is achieved with the scheme below, familiar from school:

$$
\begin{array}{rrrrr|l}
2x^4 & +x^3 & +0x^2 & -x & +2 & x^2+3 \\
\hline
2x^4 & & +6x^2 & & & 2x^2 \quad +x \quad -6 \quad = \quad q(x) \\
\hline
& +x^3 & -6x^2 & -x & +2 & \\
& +x^3 & & +3x & & \\
\hline
& & -6x^2 & -4x & +2 & \\
& & -6x^2 & & -18 & \\
\hline
r(x) \quad = & & & -4x & +20 & \\
\end{array}
$$

The division process terminates when the degree of the dividend (numerator) becomes smaller than the degree of the divisor (denominator). The Euclidean algorithm guarantees that the remainder is the first dividend satisfying said property.

With this in place, the majority of properties of the Euclidean ring $\mathbb{R}$ (such as divisibility, for instance) transfer to $\mathbb{R}[x]$. The role of the absolute value in $\mathbb{R}$ is played in $\mathbb{R}[x]$ by the degree map $\deg \colon \mathbb{R}[x] \to \{-\infty\} \cup \mathbb{N}$.

When $r = 0$ we say b divides a, or a is a multiple of b: $b(x)\big| a(x)$.

Corollary (Polynomial Remainder Theorem) *The remainder of dividing $a(x)$ by $x - x_0$ equals $a(x_0)$.*

Proof We know $a(x) = q(x)(x - x_0) + r$ with $\deg(r) \leqslant 0$, i.e. $r \in \mathbb{R}$. Hence $a(x_0) = q(x_0)(x_0 - x_0) + r = r$, as required. $\square$

Example Determine m so that $x^4 + mh^2x^2 - 5hx^2 + h^4$ be divisible by $x - h$, as $h \in \mathbb{R}$ varies.

The division's remainder is $(2 + m)h^4 - 5h^3 = h^3\big((2 + m)h - 5\big)$. So, for $h = 0$ any number m answers the question. For $h \neq 0$, we have $(2 + m)h - 5 = 0$, i.e. $m = -\dfrac{5}{h} - 2$.

More familiar for dividing $a(x) = \sum_{i=0}^{n} a_i x^i$ by $x - x_0$ might be *Ruffini's algorithm*. It's nothing more than a compact version of Euclidean division that determines the coefficients q_j of the quotient $\dfrac{a(x)}{x - x_0}$ in a recursive manner:

$$
q_{n-1} = a_n, \qquad q_{j-1} = x_0 q_j + a_j, \ \forall j = n - 1, \ldots, 1,
$$

and also the remainder $r = x_0 q_0 + a_0$. Ruffini's method can be assembled in a table of this sort:

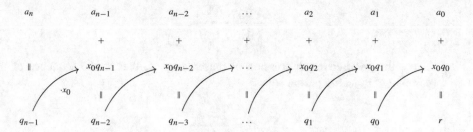

From the drill it's clear that calculating quotient and remainder has the same computational complexity of the standard division. It's only slightly quicker because with Ruffini we don't have to write the intermediate steps (in grey in the Euclidean recipe) nor the variable.

Examples 9.2

i) Divide $p(x) = x^7 - 4x^3 + 2x - 1$ by $x - 2$.
   Let's use Ruffini (but dropping the middle line and indicating just the coefficients of p and the quotient):

a_i	1	0	0	0	-4	0	2	-1
q_i	1	2	4	8	12	24	50	99

Therefore $\dfrac{x^7 - 4x^3 + 2x - 1}{x - 2} = x^6 + 2x^5 + 4x^4 + 8x^3 + 12x^2 + 24x +$

$50 + \dfrac{99}{x - 2}$.

ii) Determine the biggest exponent β such that $g(x) = 2x^5 - 11x^4 + 18x^3 - 8x^2 - 4x + 3$ be divisible by $(x - 1)^\beta$.
    First of all $g(1) = 0$, so $\beta \geqslant 1$. We'll apply the algorithm over and over again until we reach the first non-zero remainder:

$$
\begin{array}{r|rrrrrr l}
(a_n) & 2 & -11 & 18 & -8 & -4 & 3 & \\
\hline
& 2 & -9 & 9 & 1 & -3 & 0 & \rightsquigarrow \beta \geqslant 1 \\
& 2 & -7 & 2 & 3 & 0 & 0 & \rightsquigarrow \beta \geqslant 2 \\
& 2 & -5 & -3 & 0 & 0 & 0 & \rightsquigarrow \beta \geqslant 3 \\
& 2 & -3 & -6 & -6 & -6 & -6 \neq 0 & \rightsquigarrow \beta < 4
\end{array}
$$

So the answer is $\beta = 3$.

iii) Prove that $\dfrac{x^{n+1} - 1}{x - 1} = x^n + x^{n-1} + \ldots + x + 1$ for any $n \in \mathbb{N}$.
     Ruffini's rule tells

$$\frac{1\ 0\ldots 0-1}{1\ 1\ldots 1\ \ 0}.$$

Hence the **geometric progression** of reason $\rho \neq 1$, that is, the sequence of powers $(\rho^k)_{k\in\mathbb{N}}$, has partial sum

$$\sum_{k=0}^{n}\rho^k = \frac{1-\rho^{n+1}}{1-\rho}.$$

Let's move to the situation in which $\deg a < \deg b$ in the quotient $\dfrac{a(x)}{b(x)} \in \mathbb{R}(x)$. Although we cannot divide, we can still simplify the expression by transforming it in a sum of simpler fractions. The latter are called decompositions in 'partial fractions'. We will show (Propositions 13.12, 13.13) $a(x)/b(x)$ admits a unique splitting

$$\frac{a(x)}{b(x)} = \sum_{n_i=1}^{N_i}\frac{A_i}{(x-a_i)^{n_i}} + \sum_{m_j=1}^{M_j}\frac{B_j x + C_j}{(b_j x^2 + c_j x + d_j)^{m_j}} \tag{9.1}$$

where A_i, B_j, C_j are real numbers, and the denominators are determined by the decomposition

$$b(x) = \prod_{i=1}^{k}(x-a_i)^{N_i}\prod_{j=1}^{h}(b_j x^2 + c_j x + d_j)^{M_j}$$

in irreducible factors.

Example Let's decompose $\dfrac{1-x}{x^3+x^2}$ in partial fractions. For starters we have to factor the denominator completely:

$$x^3 + x^2 = x^2(x+1).$$

As x is a double factor (multiplicity 2), while $x+1$ is simple (multiplicity 1), and both are of degree one, the decomposition looks like this:

$$\frac{1-x}{x^3+x^2} = \frac{A}{x} + \frac{B}{x^2} + \frac{C}{x+1}.$$

To find A, B, C we add the terms:

$$\frac{A}{x} + \frac{B}{x^2} + \frac{C}{x+1} = \frac{Ax(x+1)+B(x+1)+Cx^2}{x^2(x+1)} = \frac{(A+C)x^2+(A+B)x+B}{x^2(x+1)}.$$

Comparing numerators, $1 - x = 0x^2 - 1x + 1$ and $(A+C)x^2 + (A+B)x + B$ give

$$\begin{cases} A + C = 0 \\ A + B = -1 \\ B = 1 \end{cases} \implies \begin{cases} C = 2 \\ A = -2 \\ B = 1. \end{cases}$$

Hence, $\dfrac{1-x}{x^3 + x^2} = -\dfrac{2}{x} + \dfrac{1}{x^2} + \dfrac{2}{x+1}$.

The presence of exponents at the denominator shouldn't come as a surprise. In the previous example it wouldn't be possible to have $\dfrac{1-x}{x^3 + x^2} = \dfrac{A}{x} + \dfrac{C}{x+1}$ for any $A, C \in \mathbb{R}$, as can be verified heuristically.

Think about numbers: take the quotient $\dfrac{5}{9}$ of $5, 9 \in \mathbb{N}$. The denominator factors as $9 = 3^2$, but there exists no $A \in \mathbb{N}$ for which $\dfrac{5}{9}$ can be written as $\dfrac{A}{3}$. If we ignore the trivial solution $A = 0$, $B = 5$ (giving no splitting), there exist only one pair (A, B) such that $\dfrac{5}{9} = \dfrac{A}{3} + \dfrac{B}{9}$, namely $A = 1$, $B = 2$.

This type of decomposition has an important application in calculus when one is looking for primitives of rational maps $\displaystyle\int \dfrac{a(x)}{b(x)} dx$. As the integration operator is linear, the integral breaks up in a sum of terms whose primitives are known elementary functions: $\dfrac{1}{x+c}$, $\log|x + c|$, $\log|x^2 + px + q|$, $\arctan(x + c)$.

The partial-fraction decomposition is an easy consequence of the Taylor series expansion of the denominator $b(x)$. Using (9.1) one can prove another formula (*Hermite decomposition*):

$$\frac{a(x)}{b(x)} = \frac{d}{dx}\left(\frac{a_1(x)}{b_1(x)}\right) + \frac{a_2(x)}{b_2(x)}.$$

The above is useful when integrating because both b_1, b_2 have the same roots as b, but those of b_2 have multiplicity 1, those of b_1 have multiplicity one less. Apart from being simpler than (9.1), Hermite's decomposition highlights the obvious rational part of the primitive.

Before we turn page let's make a short detour in several variables. Polynomials in n variables $x_1, \ldots, x_n$ are defined recursively as elements of $\mathbb{R}[x_1, \ldots, x_n] := \big(\mathbb{R}[x_1, \ldots, x_{n-1}]\big)[x_n]$. In other words, they are polynomials in x_n with polynomial coefficients in the algebra $\mathbb{R}[x_1, \ldots, x_{n-1}]$:

$$p(x_1, \ldots, x_n) = \sum_{i=1}^{r} a_{i_1 \cdots i_r} x_1^{i_1} \cdots x_r^{i_r}, \qquad a_{i_1 \cdots i_r} \in \mathbb{R}, \ i_1, \ldots, i_r \in \mathbb{N}.$$

Example: the polynomial $p(x, y, z) = 7x^3 y^4 z - \sqrt{5} y^2 z^2 + xyz^2 \in \mathbb{R}[x, y, z]$, of total degree $3 + 4 + 1$, can be viewed as:

$$p(x, y, z) = \left(7y^4z\right)x^3 + \left(yz^2\right)x - \sqrt{5}y^2z^2 \in (\mathbb{R}[y, z])[x] \quad \text{of degree 3 in } x$$

$$= \left(7zx^3\right)y^4 - \left(\sqrt{5}z^2\right)y^2 + \left(xz^2\right)y \in (\mathbb{R}[x, z])[y] \quad \text{of degree 4 in } y$$

$$= \left(yx - \sqrt{5}y^2\right)z^2 + \left((7y^4x^3\right)z \in (\mathbb{R}[y, x])[z] \quad \text{of degree 2 in } z.$$

In analogy to the case $n = 1$ it can be shown $\mathbb{R}[x_1, \ldots, x_n]$ is a Euclidean domain.

When all monomials have the same degree $i_1 + \ldots + i_r = s$ the polynomial is called **homogeneous** of degree s.

Exercises 9.3 Prove

i) every polynomial of degree s in $\mathbb{R}[x_1, \ldots, x_n]$ can be uniquely written as a sum $p_s + p_{s-1} + \ldots + p_0$, where p_i is a homogeneous polynomial of degree i, $i = 0, \ldots, s$;

ii) $p \not\equiv 0$ is homogeneous of degree $s \iff p(\lambda x_1, \ldots, \lambda x_r) = \lambda^s p(x_1, \ldots, x_r)$ for all $\lambda \in \mathbb{R}$;

iii) for $q(x) = \sum_{j=0}^{n} a_j x^j \in \mathbb{R}[x]$, define the polynomial $\dfrac{\partial q}{\partial x} := \sum_{j=1}^{n} j a_j x^{j-1}$. Show that any homogeneous $p(x_1, \ldots, x_r) \in \mathbb{R}[x_1, \ldots, x_r]$ of degree s satisfies

$$\sum_{i=1}^{r} x_j \frac{\partial p}{\partial x_j} = s\, p(x_1, \ldots, x_r).$$

In general, let V, V' be real vector spaces. A linear map $f : V \to V'$ is called **homogeneous of degree** s whenever $f(\lambda x) = \lambda^s f(x)$ for all $x \in V, \lambda \in \mathbb{R}$. It can be shown that a map $f : \mathbb{R}^n \overset{\mathscr{C}^1}{\to} \mathbb{R}$ is homogeneous of degree s if and only if it fulfils **Euler's identity for homogeneous functions**:

$$\mathbf{x} \cdot \operatorname{grad} f(\mathbf{x}) = s f(\mathbf{x}), \quad \forall \mathbf{x} \in \mathbb{R}^n.$$

9.2   Real Roots

Consider a polynomial $p(x) = \sum_{j=0}^{n} a_j x^j \in \mathbb{R}[x]$ and a number $x_0 \in \mathbb{R}$. We define the substitution[1] of x for x_0 by $p(x_0) = \sum_{j=0}^{n} a_j x_0^j \in \mathbb{R}$.

---

[1] Literally, a form of variable substitution known from logic. It's what distinguishes a polynomial from the corresponding polynomial function, see definition 11.1.

Definition Consider a polynomial $p(x) \in \mathbb{R}[x]$. A number $x_0 \in \mathbb{R}$ such that $p(x_0) = 0$ is a **zero** or **root** of p.

Corollary (of the remainder theorem) $x - x_0$ divides $a(x) \iff x_0$ is a root of p.

Proof Exercise. □

Definition The **multiplicity** of a root $x_0 \in \mathbb{R}$ of $a(x) \in \mathbb{R}[x]$ is the largest natural number $r \in \mathbb{N}$ such that $(x - x_0)^r$ divides $a(x)$:

$$a(x) = (x - x_0)^r q(x) \qquad \text{with } q(x_0) \neq 0.$$

The fact that x_0 is not a root of $q(x)$ is equivalent to imposing that $(x - x_0)^{r+1}$ doesn't divide $a(x)$.

Example The root 2 is **simple** for $x^2 - 4$ (i.e. it has multiplicity 1), **double** for $(x - 2)^2(x + 1)$ (multiplicity 2), **triple** for $(2x + 1)^7(x - 2)^3$ (multiplicity 3).

Definition Factoring $p(x) \in \mathbb{R}[x]$ means writing a **factorisation**, i.e. a product of polynomials of positive degree

$$p(x) = q_1(x)^{m_1} \cdots q_k(x)^{m_k}, \qquad m_j > 0, j = 1, \ldots k.$$

Consequently $\deg(p) = m_1 + \ldots + m_k$.

A polynomial that doesn't admit a factorisation in $\mathbb{R}[x]$ is called **irreducible**.

In this way any polynomial $mx + q$ of degree 1 is irreducible. Irreducible, monic polynomials in $\mathbb{R}[x]$ play the same part as prime numbers in $\mathbb{Z}$. *Gauß' theorem for unique-factorisation domains*, applied to our context, states that

Theorem $\mathbb{R}[x]$ *is a unique-factorisation domain: any $a(x) \in \mathbb{R}[x]$ can be decomposed in irreducible factors, and the decomposition is unique up to permutations of the factors.*

Proposition *Let $p(x), s(x) \in \mathbb{R}[x]$ with $\deg(s) < \deg(p)$, p irreducible and $s \not\equiv 0$. Then there exist $a(x), b(x) \in \mathbb{R}[x]$ such that $a(x)p(x) + b(x)s(x) = 1$.*

Proof This is a direct application of Euclid's algorithm. Take

$$S = \{ f(x) \in \mathbb{R}[x] \mid \exists\, a(x), b(x) \text{ such that } 0 \neq f(x) = a(x)p(x) + b(x)s(x) \}.$$

This set is non-empty as $s(x) \in S$. Consider an element $l(x) = c(x)p(x) + d(x)s(x) \in S$ of smallest-possible degree k, and let's prove $l(x)$ is constant. If it weren't, since $k \leqslant \deg(s)$ we could divide $p(x) = l(x)q(x) + r(x)$, with $\deg(r) < k$ and $r(x) = (1 - c(x)q(x))p(x) + (-d(x)q(x))s(x) \in S$. But $r(x) = 0$ due to the minimality of k, hence $p(x) = l(x)q(x)$. Since p is irreducible, $l(x) = l \neq 0$. To conclude it's enough to take $a(x) = l^{-1}c(x), b(x) = l^{-1}d(x)$. □

More concretely, in order to factor a (non-irreducible) polynomial we need to know its roots. The good news is that for degree up to 4 there are formulas (for the bad news you'll have to wait until p.216).

Proposition *The quadratic polynomial* $ax^2 + bx + c \in \mathbb{R}[x]$ $(a \neq 0)$

i) *is irreducible if and only if its* **discriminant** $\Delta := b^2 - 4ac$ *is negative;*
ii) *has roots*

$$x_{\pm} := \frac{-b \pm \sqrt{\Delta}}{2a} \tag{9.2}$$

in case $\Delta \geqslant 0$.

Proof Write

$$4a\left(ax^2 + bx + c\right) = 4a^2x^2 + 4abx + 4ac = (2ax + b)^2 - \Delta. \tag{9.3}$$

When $\Delta \geqslant 0$ this can be factored: $\left(2ax + b + \sqrt{\Delta}\right)\left(2ax + b - \sqrt{\Delta}\right) = 4a^2(x - x_-)(x - x_+)$.

When $\Delta < 0$ expression (9.3) is positive $\forall x$, hence $ax^2 + bx + c$ doesn't have roots, and is not reducible (Corollary 13.9). $\qquad\qquad\qquad\qquad\qquad\square$

Exercise In case $ac \neq 0$ prove the solutions of $ax^2 + bx + c = 0$, if they exist, can be written as $\frac{-2c}{b \pm \sqrt{\Delta}}$. *Hint*. Divide by x^2 and solve for $1/x$.

A handy application for monic polynomials is that

$$x^2 - (p + q)x + pq = (x - p)(x - q). \tag{9.4}$$

To determine the integer roots of $x^2 + bx + c \in \mathbb{Z}[x]$, therefore, it suffices to find two numbers $p, q \in \mathbb{Z}$ whose sum equals $-b$ and whose product is c.

Exercises

i) It's a basic linear-algebra fact that the eigenvalues λ_1, λ_2 of a real matrix $A = \left(\begin{smallmatrix} a & b \\ c & d \end{smallmatrix}\right)$ are invariants. Prove that the trace $\operatorname{tr}(A) := a + d$ and the determinant $\det(A) := ad - bc$ are invariants, too. *Hint*: write the characteristic polynomial of A.

Conversely, if we know $\operatorname{tr}(A), \det(A)$ are invariants, show λ_1, λ_2 are invariants.

ii) Suppose the cubic polynomial $x^3 + bx^2 + cx + d \in \mathbb{R}[x]$ factors as $(x - r_1)(x - r_2)(x - r_3)$. Show

$$b = -(r_1 + r_2 + r_3), \qquad c = r_1r_2 + r_1r_3 + r_2r_3, \qquad d = -r_1r_2r_3.$$

iii) For general monic polynomials there's a relationship between the coefficients a_j and the roots x_ℓ

$$x^n + a_{n-1}x^{n-1} + \ldots + a_1 x + a_0 = \prod_{\ell=1}^{n} (x - x_\ell)$$

given by the *elementary symmetric polynomials* (☞ *invariant theory*):

$$a_{n-1} = -\sum_{j=1}^{n} x_j, \quad a_{n-2} = \sum_{\substack{j,k=1 \\ j<k}}^{n} x_j x_k, \quad a_{n-3} = -\sum_{\substack{i,j,k=1 \\ j<k<i}}^{n} x_j x_k x_i, \quad \ldots \quad a_0 = (-1)^n \prod_{j=1}^{n} x_j.$$

Lemma 9.4 *Any cubic polynomial $ax^3 + bx^2 + cx + d \in \mathbb{R}[x]$ $(a \neq 0)$ can be put in the form $t^3 - 3pt - 2q \in \mathbb{R}[t]$.*

Proof The substitution $x = t - \dfrac{b}{3a}$ delivers the required form, with $-3p = \dfrac{3ac - b^2}{3a^2}$ and $-2q = \dfrac{2b^3 - 9abc + 27a^2d}{27a^3}$. □

The variable change used in the proof is, in parlance, known as a *Tschirnhaus transformation* $x \mapsto t = t(x, a, b, c, d)$. It consents to simplify the expression under exam and eventually factor it. In the above reduced form we have

Proposition *The polynomial $x^3 - 3px - 2q$ has roots*

$$x = \left(q + \sqrt{\Delta_3}\right)^{1/3} + \left(q - \sqrt{\Delta_3}\right)^{1/3} \tag{9.5}$$

where $\Delta_3 = q^2 - p^3$, and the cubic root produces 3 (complex) numbers.

Proof With the substitution $x = u - v$ we get $(u^3 - v^3) - 2q = 3(p + uv)(u - v)$, which is equivalent to the two equations $p + uv = 0$ and $u^3 - v^3 = 2q$. From the first of those we obtain $u^3 v^3 = -p$, so the system reduces to finding two numbers $u^3, -v^3$ with given sum and product. By remark (9.4) we must then solve the quadratic relation $\zeta^2 - 2q\zeta + p^3 = 0$, whence $\zeta_1 = u^3 = q + \sqrt{\Delta_3}$, $\zeta_2 = v^3 = q - \sqrt{\Delta_3}$. Now let $\omega = \dfrac{-1 + i\sqrt{3}}{2}$ be a cubic root of unity, and define

$$\begin{array}{lll} u_1 = \sqrt[3]{\zeta_1} & u_2 = \omega\sqrt[3]{\zeta_1} & u_3 = \omega^2\sqrt[3]{\zeta_1} \\ v_1 = \sqrt[3]{\zeta_2} & v_2 = \omega^2\sqrt[3]{\zeta_2} & v_3 = \omega\sqrt[3]{\zeta_2} \end{array}.$$

Then $x_i = u_i - v_i$, $i = 1, 2, 3$ give back expression (9.5). □

Proposition 9.5 *Every cubic polynomial $ax^3 + bx^2 + cx + q \in \mathbb{R}[x]$ $(a \neq 0)$ is reducible.*

In particular, calling $\Delta_3 := 18abcd - 4b^3d + b^2c^2 - 4ac^3 - 27a^2d^2$:

   if $\Delta_3 > 0$:   *there exist 3 distinct real roots,*
   if $\Delta_3 = 0$:   *there exists one triple real root,*
   if $\Delta_3 < 0$:   *there exists only one real root (the factorisation contains an irreducible quadratic term).*

Proof Exercise.   □

The previous lemma and formula (9.5) furnish a real root. Note the same conclusion is reached through the *intermediate-value theorem* for real functions.

Relatively to degree 4 there is an explicit formula as well, whose practicality is, alas, very limited. Just for curiosity's sake we'll lay out the recipe (without proof) to solve

$$z^4 + bz^3 + cz^2 + dz + e = 0.$$

The Tschirnhaus transformation $w = z+b$ reduces the equation to the more pleasing

$$w^4 + pw^2 + qw + r = 0,$$

where $p = c - \frac{3}{8}b^2$, $q = d - \frac{1}{2}bc + \frac{1}{8}b^3$, $r = e - \frac{1}{4}bd + \frac{1}{16}b^2c - \frac{3}{25}b^4$. By means of (9.5) we compute the roots s_1, s_2, s_3 of the cubic polynomial

$$u^3 - 2pu^2 + (p^2 - 4r)u + q^2 = (u - s_1)(u - s_2)(u - s_3).$$

Now the roots of the quartic equation are the 4 possible sign combinations in

$$\frac{\sqrt{-s_1} \pm \sqrt{-s_2} \pm \sqrt{-s_3}}{2}. \tag{9.6}$$

Remark 9.6 The solutions to the cubic and quartic equations are traditionally attached to Cardano's name. It was he who published general formulas in 1545, attributing them to del Ferro (deg $= 3$) and Ferrari (deg $= 4$). In reality he was informed about formula (9.5) in 1539 by Tartaglia, a circumstance that tied up the two in a decade-long spat (to set the record straight, del Ferro's solution precedes Tartaglia's). Crucially, the solutions (9.2), (9.5) and (9.6) are algebraic functions of the coefficients. This is idiosyncratic of algebraic equations up to fourth order. Having said that, there may appear non-algebraic expressions as well, as observed in Remark 12.10.

It is not possible, on the other hand, to solve the generic equation of degree $\geqslant 5$ by means of algebraic transformations and radicals (*Abel–Ruffini theorem*), notwithstanding the existence of criteria to determine which equations are solvable algebraically. For example, the irreducible polynomial $x^5 - 16x + 2 = 0$ has simple Galois group $\mathfrak{S}_5$, and so it cannot be solved 'by radicals'. In the opposite direction, $x^5 - 5x^4 - 10x^3 - 10x^2 - 5x - 1 = 0$ has root $x = 1 + \sqrt[5]{2} + \sqrt[5]{4} + \sqrt[5]{8} + \sqrt[5]{16}$. (☞ *Galois theory*.) When the roots can be written as functions of the equation's coefficients, the dependency will at least be transcendental, and the expectation is the functions won't be elementary. Hermite (1858), for instance, solved the quintic $x^5 + 5x = a$ using (analytic, non-algebraic) *theta functions*.

Although it's not always possible to solve explicitly, there are a number of criteria for estimating the number k of roots $r_1, \ldots, r_k$ of a polynomial $a(x) \in \mathbb{R}[x]$, which we'll quickly review. Evidently the degree of $a(x)$ gives an upper bound for k, since if $a(x) = \prod_{i=1}^{k} (x - r_i)$ has only linear factors, necessarily $k = \deg a$. But the mere existence of irreducible polynomials implies that $k \leqslant \deg(a)$ in general.

Proposition (Descartes' Sign Rule) *Let* $a(x) = \sum_{j=0}^{n} a_j x^j \in \mathbb{R}[x]$ *be a non-zero polynomial of degree n. The number of positive roots doesn't exceed the number of sign changes in the coefficient sequence* $a_0, a_1, \ldots, a_{n-1}, a_n$.

When counting sign flips we ignore zero coefficients: $-3 + x^2 + 7x^5 - 2x^8$ has 2 changes. Equivalently, if a_k and a_h are non-zero and consecutive (in the ordering by increasing degree), we count one sign change if $a_k a_h < 0$.

Proof The argument goes by induction.

When $\deg a = 0$ there are no changes, nor roots. Suppose $\deg a = n \geqslant 1$ and there are m sign changes in the coefficients. Clearly $m \leqslant n$. Consider the polynomial $a'(x) = \frac{a(x) - a_0}{x} = \sum_{j=1}^{n} a_j x^{j-1}$ of degree $n - 1$. Except for a_0, the coefficients of a' are the same as those of a, hence a' has either $m - 1$ or m sign changes. By the inductive hypothesis, in the former case a' has at most $m - 1$ positive roots, so a has at most m positive roots. In the latter case a' has no more than m positive roots. But as a' and a have the same number of changes, there are two further possibilities: $a_0 = 0$, or a_0 has the same sign as the first non-zero coefficient a_j, $j > 0$. If $a_0 = 0$ then a and a' have the same positive roots, meaning a' has at most m positive roots. If, on the contrary, $a_0 \neq 0$, we must consider the alternative:

$a(0) > 0$ and the function $x \mapsto a(x)$ is increasing on $(0, \epsilon)$, or

$a(0) < 0$ and $x \mapsto a(x)$ is decreasing on $(0, \epsilon)$,

for some sufficiently small $\epsilon > 0$. But in either case a has no roots in $(0, z]$, where z is the first positive zero of a'. Therefore a has m positive roots at most. □

Corollary *The number of real roots of a real polynomial cannot exceed twice the number of sign flips in the coefficients.*

Proof Apply the sign rule to both $a(x)$ and $a(-x)$. □

Example The polynomial $-3 + x^2 + 7x^5 - 2x^8$ has no more than 2 positive roots, and at most 4 real roots overall. (With elementary analytical techniques it can be shown that there are exactly 2 positive zeroes.)

When the polynomials have integer coefficients the next two results help the pursuit of roots:

Proposition (Descartes) *If* $a(x) = \sum_{j=0}^{n} a_j x^j \in \mathbb{Z}[x]$ *has* $\dfrac{p}{q} \in \mathbb{Q}$ *as root (assuming* $\mathrm{lcm}(p, q) = 1$*), then* $p \mid a_0$ *and* $q \mid a_n$.

Proof Let p/q ($\mathrm{lcm}(p, q) = 1$) be a zero of $a(x)$. Substituting, $0 = q^n a(p/q) = a_n p^n + a_{n-1} p^{n-1} q + \ldots + a_1 p q^{n-1} + a_0 q^n$. Isolating the appropriate terms we have

$$-a_n p^n = q \left(a_{n-1} p^{n-1} + \ldots + a_1 p q^{n-2} + a_0 q^{n-1} \right) \implies q \mid a_n;$$

$$p \left(a_n p^{n-1} + a_{n-1} p^{n-2} q + \ldots + a_1 q^{n-1} \right) = -a_0 q^n \implies p \mid a_0.$$

$\square$

In particular

Corollary *The (possible) integer roots of* $a(x) = \sum_{j=0}^{n} a_j x^j \in \mathbb{Z}[x]$ *divide* a_0.

Proposition (Eisenstein) *Take* $a(x) = \sum_{j=0}^{n} a_j x^j \in \mathbb{Z}[x]$ *and suppose there exists a prime* $p \in \mathbb{N}$ *such that*

(1) p does not divide a_n,
(2) p divides a_j for every $j = 0, \ldots, n - 1$,
(3) p^2 doesn't divide a_0.

Then $a(x)$ has no rational root.

Proof If, by contradiction, $a(x)$ had a rational root, it would have an integer root as well, so we could factor it as $a(x) = (b_r x^r + \ldots + b_0)(c_s x^s + \ldots c_0)$ with coefficients all integer. By hypotheses (3) and (2), p divides one (and only one) of the factors in $a_0 = b_0 c_0$, say $p \mid b_0$. But as $a_1 = b_0 c_1 + b_1 c_0$, (2) implies $p \mid b_1$. Then $a_2 = b_2 c_0 + b_1 c_1 + b_0 c_2$ would force $p \mid b_2$ and so on, by induction: $p \mid b_j$ for all $j = 0, \ldots, r$. But this would violate (1) ⨍. $\square$

Examples

i) $x^9 + 14x + 7$ is irreducible.
ii) For any prime p and $n \in \mathbb{N}^*$ the polynomial $x^n - p$ is irreducible over $\mathbb{Q}$. Therefore no n^{th} root $\sqrt[n]{p}$ is rational ($n > 1$).

After studying complex numbers (Chap. 13) we'll be able to show that every real polynomial of degree $\geqslant 3$ is reducible (Corollary 13.11).

Exercises

i) Compute quotient and remainder for:

$$1) \ \frac{x^4 + 2x^3 - 14x^2 + 2x - 15}{x^2 + 1} \qquad 2) \ \frac{2x^5 + 12x^3 + x^2 + 6}{x^3 - 4}$$

$$3) \ \frac{x^3 - x^2 - 52x + 160}{x - 5} \qquad 4) \ \frac{(x^2 + 5)(x^2 - 9)}{2x^2 + 3}$$

ii) Factor the following polynomials:

(1) $x^6 - y^6$

(2) $x^4 + 1 - 2x^2$

(3) $5a^4 + 5a$

(4) $a^4 - 3a^3 - 10a^2$

(5) $x^3 + x^2 - 4x - 4$

(6) $(x^3 - 8)(x + 2) + (x^3 + 8)(x - 2)$

(7) $2x^3 - 5x^2 - 28x + 15$

(8) $x^3 - 6x^2 - 2x + 7$

(9) $x^4 - 4x^3 - 4x^2 + 28x - 21$

(10) $x^2 + 17x + 16$

(11) $(x + 1)^2 - (x - 1)^2$

iii) Decompose in partial fractions:

$$\frac{x^2 + 2x - 1}{2x^3 + 3x^2 - 2x}; \qquad \frac{4x}{x^3 - x^2 - x + 1}; \qquad \frac{x}{(x - 2)(x^2 + 4)}; \qquad \frac{x^3 + x^2 + 1}{x(x^2 + 1)^2}.$$

Chapter 10
Real Functions of One Real Variable

The general subject matter of Chap. 3 is here narrowed down to **real-valued functions of one real variable** $f \in \mathbb{R}^{\mathbb{R}}$. An effective and comprehensive study of real functions is completed in analysis and differential geometry, fields that allow to appreciate both the global and the local (infinitesimal) features.

Pre-requisites: Remark 10.3 uses matrices. Examples 10.4, 10.7 rely on a tiny bit of calculus (limits+derivatives). The last part on bounded variations benefits from the knowledge of common series.

We'll adopt the convention that if $f(x) = y$ determines a functional relation, the natural domain will be denoted by $D \subseteq \mathbb{R}$ (explicitly or not), and the codomain will always be $\mathbb{R}$:

$$f : D \subseteq \mathbb{R} \to \mathbb{R}.$$

Examples 10.1 This selection exhibits functions intertwining analysis and arithmetics.

i) A function $a : D \subseteq \mathbb{N} \to \mathbb{R}$ is called **numerical sequence**. It's more common to indicate it by $\{a_n\}_{n \in \mathbb{N}}$, as a list of image elements $a_n := a(n)$. An example:

$$\{a_n\} = \left\{ \frac{n-1}{n} \,\middle|\, n \in \mathbb{N}^+ \right\} = \left\{ 0, \frac{1}{2}, \frac{2}{3}, \frac{3}{4}, \frac{4}{5}, \cdots \right\}.$$

Another example is the recursive sequence (cf. Sect. 14.4) of the **Fibonacci numbers**

$$f_0 = 0, \quad f_1 = 1, \quad f_n = f_{n-1} + f_{n-2}, \quad n \geqslant 2$$

They were introduced to Western European mathematics in 1202, although they had been known in India for donkey's years (at least 1500 years prior to Fibonacci). These numbers manifest themselves pervasively in nature, from the growth pattern of animals' horns and mollusc shells to phyllotaxis, and it is even witnessed in the

© The Author(s), under exclusive license to Springer Nature Switzerland AG 2021
S. G. Chiossi, *Essential Mathematics for Undergraduates*,
https://doi.org/10.1007/978-3-030-87174-1_10

shape of spiral galaxies. That's essentially because

$$f_n = \frac{\varphi^n - \hat{\varphi}^n}{\sqrt{5}}, \qquad \text{where } \varphi = \frac{1+\sqrt{5}}{2}, \quad \hat{\varphi} = -\frac{1}{\varphi} = \frac{1-\sqrt{5}}{2} \qquad (10.1)$$

in terms of the golden ratio (p. 173). Apropos φ, its defining equation $\varphi^2 = \varphi + 1$ gives $\varphi^n = f_n\varphi + f_{n-1}$ for $n > 1$, whence $\varphi^n = \varphi^{n-1} + \varphi^{n-2}$ (a nice mirror to $f_n = f_{n-1} + f_{n-2}$). For the proof of (10.1) see Example 14.9 ii). The Fibonacci progression appears in many contemporary works of art. The permanent installation 'The flight of numbers' by Mario Merz (1925–2003) displays the numbers in red neon light perched on the spire of Torino's landmark *Mole Antonelliana*.

ii) The **Dirichlet function** $\chi_{\mathbb{Q}} : \mathbb{R} \to \mathbb{R}$ is the characteristic function of the rationals:

$$\chi_{\mathbb{Q}}(x) = \begin{cases} 1, & x \in \mathbb{Q} \\ 0, & x \in \mathbb{R} \setminus \mathbb{Q} \end{cases}.$$

It's always brought up in calculus as the quintessential example of an everywhere discontinuous limit of continuous functions:

$$\chi_{\mathbb{Q}}(x) = \lim_{n \to \infty} \lim_{m \to \infty} \left(\cos(n!\pi x) \right)^{2m}.$$

iii) $\eta : \mathbb{R} \to \mathbb{R}, \quad \eta(x) = \dfrac{1}{\sigma\sqrt{2\pi}} e^{-x^2/2\sigma^2}$ is called *probability density* of the **Gaussian**, or **normal**, **distribution** (with expectation 0). See Example 10.7 iii). The constant σ is the *standard deviation* measuring the dispersion of the range of values relatively to the mean.

iv) The **prime-counting function** $\pi : \mathbb{R} \to \mathbb{N}$ assigns to x the number of primes nor exceeding x. For example $\pi(10) = 4$, $\pi(100) = 25$, $\pi(10^5) = 9,592$. Riemann found the exact closed-form formula for it, but more relevantly, its asymptotic growth is governed by the

prime-number theorem

$$\lim_{x \to \infty} \frac{\pi(x)}{\frac{x}{\log x}} = 1,$$

proved by Hadamard and de la Vallée-Poussin in 1896 (by way of Theorem 6.26, interestingly). The behaviour of π is also that of the integral logarithm $\mathrm{Li}(x) := \displaystyle\int_2^x \frac{dt}{\log t}$ (as shown by Schoenfeld, assuming the Riemann hypothesis).

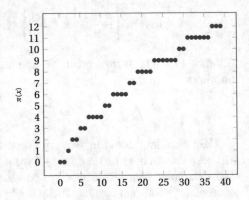

To evaluate $\pi(x)$ when x is small-ish, trial divisions such as *Eratosthenes' sieve* (step 0 in ☞ *sieve theory*) work reasonably. Legendre discovered a remarkable expression, which we mention because it involves (6.5):

$$\pi(x) = \lfloor x \rfloor - \sum_i \left\lfloor \frac{x}{p_i} \right\rfloor + \sum_{i<j} \left\lfloor \frac{x}{p_i p_j} \right\rfloor - \sum_{i<j<k} \left\lfloor \frac{x}{p_i p_j p_k} \right\rfloor + \cdots$$

for distinct primes $\{p_\alpha\}$ that don't divide x. The literature on $\pi(x)$ is more than one can read in many lifetimes.

v) In the same circle of ideas, *Euler's totient function* $\phi : \mathbb{R} \to \mathbb{N}$, also known as 'phi' function, counts positive integers $p \leqslant x$ such that $\gcd(p, \lfloor x \rfloor) = 1$. The name 'totient' was coined by Sylvester in 1879. Its explicit expression

$$\phi(x) = \lfloor x \rfloor \cdot \prod_{\substack{p \text{ prime} \\ p \mid \lfloor x \rfloor}} \left(1 - \frac{1}{p}\right) \tag{10.2}$$

can be proved via (6.6), but is only apparently simple. The totient function shows up in the RSA encryption algorithm (see Remark 6.27 for notation) because the latter is based on solving

$$x^{\phi(n)} \equiv_n 1$$

(*Fermat's little theorem* when n is prime, see Exercise 14.6 ii)). This equation implies that the inverse of $c(x) = x^e (\bmod\ n)$ is the function $m(y) = y^d (\bmod\ n)$, where the private key d is the inverse of e modulo $\phi(n)$. Hence computing $\phi(n)$ without knowing the factorisation of n is as hard as finding d. The private-key owner knows the factorisation, but only n is publicly disclosed. Given the notorious difficulty to factor large numbers it's practically guaranteed that no one else will know the factorisation, and therefore crack the encryption.

Interestingly, ϕ is *multiplicative*: $\phi(1) = 1$ and $\phi(xy) = \phi(x)\phi(y)$ whenever x, y are coprime integers (the proof goes through Theorem 6.22).

Exercise Suppose a map $f \colon \mathbb{Z}^+ \to \mathbb{R}$ is multiplicative. Show $f(a)f(b) = f(\gcd(a, b)) f(\operatorname{lcm}(a, b))$ for any $a, b \in \mathbb{Z}^+$.

After this long break let's return to general matters. Once more, it has to be stressed that the natural domain is an integral part of a function, and it is absolutely imperative to determine it in order to work with functions. There are three common conditions that must be imposed when we determine a domain:

① every denominator must be $\neq 0$

     reason: $\mathbb{R}$ is a group under multiplication, so 0 is the only non-invertible number.

② every radicand (for even roots only) must be $\geqslant 0$

     reason: the map $\zeta \mapsto \zeta^{2m}$, $m \in \mathbb{N}^*$ has image $[0, +\infty)$.

③ every logarithm's argument must be > 0

     reason: the map $\exp_a$, $a > 0, a \neq 1$ has image $\mathbb{R}^+$.

Examples

$$f(x) = \frac{1}{x-3} + x^{x^x} \qquad \text{has domain } (\mathbb{R} \setminus \{3\}) \cap (0, +\infty) = (0,3) \cup (3, +\infty);$$

$$g(x) = \sqrt{x^2 - 5} - \arctan(x/\pi) \qquad \text{has domain } (-\infty, -\sqrt{5}] \cup [\sqrt{5}, +\infty);$$

$$h(x) = \sin\left(\log_7(6-x)\right) - (x-1)^{1/2} \text{ has domain } (-\infty, 6) \cap [1, +\infty) = [1, 6)$$

Exercises

i) Find the domains of

$$f(x) = \frac{2x}{x^2 - 1}, \qquad l(x) = \sqrt{\frac{2x-1}{x^2-3}}, \qquad q(x) = \frac{1}{\sqrt{|3x-4|}}.$$

ii) Are the maps $h_1(x) = \sqrt{\dfrac{x-1}{x-2}}$, $h_2(x) = \dfrac{\sqrt{x-1}}{\sqrt{x-2}}$ equal?

Prove that h_2 is a restriction of h_1.

Whilst calculating the domain might be reasonable, determining a map's image can prove very hard even for simple expressions, as in the case of $\alpha : \mathbb{R} \to \mathbb{R}, \alpha(x) = x^4 + x^3 + x^2 + x + 1$. The good news is that analytical methods easily show that

$$\text{Im}(\alpha) = \left[\frac{-1695 - (-135 + 20\sqrt{6})^3 \sqrt[3]{135 + 60\sqrt{6}} + (-49 + 24\sqrt{6})^3 \sqrt[3]{(135 + 60\sqrt{6})^2}}{2304}, +\infty \right).$$

Along the same lines, establishing whether a map is invertible is typically rather delicate. At times analysis help answering the question, via the *implicit function theorem*.

For instance, the definition is useless for deciding whether λ : $\mathbb{R} \to \mathbb{R}$, $\lambda(x) = x - \sin x$ is 1-1. Analytical tools, though, allows to verify it in a simple way. As its image is $\mathbb{R}$ furthermore, λ is invertible. That said, inverting it explicitly is far from trivial:

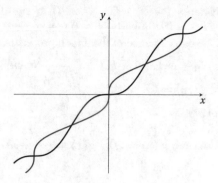

$$\lambda^{-1}(x) = \sum_{n=1}^{\infty} \frac{x^{n/3}}{n!} \lim_{\theta \to 0} \left(\frac{d^{n-1}}{d\theta^{n-1}} \left(\frac{\theta}{\sqrt[3]{\theta - \sin\theta}} \right)^n \right).$$

Writing an explicit formula for the inverse map can be challenging, even in the case of the simplest functions, like polynomial ones. Try with $f(x) = x^3 + 2x^2 + x - 3$. The issue boils down, in practice, to solving the cubic equation $x^3 + 2x^2 + x - 3 - y = 0$ for x.

10.1   The Algebra of Real Functions

The space of maps $f : D \to \mathbb{R}$ comes equipped with operations defined pointwise and inherited from the operations of real numbers.

Definition 10.2 Let $f, g \in \mathbb{R}^D$ be maps defined on the same domain and λ a real number. One calls

  sum: the map $(f + g)(x) := f(x) + g(x)$,
  product: the map $(fg)(x) := f(x)g(x)$,
  product by scalars: the map $(\lambda f)(x) := \lambda f(x)$.

By definition, then, $\mathrm{Dom}(\lambda f) = \mathrm{Dom}(f)$, whilst

$$(\dagger) \qquad \mathrm{Dom}(f + g), \ \mathrm{Dom}(fg) \subseteq \mathrm{Dom}(f) \cap \mathrm{Dom}(g).$$

Proposition *The above operations make the space $\mathbb{R}^D$ of real-valued maps of one real variable a commutative $\mathbb{R}$-module with unit.*

Proof Since operations are defined pointwise, the additive neutral element is the zero map $0(x) = 0$, the unit is the constant map $1(x) = 1$, for all $x \in \mathbb{R}$. All properties reduce to verifications over the field $\mathbb{R}$, so we leave them as exercise. $\square$

Notation: we'll write $f \equiv 0$ to signify f equals the zero map, i.e. $\forall x \in D \; f(x) = 0$. Similarly, $f \not\equiv 0$ means $\exists x \in D$ such that $f(x) \neq 0$.

One consequence of the above proposition is that if $f : \mathbb{R} \to \mathbb{R}$ is not identically zero, its reciprocal $\dfrac{1}{f}(x) := \dfrac{1}{f(x)}$ is well defined on $\mathrm{Dom}\left(\dfrac{1}{f}\right) = \mathrm{Dom}(f) \setminus \{x_0 \mid f(x_0) = 0\}$.

Examples

i) Consider $f(x) = \sqrt{x}$, $g(x) = \dfrac{x\sqrt{x}}{1-x}$. Then

$$(f+g)(x) = \frac{\sqrt{x}}{1-x} \qquad (fg)(x) = \frac{x^2}{1-x} \qquad (f/g)(x) = \frac{1-x}{x}.$$

It's clear that the domain resulting from algebraic operations cannot be determined beforehand from the knowledge of the constituent functions' domains. Note that this doesn't violate (†): the map $t(x) = \dfrac{x^2}{1-x}$ has $\mathrm{Dom}(t) = \mathbb{R} \setminus \{1\}$, while the product $(fg)(x) = \dfrac{x^2}{1-x}$ is well defined on $[0, +\infty) \setminus \{1\}$. That is to say: $t = fg$ only on $[0, +\infty) \setminus \{1\}$.

ii) Both maps $h(x) = \sqrt{x}$ and $k(x) = -\sqrt{x}$ have domain $[0, +\infty)$, as does their sum $h + k$. Clearly, though, the constant map 0 is defined everywhere, even for $x < 0$.

These examples should convince us even more that a function is not just a law/a formula.

Taking into account the composition operation:

Theorem $(\mathbb{R}^{\mathbb{R}}, +, \circ, \cdot)$ *is an associative $\mathbb{R}$-algebra with unit $\mathbb{1}_{\mathbb{R}}$, non-commutative and of infinite dimension.*

Proof The only property we'll comment on is the last one. Polynomial functions form a proper subset of real functions, and we know from Theorem 9.1 that the subspace of polynomials already is infinite dimensional. Hence *a fortiori* $\mathbb{R}^{\mathbb{R}}$ doesn't have finite bases. □

While we're counting, we also recall that $\mathrm{card}\,(\mathbb{R}^{\mathbb{R}}) = 2^{\aleph_1}$, by Proposition 7.26. Regarding orderings, the algebra of functions $\mathbb{R}^D$ is a poset:

Definition Let $g, f : D \subseteq \mathbb{R} \to \mathbb{R}$ be maps defined on a common domain D. We'll write $f \leqslant g$ whenever $f(x) \leqslant g(x)$ for every $x \in D$.

Geometrically, the graph of f will stay below the graph of g at each point.

There is no significant ordering on $\mathbb{R}^{\mathbb{R}}$ since maps defined over different domain cannot be reasonably compared.

Exercises Let $f : D = \mathbb{R} \to \mathbb{R}$ be a map. Prove the following facts.

i) If $f \circ g = g \circ f$ for any $g \in \mathbb{R}^{\mathbb{R}}$, show $f = \mathbb{1}_{\mathbb{R}}$. (*Hint*: take g constant.)

ii) (challenge) Suppose f is *additive* and *multiplicative*, i.e.

$$f(x + y) = f(x) + f(y), \quad f(xy) = f(x)f(y) \quad \forall x, y \in \mathbb{R}.$$

Prove f is either the zero map or the identity $\mathbb{1}_{\mathbb{R}}$.

(*Hint*: if $f \not\equiv 0$, show $f(1) = 1$ and $f\big|_{\mathbb{Q}} = \mathbb{1}_{\mathbb{Q}}$. Then prove that $f > 0$ if $x > 0$, and that f is increasing. Conclude using the density of $\mathbb{Q} \subseteq \mathbb{R}$.)

10.2   Graphs and Curves

By identifying $\mathbb{R}$ with a **real line** (which is **oriented**, since $\mathbb{R}$ is ordered)

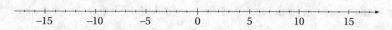

we may then identify the Cartesian product $\mathbb{R} \times \mathbb{R} =: \mathbb{R}^2$ with the **Cartesian plane Oxy**, determined by a pair of orthogonal real lines crossing each other at the 0 of both. In this way a pair $(x_P, y_P) \in \mathbb{R}^2$ corresponds to a point P on the plane, and the numbers x_P, y_P are called the **Cartesian coordinates** of P: x_P is the aptly named **x-coordinate** (or *abscissa*), y_P the **y-coordinate** (or *ordinate*), cf. Theorem 18.6. The values of variable x belong to the horizontal line (**x-axis**), while the values of variable y are placed on the vertical line (**y-axis**) (Fig. 10.1).

A thorough explanation will come immediately after Thales' theorem 18.5. Geometrically, the coordinates of P are obtained by orthogonal projection of P

Fig. 10.1 Cartesian plane

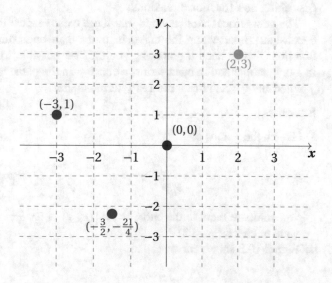

Fig. 10.2 Graph of
$\omega(x) = 2x^3 - 3x^2 - 3x + 2$

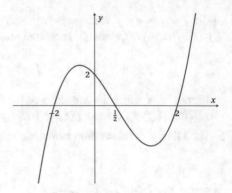

on the axes: $x_P = \mathrm{pr}_1(P)$, $y_P = \mathrm{pr}_2(P)$, where $\mathrm{pr}_i : \mathbb{R}_1 \times \mathbb{R}_2 \to \mathbb{R}_i$, $i = 1, 2$ are
the projections on the factors, cf. (3.3).

To understand the behaviour of a map $f : D \to \mathbb{R}$ we can visualise it as a subset
of the plane (a curve, in the cases of concern). Although this curve doesn't prove
anything (after all, it's only a drawing), it still is quite useful for indications.

Definition The **graph** of a map $f : D \to \mathbb{R}$ is the subset

$$\mathrm{graph}(f) := \left\{ (x, f(x)) \in \mathbb{R}^2 \mid x \in D \right\} \subseteq D \times f(D).$$

For instance: $\omega : \mathbb{R} \to \mathbb{R}$, $\omega(x) = 2x^3 - 3x^2 - 3x + 2$ (see Fig. 10.2) has
$\mathrm{Dom}(\omega) = \mathbb{R}$ (the entire x-axis). The point $(0, 2)$ belongs to $\mathrm{graph}(\omega)$ because
$\omega(0) = 2$, while $(1, 0) \notin \mathrm{graph}(\omega)$ since $\omega(1) \neq 0$.

It was mentioned in Example 3.1, in passing, that not every curve in the plane is
the graph of a function, but certainly it is the graph of a relation. Here are curves
describing non-functional relations:

The geometrical explication is that there exist vertical lines $\{(x, y) \in \mathbb{R}^2 \mid x = c \text{ constant}\}$ intersecting the curve in more than one point, so if there existed a
function f with such a graph, it wouldn't be possible to define the image $f(c)$
in a unique manner. In other terms, a curve γ in the plane is the graph of a function
f when any vertical line r crosses γ at most once.

Exercises

i) Sketch the graphs of

$$y = \sqrt{1 - x^2}, \qquad y = -\sqrt{1 - x^2}$$

and compare them to the curve $\gamma = \{(x, y) \in \mathbb{R}^2 \mid x^2 + y^2 = 1\}$. (You might
want to check Fig. 3.2).

ii) Figure 10.3 shows curves called:

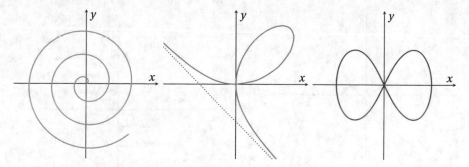

Fig. 10.3 Curves that don't represent functions globally

$$x(t) = \tfrac{1}{3}t\cos(t), \ y(t) = \tfrac{1}{3}t\sin(t), \ t > 0 \qquad \text{Archimedes' spiral}$$
$$x^3 + y^3 - 3xy = 0 \qquad\qquad\qquad\qquad \text{Descartes' folium}$$
$$\big((x-a)^2 + y^2\big)\big((x+a)^2 + y^2\big) = a^4, \ a \neq 0 \ \text{Bernoulli's lemniscate.}$$

Show that each curve doesn't define a unique function $y(x)$ defined globally (on the entire $\mathbb{R}$).

The issue 'local vs global' is sorted out by the *implicit function theorem*, which dictates how and when, for equation $F(x, y) = 0$, there exists a function $y = f(x)$ such that $F(x, f(x)) = 0$.

For example, the locus of points $(x, y) \in \mathbb{R}^2$ satisfying $F(x, y) = y^5 + 16y - 32x^3 + 32x = 0$ is a curve in the plane. If we plot it, we might suspect that the locus in question is the graph of some function $y = f(x)$. This turn out to be indeed the case (and almost a miracle), for using infinitesimal calculus one proves that the polynomial map $y \mapsto y^5 + 16y - 32x_0^3 + 32x_0$ has a unique zero y_0, for any chosen $x_0 \in \mathbb{R}$. The required function is then defined by the assignment $y_0 = f(x_0)$.

Starting now we'll make a number of considerations regarding graphs.

- The naive idea of curve shouldn't make us think there cannot be interruptions (i.e., that the curve is 'continuous'). A graph may well be disconnected, which often happens when the domain D is not a single interval, or if we are considering piecewise-defined maps, as in Fig. 10.4.
- The intersections of graph(f) with the x-axis are given by the roots of f. The equation $f(x) = 0$, in fact, can be written as a system $\begin{cases} y = f(x) \\ y = 0 \end{cases}$.
- The graph might reveal whether a map in injective: if graph(f) intersects an arbitrary horizontal line (the graph for the constant map $y = a, a \in \mathbb{R}$) in 1 point at most then f is 1-1. If there are more intersections, f is not one-to-one.

Examples

i) The absolute value map abs: $x \mapsto |x|$ is not 1-1 (Fig. 10.5): any straight line $y = c > 0$ meets the graph in the two points $(c, c) \neq (-c, c)$

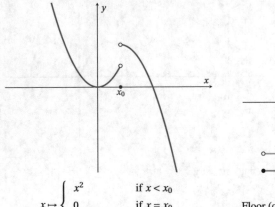

$$x \mapsto \begin{cases} x^2 & \text{if } x < x_0 \\ 0 & \text{if } x = x_0 \\ 2-(x-1)^2 & \text{if } x > x_0 \end{cases}$$

Fig. 10.4 Piecewise maps

Fig. 10.5 abs: $x \mapsto |x|$

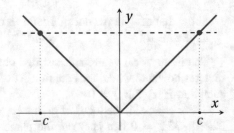

Floor (orange) & ceiling (cyan) functions (3.4).

$$\text{abs}^{-1}(c) = \{c, -c\} \quad \text{i.e.} \quad |c| = |-c|.$$

ii) Polynomial functions $f(x) = c_n x^n + \ldots + c_1 x + c_0$ of degree $n > 1$ are expected to be non-injective, since the graph will in general have crests and troughs (Fig. 11.1 or Fig. 10.3).

• The effect of translations and reflections on graphs can be pictured easily (Fig. 10.6). Starting from graph(f) we may plot the graphs of:

$y = f(x - p)$ by shifting horizontally by p (to the left if $p > 0$)
$y = f(x) + q$ by shifting vertically by q (above, if $q > 0$)
$y = f(x - p) + q$ by combining the previous operations
$y = f(-x)$ by reflecting about the y-axis
$y = -f(x)$ by reflecting about the x-axis
$y = -f(-x)$ by reflecting about the origin (combination of previous two).

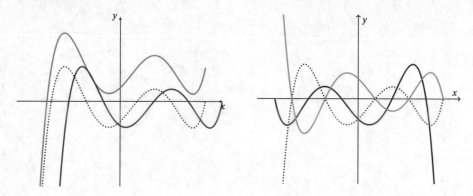

Fig. 10.6 (Left) Translations along the axes: vertically (brown), horizontally (blue). (Right) Axial reflections: in y (purple), in x (cyan)

Remark 10.3 All of the above are 'rigid' motions of the Euclidean plane $\mathbb{R}^2$ (cf. Definition 18.10), and correspond to linear transformations $\binom{x}{y} \mapsto A\binom{x}{y} + b$ of the coordinate system:

Movement	A	b
General translation	$\begin{pmatrix} 1 & 0 \\ 0 & 1 \end{pmatrix}$	$\begin{pmatrix} p \\ q \end{pmatrix}$
Reflection about y-axis	$\begin{pmatrix} -1 & 0 \\ 0 & 1 \end{pmatrix}$	$\begin{pmatrix} 0 \\ 0 \end{pmatrix}$
Reflection about x-axis	$\begin{pmatrix} 1 & 0 \\ 0 & -1 \end{pmatrix}$	$\begin{pmatrix} 0 \\ 0 \end{pmatrix}$
Reflection about origin	$\begin{pmatrix} -1 & 0 \\ 0 & -1 \end{pmatrix}$	$\begin{pmatrix} 0 \\ 0 \end{pmatrix}$

Example Take the parabola of equation $y = x^2$ and shift it horizontally by -1: the graph moves leftwards, and the equation becomes $y = (x + 1)^2$.

Move the latter upwards by $q = 2$: the new equation reads $y = (x+1)^2 + 2 = x^2 + 2x + 3$.

Now symmetrise the graph about the origin to obtain $y = -\left((-x)^2 + 2(-x) + 3\right) = -x^2 + 2x - 3$.

Exercise Describe the motions transforming the parabola of equation $y = x^2 + bx + c$ into the parabola of equation $y = x^2$, for generic numbers $b, c \in \mathbb{R}$.

- It's immediate to plot the graphs of composites with the absolute value map, by exploiting the latter's properties. Operationally (Fig. 10.7):

  a) $f \circ \text{abs} : x \mapsto f(|x|)$ coincides with $y = f(x)$ on the positive x-semi-axis, and with its reflection (about y) on the negative semi-axis. Therefore it turns out being symmetric about the vertical axis ($f \circ \text{abs}$ is an even map, see Definition 10.8).

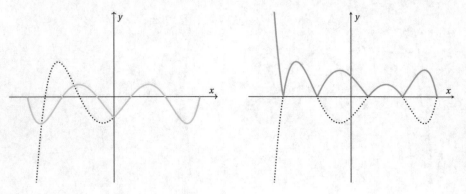

Fig. 10.7 Composing with the absolute value: $y = f(|x|)$ and $y = |f(x)|$

Fig. 10.8 The graph (cyan)
of $f(x) = x^3 + 2x^2 - 4$ and
its reflection (red),
corresponding to the *curve*
$x = y^3 + 2y^2 - 4$

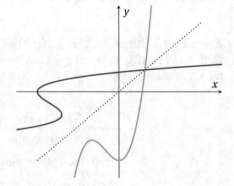

b) $|f| := \mathrm{abs} \circ f : x \mapsto |f(x)|$: the graph's positive part (above the x-axis) stays
the same, whilst the negative part (below the x-axis) gets reflected above.

- Suppose f is invertible, but we're not able to determine f^{-1} explicitly in order
  to plot it. Still, we can sketch the graph of f^{-1} from graph(f), based on the
  following observation:

$$y = f(x) \iff x = f^{-1}(y)$$
$$(x, y) \in \mathrm{graph}(f) \iff (y, x) \in \mathrm{graph}(f^{-1})$$

In the plane, therefore, the graphs of f and f^{-1} are symmetric about the straight
line $y = x$ (graph of $\mathbb{1}_{\mathbb{R}}$).

The function $f(x) = x^3 + 2x^2 - 4$ isn't injective, for instance because $f(0) = -4 = f(-2)$. This can be guessed from the reflected red curve in Fig. 10.8, which
does *not* represent a function g: the value $g(-4)$ cannot be defined, for it would
have to satisfy $0 = g(f(0)) = g(-4) = g(f(-2)) = -2$ ⚡.

The problem disappears if we restrict f to a smaller domain, in particular to
the intervals $(-\infty, -4/3]$, $[-4/3, 0]$ and $[0, +\infty)$ separately, on which f becomes
monotone. Thus the restrictions become invertible, cf. Definition 10.5.

Fig. 10.9 Graphs of $y = \dfrac{2x+3}{x+4}$ (brown) and its inverse (blue)

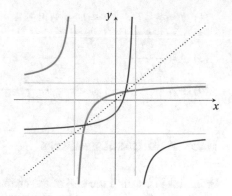

Example 10.4 Take real numbers $\alpha, \delta, \beta, \gamma$ such that $\gamma \neq 0$ and $\Xi := \alpha\delta - \beta\gamma \neq 0$. The graph of

$$h(x) = \frac{\alpha x + \beta}{\gamma x + \delta}$$

is called a **equilateral hyperbola**. Imposing $\gamma \neq 0$ excludes the case where h is affine (Fig. 10.9).

Some of its features are:

- the domain $\mathbb{R} \setminus \{-\delta/\gamma\}$ suggest the presence of a vertical asymptote $x = -\dfrac{\delta}{\gamma}$;

- it can be proved that h is monotone:

$$\text{increasing} \iff \Xi > 0, \quad \text{decreasing} \iff \Xi < 0;$$

- polynomial division

$$\frac{\alpha x + \beta}{\gamma x + \delta} = \frac{\alpha}{\gamma} - \frac{\Xi}{\gamma(\gamma x + \delta)} \tag{10.3}$$

reveals graph(h) is the shift of the hyperbola $xy = -\dfrac{\Xi}{\gamma^2}$ under the translation $\begin{pmatrix} x \\ y \end{pmatrix} \mapsto \begin{pmatrix} x + \delta/\gamma \\ y - \alpha/\gamma \end{pmatrix}$. This motion is determined by saying the asymptotes' intersection is the origin of the translated coordinate system.

Exercises

i) Prove that $h(x) = \dfrac{\alpha x + \beta}{\gamma x + \delta}$ is invertible (on its domain) if and only if $\Xi \neq 0$.

   Compute the inverse map explicitly and conclude the image of h is $\mathbb{R} \setminus \{\alpha/\gamma\}$.

   Show that $y = \dfrac{\alpha}{\gamma}$ is the horizontal asymptote of h (also confirmed by (10.3)).

ii) Sketch the graphs of the following maps

(1) $y = |-x|$ $y = |-x + 3| - 2$ $y = -|-x + 3| + 2$

(2) $y = |2x - 1|$ $y = 2|x| - 1$ $y = |2|x| - 1|$

(3) $y = \dfrac{1}{|x - 4|}$ $y = \dfrac{x - 3}{x - 4} = 1 + \dfrac{1}{x - 4}$

(4) $y = \left| \dfrac{-2x + 3}{x - 5} \right|$ $y = \dfrac{-2|x| + 3}{|x| - 5}$.

10.3   Additional Features

From this point until the end of the chapter we will explore some properties of real functions, without the pretence of being exhaustive.

10.3.1   Monotonicity Conditions

Definition 10.5 A map $f: D \to \mathbb{R}$ is said

Increasing	if $x_1 < x_2 \Longrightarrow f(x_1) < f(x_2)$	(f preserves the strict order)
Decreasing	if $x_1 < x_2 \Longrightarrow f(x_1) > f(x_2)$	(inverts the strict order)
Non-decreasing	if $x_1 < x_2 \Longrightarrow f(x_1) \leqslant f(x_2)$	(weakly preserves the order)
Non-increasing	if $x_1 < x_2 \Longrightarrow f(x_1) \geqslant f(x_2)$	(weakly inverts the order)

for all $x_1, x_2 \in D$. When either increasing or decreasing, f is called **monotone**.

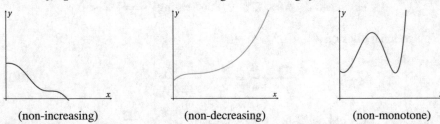

(non-increasing)   (non-decreasing)   (non-monotone)

Corollary *A monotone function is one-to-one.*

Proof Exercise.   □

Exercise Explain why the definition of increasing map f can be taken to be

$$\forall x_1, x_2 \in D \quad x_1 < x_2 \iff f(x_1) < f(x_2).$$

Ditto for decreasing.

Examples

i) As already mentioned, a polynomial map $f(x) = c_n x^n + \ldots + c_1 x + c_0$ of degree $n > 1$ will, in general, be increasing on certain subintervals of the domain and decreasing on others.

   Rare examples of monotone polynomial functions are those of degree one (whose graphs are non-horizontal straight lines), or $f(x) = x^{2k+1} + cx$, for $k \in \mathbb{N}$ and $c > 0$.

ii) Monotonicity doesn't imply invertibility; and vice-versa, invertibility doesn't force monotonicity: the map $f(x) = \begin{cases} 1 - x, & 0 \leqslant x \leqslant 1 \\ x, & 1 < x \leqslant 2 \end{cases}$ is bijective but not monotone.

   Another example is $g : [0, 1] \to [0, 1]$ defined by

$$x \mapsto \begin{cases} x, & x \in \mathbb{Q} \cap [0, 1] \\ 1 - x, & x \in [0, 1] \setminus \mathbb{Q} \end{cases}.$$

iii) The sum of monotone maps may not be monotone: $f(x) = \sin x + 2x$ is increasing, $g(x) = \sin x - 2x$ is decreasing, but $f + g$ is neither.

Exercises Prove that

i) f is increasing if and only if $\dfrac{f(x_1) - f(x_2)}{x_1 - x_2} > 0$ for any $x_1 \neq x_2 \in \text{Dom}(f)$;

ii) f increasing $\Longleftrightarrow -f$ decreasing;

iii) f increasing $\Longleftrightarrow f^{-1}$ increasing (assuming it's bijective);

iv) f, g increasing $\Longrightarrow f \circ g$ increasing (where defined);

v) Prove the corresponding statements for decreasing maps.

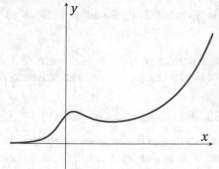

Establishing the monotonicity might not be immediate, as in the case of

$$f(x) = \frac{2^x}{x^2 + 1}.$$

Calculus provides some tools that eschew having to rely on the definition.

Definition 10.6 A map $f : D \to \mathbb{R}$ is called **convex** if

$$f(\lambda x_1 + \mu x_2) \leqslant \lambda f(x_1) + \mu f(x_2)$$

for all $x_1, x_2 \in D$ and every $\lambda, \mu \geqslant 0$ with $\lambda + \mu = 1$. We call it **concave** (Fig. 10.10) if $-f$ is convex.

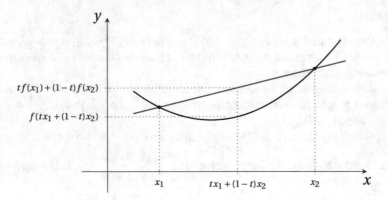

Fig. 10.10 A convex map

Geometrically the graph of a convex map always lies below the segment $t P_1 + (1 - t) P_2$, $t \in [0, 1]$ joining any two points $P_1, P_2 \in \text{graph}(f)$. Roughly, convexity conveys the idea of a smooth depression. (Elsewhere, we'd call a convex map 'sublinear'.)

Examples The maps $x \mapsto |x|$, $x \mapsto e^x$, $x \mapsto -\log(x)$ are convex. A quadratic map $x \mapsto ax^2 + bx + c$ is convex if and only if $a \geqslant 0$.

Exercises If f, g are convex maps, prove that

i) $f + g$ is convex, αf is convex for any $\alpha \geqslant 0$
    (the subspace of convex maps is a cone inside $\mathbb{R}^D$);
ii) the mirror $y = f(-x)$ is convex;
iii) the composite $f \circ g$ (where defined) is convex;
iv) f convex $\iff$ $x \mapsto \dfrac{f(x) - f(x_0)}{x - x_0}$ is non-decreasing for any $x \in \text{Dom}(f) \setminus \{x_0\}$.

Definition Consider a map $f : D \to \mathbb{R}$ and a point $x_0 \in D$ in its domain. If f is convex on the interval $(x_0 - \epsilon, x_0)$ and concave on $(x_0, x_0 + \epsilon)$ for a suitable $\epsilon > 0$, we say x_0 is an **inflection** (point).

Examples 10.7 Using infinitesimal methods the following can be proved.

i) Every parabola is either convex or concave (hence it admits no inflection).
ii) A generic polynomial map of degree > 2 has at least one inflection (here, 'generic' means the expectation is for inflections to exist. But there are convex polynomials, like $x^4 - x$ in Fig. 10.12).
iii) The map $\eta(x) = \dfrac{1}{\sigma\sqrt{2\pi}} e^{-x^2/2\sigma^2}$, $\sigma > 0$, has two symmetric inflections at $x = \pm\sigma$. Its graph is called **bell curve** due to its suggestive profile, Fig. 10.11. Of the values drawn from a normal distribution (cf. Examples 10.1), about 2/3 lie in the concavity interval $[-\sigma, \sigma]$, and about 95% lie in $[-2\sigma, 2\sigma]$. For $|x| > 2\sigma$ the curve tails off (☞ *probability theory*).

Fig. 10.11 For whom the bell tolls (it tolls for thee)

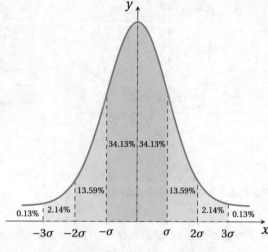

Fig. 10.12 The black map $y = -x^3 + 3x^2 - 2x + 1$ inflects at $x = 1$, while the red map $y = x^4 - x$ is everywhere convex

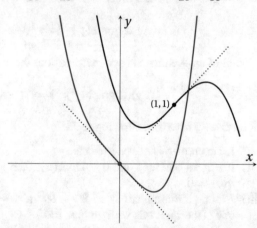

iv) A non-critical point is an inflection if (and only if) the intersection multiplicity between the graph and the tangent line (at the inflection) is at least 3 and odd (black curve in Fig. 10.12).

v) The trigonometric functions sine, cosine, tangent (Sect. 12.4) have infinitely many inflections, located at the crossings with the x-axis.

10.3.2 Symmetry Conditions

Definition 10.8 A map $f : D \to \mathbb{R}$ is called **odd** if $f(x) = -f(-x)$, and **even** if $f(x) = f(-x)$, for all $x \in D$.

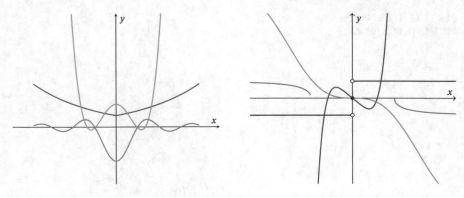

Fig. 10.13 Even and odd functions

The graph of an odd function is therefore symmetric about the origin

$$(x, y) \in \text{graph}(f) \iff (-x, -y) \in \text{graph}(f),$$

that of an even map is symmetric about the y-axis (Fig. 10.13)

$$(x, y) \in \text{graph}(f) \iff (-x, y) \in \text{graph}(f).$$

Exercises Prove the following facts.

i) An even map is not one-to-one.
ii) If u is an odd map and $0 \in \text{Dom}(u)$, then $\text{graph}(u)$ passes through the origin: $u(0) = 0$.
iii) Among affine functions $a : \mathbb{R} \to \mathbb{R}$, $a(x) = mx + q$, which are even? Which odd? (Interpret this with graphs, too.)
      Can a map be both even and odd simultaneously?

Examples Even maps include: (examples)

- polynomial functions only containing terms
  of even degree x^{2k}: $y = x^4 - 3x^2 + 2$
- functions in the variable $|x|$: $y = e^{|x|/3}$
- products of two odd maps: $y = -\frac{1}{x} \sin(3x)$
- products of two even maps: $y = (x^2 - 5x^6) \log |x|$
- composites of two even maps, or one
  odd and one even: $y = 3 \sin(x^2)$
- $y = \cosh x$, $y = \cos x$
- the *sign function* $y = \text{sgn}(x)$.

Odd examples:

- polynomial functions only containing terms
  of odd degree x^{2k+1}: $y = \dfrac{1}{2}x^5 - x$

- products of an odd map and an even one: $y = x^{-2}\sin\left(\dfrac{10}{x}\right)$
- inverses of odd maps: $y = \arctan(x)$
- composites of odd maps: $y = \frac{1}{3}\sinh\left(\frac{x^3}{2}\right)$
- $y = \sinh x$, $y = \sin x$.

But it should be evident that the vast majority of functions won't be neither even nor odd (a generic polynomial, for instance).

Proposition 10.9 *For any function f*

$$f^e(x) := \frac{f(x) + f(-x)}{2} \quad \text{is even and} \quad f^o(x) := \frac{f(x) - f(-x)}{2} \quad \text{is odd.}$$

Therefore, any map f can be written as a sum of an even map and an odd map $f = f^e + f^o$, in a unique way.

Proof The expressions for f^e, f^o add up to f (proving the existence of the decomposition). As regards uniqueness, suppose $f = g + h$ is another even/odd sum. Then $2f^e(x) = f(x) + f(-x) = 2g(x)$ because g is even, and so $h = f - g = f - f^e = f^o$. $\qquad\square$

Example A polynomial map $x \mapsto q(x) = \sum_{j=0}^{n} a_j x^j$ splits according to the above proposition if we separate monomials with even/odd exponent.
    Concretely, for $q(x) = 3x^5 - 14x^4 - x^2 + 7x - 3$ we have

$$q^e(x) = -14x^4 - x^2 - 3, \qquad q^o(x) = 3x^5 + 7x.$$

Exercises

i) Prove: f even and odd $\Longrightarrow f \equiv 0$.
ii) Show: f odd $\Longrightarrow |f|$ even.
iii) Show that the composite $f \circ \beta$ of an arbitrary function f and an even map β is even.
iv) Prove an even map f can be put in the form $f = g \circ \text{abs}$, i.e. $f(x) = g(|x|)$, $\forall x \in \text{Dom}(f)$, for some function g.
v) According to Proposition 10.9 we have

$$2\sqrt{x} = \underbrace{\left(\sqrt{x} + \sqrt{-x}\right)}_{a(x)} + \underbrace{\left(\sqrt{x} - \sqrt{-x}\right)}_{b(x)}.$$

The maps a, b have domain $\{0\}$, while $2\sqrt{x}$ is defined on $[0, +\infty)$. Does this violate the proposition?

Fig. 10.14 Periodic functions

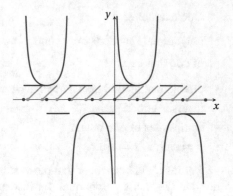

What about $\underbrace{2\log(x)}_{\text{Dom}=\mathbb{R}^+} = \underbrace{\left(\log(x) + \log(-x)\right)}_{\text{Dom}=\varnothing} + \underbrace{\left(\log(x) - \log(-x)\right)}_{\text{Dom}=\varnothing}$?

(The justification is elementary, but there is a deep explanation in ☞ *complex analysis*).

Definition A (non-constant) map $f \colon \mathbb{R} \to \mathbb{R}$ is said to be **T-periodic** if $T > 0$ is the smallest real number such that

$$f(x) = f(x + T) \quad \text{for all } x.$$

A T-periodic function then satisfies $f(x) = f(x + \mathbb{Z}T)$ for all x. Necessarily, the graph is invariant under horizontal translations $x \mapsto x + kT, k \in \mathbb{Z}$.

The study of periodic functions can therefore be confined to a domain interval of length T.

Examples

- Every trigonometric function:

  $T = 2\pi$ for: $\quad y = \sin x, \; y = \cos x, \; y = \dfrac{1}{\cos x}, \; y = \dfrac{1}{\sin x}$ (blue, Fig. 10.14)

  $T = \pi$ for: $\quad y = \tan x, \; y = \cot x$.

- The integer part $y = \lfloor x \rfloor$ ($T = 1$), or the *sawtooth wave* $y = x - \lfloor x \rfloor$ (olive, Fig. 10.14).
- *Square waves*, e.g. $y = \mathrm{sgn}(\sin 2x)$ (red, Fig. 10.14).
- The triangle wave $y = \left| x \pmod 4 - 2 \right| - 1$.
- *Cycloids*, described parametrically by

  $$x(\tau) = a\tau - b\sin(\tau), \quad y(\tau) = a - b\cos(\tau), \quad \tau \in \mathbb{R}$$

  with $a \geqslant b > 0$, Fig. 10.15.

Fig. 10.15 Cycloids and their ilk

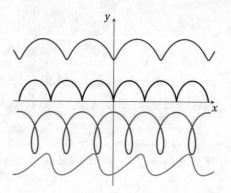

The black cycloid has $a = b = 1$, the red curve is a minor generalisation: $x(\tau) = 3\tau - \sin(2\tau)$, $y(\tau) - 2 = 3 - \cos(2\tau)$. The orange curve is not a cycloid, but is still described by similar equations: $x(\tau) = 3\tau - \sin(2\tau + \pi/2)$, $y(\tau) = -6 - \cos(2\tau)$.

The restriction $a \geqslant b$ is fundamental. In case $a < b$ the curve, called trochoid, doesn't represent a function: the blue trochoid has equations $x(\tau) = -\tau - 2\sin(\tau)$, $y(\tau) = -3 + 2\cos(\tau)$ ☞ *differential geometry*.

Exercises Suppose f has period T. Show that

i) f is not 1-1;

ii) $x \mapsto f(\omega x)$ has period $\dfrac{T}{\omega}$, $\forall \omega \neq 0$.

Observe that the sum of periodic functions isn't always periodic [39, p.369]. Take $\sin(x) + \sin(ax)$ with a irrational. If it had period T,

$$\sin(x + T) - \sin x = -\big(\sin(ax + aT) - \sin(ax)\big) \quad \Longrightarrow$$

$$\cos(x + T/2)\,\sin(T/2) = -\cos(ax + aT/2)\,\sin(aT/2) \quad \Longrightarrow$$

$$\cos x\,\sin(T/2) = -\cos(ax)\,\sin(aT/2).$$

Evaluating at $x = \pi/2$ and after at $ax = \pi/2$ we'd obtain T is an integer multiple of $\frac{2\pi}{a}$ and 2π. But this is nonsense since $a \notin \mathbb{Q}$.

Another definition, relating back to Definition 7.4:

Definition For two maps $f, g \in \mathbb{R}^D$ one indicates by $\max(f, g)$, $\min(f, g)$ the pointwise **maximum/minimum** functions:

$$\max(f, g)(x) := \max_{x \in D}\{f(x), g(x)\}, \quad \min(f, g)(x) := \min_{x \in D}\{f(x), g(x)\}.$$

Then

$$f^+ := \max(f, 0), \quad f^- := \min(f, 0)$$

are the **positive/negative part** of f.

Exercises Choose an arbitrary real function f.

i) Sketch the graphs of f^+, f^- starting from graph(f).
ii) Show $|f| = f^+ - f^-$.
iii) Prove that $f = g - h$ for certain non-negative maps g, h.

10.3.3   Boundedness Conditions

Definition A function $f: D \to \mathbb{R}$ is **bounded** when

$$\exists\, M \in \mathbb{R} \text{ such that } |f(x)| < M \text{ for every } x \in D.$$

Being bounded means the image is a bounded subset, i.e. contained in some interval: $\mathrm{Im}(f) \subseteq (-M, M)$. Hence the graph is confined to the horizontal strip determined by the straight lines $y = \pm M$.

A map f is called **unbounded** if, on the contrary, $\forall M \in \mathbb{R} \quad \exists x \in D$ such that $|f(x)| > M$, meaning its graph goes off to infinity.

Examples 10.10 Bounded maps include:

i) constants $y = c$ (Fig. 11.2);
ii) $y = \dfrac{3x - 2}{x^2 + 1} + 1$. In fact, $-3 < y < 2$ for all $x \in \mathbb{R}$;
iii) $\xi(x) = e^{-x^2} + 2$, since $2 < \xi(x) \leqslant 3$ for every $x \in \mathbb{R}$;
iv) $y = \sin(x)$, $y = \cos(x)$ have image $[-1, 1]$ (Fig. 12.10);
v) $y = \arctan(x)$ has image $[-\pi/2, \pi/2]$ (Fig. 12.17);
vi) $y = -4xe^{-x^2/2}$ has image $[-4, 4]$.

More generally, Hermite functions $y_n(x) := (-1)^n e^{x^2/2} \dfrac{d^n}{dx^n}\left(e^{-x^2}\right)$, which solve the Schrödinger equation $y'' + (2n + 1 - x^2)y = 0$ for the quantum harmonic oscillator (☞ *quantum mechanics, differential equations*).

Exercise

i) Referring to Fig. 10.16, can you tell which graphs correspond to which functions in the above list?
ii) Consider polynomial maps $p(x), q(x)$ where q has no zeroes and $\deg(p) = \deg(q)$. Prove the rational map $y = \dfrac{p(x)}{q(x)}$ is bounded.

Fig. 10.16 Bounded maps
from Example 10.10

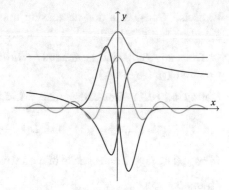

Examples In the opposite direction, the following are unbounded on their respective natural domains:

i) $y = mx + q$ $(m \neq 0)$
ii) any non-constant polynomial function p (i.e., $\deg(p) \geqslant 1$)
iii) $y = x^{-n}$, $y = \sqrt[n]{x}$ $(n \in \mathbb{N} \setminus \{0\})$
iv) $y = \tan(x)$
v) $y = a^x$, $y = \log_a(x)$ $(a > 0, a \neq 1)$
vi) the reciprocal $1/f$ of a map f with zeroes.

More frequent is the presence of one-sided bounds:

Definition A function $f \colon D \to \mathbb{R}$ is **upper bounded** (or **bounded above**) if

$$\sup_D f := \sup \operatorname{Im}(f) < +\infty,$$

and **lower bounded** (or **bounded below**) if

$$\inf_D f := \inf \operatorname{Im}(f) > -\infty.$$

Example The map $y = x^{-2}$ is well defined on $\mathbb{R}^*$: it's lower bounded by $0 = \inf\limits_{x \neq 0} y$,

and unbounded above since $\sup \dfrac{1}{x^2} = +\infty$.

Exercises Let f, g be real maps and fix $0 < \lambda$. Prove that

i) $\sup(\lambda f) = \lambda \sup f$, $\quad \sup(\lambda f) = \lambda \inf f$
ii) $\sup(f + g) \leqslant \sup f + \sup g$
iii) $\sup(f + \lambda) = \sup(f) + \lambda$
iv) $\sup(fg) \leqslant \sup(f) \sup(g)$, assuming $f, g \geqslant 0$
v) $\sup |f + g| \leqslant \sup |f| + \sup |g|$.

State the analogue properties involving inf.

Exercise Consider a non-decreasing map $\phi\colon \mathbb{R} \to \mathbb{R}$ and an upper-bounded map $f\colon D \to \mathbb{R}$. Show $\phi(\sup_D f) \geqslant \sup_D(\phi \circ f)$.

Does equality stand in general? (*Hint*: take $f(x) = \frac{x}{x+1}$ and the characteristic function $\chi_{[1,+\infty)}$ as ϕ.)

Definition 10.11 One calls a map f **L-Lipschitz** if there exists an $L \geqslant 0$ such that

$$|f(x_1) - f(x_2)| \leqslant L|x_1 - x_2|, \quad \forall x_1, x_2 \in \mathrm{Dom}(f).$$

In case $0 \leqslant L < 1$ one speaks of a **contraction**, cf. Definition 19.24.

Examples

i) Affine maps $a(x) = mx + q$ are Lipschitz, and realise equality: $|a(x_1) - a(x_2)| = |m|\,|x_1 - x_2|$.
ii) $y = x^2$ is not Lipschitz:
     if it were, $\left|x_1^2 - x_2^2\right| \leqslant L|x_1 - x_2| \implies |x_1 + x_2| \leqslant L$ for any $x_2 \neq x_1$, but $x_1 + x_2$ is unbounded as the x_i vary. $\not{\mathit{z}}\mathit{z}$
iii) Sine is 1-Lipschitz: for any $t_1, t_2 \in \mathbb{R}$ set $h = t_1 - t_2$, so

$$\sin t_1 - \sin t_2 = \sin(t_2 + h) - \sin t_2 = (\cos h - 1)\sin t_2 + \cos t_2 \sin h =$$

$$= -2\sin^2\frac{h}{2}\sin t_2 + 2\cos y \sin\frac{h}{2}\cos\frac{h}{2} = 2\sin\frac{h}{2}\cos\left(t_2 + \frac{h}{2}\right).$$

Hence

$$|\sin t_1 - \sin t_2| = 2\left|\sin\frac{h}{2}\right|\left|\cos\left(t_2 + \frac{h}{2}\right)\right| \leqslant 2\left|\sin\frac{h}{2}\right| \overset{12.8}{\leqslant} 2\left|\frac{h}{2}\right| = |t_1 - t_2|.$$

iv) $y = \sqrt{x}$ is not Lipschitz.
     If, by contradiction, there were an $L > 0$ such that $\left|\sqrt{x_1} - \sqrt{x_2}\right| \leqslant L|x_1 - x_2|$ for any $x_i \geqslant 0$, in particular we'd have (put $x_2 = 0$) $1 \leqslant L\sqrt{x_1}$ for any $x_1 > 0$. But this becomes false if we take $x_1 = \frac{1}{4L^2}$. $\not{\mathit{z}}\mathit{z}$
v) Fix a subset $\varnothing \neq Z$ in a metric space (X, d). The distance map $\mathrm{dist}(Z, \cdot)$ is 1-Lipschitz, see Proposition 19.10.

Exercises

i) Decide whether abs, or an isometry (see Definition 19.14), are Lipschitz.
ii) Prove that generic polynomial functions aren't Lipschitz.
iii) Prove $y = x + \sin x$ is Lipschitz.
iv) Find examples showing that: bounded $\not\Longrightarrow$ Lipschitz.   (*Hint*: $y = \sin(x^2)$.)
v) Find examples showing that: Lipschitz $\not\Longrightarrow$ bounded.

The final notion we'll present is inescapable in any theory of integration (Riemann, Lebesgue, ...).

Definition Let $f \colon D \to \mathbb{R}$ be a map and $t_0, \ldots, t_n \in D$ points in the domain. We say f has **bounded variation** over D if

$$\mathrm{var}(f) := \sup_{n \in \mathbb{N}} \sum_{i=1}^{n} |f(t_i) - f(t_{i-1})| < +\infty.$$

This notion of variation *a priori* depends on the choice of partition $\{t_i\}$, but we won't be concerned about it and shamelessly sweep the matter under the rug (in an analysis course this matter is addressed properly).

Example The sine function has bounded variation on $[0, 2\pi]$. To prove it we subdivide the interval into $\underbrace{[0, \pi/2]}_{I_1} \cup \underbrace{[\pi/2, 3\pi/2]}_{I_2} \cup \underbrace{[3\pi/2, 2\pi]}_{I_3}$, on each of which sine is monotone. By the supremum's properties

$$\mathrm{var}(f) \leqslant \mathrm{var}(f \mid_{I_1}) + \mathrm{var}(f \mid_{I_2}) + \mathrm{var}(f \mid_{I_3})$$
$$= \big| \sin(0) - \sin(\pi/2) \big| + \big| \sin(\pi/2) - \sin(3\pi/2) \big|$$
$$+ \big| \sin(3\pi/2) - \sin(2\pi) \big| = 4.$$

Over the entire real line, though, it can be proved that $\mathrm{var}(\sin) = +\infty$.

Exercise Show that a real map is constant if and only if its variation is 0.

Examples (*maps with unbounded variation*)

i) The characteristic function of $Q := \mathbb{Q} \cap [0, 1]$

$$\chi_Q(x) = \begin{cases} 1 & x \in [0, 1] \cap \mathbb{Q} \\ 0 & x \in [0, 1] \setminus \mathbb{Q} \end{cases}$$

doesn't have bounded variation. Partition $[0, 1]$ so that t_{2k} is rational and t_{2k+1} irrational. Then $\sum_{i=1}^{n} |\chi_Q(t_i) - \chi_Q(t_{i-1})| = n$. By increasing the number of points we obtain $\mathrm{var}(\chi_Q) = +\infty$.

ii) Consider, on $D = [0, 2/\pi]$,

$$g(x) = \begin{cases} x \sin \dfrac{1}{x} & x \in (0, 2/\pi] \\ 0 & x = 0 \end{cases}$$

The claim is that its total variation is infinite. Fix $n \in \mathbb{N}$ and choose points $t_k = \frac{1}{(n-k+1/2)\pi}$, $k = 1, \ldots, n$. From

$$|g(t_k) - g(t_{k-1})| = \left| t_k(-1)^{n-k} - t_{k-1}(-1)^{n-k+1} \right|$$

$$= \left| (-1)^{n-k}(t_k + t_{k-1}) \right| \geqslant t_k > \frac{1}{(n-k+1)\pi}$$

we deduce

$$\sum_{k=1}^{n} |g(t_k) - g(t_{k-1})| \geqslant \frac{1}{\pi} \sum_{k=1}^{n} \frac{1}{(n-k)+1} = \frac{1}{\pi} \left(\frac{1}{n} + \cdots + \frac{1}{2} + 1 \right).$$

The term within brackets is the partial sum of the harmonic series, so letting $n \in \mathbb{N}$ vary results in $\mathrm{var}(g) = +\infty$.

Exercise Using a similar argument, show that over $D = [0, 1]$ the map

$$f(x) = \begin{cases} \sin \dfrac{1}{x} & x > 0 \\ 0 & x = 0 \end{cases}$$

is bounded, but $\mathrm{var}(f) = +\infty$, see Fig. 19.1.

To better explain the relationship between boundedness and bounded variations,

Proposition *Any map* $f : [a, b] \mapsto \mathbb{R}$ *with bounded variation is bounded.*

Proof For any $x \in [a, b]$ and suitable choice of partition points, we compute

$$|f(x)| - |f(a)| \leqslant |f(x) - f(a)| \leqslant \sum_{i=1}^{n} |f(t_i) - f(t_{i-1})| \leqslant \mathrm{var}(f) < \infty.$$

Therefore $|f(x)| \leqslant |f(a)| + \mathrm{var}(f) < +\infty$. $\square$

Example Slightly modifying the previous examples may alter the picture completely. Over $[0, 1]$

$$v(x) = \begin{cases} x^2 \sin \dfrac{1}{x} & x \in (0, 1] \\ 0 & x = 0 \end{cases}$$

has finite variation. As before, fix $n \in \mathbb{N}$ and take $t_k = \frac{1}{(n-k)+1}$, $k = 1, \ldots, n$. For any $x \in [0, 1]$

$$v(t_k) \leqslant (-1)^{n-k} t_k^2 = \begin{cases} t_k^2 & \text{for } n-k \text{ even} \\ -t_k^2 & \text{for } n-k \text{ odd} \end{cases}$$

and thus

$$|v(t_k) - v(t_{k-1})| \leqslant \left| (-1)^{n-k} t_k^2 - (-1)^{n-k+1} t_{k-1}^2 \right| = \left| (-1)^{n-k} (t_k^2 + t_{k-1}^2) \right| \leqslant 2t_k^2.$$

Consequently

$$\sum_{k=1}^{n} |v(t_k) - v(t_{k-1})| \leqslant 2 \sum_{k=1}^{n} \left(\frac{1}{n-k+1} \right)^2 = 2 \left(\frac{1}{n^2} + \cdots \frac{1}{2^2} + 1 \right).$$

The last sum converges to $\pi^2/6$, so in the end $\mathrm{var}(v) \leqslant \pi^2/3$.

Exercises Take $f: [a, b] \to \mathbb{R}$ and show

i) f Lipschitz $\Longrightarrow \mathrm{var}(f) < +\infty$;
ii) f monotone $\Longrightarrow \mathrm{var}(f) = |f(b) - f(a)|$ ($\Longrightarrow$ bounded variation);
iii) if f is monotone, $\mathrm{var}(f) < +\infty$ if and only if f is bounded;
iv) for any $\alpha \in \mathbb{R}$ and $f, g: [a, b] \to \mathbb{R}$ with bounded variation, $|f|$, αf, $f + g$ all have bounded variations.

At last we mention, without proof, a converse to exercises iii)–iv).

Proposition (Jordan) *A map $f : [a, b] \mapsto \mathbb{R}$ has bounded variation if and only if it is the difference of two increasing maps.*

Chapter 11
Algebraic Functions

After setting up the general theory in Chaps. 3 and 9, we are fully fledged for studying concrete cases in detail. The present chapter might seem a gallery of easy, special examples, but in the true spirit of a reference text it is justified in order to provide the background for solving algebraic equations and inequalities.

Put informally, an algebraic function is one obtained composing a finite number of arithmetic operations (sums, products, divisions, root extractions). More formally,

Definition A function $f : D \subseteq \mathbb{R} \to \mathbb{R}$ is said **algebraic** when there exists a polynomial $p(y) = \sum_{j=0}^{n} a_j(x) y^j$, whose coefficients $a_j(x) \in \mathbb{R}[x]$ are themselves polynomials (not all zero), such that $p\big(f(x)\big) = 0$ for every $x \in D$.

The map $y = x + \sqrt[3]{2x^2 + \frac{1}{2x}} - 1$ solves the polynomial equation

$$(2x)y^3 + \big(6x(1-x)\big)y^2 + \big(6x(1-x)^2\big)y + \big(-2x^4 + 2x^3 - 6x^2 + 2x + 1\big) = 0$$

in the variable y, and as such is algebraic (Fig. 11.1).

The first class we wish to examine is made of functions defined by polynomials. One prominent among a plethora of reasons, even more relevant than simplicity, is the *Weierstraß approximation theorem*, according to which any continuous map $f : [a, b] \subseteq \mathbb{R} \to \mathbb{R}$ can be approximated uniformly, and as well as we want, by a polynomial function:

$$\forall \epsilon > 0 \, \exists \, p(x) \in \mathbb{R}[x] \text{ such that } \sup_{x \in [a,b]} \big|f(x) - p(x)\big| < \epsilon.$$

Since polynomials can be evaluated directly by computer programs, the theorem has both a practical and a theoretical relevance, especially if we think of polynomial interpolation or series. It suggests that polynomial maps are dense in $\mathscr{C}\big([a, b]\big)$ (☞

© The Author(s), under exclusive license to Springer Nature Switzerland AG 2021   249
S. G. Chiossi, *Essential Mathematics for Undergraduates*,
https://doi.org/10.1007/978-3-030-87174-1_11

Fig. 11.1 Polynomial map

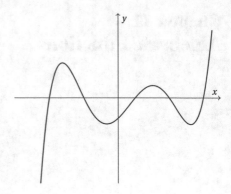

functional analysis). The same idea lurks in the background of the *Taylor expansion*, see Fig. 12.11.

11.1   Polynomial Functions

Definition 11.1 Every real polynomial $p(t) = \sum_{j=0}^{n} a_j t^j \in \mathbb{R}[t]$ determines a unique **polynomial function**

$$p : \mathbb{R} \to \mathbb{R}, \ x \mapsto p(x) = \sum_{j=0}^{n} a_j x^j$$

defined on the entire real line: $\mathrm{Dom}(p) = \mathbb{R}$.

The converse holds as well: it's known that a real polynomial map defines a polynomial uniquely, because $\mathbb{R}$ is an infinite domain with unit. It's not uncommon to mix up the polynomial $p(t)$ with the polynomial function $y = p(x)$, since—truth be told—we've glossed over the difference (see footnote on p. 212).

But the above correspondence is false if the coefficients are taken in a finite integral domain. For instance the two polynomials t^3, $t \in \mathbb{Z}_3[t]$ are distinct, yet they induce the same polynomial function because $f^3 - f \equiv_3 0$ for $f = -1, 0, 1$. (This is a manifestation of *Fermat's little theorem*, see p. 327.)

To set things in motion let's take a look at the graphs of some simple polynomial functions. We'll group them according to the degree.

Degree 0: the graph of a **constant map** $y = q$ is a horizontal line (parallel to the x-axis) passing through the point $(0, q)$.

Degree 1: the graph of an **affine function** $y = mx + q$ is a straight line through $(0, q)$ with **slope**, or **angular coefficient**, $m = \tan \theta$ (θ is the oriented angle between the positive x-axis and the line) (Fig. 11.2).

Fig. 11.2 Affine functions

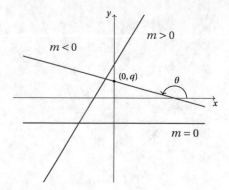

The map is increasing whenever $m > 0$, i.e. $\theta \in (0, \pi/2)$, decreasing for $m < 0$, that is $\theta \in (\pi/2, \pi)$. For $m = 0$ ($\theta = 0$) we obtain the constant map $y = q$ (degree 0).

Degree 2: the graph of a **quadratic map** $y = ax^2 + bx + c$ is called **parabola** (Fig. 11.3).

If $\Delta = b^2 - 4ac \geqslant 0$, the polynomial's roots $x_\pm := \dfrac{-b \pm \sqrt{\Delta}}{2a}$ (see (9.2)) give the intersections with the horizontal axis.

The sign of a determines the convexity.

Since $a\left(x - \frac{b}{a}\right)^2 + b\left(x - \frac{b}{a}\right) + c = ax^2 + bx + c$, the quadratic polynomial is invariant under the transformation $x \rightsquigarrow x - \dfrac{b}{a}$, making the parabola symmetric about the line $x = -\dfrac{b}{2a}$. Hence the extremum is attained at $\left(-\dfrac{b}{2a}, \dfrac{\Delta}{4a}\right)$, called the **vertex**. This is a minimum if $a > 0$, a maximum if $a < 0$.

Examples

i) $y = 2x^2 - 12x + 16 = 2(x^2 - 6x + 8) = 2(x - 4)(x - 2)$, $\Delta = 4^2 > 0$
ii) $y = x^2 + 2 \geqslant 2 > 0$ for all x, $\Delta = -8 < 0$ and $a = 1 > 0$
iii) $y = x^2 - \frac{4\sqrt{2}}{8}x + \frac{1}{8} = \left(\frac{1}{2\sqrt{2}} - x\right)^2$, $\Delta = 0$.

The set $\{x \in \mathbb{R} \mid f(x) > 0\}$ where the quadratic map is positive can be read off this table:

	$a > 0$	$a < 0$
$\Delta > 0$	$(-\infty, x_-) \cup (x_+, +\infty)$	(x_-, x_+)
$\Delta = 0$	$\mathbb{R} \setminus \{x_+\}$	$\varnothing$
$\Delta < 0$	$\mathbb{R}$	$\varnothing$

For quadratic maps with zeroes this set is always either an interval bounded by the roots, or the complement (first row).

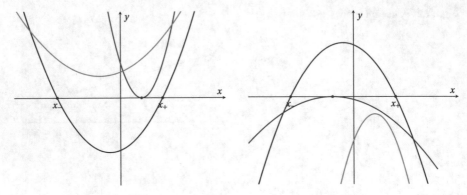

Fig. 11.3 Parabolas: $a > 0$ (left), $a < 0$ (right); $\Delta > 0$, $\Delta = 0$, $\Delta < 0$

Example We want to solve the following problem: choose values for the coefficients a, b, c in $f(x) = ax^2 + bx + c$ so that the set $\{x \in \mathbb{R} \mid f(x) > 0\}$ equals:

(1) $\mathbb{R}$ (3) $\varnothing$ (5) $(-\infty, 2) \cup (2, \infty)$
(2) $(-2, 3)$ (4) $(-\infty, 1) \cup (5, \infty)$ (6) $(3, +\infty)$.

Let's remind preliminarily that graf(f) is a parabola exactly when $a \neq 0$.

(1) It suffices to choose a triple (a, b, c) such that $b^2 - 4ac < 0$: there are infinitely
   many solutions, e.g. $(1, 0, 1)$ giving $y = x^2 + 1$, or $(1, 1, 500)$ giving $y = x^2 + x + 500$. The simplest solutions are $(0, 0, c > 0)$, corresponding to horizontal
   lines.
(2) The expression $ax^2 + bx + c$ must necessarily factor as $a(x + 2)(x - 3)$ for
   some $a < 0$, where $-2, 3$ are the roots. The possible triples satisfy

$$\begin{cases} ax^2 + bx + c = a(x + 2)(x - 3), \ \forall x \\ a < 0 \end{cases} \implies \begin{cases} b = -a \\ c = -6a \end{cases},$$

   so they must have the form $(a, -a, -6a)$, with $a \in \mathbb{R}^-$ arbitrary.
(3) Inspecting the previous table, we may pick (a, b, c) such that $a < 0$, $b^2 - 4ac <$
   0; for instance $(-1, 0, -1)$.
(4) We need (a, b, c) such that $ax^2 + bx + c = a(x - 1)(x - 5)$ and $a > 0$. The
   possibilities are triples of the form $(a, -6a, 5a)$, for $a > 0$.
(5) In analogy to the previous case,

$$\begin{cases} ax^2 + bx + c = a(x - 2)^2, \ \forall x \\ a > 0 \end{cases} \implies \begin{cases} a > 0 \\ c = 4a \\ b = -4a \end{cases}.$$

(6) Here the graph cannot be a parabola, because a has to vanish. The function then will be affine. Condition $bx + c > 0 \iff x > 3$ forces $b > 0$ and $c = -3b$. Summing up, we can choose $(0, b, -3b)$ for any $b > 0$, corresponding to lines $y = b(x - 3)$ with positive slope.

Remark (Ballistics) The name parabola (see the appendix) points directly to a physics link. The motion experienced by a point-particle thrown near the Earth's surface under the action of gravity only (no air resistance or the like) is planar, in particular parabolic.

In a suitable frame system with initial point of motion s_0 and initial velocity $v_0 \neq 0$ at angle $\theta \neq \pm\frac{\pi}{2}$, the acceleration $a(t) = -g\mathbf{j}$ integrates to the trajectory's displacement $t \mapsto s_0 + v_0 t - \frac{1}{2}gt^2\mathbf{j}$. Eliminating the time parameter produces the quadratic equation

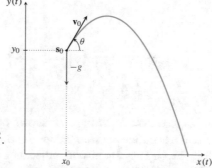

$$y = y_0 + \tan(\theta)(x - x_0) - \frac{g}{2\|v_0\|^2 \cos^2 \theta}(x - x_0)^2.$$

Examples Let's determine all numbers $\lambda \in \mathbb{R}$ for which:

i) equation $\dfrac{\lambda}{x - 2} + x = \lambda$ has exactly two distinct real solutions.

Easy algebraic manipulations give us $0 = \dfrac{x^2 - (2 + \lambda)x + 3\lambda}{x - 2}$. Now, the numerator has distinct roots if and only if $(2 + \lambda)^2 - 12\lambda > 0$, i.e. whenever $\lambda < 4 - 2\sqrt{3}$ or $\lambda > 4 + 2\sqrt{3}$.

ii) $2x^2 - \lambda x + 2\lambda \geqslant 0$ for every $x \in \mathbb{R}$.

For any chosen λ, the parabola $y = 2x^2 - \lambda x + 2\lambda$ intersect the x-axis in one point at most when $\lambda^2 - 16\lambda \leqslant 0$, i.e. if $\lambda \in [0, 16]$.

iii) $\dfrac{4}{\sqrt{\lambda}} - \dfrac{1}{\lambda}$ attains its maximum value.

Write $\dfrac{4}{\sqrt{\lambda}} - \dfrac{1}{\lambda} = -\left(\dfrac{1}{\sqrt{\lambda}}\right)^2 + \dfrac{4}{\sqrt{\lambda}} - 4 + 4 = -\left(\dfrac{1}{\sqrt{\lambda}} - 2\right)^2 + 4$. As $-(x - 2)^2 + 4 \leqslant 4$ for any x, and equality is attained for $x = 2$, the value we're looking for is $\dfrac{1}{\sqrt{\lambda}} = 2 \implies \lambda = 1/4$.

Geometrically, in fact, the highest point on the parabola $y = -(x - 2)^2 + 4$ is the vertex, which is symmetrical with respect to the axis intersections: since $-(x - 2)^2 + 4 = x(4 - x)$, the vertex has first coordinate $x = \dfrac{0 + 4}{2} = 2$.

Fig. 11.4 Reduced cubics

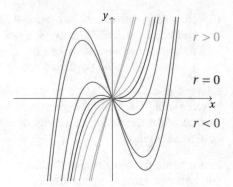

Alternatively, $\max\left(\dfrac{4}{\sqrt{\lambda}} - \dfrac{1}{\lambda}\right) = \min\left(\dfrac{1}{\sqrt{\lambda}} - 2\right)^2 + 4$ is reached when $\dfrac{1}{\sqrt{\lambda}} - 2 = 0.$

Degree $n = 3$: Lemma 9.4 shows how cubics reduce, under rigid motions and rescalings (affine transformations), to three 'normal forms' $y = x^3 + rx = x(x^2 + r)$ (Fig. 11.4). They are all odd, inflect at the symmetry point 0 and depend on the sign of the surviving coefficient. When $r < 0$ there are two further roots $\pm\sqrt{-r}$. When $r < 0$, instead, the map is increasing: $a > b > 0 \implies a(a^2 + r) > b(b^2 + r)$ (and then use the oddity). We won't pursue this case, which requires more tools than we have provided in order to give general recipes. The general theory of *plane cubic curves* is studied in ☞ *algebraic geometry, number theory*.

Degree $n > 3$: basic calculus techniques allow to handle polynomial maps of any degree in one go, in that they almost effortlessly give us the asymptotic behaviour and the position of peaks and troughs. In very special situations those tools might even have a say on the position of the roots, though this is rather rare an occasion.

Higher degrees can be tackled in an elementary way if, instead of looking at polynomials, we consider the simple case of monomials, as we set out to do next.

11.2 Algebraic Equations and Inequalities

As a warm-up to equations we wish to review functions of the form

$$y = x^{\alpha}, \quad \alpha \in \mathbb{Q}.$$

We'll mostly concentrate on α being an integer or the inverse of one. In these cases the map's behaviour is mostly determined by the parity of α, and there are only finitely many blueprints, irrespective of the specific α. Rational exponents $\alpha = p/q$ will receive a brief treatment, with no details, since properties of the relative maps refer back to previous cases. Instead, choosing α irrational produces transcendental functions (Definition 12.4).

Let us start by looking at $y = x^n$ over $\mathbb{R}^+$ for various values of natural exponent. These maps are essentially controlled by the fact that for any $m > n > 1$

$$0 < x < 1 \iff x^m < x^n, \qquad x > 1 \iff x^m > x^n.$$

The larger the exponent, the 'flatter' the graph becomes near the origin, and the faster the function rockets to infinity as x grows Fig. 11.5.

Case $\alpha \in \mathbb{Z}^+$: these are polynomial maps of degree α, hence defined on $\mathbb{R}$ (Fig. 11.6):

- $y = x^{2n+1}$ is odd, increasing and its image is $\mathbb{R}$. It is convex for $x \geqslant 0$, with a unique inflection at the origin.
- $y = x^{2n}$ is even, convex, increasing for $x \geqslant 0$ and with image $[0, +\infty)$.

Case $\alpha \in \mathbb{Z}^-$: these are rational maps $y = \dfrac{1}{x^{|\alpha|}}$, with domain $\mathbb{R} \setminus \{0\}$ (Fig. 11.7):

- $y = x^{-(2n+1)}$ is odd, decreasing, convex for $x > 0$ and with image $\mathbb{R} \setminus \{0\}$.
- $y = x^{-2n}$ is even, convex, decreasing for $x > 0$ and with image $(0, +\infty)$.

Case $\alpha = \dfrac{1}{m} > 0$ ($m \in \mathbb{N}^*$): these are **mth roots** (Fig. 11.8)

- $y = \sqrt[2n+1]{x} = x^{1/(2n+1)}$ has domain and image $\mathbb{R}$, it is increasing and odd.
- $y = \sqrt[2n]{x} = x^{1/2n}$, $n \in \mathbb{N}^*$, has domain and image $[0, +\infty)$ and is increasing.

Fig. 11.5 Comparing powers x^n

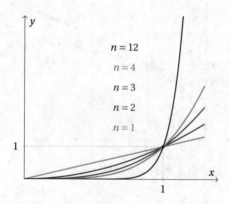

$n = 12$

$n = 4$

$n = 3$

$n = 2$

$n = 1$

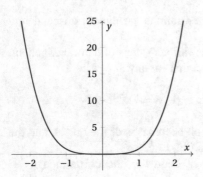
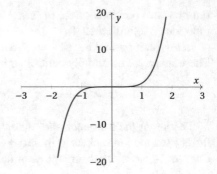

Fig. 11.6 Positive integer exponent: even and odd

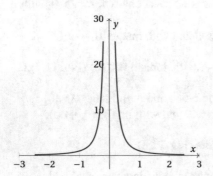
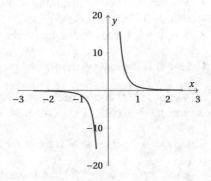

Fig. 11.7 Negative integer exponent: even and odd

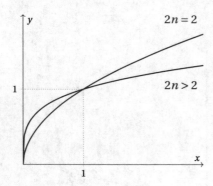
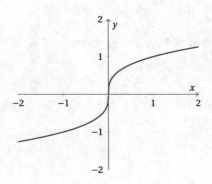

Fig. 11.8 Even and odd roots

Case $\alpha = \dfrac{1}{m} < 0$ ($m \in \mathbb{N}^*$): this defines reciprocals of roots (Fig. 11.9)

- $y = \dfrac{1}{\sqrt[2n+1]{x}}$ has domain and image $\mathbb{R} \setminus \{0\}$, it is odd and decreasing.

- $y = \dfrac{1}{\sqrt[2n]{x}}$ has domain and image $\mathbb{R}^+$ and is decreasing.

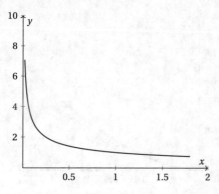

Fig. 11.9 Reciprocals of roots: even and odd

Exercise Relative to the previous cases, prove the statements concerning domains & images, evenness & oddness, monotonicity.

Case $\alpha \in \mathbb{Q}$: regarding $y = x^{p/q} = \left(\sqrt[q]{x}\right)^{p}$ we shall treat rapidly only positive exponents $\dfrac{p}{q} > 0$ that are reduced: $\gcd(p, q) = 1$ (and exclude previously considered situations).

First, if q is even the domain equals $[0, +\infty)$, whilst for q odd it is the entire line $\mathbb{R}$. The behaviour of y depends on the relative parity of p, q and whether p is smaller or larger than q:

- if both p, q are odd, the map is odd: the profile is similar to that in Fig. 11.6 left for $p/q > 1$, and Fig. 11.8 right for $p/q < 1$.
- if p is even the map is even and increasing on $\mathbb{R}^{+}$: when $\dfrac{p}{q} = \dfrac{2m}{2n+1} > 1$ the profile is as in Fig. 11.6; when $\dfrac{p}{q} = \dfrac{2m}{2n+1} < 1$ the convexity changes due to the singular point at the origin, see Fig. 11.10 left.
- if p is odd the map is increasing: when $\dfrac{p}{q} = \dfrac{2m+1}{2n} < 1$ we'll obtain Fig. 11.8 right, and when $\dfrac{p}{q} = \dfrac{2m+1}{2n} > 1$ Fig. 11.10 right.

All this information is useful to solve algebraic equations and inequalities. An **algebraic equation** is an equation $F(x) = 0$ where F is an algebraic function.

Anytime we have to do with an even root $\sqrt[2n]{f(x)}$ in an equation it is necessary to impose what we call the *existence condition* for solutions: $f(x) \geq 0$.

Equations of type

$$\sqrt[n]{f(x)} = \sqrt[m]{h(x)}$$

for $n > 1$, $m \geq 1$ can be solved, provided we assume $x \in \text{Dom}(\sqrt[n]{}) \cap \text{Dom}(\sqrt[m]{})$, by raising to the appropriate power to eliminate the roots.

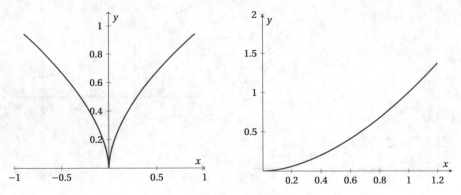

Fig. 11.10 Graphs of $y = \sqrt[7]{x^4}$ and $y = \sqrt[4]{x^7} = |x|\sqrt[4]{x^3}$

Examples

i) $\sqrt[3]{x^3 + 4} - 1 = x$.

    The domain of existence for the solutions is $\mathbb{R}$. As the map $t \mapsto \sqrt[3]{t}$ is invertible,

$$\sqrt[3]{x^3 + 4} = x + 1 \iff \left(\sqrt[3]{x^3 + 4}\right)^3 = (x + 1)^3 \iff x = \frac{-1 \pm \sqrt{5}}{2}. \quad (\checkmark)$$

ii) $\sqrt{6 - x} = \sqrt{x - 1} + \sqrt{x + 1}$.

    The common domain of the 3 root functions is $(-\infty, 6] \cap [1, +\infty) \cap [-1, +\infty) = [1, 6]$. Both sides are non-negative, so squaring we obtain the equivalent equation $6 - 3x = 2\sqrt{(x - 1)(x + 1)}$. Now, the expression on the right is non-negative, so we must additionally impose $x \leqslant 2$. Therefore the solutions must belong to $[1, 2]$. Squaring once more produces $4x^2 - 4 = 36 - 36x + 9x^2$, i.e. $x = \frac{18 - 2\sqrt{31}}{5} \in [1, 2]$ ($\checkmark$) or $x = \frac{18 + 2\sqrt{31}}{5} \notin [1, 2]$ ($\boldsymbol{X}$). So, there is only one solution.

    Inequalities involving roots

$$\sqrt[n]{f(x)} > \sqrt[m]{h(x)}$$

with $n > 1, m \geqslant 1$ are a bit more delicate.

    In case both roots are odd it suffices to raise everything to $\mathrm{lcm}(n, m)$. For example

$$\sqrt[3]{f(x)} > \sqrt[5]{h(x)} \iff \sqrt[3]{f(x)}^{15} > \sqrt[5]{h(x)}^{15} \iff f(x)^5 > h(x)^3;$$
$$\sqrt[2n+1]{f(x)} > g(x) \iff f(x) > g(x)^{2n+1}.$$

In presence of an even root we should watch out for 'implicit' constraints:

$$\sqrt[2n]{f(x)} < g(x) \iff \begin{cases} f(x) \geqslant 0 \\ g(x) > 0 \\ f(x) < (g(x))^{2n} \end{cases}.$$

Specifically, the non-obvious requirement $g(x) > 0$ is necessary for solutions to exist.

On the other hand, we have to treat two cases when

$$\sqrt[2n]{f(x)} > g(x) \iff \begin{cases} f(x) \geqslant 0 \\ g(x) < 0 \end{cases} \quad \vee \quad \begin{cases} g(x) \geqslant 0 \\ f(x) > (g(x))^{2n} \end{cases}.$$

Example

i) $\sqrt{x-1} > 12 - 2x$.

We need only consider $x \geqslant 1$.

The inequality is true for every x whenever the right-hand side is negative, i.e. $x > 6$.

Instead, when $x \leqslant 6$ (and $x \geqslant 1$) we may square to obtain $x - 1 > 144 + 4x^2 - 48x$, that is $5 < x < 29/4$.

In the end, the inequality is solved by any $x \in (5, 29/4) \cup (6, +\infty) = (5, +\infty)$.

ii) $\sqrt{x} - 3 \geqslant \dfrac{2}{\sqrt{x} - 2}$.

First of all, $x \geqslant 0$ and $\sqrt{x} \neq 2$.

The first case is $\begin{cases} \sqrt{x} - 3 \geqslant 0 \\ \dfrac{2}{\sqrt{x} - 2} < 0 \end{cases} \iff \begin{cases} \sqrt{x} \geqslant 3 \\ \sqrt{x} < 2 \end{cases}$ which

has no solution. The other possibility $\begin{cases} \sqrt{x} - 3 \geqslant \dfrac{2}{\sqrt{x} - 2} \\ \dfrac{2}{\sqrt{x} - 2} > 0 \end{cases} \iff$

$\begin{cases} (\sqrt{x} - 3)(\sqrt{x} - 2) \geqslant 2 \\ \sqrt{x} > 2 \end{cases} \iff \begin{cases} (\sqrt{x} - 1)(\sqrt{x} - 4) \geqslant 0 \\ \sqrt{x} > 2 \end{cases}$ reduces to

$\sqrt{x} \geqslant 4$, so eventually $x \geqslant 16$.

Exercises

i) Solve

(1) $2\sqrt{x - 1} - x = 0$

(2) $\sqrt{x + 3} = 1 - 3x$

(3) $\sqrt{2x + 6} - x + 1 = 0$

(4) $3\sqrt{x + 2} - x - 4 = 0$

(5) $\sqrt[3]{x + 4} = 3$

(6) $\sqrt{x - 1} - \sqrt{2x - 3} = 0$

(7) $\sqrt{5x-6} > x$

(8) $\sqrt{x+2} + \sqrt{3x-1} > 0$

(9) $\sqrt{\dfrac{x-4}{x+2}} < 2.$

ii) For every given instance, find the domain and where the map is positive:

(1) $y = \sqrt{x-2} + 1$

(2) $y = \sqrt{x+3} + \sqrt{x^2+9}$

(3) $y = \sqrt[3]{x^2-1}$

(4) $y = \sqrt[3]{x-1} + \sqrt[3]{x-2}$

(5) $y = \dfrac{\sqrt{x}-1}{\sqrt{|x|}-2}$

(6) $y = \sqrt[3]{x+2} - \sqrt{x+2}$

(7) $y = \dfrac{\sqrt{x-5} + \sqrt{2x+1}}{\sqrt[3]{1-x}}$

(8) $y = \sqrt[4]{x^4-1} - x^2$

(9) $y = \dfrac{\sqrt{x-1}\sqrt{x+2}}{\sqrt{6x^2+x-2}}.$

iii) Determine domains, the possible symmetries and prove the property indicated:

(1) $y = x^{8/3}$ is convex

(2) $y = x^{7/3}$ is convex on $\mathbb{R}^+$

(3) $y = x^{19/4}$ is convex

(4) $y = x^{4/7}$ is concave

(5) $y = x^{5/13}$ is convex on $\mathbb{R}^-$

(6) $y = x^{5/12}$ is concave.

11.3 Rational Functions

Definition A **rational function** is an element of the field of fractions $\mathbb{R}(x)$ of the polynomial ring $\mathbb{R}[x]$, cf. Remark 6.28. More prosaically, it is the quotient map of two polynomials $a(x)$ and $b(x) \not\equiv 0$

$$x \mapsto \frac{a(x)}{b(x)}.$$

We'll assume $\deg b > 0$ so not to consider polynomial maps again. The domain is always

$$\mathrm{Dom}\left(\frac{a}{b}\right) = \Big(\mathrm{Dom}(a) \cap \mathrm{Dom}(b)\Big) \setminus \{x_0 \mid b(x_0) = 0\}.$$

Without infinitesimal tools it's not possible to say a lot that is useful in general: only those tools allow to complete the study of these functions, and as a matter of fact they make it rather simple. With the present knowledge we may establish the sign, the intersections with the axes, certain symmetries and perhaps divine the asymptotic behaviour.

Supposing $a(x)$, $b(x)$ don't have non-constant common factors, there will be a vertical asymptote $x = x_0$ for each zero x_0 of the denominator b.

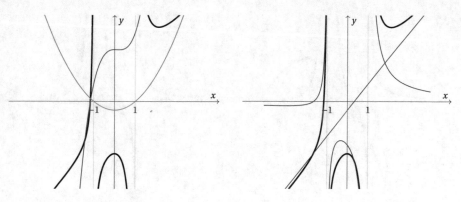

Fig. 11.11 Constructing a rational graph

When $\deg(a) \geqslant \deg(b)$: $y = \dfrac{a(x)}{b(x)} = q(x) + \dfrac{r(x)}{b(x)}$ is asymptotic to $y = q(x)$,

because as x grows indefinitely, $\deg(r) < \deg(b)$ forces the fraction $\dfrac{r(x)}{b(x)}$ to become

infinitesimal.

Examples

i) $f(x) = \dfrac{3x^3 - x^2 + 6}{x^2 - 1}$: using the graphs of numerator (blue, in Fig. 11.11 left)
and denominator (orange) it's not so easy to guess the behaviour of f (black).

   Dividing, though, we obtain $f(x) = 3x - 1 + \dfrac{3x + 5}{x^2 - 1}$: the quotient (straight

   line in Fig. 11.11 right) and $\dfrac{r(x)}{b(x)}$ (red) are more effective for inferring the

   rational graph.

ii) We saw in (10.3) the graph of $f(x) = \dfrac{\alpha x + \beta}{\gamma x + \delta}$, with $\alpha\delta - \beta\gamma \neq 0$ and $\gamma \neq 0$, is

   an equilateral hyperbola with asymptotic lines $y = \dfrac{\alpha}{\gamma}$, $x = -\dfrac{\delta}{\gamma}$, in agreement

   with $\lim\limits_{x \to -\delta/\gamma} f(x) = \infty$, $\lim\limits_{|x| \to \infty} f(x) = \alpha/\gamma$.

When $\deg(a) < \deg(b)$: we might attempt to sketch the graph using the partial
fraction decomposition (9.1).

Example To sketch $f(x) = \dfrac{-x^2 + 4x + 1}{x^3 - x^2 - x + 1}$ it's virtually hopeless to use the
numerator (blue in Fig. 11.12 left) and the denominator (orange).

   The situation is slightly better if we decompose $f(x) = -\dfrac{1}{x + 1} + \dfrac{2}{(x - 1)^2}$

(Fig. 11.12 right: the two partial fractions are blue and red).

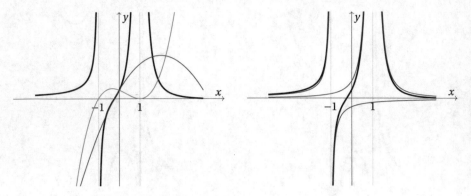

Fig. 11.12 Constructing another rational graph, II

At the cost of being pedantic, remember that when handling a rational expression $\dfrac{a(x)}{b(x)}$ in an equation/inequality we must impose the *existence condition* for the solutions: $b(x) \neq 0$.

Exercises

i) Solve

$$\frac{x^2 + 5x + 4}{x^4 + 1} > 0 \qquad \frac{x^3 + 8}{x^2 - 1} > 0 \qquad \frac{x - 2}{|2x + 1|} > -\frac{1}{3}x \qquad \frac{|x - 1|}{|3x + 1|} \leqslant 1.$$

ii) Find the expression and sketch the symmetric graphs to

$$y_1 = |x|, \quad y_2 = x^2 - x, \quad y_3 = x^2 - 2x + 3$$

about the x-axis, then the y-axis.

iii) Sketch the graphs of

(1) $y = \dfrac{1}{(x - 1)^2} + 3$ (3) $y = \left| x^2 - 5x + 4 \right|$ (6) $y = \sqrt{x + 2}$

                        (4) $y = \left| x^2 - 5|x| + 4 \right|$ (7) $y = -\sqrt{|3 - x|}$

(2) $y = \dfrac{1}{(2 - x)^5}$ (5) $y = \sqrt{x} + 2$ (8) $y = x|x| - |x| + x - 1.$

Chapter 12
Elementary Transcendental Functions

Pre-requisites: calculus for proving Theorem 12.2 and the properties of e. All this is optional, and may be skipped without consequences.

Definition A function that is not algebraic is called **transcendental**.

This chapter is devoted to the simplest transcendental functions: exponential maps (Sects. 12.1 and 12.3), logarithmic maps (Sect. 12.2) and trigonometric functions (Sect. 12.4). Also falling in this class are functions of type $y = x^r$ for $r \in \mathbb{R} \setminus \mathbb{Q}$, or $y = x^x$ etc.

Every algebraic function and the above transcendental ones are called **elementary** functions (☞ *differential algebra*).

Examples of non-elementary functions, for the record, include:

- the primitives of $\log(\log t)$, $\dfrac{1}{\log t}$, $\dfrac{e^t}{t}$, $\dfrac{\sin t}{t}$, ...

- Fresnel integrals $\displaystyle\int_0^x \sin(t^2)\,dt$, $\displaystyle\int_0^x \cos(t^2)\,dt$

- the error function $\operatorname{erf}(x) = \dfrac{2}{\sqrt{\pi}}\displaystyle\int_0^x e^{-t^2}\,dt$

- elliptic integrals, like $\displaystyle\int_0^x \sqrt{1-t^4}\,dt$, $\displaystyle\int_0^x \dfrac{dt}{\sqrt{1-k^2\sin^2 t}}$

- elliptic functions (Weierstraß' $\wp$, Jacobi's θ, ...)

- the gamma function $\Gamma(t) = \displaystyle\int_0^\infty x^{t-1}e^{-x}\,dx$,

- Bessel functions $J_\alpha(x) = \displaystyle\sum_{m=0}^\infty \dfrac{(-1)^m}{m!\,\Gamma(m+\alpha+1)}\left(\dfrac{x}{2}\right)^{2m+\alpha}$.

Non-elementary functions require the tools of ☞ *algebraic geometry, complex analysis* or *harmonic analysis* to be studied. A profusion of non-elementary maps can be found in the encyclopaedia [76]. As the examples show, the integration of

S. G. Chiossi, *Essential Mathematics for Undergraduates*,
https://doi.org/10.1007/978-3-030-87174-1_12

an elementary function typically results being non-elementary. In fact, the problem of integration represented a crucial incentive propelling the inception of one of the most glamorous mathematical theories, for reach and breadth: the theory of holomorphic maps.

12.1   Exponential Functions

Definition Let $a > 0, a \neq 1$ be a real number. One calls **exponential function in base** a

$$\exp_a : \mathbb{R} \longrightarrow \mathbb{R}$$
$$x \longmapsto a^x .$$

The domain is $\mathbb{R}$, the image $\mathbb{R}^+$. From Proposition 7.12 we know that $\exp_a$ is increasing for $a > 1$, decreasing for $a < 1$. Furthermore, the graphs of the exponential maps in base a and $1/a$ are symmetric about the y-axis, since

$$\left(\frac{1}{a}\right)^x = a^{-x} \tag{12.1}$$

as shown in Fig. 12.1.

Corollary 12.1 *The restriction* $\exp_a : \mathbb{R} \to (0, +\infty)$ *is bijective, hence invertible.*

Proof Exercise.   □

A topologically flavoured observation is the following:

Theorem 12.2 $\exp_a$ *is the unique continuous isomorphism between the additive group* $(\mathbb{R}, +)$ *and the multiplicative group* $(\mathbb{R}^+, \cdot)$ *such that* $\exp_a(1) = a$.

We'll need analytical tools without further explanation, starting from a lemma.

Fig. 12.1 Exponential maps

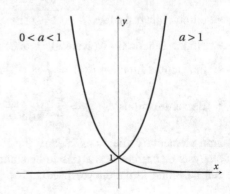

Lemma *For any $a > 0$ there exists a unique homomorphism $f : (\mathbb{Q}, +) \to (\mathbb{R}^+, \cdot)$ such that $f(1) = a$. In case $a > 1$ the homomorphism f is increasing.*

Proof Assuming said homomorphism f exists, it must act on $\mathbb{Z}$ as follows:

$$f(n) = f(1)^n = a^n \text{ for all } n \in \mathbb{N}, \quad f(0) = 1, \quad f(m) = 1/a^{-m} \text{ for } m < 0.$$

Therefore $f(m) = a^m$ for all $m \in \mathbb{Z}$. Now we take the latter as definition, and we extend f to $\mathbb{Q}$. First, we must have $f(1/n) = \sqrt[n]{a}$ for $n \in \mathbb{Z}^+$ (because the root in unique). Consequently $f(m/n) = (\sqrt[n]{a})^m$ for $n > 0, m \in \mathbb{Z}$. The last fact doesn't depend on the fraction representing the rational number m/n since

$$m/n = m'/n' \implies \left((\sqrt[n]{a})^m\right)^{nm'} = a^{mm'} = \left((\sqrt[n']{a})^{m'}\right)^{n'm}.$$

This shows $f(x + y) = f(x)f(y)$ for every $x, y, \in \mathbb{Q}$.

When $q = m/n > 0$ we have $f(q) = a^{m/n} > 1$. Take $p < p' \in \mathbb{Q}$ and set $p' = p + q$. Then $a^{p'} = a^p a^q > a^p$. □

Proof of Theorem 12.2. Let $f : (\mathbb{Q}, +) \to (\mathbb{R}^+, \cdot)$ denote the homomorphism built in the lemma. The idea is to extend it to a continuous homomorphism $(\mathbb{R}, +) \to (\mathbb{R}^+, \cdot)$. If such exists, the extension f is unique in view of continuity and the density of $\mathbb{Q} \subseteq \mathbb{R}$.

In case $a = 1$, we can clearly extend the constant map f.

Suppose $a > 1$. For any $m \in \mathbb{Z}^+$ one can prove the restriction $f\big|_{\mathbb{Q} \cap [-m,m]}$ is uniformly continuous, and $f(\mathbb{Q} \cap [-m, m]) \subseteq [a^{-m}, a^m]$. Therefore f extends to the intervals $[-m, m]$. These are nested, and any real number x belongs to one $[-m, m]$ as soon as $m > |x|$. Hence f is defined for any $x \in \mathbb{R}$, and is continuous. Taking the limit one sees f is still an increasing surjective homomorphism.

In case $0 < a < 1$ we consider the homomorphism $x \mapsto \left(\dfrac{1}{a}\right)^{-x}$: it is continuous, it maps 1 to a and it is unique. But $x \mapsto f(-x) = 1/f(x)$ sends 1 to $1/a > 1$ (it's unique), so $f(-x) = (1/a)^x$ i.e. $f(x) = (1/a)^{-x}$. □

Among all possible choices for a there is one base of paramount importance in mathematics.

Proposition *The rational sequence $a_n := \left(1 + \dfrac{1}{n}\right)^n$ is increasing and bounded:* $2 < a_n < 3$ *for all $n \in \mathbb{N}^*$. Therefore $\lim\limits_{n \to \infty} a_n = \sup\limits_{\mathbb{N}} a_n < 3$.*

Proof Using the binomial formula (Sect. 14.3)

$$a_n = 1 + \binom{n}{1}\frac{1}{n} + \binom{n}{2}\frac{1}{n^2} + \ldots + \binom{n}{n}\frac{1}{n^n}$$

$$= 2 + \frac{1}{2!} \frac{n(n-1)}{n^2} + \frac{1}{3!} \frac{n(n-1)(n-2)}{n^3} + \ldots + \frac{1}{n!} \frac{n!}{n^n}$$

$$= 2 + \frac{1}{2!} \left(1 - \frac{1}{n}\right) + \frac{1}{3!} \left(1 - \frac{1}{n}\right) \left(1 - \frac{2}{n}\right) + \ldots$$

$$+ \frac{1}{n!} \left(1 - \frac{1}{n}\right) \left(1 - \frac{2}{n}\right) \cdots \left(1 - \frac{n-1}{n}\right)$$

$$= a_{n-1} + \frac{1}{n!} \prod_{i=1}^{n-1} \left(1 - \frac{i}{n}\right) > a_{n-1}.$$

Moreover $a_1 = 2$, and since $j! \geqslant 2^{j-1}$, we have

$$a_n \leqslant 2 + \sum_{j=2}^{n} \frac{1}{j!} \leqslant 2 + \sum_{j=1}^{n-1} \frac{1}{2^j} = 2 + \frac{1}{2} \sum_{j=0}^{n-2} \frac{1}{2^j} = 2 + \frac{1}{2} \frac{1 - 1/2^{n-1}}{1 - 1/2} = 3 - \frac{1}{2^{n-1}} < 3.$$

Hence $\sup_{\mathbb{N}} a_n < \infty$, and by Proposition 19.16 the sequence converges to its least upper bound.   □

Definition (Bernoulli 1683) The limit

$$e := \lim_{n \to \infty} \left(1 + \frac{1}{n}\right)^n$$

is a real constant, with approximate value 2.71828, universally denoted by the letter '**e**'.

The discovery this number is not rational is due to Euler (who used the continuous fraction on p. 174). We shall present another classical argument:

Proposition (Fourier) *The number e is irrational.*

Proof We'll invoke a known fact:

$$e = \sum_{n=0}^{\infty} \frac{1}{n!}, \tag{12.2}$$

which we accept without further ado. Suppose by contradiction that $e = a/b$ for some coprime $a, b \in \mathbb{Z}$. Then $n!e$ is integer for every $n > b$. Because $n!/k!$ is an integer for all natural numbers $k \leqslant n$, then $N := n! \left(e - \sum_{k=0}^{n} \frac{1}{k!}\right)$ is integer too. But

$$N = \frac{1}{n+1} + \frac{1}{(n+1)(n+2)} + \frac{1}{(n+1)(n+2)(n+3)} + \ldots$$

$$< \frac{1}{n+1} + \frac{1}{(n+1)^2} + \frac{1}{(n+1)^3} + \cdots = \sum_{k=1}^{\infty} \frac{1}{(n+1)^k} = \frac{1}{1 - \frac{1}{n+1}} - 1 = \frac{1}{n}. \;\; \sharp\sharp$$

$\square$

In reality e is a transcendental number (cf. Definition 13.14): it is not the root of any polynomial in $\mathbb{Q}[x]$. To justify this we'd need more refined tools than what those furnished. You can look in [66].

The majority of mathematicians in all likelihood agrees on crowning the exponential function e^x in base e the most important function in mathematics and applications, by far. One of the many reasons is that $y = e^x$ is the basic solution to the differential equation $y' = y$, i.e. an eigenfunction of the differential operator $\dfrac{d}{dx}$ with eigenvalue 1 ($\text{\tiny ☞}$ differential equations).

Theorem 12.3 *For every* $x \in \mathbb{R}$

$$e^x := \sum_{n=0}^{\infty} \frac{x^n}{n!} = 1 + x + \frac{1}{2}x^2 + \frac{1}{6}x^3 + \frac{1}{24}x^4 + \frac{1}{120}x^5 + \cdots$$

Proof We will use (without proving) the fact that the series $E(x) := \sum_{n=0}^{\infty} \dfrac{x^n}{n!}$ converges (uniformly) for all x in bounded intervals of the real line. In this way $E : \mathbb{R} \to \mathbb{R}$ defines a (continuous) function. What we can show is that E is a homomorphism:

$$E(x)E(y) = \left(\sum_{n=0}^{\infty} \frac{x^n}{n!} \right) \left(\sum_{n=0}^{\infty} \frac{y^n}{n!} \right) = \sum_{n=0}^{\infty} \sum_{k=0}^{n} \frac{x^k y^{n-k}}{k!\,(n-k)!}$$

$$= \sum_{n=0}^{\infty} \frac{1}{n!} \sum_{k=0}^{n} \frac{n!}{k!\,(n-k)!} x^k y^{n-k} = \sum_{n=0}^{\infty} \frac{1}{n!} \sum_{k=0}^{n} \binom{n}{k} x^k y^{n-k}$$

$$= \sum_{n=0}^{\infty} \frac{1}{n!} (x+y)^n = E(x+y).$$

Formula (12.2) implies $E(1) = e$, so Theorem 12.2 allows to conclude $E = \exp_e$.

$\square$

The following table summarises common exponential equations and inequalities, and the standard solution technique.

Equation/inequality	Method
$a^x = k$ $(a > 0, \neq 1)$ and $k \in \mathbb{R}$	if $k > 0$, $x = \log_a k$.
$a^{f(x)} = a^{g(x)}$	$f(x) = g(x)$
$a^{f(x)} = b^{g(x)}$	set $b^{g(x)} = a^{g(x) \log_a b}$
$f(a^x) = 0$	set $a^x = t$ and solve $f(t) = 0$
$a^{f(x)} > a^{g(x)}$	if $a > 1$, $f(x) > g(x)$ if $a < 1$, $f(x) < g(x)$
$f(a^x) > c$	set $a^x = t$ and solve $f(t) > c$

Examples

(1) $2^{2x^2+x} - 2^{x^3+2x} = 0$.

As $\exp_2$ is one-to-one,

$$2^{2x^2+x} = 2^{x^3+2x} \iff x(2x+1) = x(x^2+2).$$

This is great news because we've reduced a transcendental equation to an algebraic (polynomial) one, so there remains to solve

$$0 = x(2x+1) - x(x^2+2) = x(-x^2+2x-1) = -x(x-1)^2 \iff x = 0, 1.$$

(2) $\left(\left(\frac{1}{7}\right)^{x+1}\right)^x > \frac{1}{49}$.

Let's rewrite as $\left(\frac{1}{7}\right)^{(x+1)x} > \left(\frac{1}{7}\right)^2$. The exponential in base $1/7$ is decreasing, so again we transform the problem into the algebraic inequality $(x+1)x < 2$, solved by $x \in (-2, 1)$.

(3) $4^x - 2 \cdot 2^x - 8 \leqslant 0$.

$$4^x - 2 \cdot 2^x - 8 \leqslant 0 \iff 2^{2x} - 2 \cdot 2^x - 8 \leqslant 0 \overset{2^x = t}{\iff} t^2 - 2t - 8 \leqslant 0$$
$$\iff -2 \leqslant t \leqslant 4 \qquad\qquad \iff -2 \leqslant 2^x \leqslant 8$$
$$\iff 2^x \leqslant 8 \qquad\qquad \iff x \leqslant 3.$$

(4) How many solutions does $\left(\sqrt{2+\sqrt{3}}\right)^x + \left(\sqrt{2-\sqrt{3}}\right)^x = 4$ have ?

First, set $r := \sqrt{2+\sqrt{3}} > 1$ so $\sqrt{2-\sqrt{3}} = \dfrac{1}{r}$, and the equation reads $r^x + r^{-x} = 4$. The map $y = r^x + r^{-x}$ has domain $\mathbb{R}$, is even and for $x \geqslant 0$ increasing:

$$x_1 > x_2 \geqslant 0 \implies 0 < \frac{1}{r^{x_2}} - \frac{1}{r^{x_1}} = \frac{r^{x_1} - r^{x_2}}{r^{x_1} r^{x_2}} < r^{x_1} - r^{x_2}$$

$$\implies r^{x_1} + r^{-x_1} > r^{x_2} + r^{-x_2}.$$

Hence the minimum value is $y(0) = 2 < 4$, which implies that the intersection between the graph of y and the line $y = 4$ consists of two symmetric points with respect to the y-axis.

This is also confirmed by solving directly

$$r^x + r^{-x} = 4 \iff r^{2x} - 4r^x + 1 = 0 \iff r^x = 2 \pm \sqrt{3} = r^{\pm 2} \iff x = \pm 2.$$

12.2 Logarithmic Functions

Choose $a \in (0, 1) \cup (1, +\infty)$ and consider the equation

$$a^x = y_0.$$

When $y_0 \leqslant 0$ there are no solutions because the exponential's graph doesn't meet $y = y_0$. But when $y_0 > 0$, by Corollary 12.1 the equation admits exactly one solution.

Definition Fix numbers $a > 0, a \neq 1$ and $y_0 > 0$. The number $x_0 =: \log_a y_0$, called **logarithm in base a of** y_0, is the exponent x_0 such that $a^{x_0} = y_0$.

Logarithms were invented by Bürgi around 1600 and Napier in 1614 in independent work.

Since

$$a^{\log_a y_0} = y_0 \quad \text{for all } y_0 > 0,$$

$$\log_a \left(a^{x_0} \right) = x_0 \quad \text{for all } x_0 \in \mathbb{R},$$

we are to all effects defining a 1-1 correspondence associating a unique number x_0 to each positive y_0. In other words,

Definition One calls **logarithmic function in base a**,

$$\log_a : (0, +\infty) \longrightarrow \mathbb{R}$$
$$x \longmapsto \log_a x,$$

the inverse map to $\exp_a$.

As inverse to the exponential, it satisfies

$$\text{Dom}(\log_a) = (0, +\infty) = \text{Im}(\exp_a), \qquad \text{Im}(\log_a) = \mathbb{R} = \text{Dom}(\exp_a).$$

Exercises Fix $0 < a \neq 1$, $x, y \in (0, +\infty)$, and z arbitrary. Prove

i) $\log_a(xy) = \log_a x + \log_a y$, $\log_a\left(\dfrac{x}{y}\right) = \log_a x - \log_a y$, $\log_a(x^z) = z \log_a x$.

In words, the log function transforms products in sums, powers in products. (Abstractly this is clear: $\exp_a$ isomorphism $\Longrightarrow \log_a = (\exp_a)^{-1}$ isomorphism).

ii) Suppose b is a positive number different from 1. Prove

$$\log_b x = \frac{\log_a x}{\log_a b}$$

for all $x > 0$, called *base-change formula* for logarithms.

iii) Deduce $\log_a(b) = -\log_{1/a}(b) = \log_{1/a}(1/b)$ for all $b > 0$.

Warning: $\log_a(x + y)$ *cannot* be expressed in terms of $\log_a x$ and $\log_a y$ (in an elementary way).

As $\exp_a$ and $\log_a$ are inverse, their graphs are mirrors with respect to the line $y = x$.

Corollary *The map $\log_a$ is*

when $a > 1$: increasing, negative on $(0, 1)$, positive on $(1, \infty)$ (Fig. 12.2 red);
when $a < 1$: decreasing, positive on $(0, 1)$ and negative on $(1, \infty)$ (Fig. 12.2 purple).

Proof Exercise.   □

The patent horizontal symmetry between the above curves is a manifestation of the fact $\log_a(x) = -\log_{1/a}(x)$, as a consequence of (12.1).

When the base is the number e, we'll omit the symbol and simply write $y = \log x$. (Some authors like to call this 'natural logarithm' and use the symbol ln). From an analytical point of view the definition is $\log x := \displaystyle\int_1^x \frac{dt}{t}$.

Fig. 12.2 Logarithmic
functions

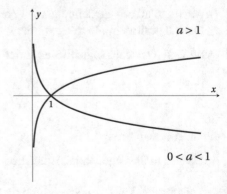

With this in place we can interpret the operation of raising to any power—even irrational, so that expression like $2^{\sqrt{3}}$, 5^{π} etc. finally make sense:

Definition 12.4 Consider real numbers $x > 0$ and $r \in \mathbb{R}$. Define

$$x^r := e^{r \log x} = \sum_{n=0}^{\infty} \frac{(r \log x)^n}{n!},$$

in agreement with Theorem 12.3.

Examples By this definition

$$2^{\sqrt{3}} = e^{\sqrt{3}\log 2} = 1 + \sqrt{3}\log 2 + \frac{3\log^2 2}{2} + \frac{\sqrt{3}\log^3 2}{2} + \frac{3\log^4 2}{8} + \cdots$$

$$\frac{1}{5^{\pi}} = e^{-\pi \log 5} = 1 - \pi \log 5 + \frac{\pi^2 \log^2 5}{2} - \frac{\pi^3 \log^3 5}{6} + \frac{\pi^4 \log^4 5}{24} - \cdots$$

The sum of infinitely many terms is indicative of the approximating process carried out as we compute said quantities.

Each time we have the expression $\log_a f(x)$ in an equation or inequality we must impose the *existence condition* for the solutions: $f(x) > 0$.

The table below displays a few customary logarithmic problems and the technique for solving.

Equation/inequality	Method
$\log_a x = b \quad (b \in \mathbb{R})$	$x = a^b$
$\log_a f(x) = b \quad (b \in \mathbb{R})$	if $f(x) > 0$, $f(x) = a^b$
$\log_a f(x) = \log_a g(x)$	if $f(x) > 0$ and $g(x) > 0$, $f(x) = g(x)$
$f(\log_a x) = 0$	set $\log_a x = t$ and solve $f(t) = 0$
$\log_a f(x) > \log_a g(x)$	if $a > 1, \quad f(x) > g(x)$ if $a < 1, \quad f(x) < g(x)$
$f(\log_a x) > c$	set $\log_a x = t$ and solve $f(t) > c$

Examples

(1) $\log_4 (x + 6) + \log_4 x = 2$.

The equation only makes sense when $\begin{cases} x + 6 > 0 \\ x > 0 \end{cases}$, so for $x \in (0, +\infty)$. In that case

$$\log_4 (x + 6) + \log_4 x = 2 \iff \log_4 \left(x^2 + 6x\right) = 2$$

$$\iff \log_4 \left(x^2 + 6x\right) = \log_4 16$$

$$\iff x^2 + 6x = 16$$

$$\iff x_1 = -8 \notin \mathbb{R}^+ \ (\mathbf{X}) \quad \vee \quad x_2 = 2 \ (\checkmark).$$

(2) $\log_2 (x + 1) = \log_4 (2x + 5)$.

The solutions' existence domain is $\{x > -1\} \cap \{x > -5/2\} = (-1, +\infty)$. As it easier to work in a single base, we put everything in base 2, so $\log_2 (x + 1) = \dfrac{\log_2 (2x + 5)}{\log_2 4}$. Then

$$\log_2(x + 1) = \tfrac{1}{2} \log_2(2x + 5) \iff \log_2(x + 1) = \log_2(2x + 5)^{\frac{1}{2}}$$

$$\iff x + 1 = \sqrt{2x + 5} \iff x^2 - 4 = 0$$

$$\iff x_1 = -2 \ (\mathbf{X}) \quad \vee \quad x_2 = 2 \ (\checkmark).$$

(3) $\log_2 x - \log_2 3 < \log_2(x + 2)$.

The domain is $(0, \infty)$. Immediately,

$$\log_2 \frac{x}{3} < \log_2(x + 2) \iff \frac{x}{3} < x + 2 \iff x > -\frac{1}{3},$$

and so $x \in (0, \infty)$.

(4) $\log_2^3 x - 2 \log_2 x > 0$.

There might be solutions only in $(0, +\infty)$.

Set $t = \log_2 x$ to get $t^3 - 2t = t(t^2 - 2) > 0 \iff t > \sqrt{2}$, or $-\sqrt{2} < t < 0$.

Substituting back to x: $\log_2 x > \sqrt{2}$ or $-\sqrt{2} < \log_2 x < 0$.

In conclusion the solution is $x \in \left(2^{\sqrt{2}}, +\infty\right) \cup \left(\frac{1}{2^{\sqrt{2}}}, 1\right)$.

(5) $\dfrac{2^{x+1} 5^{x-1}}{3^x} = 2$.

Multiplying by $3^x > 0$ both sides we obtain

$$2^{x+1} 5^{x-1} = 2 \cdot 3^x \iff 2^x 5^{x-1} = 3^x \overset{*}{\iff} \log 2^x + \log 5^{x-1} = \log 3^x$$

$$\iff x \log 2 + x \log 5 - x \log 3 = \log 5 \iff x = \frac{\log 5}{\log 2 + \log 5 - \log 3}.$$

Had we chosen to apply $\log_{10}$ at step $*$, instead of using base e, we would have obtained $x = \dfrac{\log_{10} 5}{\log_{10} 2 + \log_{10} 5 - \log_{10} 3}$. Is it the same solution?

Exercises To finish off here is an assortment of problems on exponentials and logarithms.

i) For each case find a polynomial $p(x)$ so that the expression does *not* define a function:

$$\sqrt[4]{p(x)} \qquad \sqrt[7]{p(x)} \qquad e^{p(x)} \qquad \frac{1}{e^{p(x)}}$$

$$\log\left(p(x)\right) \qquad \frac{1}{\log|p(x)|} \qquad \log_{p(x)} p(x) \qquad p(x)^{p(x)}.$$

ii) Solve

$$(3^x)^{x+3} = 1 \qquad\qquad \log_3(x-8) > 0 \qquad\qquad \log_3|x| = 2$$

$$5^x - 4 = 5^{1-x} \qquad\qquad 2^{3x+1} + 2^{3x+2} - 3 \cdot 8^x = 32 \qquad 7^{2x} - 5 \cdot 7^x + 6 < 0$$

$$\left(\frac{1}{3}\right)^{x+2} > 1 \qquad\qquad |e^x - 2| = 1 \qquad\qquad \frac{\log_2(1-x)}{\log_2(x^2+x)} = \frac{1}{3}$$

$$\log_{10}(x+5) = 1 \qquad\qquad \log_2 \frac{x-1}{x+1} = \log_2 x \qquad\qquad \log(9-x) < 0$$

$$\log_a^2 x - 2\log_a x + 1 = 0 \qquad \log_2(\log_3(6x+1)) = 0 \qquad \frac{\log_3(x+2) - 1}{2 + \log_3 x} = 1$$

$$\log_3(x^2+1) > 0 \qquad\qquad 5^{x^2-x} > 0 \qquad\qquad \log_2 x - 3 > -\frac{2}{\log_2 x}$$

$$\frac{e^{2x} - 8e^x + 7}{e^x - 4} < 0 \qquad\qquad e^{|x|} - 3 < 0 \qquad\qquad \frac{2e^{-x} + e^x - 3}{1 - e^x} > 0.$$

iii) Solve the equations

$$e^{-|x+2|} = k \qquad\qquad |\log|x|| = k$$

as $k \in \mathbb{R}$ varies.

iv) Depending on the choice of real number α, estimate graphically when equations

$$2^x = \alpha - x, \qquad e^x = \alpha x, \qquad \log x = \alpha - x, \qquad \log x = x - \alpha$$

admit solutions, and how many.

12.3 Hyperbolic Functions

Hyperbolic functions are traditionally considered similar to trigonometric ones, because of the formal and practical kinship. I prefer—in this elementary, introductory context—to discuss them after exponential maps to emphasise that their

Fig. 12.3 Hyperbolic sine and cosine

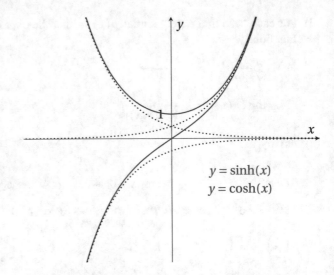

$$y = \sinh(x)$$
$$y = \cosh(x)$$

properties descend from the previous sections. Besides, we shall only talk about three hyperbolic functions, namely the hyperbolic sine, cosine and tangent.

Definition The **hyperbolic cosine** is the map $\cosh : \mathbb{R} \to \mathbb{R}$

$$x \mapsto \cosh x := \frac{e^x + e^{-x}}{2}.$$

The graph of cosh is a curve called *catenary*, because it is the shape assumed by an idealised chain (*catena*, in Latin) hanging by the ends under its own weight. As it is only subject to the force of gravity, physically the profile is that of least potential energy (Fig. 12.3).

Proposition *The map* cosh *is positive, even, and for* $x \geqslant 0$ *increasing.*

Proof By definition $\cosh(x)$ is the even part of e^x. It's positive as sum of the positive maps $e^x/2$ and $e^{-x}/2$.
Monotonicity: taking $x_1 > x_2 \geqslant 0$ we obtain $e^{x_1} > e^{x_2} \geqslant 1 \implies e^{-x_2} > e^{-x_1}$ and so

$$0 < e^{-x_2} - e^{-x_1} = \frac{e^{x_1} - e^{x_2}}{e^{x_1} e^{x_2}} < e^{x_1} - e^{x_2} \implies e^{x_1} + e^{-x_1} > e^{x_2} + e^{-x_2}.$$

$\square$

The image of the map is the interval $[1, +\infty)$. In fact, both $e^x/2$ and $e^{-x}/2$ have image $\mathbb{R}^+$, so $\cosh(\mathbb{R}) \subseteq [1, +\infty)$. But cosh is increasing on $[0, +\infty)$, and $1 = \cosh(0)$ is its minimum value. In a short while we'll settle this matter, too.

Besides, using the Taylor expansion of the exponential map it becomes clear that cosh is equal to the sum

$$\cosh x = \sum_{n=0}^{\infty} \frac{x^{2n}}{(2n)!} = 1 + \frac{x^2}{2!} + \frac{x^4}{4!} + \frac{x^6}{6!} + \cdots$$

of even powers (all increasing if $x > 0$).

Definition The **hyperbolic sine** is the map $\sinh : \mathbb{R} \to \mathbb{R}$

$$x \mapsto \sinh x := \frac{e^x - e^{-x}}{2}.$$

Proposition $\sinh$ *is odd, increasing, and positive for* $x > 0$. *Its image is* $\mathbb{R}$.

Proof First of all, $\sinh$ is the odd part of e^x. It is increasing as sum of the increasing maps $e^x/2$ and $-e^{-x}/2$. The fact that

$$x > 0 \implies e^x > 1 > e^{-x} > 0$$

forces it to be positive on one hand, on the other it guarantees upper unboundedness (like exp). And being odd it must be unbounded below as well. Hence $\mathrm{Im}(\sinh) = \mathbb{R}$. (Another explanation will come when we invert it.) □

These facts are clear once we know the Taylor series

$$\sinh x = \sum_{n=0}^{\infty} \frac{x^{2n+1}}{(2n+1)!} = x + \frac{x^3}{3!} + \frac{x^5}{5!} + \frac{x^7}{7!} + \cdots$$

The names come from the analogy with relation (12.6): just as $\sin, \cos$, the hyperbolic functions obey an equation of triangular flavour:

Proposition *For every* $x \in \mathbb{R}$

$$\cosh^2 x - \sinh^2 x = 1.$$

Proof Exercise. □

The above relationship, called hyperbolic Pythagoras theorem lives naturally on the equilateral hyperbola $x^2 - y^2 = 1$, whose representation as a parametric curve (Fig. 12.4) reads $x(t) = \cosh t$, $y(t) = \sinh t$.

Other symbols appearing in the literature are $\mathrm{sh}\, x := \sinh x$, $\mathrm{ch}\, x := \cosh x$.

Exercise Prove the addition formula $\sinh(a + b) = \sinh(a)\cosh(b) + \cosh(a)\sinh(b)$ for $\sinh$, valid for all $a, b \in \mathbb{R}$.

Example Let's practice a little the basic properties to prove that $\sinh(\cosh x) \geqslant \cosh(\sinh x)$ for all $x \in \mathbb{R}$. From the above exercise

Fig. 12.4 Hyperbolic
pythagoras theorem

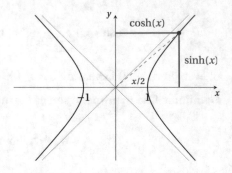

$$\sinh^2(a+b) - \sinh^2(a-b) = (\sinh a \cosh b + \cosh a \sinh b)^2$$

$$- (\sinh a \cosh b - \cosh a \sinh b)^2$$

$$= 4\sinh(a)\cosh(a)\sinh(b)\cosh(b) = \sinh(2a)\sinh(2b).$$

Looking at the series of sinh we see $\sinh t - t$ is a sum of non-negative terms for $t \geqslant 0$, so $\sinh t \geqslant t$ whenever $t \geqslant 0$. Therefore

$$\sinh^2(\cosh x) - \sinh^2(\sinh x) = \sinh^2\left(\tfrac{e^x}{2} + \tfrac{e^{-x}}{2}\right) - \sinh^2\left(\tfrac{e^x}{2} - \tfrac{e^{-x}}{2}\right)$$

$$= \sinh(e^x)\sinh(e^{-x}) \geqslant e^x e^{-x} = 1$$

$$= \cosh^2(\sinh x) - \sinh^2(\sinh x),$$

implying the claim.

Exercise Show that

i) $y - \sqrt{y^2+1} < 0$ for all real numbers y,

ii) $y + \sqrt{y^2-1} = \dfrac{1}{y-\sqrt{y^2-1}}$ for every $|y| \geqslant 1$.

Now we shall address the invertibility of sinh and cosh.

Inverting the hyperbolic sine is no problem: by the aforementioned exercise $\sinh x = y \iff e^x = y + \sqrt{y^2+1} \iff x = \log\left(y + \sqrt{y^2+1}\right)$, which proves sinh is bijective, and that it has image $\mathbb{R}$.

As for the hyperbolic cosine things are slightly more complex, because the algebraic manipulations give $\cosh x = y \iff e^x = y \pm \sqrt{y^2-1} \iff x = \pm\log\left(y + \sqrt{y^2-1}\right)$. The sign ambiguity is no surprise, given that cosh is even, but monotone on $x \geqslant 0$ and $x \leqslant 0$ separately. This fact reflects the possible choices for restricting the domain if we expect it to become bijective.

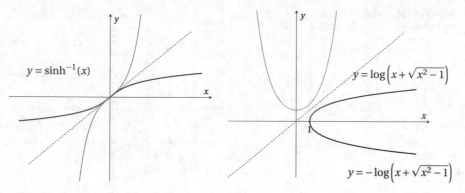

Fig. 12.5 (Left) Hyperbolic arcsine. (Right) 'Partial inverses' to cosh

By the previous exercise,

$$-\log\left(y + \sqrt{y^2 - 1}\right) = \log\left(y + \sqrt{y^2 - 1}\right)^{-1} = \log\left(y - \sqrt{y^2 - 1}\right).$$

When we restrict cosh to $[0, +\infty)$ or $(-\infty, 0]$ we are implicitly choosing the corresponding logarithm branch (☞ *complex analysis*):

> if $x \in [0, +\infty)$ we must take $x = \log\left(y + \sqrt{y^2 - 1}\right)$ red in Fig. 12.5;
>
> if $x \in (-\infty, 0]$ necessarily $x = \log\left(y - \sqrt{y^2 - 1}\right)$ purple in Fig. 12.5.

Definition The inverse functions of sinh: $\mathbb{R} \to \mathbb{R}$ and cosh: $[0, +\infty) \to [1, +\infty)$ are called, respectively:

hyperbolic arcsine $\operatorname{arcsinh}(x) = \sinh^{-1}(x) = \log\left(x + \sqrt{x^2 + 1}\right), x \in \mathbb{R}$

hyperbolic arccosine $\operatorname{arccosh}(x) = \cosh^{-1}(x) = \log\left(x + \sqrt{x^2 - 1}\right) x \geqslant 1.$

The third interesting hyperbolic function is the quotient of sinh and cosh:

Definition The **hyperbolic tangent** is the map tanh : $\mathbb{R} \to (-1, 1)$

$$x \mapsto \tanh x := \frac{\sinh x}{\cosh x} = \frac{e^{2x} - 1}{e^{2x} + 1}.$$

Fig. 12.6 Hyperbolic tangent
and arctangent

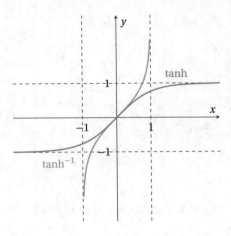

Exercises 12.5 Prove that

i) one can express $\cosh x = \dfrac{1+t^2}{1-t^2}$, $\sinh x = \dfrac{2t}{1-t^2}$ as rational functions, in
terms of the variable $t = \tanh \dfrac{x}{2}$.

ii) $\tanh$ is a 1-1 correspondence with inverse $\tanh^{-1} : (-1, 1) \to \mathbb{R}$

$$\tanh^{-1}(x) = \frac{1}{2} \log \frac{1+x}{1-x} ,$$

called **hyperbolic arctangent** (Fig. 12.6).

iii) $\tanh^{-1}(s) + \tanh^{-1}(t) = \tanh^{-1} \left(\dfrac{t+s}{1+ts} \right)$ for all $|t|, |s| < 1$.

iv) Use i) or iii) to deduce $\tanh(2x) = \dfrac{2 \tanh(x)}{1 + \tanh^2(x)}$.

The map $\tanh^{-1}$ is an important player to describe natural distances on spaces
with negative curvature, see (17.7), (18.5) and Proposition 19.3 (☞ *Riemannian
geometry*).

Other, less relevant, hyperbolic functions include the reciprocals

$$\coth x = \frac{1}{\tanh x} \qquad \operatorname{sech} x = \frac{1}{\cosh x} \qquad \operatorname{cosech} x = \frac{1}{\sinh x}$$

respectively called hyperbolic cotangent, secant and cosecant.

12.4   Trigonometric Functions

Defining precisely the trigonometric functions is trickier than one might imagine. For this reason the section launches with intuitive and informal considerations, which should be taken for what they are (see [88] for an elementary approach).

An **angle** is made by two rays emanating from a common initial point (cf. p. 403).

On the Cartesian plane angles θ are measured anti-clockwise, starting from the x-axis. Defining $\rho = \text{dist}(Q, O)$, and indicating by $\widehat{QE}$ the length of the circular arc joining Q and E, we set

$$\theta := \frac{\widehat{QE}}{\rho}$$

to denote the angle at the centre subtended by the arc.

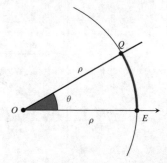

The unit is the **radian**, the measure of an angle subtended by an arc equal in length to the radius: $\widehat{QE} = \rho \rightsquigarrow \theta = 1$ rad.

By declaring the circle $S^1(\rho)$ of radius $\rho > 0$ centred at the origin has length $2\pi\rho$, to all effects we set up the conversion between degrees and radians: $360° : 2\pi = \theta° : \theta$, so that the measure of angles (mod 2π) is a real number belonging in $[0, 2\pi)$ (Fig. 12.7).

Remark In this set-up the transcendental number π is by definition half the length of the unit circle:

$$\pi := 4 \arctan(1) = \int_{-1}^{1} \frac{dt}{\sqrt{1 - t^2}}. \tag{12.3}$$

There exist scores of expressions that may be adopted as definitions of π: some originate in trigonometry (the above formula), others in ☞ *number theory*, $\pi =$

Fig. 12.7 Angles in radians

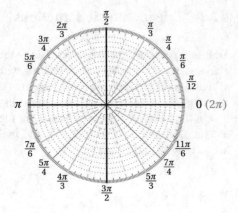

Fig. 12.8 Pythagoras
theorem: linking Cartesian
and polar coordinates

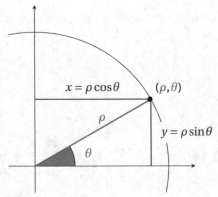

$\sqrt{6 \sum\limits_{k=0}^{\infty} \dfrac{1}{k^2}}$. The bibliography on the subject is monumental, see for example [5, 6, 14, 19].

Take a point $Q = (x, y)$ on $S^1(\rho)$, and let θ be the corresponding angle. Our first definition is informal, and is only meant to set things in motion and allow us to start working:

$$\sin \theta = \frac{y}{\rho}, \qquad \cos \theta = \frac{x}{\rho}. \tag{12.4}$$

This reveals $\sin \theta, \cos \theta$ are the lengths of the projections onto the axes of the segment OQ (Fig. 12.8).

The universally known *Pythagoras theorem* 18.2

$$x^2 + y^2 = \rho^2, \tag{12.5}$$

guarantees immediately that

$$\sin^2\theta + \cos^2\theta = 1, \tag{12.6}$$

which is considered the fundamental relationship between the trigonometric functions.

Exercise Deduce from (12.6) (or (12.4)) that

$$\left|\sin\theta\right| \leqslant 1, \quad \text{and} \quad \left|\cos\theta\right| \leqslant 1, \quad \forall\theta \in \mathbb{R}.$$

Definition The numbers $\rho > 0$ and $\theta \in [0, 2\pi)$ defined by (12.5), (12.4) are called **polar coordinates** of the point $Q = (x, y)$ on the plane.

Whilst in the Cartesian frame Oxy the coordinate curves $\{x = x_0\}, \{y = y_0\}$ form two pencils of parallel lines that are orthogonal to one another, (ρ, θ) define a polar frame system polar in which the level sets are

$$\text{circles}: \ S^1(\rho_0) = \{(\rho_0, \theta) \mid \rho_0 > 0, \theta \in \mathbb{R}\} \quad \text{and} \quad \text{rays}: \ \{(\rho, \theta_0) \mid \theta_0 \in \mathbb{R}, \rho > 0\} \approx (0, +\infty).$$

The two families of curves are automatically perpendicular, making the polar frame system orthogonal (Fig. 12.9).

In this way the point $Q \in S^1(\rho)$ will have Cartesian coordinates given by (12.4). For the time being (until Definition 13.3) we indicate with $\theta_0 \in [0, 2\pi)$ the unique real number solution to

$$\cos\theta_0 = \frac{x}{\sqrt{x^2 + y^2}}, \qquad \sin\theta_0 = \frac{y}{\sqrt{x^2 + y^2}},$$

i.e. the anti-clockwise angle between the x-axis and the half-line OQ. No angle is associated to the origin $O = (0, 0)$, because of (12.4). As a matter of fact:

Fig. 12.9 Level curves in polar coordinates

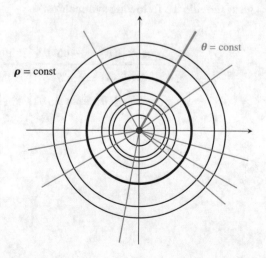

Corollary 12.6 *Polar coordinates* (ρ, θ) *define a 1-1 correspondence between* $\mathbb{R}^+ \times [0, 2\pi)$ *and* $\mathbb{R}^2 \setminus \{(0, 0)\}$.

Proof The required map is (12.4), whose inverse associates with any $(x, y) \neq (0, 0)$ the polar pair $(\sqrt{x^2 + y^2}, \theta_0)$.   □

Below we have angles and the corresponding values of sine and cosine, deduced from equations (12.4):

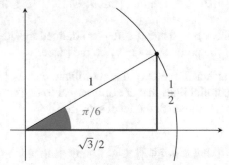

θ	0	$\dfrac{\pi}{6}$	$\dfrac{\pi}{4}$	$\dfrac{\pi}{3}$	$\dfrac{\pi}{2}$	π	$\dfrac{3\pi}{2}$
sin	0	$\dfrac{1}{2}$	$\dfrac{\sqrt{2}}{2}$	$\dfrac{\sqrt{3}}{2}$	1	0	-1
cos	1	$\dfrac{\sqrt{3}}{2}$	$\dfrac{\sqrt{2}}{2}$	$\dfrac{1}{2}$	0	-1	0

The numbers in column two/four are the legs of a right triangle of hypotenuse 1 and one angle $\pi/6$ (itself the half of an equilateral triangle).

Similar considerations from elementary geometry (congruence of right triangles) provide the following symmetries:

$\cos(\pi \pm \theta) = -\cos\theta$	$\sin(\pi \pm \theta) = \mp\sin\theta$
$\cos(-\theta) = \cos\theta$	$\sin(-\theta) = -\sin\theta$
$\cos\left(\dfrac{\pi}{2} \pm \theta\right) = \mp\sin\theta$	$\sin\left(\dfrac{\pi}{2} \pm \theta\right) = \cos\theta$

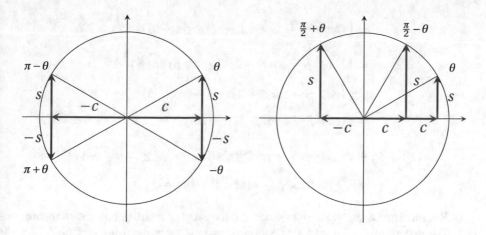

Exercises

i) Evaluate sin, cos on the angles $\dfrac{5\pi}{6}$, $\dfrac{2\pi}{3}$, $\dfrac{7\pi}{2}$, $\dfrac{5\pi}{4}$, 5π and plot them over the unit circle in the Cartesian plane.

ii) Interpret the last table's second row in terms of symmetry (evenness / oddness) of sin and cos.

iii) Explain why negative numbers $\theta < 0$ correspond to clockwise angles.

The addition formulas

$$\cos(\theta_1 + \theta_2) = \cos\theta_1 \cos\theta_2 - \sin\theta_1 \sin\theta_2$$
$$\sin(\theta_1 + \theta_2) = \cos\theta_1 \sin\theta_2 + \sin\theta_1 \cos\theta_2$$

(12.7)

too, can be proved elementarily. We shall not do it because they fall out of the multiplication law of complex numbers (13.2). A useful special case are the duplication formulas

$$\sin 2\theta = 2\sin\theta \cos\theta, \qquad \cos 2\theta = \cos^2\theta - \sin^2\theta = 2\cos^2\theta - 1 = 1 - 2\sin^2\theta.$$

The above relations confirm our first hunch that the trigonometric functions are very far from being linear.

Examples

i) Let's express $\dfrac{\cos 6\beta + 1}{2}$ and $\sin\left(2\beta + \frac{\pi}{6}\right)$ in terms of $\sin\beta$, $\cos\beta$ only.

$$\frac{\cos 6\beta + 1}{2} = \cos^2 3\beta = \left(\cos(2\beta + \beta)\right)^2 = (\cos 2\beta \cos\beta - \sin 2\beta \sin\beta)^2$$

$$= \left((2\cos^2\beta - 1)\cos\beta - (2\sin\beta\cos\beta)\sin\beta\right)^2$$

$$= \left(2\cos^3\beta - \cos\beta - 2(1 - \cos^2\beta)\cos\beta\right)^2$$

$$= (2\cos^3\beta - \cos\beta - 2\cos\beta + 2\cos^3\beta)^2$$

$$= (4\cos^3\beta - 3\cos\beta)^2 = 16\cos^6\beta - 24\cos^4\beta + 9\cos^2\beta.$$

$$\sin\left(2\beta + \tfrac{\pi}{6}\right) = \sin 2\beta\cos\tfrac{\pi}{6} + \cos 2\beta\sin\tfrac{\pi}{6} = \tfrac{\sqrt{3}}{2}\sin 2\beta + \tfrac{\sqrt{3}}{2}\cos 2\beta$$

$$= \sqrt{3}\sin\beta\cos\beta + \tfrac{1}{2}(\cos^2\beta - \sin^2\beta)$$

ii) Several among the ways to approximate the number π are geometric in nature. For instance from Eq. (12.3) it follows that π is the area of the unit disc. The area of a regular polygon of m sides inscribed in the unit circle equals $A(m) = m\cos\tfrac{\pi}{m}\sin\tfrac{\pi}{m} = \tfrac{m}{2}\sin\tfrac{2\pi}{m}$. Now set

$$a_n := A(2^n) = 2^{n-1/2}\sqrt{1 - \sqrt{1 - 4^{1-n}a_{n-1}^2}}.$$

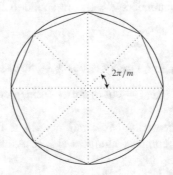

This sequence is increasing and bounded ($a_n < 4 =$ area of the circumscribed square), and Catalan (1842) demonstrated that

$$\pi = \lim_{n\to\infty} a_n$$

whilst studying continuous fractions [48]. Nowadays we phrase this as the fundamental relation $\lim\limits_{t\to 0}\dfrac{\sin t}{t} = 1$.

Although (12.4) is a legitimate definition, it has the disadvantage of depending on certain choices. If we want to be more formal, the aproblematic (true) definitions go as follows:

Definition 12.7 One calls **cosine** the function $\cos \colon \mathbb{R} \to [-1, 1]$

$$x \mapsto \cos x := \sum_{n=0}^{\infty} (-1)^n \frac{x^{2n}}{(2n)!} = 1 - \frac{x^2}{2!} + \frac{x^4}{4!} - \frac{x^6}{6!} + \frac{x^8}{8!} - \cdots$$

and **sine** the function $\sin \colon \mathbb{R} \to [-1, 1]$

$$x \mapsto \sin x := \sum_{n=0}^{\infty} (-1)^n \frac{x^{2n+1}}{(2n+1)!} = x - \frac{x^3}{3!} + \frac{x^5}{5!} - \frac{x^7}{7!} + \cdots$$

We won't show the above series converge on $\mathbb{R}$, nor that these objects coincide with the functions seen earlier (it is a standard chapter in any analysis course) (Fig. 12.10).

Regarding the uncanny resemblance between the two graphs, the last row of the earlier table tells that the graph of sin is the translate of the graph of cos, with a shift (a.k.a. phase difference) of $\pi/2$.

The idea behind the infinite sum in Definition 12.7 is that successive truncations $x - \frac{x^3}{6} + \frac{x^5}{120} - \cdots + (-1)^k \frac{x^{2k+1}}{(2k+1)!}$, called *Taylor polynomials*, approximate increasingly better the function in question as k grows, especially away from the origin (Fig. 12.11).

Proposition *The function* $\cos \colon \mathbb{R} \to [-1, 1]$ *is*

- *2π-periodic, onto, even*
- *positive on $(-\pi/2, \pi/2)$, negative on $(\pi/2, 3\pi/2)$*
- *decreasing on $[0, \pi]$ and increasing on $[\pi, 2\pi]$.*

The function $\sin \colon \mathbb{R} \to [-1, 1]$ *is*

- *2π-periodic, onto, odd*
- *positive on $(0, \pi)$, negative on $(\pi, 2\pi)$*
- *increasing on $[-\pi/2, \pi/2]$, decreasing on $[\pi/2, 3\pi/2]$.*

Proof Exercise (use the previous results). □

The next inequality has countless applications. Its proof depends on elementary arguments (sketched below) or analytical ones.

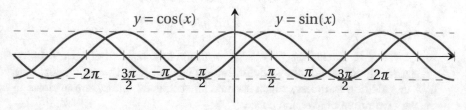

Fig. 12.10 Cosine and sine

Fig. 12.11 'Tayloring' polynomials to fit $y = \sin x$

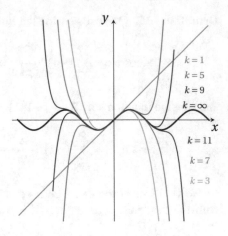

Proposition 12.8 *For every $x \in \mathbb{R}$, $\left|\sin x\right| \leqslant \left|x\right|$.*

Proof Since sine is odd it suffices to prove $\sin(x) \leqslant x$ when $x > 0$. Actually, just for $0 < x < 1$, since sine is bounded above by 1.

On the circle of radius $\rho = 1$ we identify the angle x with the arc x, Fig. 12.12. The segment $\overline{EQ}$ is shorter than the arc x (the shortest distance between points in the plane is realised by a straight line). The same segment is also the hypotenuse of a right triangle $Q\overset{\triangle}{E}T$ with leg $\overline{QT} = \sin(x)$. Therefore $\sin(x) \leqslant \overline{EQ} \leqslant x$.   □

Remark (Alternative Definitions)

i) sin, cos can be defined equivalently by means of the complex exponential map

$$e^{ix} = \cos x + i \sin x, \quad \forall x \in \mathbb{R},$$

see Remark 13.4. Starting from the observation that $h(x) = e^{ix}$ is the unique non-trivial continuous homomorphism $h \colon \mathbb{R} \to S^1 \subseteq \mathbb{C}$, hence periodic (this fact generalises Theorem 12.2), it can be proved that its real and imaginary parts $f = \operatorname{Re} h$, $g = \operatorname{Im} h$ satisfy

$$\begin{cases} f(t)^2 + g(t)^2 = 1 \\ f(t+s) = f(t)f(s) - g(t)g(s) \\ g(t+s) = f(t)g(s) + f(s)g(t) \end{cases},$$

for all $s, t \in \mathbb{R}$. Excluding $f \equiv 1$, $g \equiv 0$, the system has only one solution (f, g), which one could take as definition for $\cos := f$, $\sin := g$. (Aside: note how the addition formulas, which seemed an accidental oddity, are intrinsic to the very definition of sine and cosine.)

Fig. 12.12 $|\sin x| \leqslant |x|$

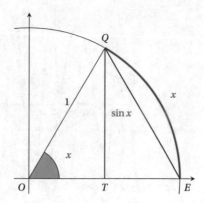

ii) A further characterisation (secretly, the same one) arises through differential equations. The initial-value problem $y''(x) + y(x) = 0, \quad y(0) = 0, \quad y'(0) = 1$ admits an analytical solution $y(x) = \sum_{n=0}^{\infty} c_n x^n$ on a neighbourhood of 0. By substitution, the coefficients fulfil $c_0 = 0, c_1 = 1$ and the recursive relation $(n+1)(n+2)c_{n+2} + c_n = 0$ for $n > 1$, so $c_{2k} = 0, c_{2k+1} = \dfrac{(-1)^k}{(2k+1)!}$ for all $k \in \mathbb{N}$. But these are nothing more than the Taylor coefficients in the sine series of Definition 12.7.

The cosine arises similarly, as the unique solution to the Cauchy problem $y'' + y = 0, \quad y(0) = 1, \quad y'(0) = 0$.

Example We wish to study the map

$$y = A \sin(\omega x + \phi)$$

where $\omega > 0$ and $A, \phi \in \mathbb{R}$. The graph is an **elementary wave** whose behaviour is sinusoidal (Fig. 12.13).

- $|A|$ is called **amplitude** and controls the width of the wave's range $\big[-|A|, |A|\big]$;
- ω is the **angular frequency**:
   when $\omega > 1$ the wave is compressed (compared to the profile of sin),
   when $0 < \omega < 1$ it is stretched.
   This quantity determines the wave's **period** $T = 2\pi/\omega$, a.k.a. *wavelength*, namely the distance between successive crests or successive troughs. The **frequency** $\nu = 1/T$ indicates how fast the curve oscillates, representing the number of wave crests passing through a point per unit of time.
- ϕ is the **phase** and detects a horizontal displacement with respect to the origin.

The names of A, ω, ϕ are dictated by physics' tradition (the simple pendulum), since the function is a solution to the **simple harmonic motion** $y'' + \omega^2 y = 0$. ☞

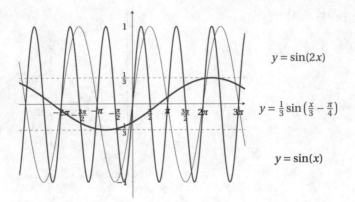

$$y = \sin(2x)$$

$$y = \tfrac{1}{3}\sin\left(\tfrac{x}{3} - \tfrac{\pi}{4}\right)$$

$$y = \sin(x)$$

Fig. 12.13 Elementary waves

Fig. 12.14 The periodic signal $y = \cos(3x) - \sin(x) + \sin(3x) + \sin(5x)$

Fourier theory explains that any sufficiently regular periodic function decomposes as an infinite sum of elementary waves, see Fig. 12.14.

Take for instance $s(x) = \dfrac{x}{\pi}$, with $|x| < \pi$, and extend it periodically to the entire line $\mathbb{R}$. The resulting periodic map is called 'sawtooth wave' (Fig. 10.14), and can be written as the series $\dfrac{2}{\pi} \displaystyle\sum_{n=1}^{\infty} \dfrac{(-1)^{n+1}}{n} \sin(nx)$, for $x \notin (2\mathbb{Z} + 1)\pi$.

Exercise Compute the period of the wave in Fig. 12.14.

The term 'simple' differentiates the above situation from complex harmonic motions (compound pendulums et al.), whose trajectories are the Bowditch–Lissajous curves: $x(t) = A\sin(\omega_1 t + \phi)$, $y(t) = B\sin(\omega_2 t)$, where $\omega_1/\omega_2 \in \mathbb{Q}$ (☞ *differential geometry*) (Fig. 12.15).

Definition The function $\tan : \left\{ x \in \mathbb{R} \;\middle|\; x \neq \dfrac{\pi}{2} + k\pi, \; k \in \mathbb{Z} \right\} \to \mathbb{R}$

$$\tan x := \frac{\sin x}{\cos x}$$

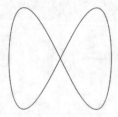 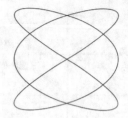 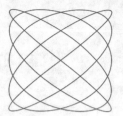

Fig. 12.15 Lissajous curves with ratios $1/2$, $3/2$, $5/4$

Fig. 12.16 The tangent function, with asymptotic lines $x = \frac{\pi}{2} + k\pi$, $k \in \mathbb{Z}$

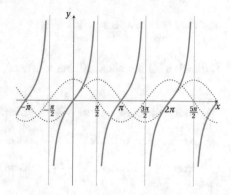

is called **tangent**. Its graph is plotted in Fig. 12.16.

Proposition *The function* tan *is*

- *π-periodic, odd and onto,*
- *positive on $(0, \pi/2)$, negative on $(-\pi/2, 0)$,*
- *increasing and unbounded.*

Proof Exercise. $\qquad\qquad\qquad\qquad\qquad\qquad\qquad\qquad\qquad\qquad\qquad\qquad$ □

Example If x is an angle in the second quadrant ($\pi/2 \leqslant x \leqslant \pi$), compute $\cos x$ and $\sin(2x)$ knowing that $\tan x = -\sqrt{6}$.

We begin observing, since the tangent is bijective on $[-\pi, -\pi/2)$, that x will be unique (mod 2π). There are various ways to solve this (cf. p. 291) and here we'll rely on elementary facts. The zeroes of cosine aren't solutions, so we can multiply by $\cos x \neq 0$ to obtain $\sin x = -\sqrt{6}\cos x$. Substituting in (12.6) gives $\cos^2 x = 1/7$. As x belongs in the second quadrant, where cosine is negative and sine positive, $\cos x = -1/\sqrt{7}$ and $\sin x = \sqrt{6/7}$. Hence $\sin(2x) = 2\cos x \sin x = -2\sqrt{6}/7$.

Exercises 12.9 Prove the following formulas:

i) $\tan 2x = \dfrac{2\tan x}{1 - \tan^2 x}$ for all $x \notin \left(\dfrac{\pi}{2} + \mathbb{Z}\pi\right) \cup \left(\dfrac{\pi}{4} + \dfrac{\pi}{2}\mathbb{Z}\right)$;

ii) $\sin x = \dfrac{2t}{1 + t^2}$, $\cos x = \dfrac{1 - t^2}{1 + t^2}$ where $t = \tan\dfrac{x}{2}$, for $x \notin (2\mathbb{Z} + 1)\pi$.

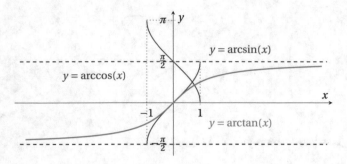

Fig. 12.17 Arc-functions

Definition Other classical, but less used, trigonometric functions are:

$$
\begin{array}{llll}
cotangent & x \mapsto \cot x := \dfrac{1}{\tan x} & \mathrm{Dom} = \mathbb{R} \setminus \mathbb{Z}\pi & \mathrm{Im} = \mathbb{R}; \\[2ex]
cosecant & x \mapsto \operatorname{cosec} x := \dfrac{1}{\sin x} & \mathrm{Dom} = \mathbb{R} \setminus \mathbb{Z}\pi & \mathrm{Im} = \{x \in \mathbb{R} \mid |x| \geqslant 1\}; \\[2ex]
secant & x \mapsto \sec x := \dfrac{1}{\cos x} & \mathrm{Dom} = \mathbb{R} \setminus \dfrac{\pi}{2} + \mathbb{Z}\pi & \mathrm{Im} = \{x \in \mathbb{R} \mid |x| \geqslant 1\},
\end{array}
$$

whose properties descend from the respective reciprocals, e.g.: $\cot\left(\frac{\pi}{2} - x\right) = \tan x$.

Exercise Sketch the graphs of cotangent, cosecant and secant using only information from the graphs of tangent, sine and cosine.

Suitably restricting domains, the functions sin, cos, tan become injective and can therefore be inverted (Fig. 12.17).

Definition The inverse trigonometric functions are called:

arctangent	$\arctan x := \tan^{-1} x$	$\mathrm{Dom} = \mathbb{R}$	$\mathrm{Im} = (-1, 1)$
arcsine	$\arcsin x := \sin^{-1} x$	$\mathrm{Dom} = [-1, 1]$	$\mathrm{Im} = [-\pi/2, \pi/2]$
arccosine	$\arccos x := \cos^{-1} x$	$\mathrm{Dom} = [-1, 1]$	$\mathrm{Im} = [0, \pi)$.

The identification between angles and circular arcs justifies these names. For instance, $\pi/6 = \arcsin(1/2)$ means the arc on the unit semicircle between $-\pi/2$ and $\pi/2$ whose sine equals $1/2$ has length $\pi/6$.

Exercises

i) Solve the equation $\arcsin(x) = \arccos(x)$.

ii) Show that $\dfrac{\arcsin(x) + \arccos(x)}{2} = \dfrac{\pi}{4}$ for any $-1 \leqslant x \leqslant 1$.

   Hint: look at the graphs, or remember that $\sin(x - \frac{\pi}{2}) = -\cos(x)$.

iii) Suppose $ax^2 + bx + c = 0$, $ac \neq 0$, has real solutions $x_{\pm}$. Prove that they may be computed as follows:

if $ac > 0$ define the angle α by $\tanh 2\alpha = -\frac{2\sqrt{ac}}{b}$. Then $x_{\pm} = \sqrt{\frac{c}{a}}(\tanh\alpha)^{\pm 1}$;

if $ac < 0$ define α by $\tan 2\alpha = \frac{2\sqrt{-ac}}{b}$. Then $x_{\pm} = \sqrt{-\frac{c}{a}}(\tan\alpha)^{\pm 1}$.

Remark 12.10 The trigonometric functions allows to write the general solution to the cubic equation.

Without loss of generality consider $t^3 - 3pt - 2q = 0$ (Lemma 9.4), with $p \neq 0$. Change variable $t = 2\sqrt{p}\cos\theta$, and use the relation $\cos(3\theta) = 4\cos^3\theta - 3\cos\theta$, to obtain $\cos(3\theta) = -\dfrac{q}{p\sqrt{p}}$. Its solutions read

$$t_k = 2\sqrt{p}\cos\left(\frac{1}{3}\arccos\left(-\frac{q}{p\sqrt{p}}\right) - \frac{2k\pi}{3}\right) \qquad k = 0, 1, 2.$$

The formula produces three real numbers in case $\dfrac{q}{p\sqrt{p}} \in [-1, 1]$, i.e. when $q^2 \leqslant p^3$ (which implies $p > 0$), in agreement with Proposition 9.5.

Examples There are no universal, catch-all recipes to solve trigonometric equations. The method is *ad hoc*, that is, it depends on the specific case at hand. Some general tips:

- write all expression in terms of one angle/variable x;
- write all expression in terms of $\sin x$, $\cos x$ only;
- in absence of constant terms, try factoring.

Let's examine a few types of equations:

(1) $\sin x = c$, $\cos x = c$:
  if $|c| > 1$ there's no solution;
  if $|c| \leqslant 1$ there are infinitely many solutions:
    (sin) $x_1 = \arcsin(c) \in \left[-\frac{\pi}{2}, \frac{\pi}{2}\right]$ and $x_2 = \pi - x_1$. The others are $x_1 + 2\pi\mathbb{Z}$, $x_2 + 2\pi\mathbb{Z}$.
    (cos) $x_1 = \arccos(c) \in [0, \pi]$ and $x_2 = -x_1$. The others are $\pm x_1 + 2\pi\mathbb{Z}$.
(2) $\tan x = c$:
  there is exactly one solution $x_1 = \arctan(c)$ in $\left(-\frac{\pi}{2}, \frac{\pi}{2}\right)$; any other has the form $x_1 + \pi\mathbb{Z}$.
(3) $a\sin x + b\cos x = c$ ('linear equations' in sin, cos):

  - using the fundamental relation (12.6) we solve the quadratic system in the auxiliary variables $C := \cos x$, $S := \sin x$

$$\begin{cases} aS + bC = c \\ S^2 + C^2 = 1 \end{cases}$$

and then we recover x from C, S.

Example: $\sin x + \cos x = 1$. Solving $C + S = 1, C^2 + S^2 = 1$ gives $(C, S) = (1, 0)$ or $(0, 1)$, so $x = 2\pi k$ or $\pi/2 + 2\pi k, k \in \mathbb{Z}$.

- A second possibility is to change variable with the rational parametrisation

$$C = \frac{1 - t^2}{1 + t^2}, \qquad S = \frac{2t}{1 + t^2},$$

where $t = \tan(x/2)$ (valid for $x \notin (2\mathbb{Z} + 1)\pi$).

Example: $\sqrt{3}\sin x + \cos x = 1$. The parametric method furnishes $t(t - \sqrt{3}) = 0$, i.e. $x \in 2\pi\mathbb{Z}, 2\pi/3 + 2\pi\mathbb{Z}$.

- (If $c = 0$) we may divide by C (carefully) and reduce to the variable $T = S/C$.

Example: $\sqrt{3}\sin x - 3\cos x = 0$. As $\cos x \neq 0$, dividing we find $T = \sqrt{3}$, so $x \in \pi/3 + \mathbb{Z}\pi$.

- Let ϕ be the angle such that $\tan\phi = a/b$. Substituting $a = \sqrt{a^2 + b^2}\sin\phi$, $b = \sqrt{a^2 + b^2}\cos\phi$ produces the simpler equation $\sqrt{a^2 + b^2}\cos(x - \phi) = c$.

Example: for $\sin x + \sqrt{3}\cos x = 1$ we take $\tan\phi = \sqrt{3}/3$, i.e. $\phi = \pi/6$. Then we must solve $\cos(x - \pi/6) = 1/2$, whose solutions are $\pi/2 + 2\pi\mathbb{Z}, -\pi/6 + 2\pi\mathbb{Z}$.

- If the equation is 'symmetric' in $\cos x, \sin x$, we can substitute $x = y + \pi/4$ so that $\sin x = \dfrac{1}{\sqrt{2}}(\sin y + \cos y)$, $\cos x = \dfrac{1}{\sqrt{2}}(\cos y - \sin y)$, $\sin x \cos x = \cos^2 y - \dfrac{1}{2}$. Eventually we have an equation in only $C = \cos y$ or $S = \sin y$.

Example: $\sin x + \sin x \cos x + \cos x = 1$. Switching to y gives $2C^2 + 2\sqrt{2}C - 3$, solved by $y \in \pm\pi/4 + 2\pi\mathbb{Z}$, i.e. $x \in \pi/2 + 2\pi\mathbb{Z}, 2\pi\mathbb{Z}$.

(4) $a\sin^2 x + b\cos^2 x + c\sin x \cos x = d$ ('quadratic homogeneous' in sin, cos): using (12.6) we arrive at $(a - d)S^2 + (b - d)C^2 + cSC = 0$. In case $a = d$ or $b = d$ the solution is immediate by factorisation. In case $a \neq d$, we may divide by C^2 (the cosine zeroes are not solutions) and thus obtain the quadratic relation $(a - d)T^2 + cT + (b - d) = 0$ in $T := \tan x$.

Example: $\sin^2 x - (1 + \sqrt{3})\sin x \cos x + \sqrt{3}\cos^2 x = 0$. Since $C = 0$ doesn't give solutions, transforming we find $T^2 - (1 + \sqrt{3})T + \sqrt{3} = 0$. Hence $T = 1, \sqrt{3}$, and so $x = \pi/4, \pi/3 \mod \mathbb{Z}\pi$.

Regarding (simple) inequalities, in this context it is convenient to rely on the geometrical visualisation, sketching graphs or inspecting the unit circle. Here are some examples:

(5) One way is to intersect the graphs of the maps in question. Another is to look at the unit circle. We'll write the solutions without details, and the task is to understand how they have been achieved.

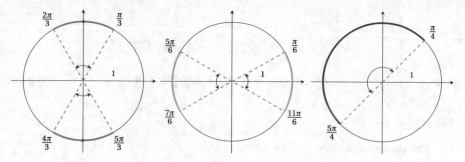

Fig. 12.18 Solving trigonometric inequalities, graphically

$|\tan x| \geqslant \sqrt{3}$: $\quad x \in \left[\dfrac{\pi}{3}, \dfrac{2\pi}{3}\right] \setminus \left\{\dfrac{\pi}{2}\right\}$ (mod π)   (Fig. 12.18 left)

$|\sin x| < \dfrac{1}{2}$: $\quad x \in \left(-\dfrac{\pi}{6}, \dfrac{\pi}{6}\right)$ (mod π)   (Fig. 12.18 middle)

$\sin x \geqslant \cos x$: $\quad x \in \left[\dfrac{\pi}{4}, \dfrac{5\pi}{4}\right]$ (mod 2π)   (Fig. 12.18 right)

(6) For $\sin^2 x - \sin x \cos x \geqslant 0$ we factor

$$\sin x(\sin x - \cos x) \geqslant 0 \iff \begin{cases} \sin x \geqslant 0 \\ \sin x \geqslant \cos x \end{cases} \quad \vee \quad \begin{cases} \sin x \leqslant 0 \\ \sin x \leqslant \cos x \end{cases}.$$

The systems are solved by $x \in \left[\dfrac{\pi}{4} + 2\pi\mathbb{Z}, \pi + 2\pi\mathbb{Z}\right] \cup \left[\dfrac{5\pi}{4} + 2\pi\mathbb{Z}, 2\pi\mathbb{Z}\right]$.

(7) Solve $\cos(\pi(x^2 - 10x)) - \sqrt{3}\sin(\pi(x^2 - 10x)) > 1$.

Setting $t = \pi(x^2 - 10x)$ gives $\cos t - \sqrt{3}\sin t > 1$. Writing the latter as $2(\cos t - \sqrt{3}\sin t)\cos(t + \pi/3) > 1/2$ gives solutions $-2\pi/3 + 2k\pi < t < 2k\pi$, $k \in \mathbb{Z}$. So we have the system

$$\begin{cases} x^2 - 10x - 2k < 0 \\ x^2 - 10x - 2k + 2/3 > 0 \end{cases}.$$

The first inequality admits solutions if and only if $\kappa := 25 + 2k \geqslant 0 \iff k \geqslant -12$ (integer). For these values the second relation has positive discriminant. Therefore

$$\begin{cases} 5 - \sqrt{\kappa} < x < 5 + \sqrt{\kappa} \\ x > 5 + \sqrt{\kappa - 2/3} \ \lor \ x < 5 - \sqrt{\kappa - 2/3} \end{cases}$$

i.e. $5 - \sqrt{\kappa} < x < 5 - \sqrt{\kappa - 2/3}$ and $5 + \sqrt{\kappa - 2/3} < x < 5 + \sqrt{\kappa}$, with $\kappa \geqslant 1$ integer.

Exercises

i) The solutions of $\sin x = \cos x$ are:

(1) $x = 2k\pi$, k integer   (4) $x = \pi + 2k\pi$, k integer

(2) $x = \frac{\pi}{4} + k\pi$, $k \in \mathbb{Z}$ (5) none of (1)–(4) are correct.

(3) $x = \frac{\pi}{4} + 2k\pi$, $k \in \mathbb{Z}$

ii) Assuming $0 \leqslant t < 2\pi$, solve:

$\sqrt{2} \sin t + 1 = 0$ $\sin\left(t + \frac{\pi}{3}\right) = \sin(2t)$

$\tan t = 1$ $\tan\left(\frac{\pi}{6} - 2t\right) = \tan(3t)$

$3 \tan^2 t - 1 = 0$ $\sin t - 2 \sin t \cos t = 0$

$\left| \sin|t| \right| = 1$ $2 \cos^2 t - 1 = 0$

     $2 \cos^2 t - \cos t = 1$

$\cos\left(2t + \frac{\pi}{4}\right) = \frac{\sqrt{2}}{2}$ $2 \sin^2\left(\frac{t}{2}\right) - 5 \sin\left(\frac{t}{2}\right) + 2 = 0$.

iii) Solve in $[0, 2\pi]$:

$\sin x + 2 \sin^2 x < 1$ $\tan x < \frac{\sqrt{3}}{3}$

$3 \tan^2 x - 1 < 0$ $2 \cos\left(x - \frac{\pi}{3}\right) - 1 < 0$

$\sqrt[3]{\tan\left(x - \frac{\pi}{4}\right)} < 1$ $2^{\sin x - \cos x} > 1$

$2 \cos^2 x + \cos x - 1 > 0$ $2 \cos^2 x - 5 \cos x - 3 > 0$

$\sqrt{3 + 2 \tan x - \tan^2 x} \geqslant \frac{1 + 3 \tan x}{2}$ $\log_{\frac{2x}{\sqrt{3}}} \sqrt{1 + 2 \cos 2x} \geqslant 1$

$\log\left(\frac{1}{2} - |\sin x|\right) < 0$ $\log_{x^2} \frac{4x - 5}{|x - 2|} \geqslant \frac{1}{2}$

$\sqrt{4 \sin^2 x - 1} \log_{\sin x} \frac{x - 5}{2x - 1} \geqslant 0$ $\left(\log_{\sin x} 2\right)^2 < \log_{\sin x}\left(4 \sin^3 x\right)$

$\log_{\frac{2x+1}{x^2-5}} 2 \leqslant \frac{1}{2} \log_{\sin\left(\frac{\pi}{3}\right)} \frac{4}{3}$ $\left| 3^{\tan(\pi x)} - 3^{1 - \tan(\pi x)} \right| \geqslant 2$.

Chapter 13
Complex Numbers

Pre-requisites: two-variable calculus for Remark 13.2.

Complex numbers started to be used formally in the formulas to solve cubic and quartic equations in the XVI century. Cardano recognised the existence of what we would call imaginary numbers, although he didn't understand their properties. The basic working rules of $\pm i$ were described for the first time by his contemporary Bombelli, who in 1572 solved equations with the del Ferro/Tartaglia method (cf. Remark 9.6).

For a long time complex numbers were considered more 'complicated' than real numbers (a misconception that persists today as well), and at least 200 years[1] had to pass before mathematicians understood that, in reality, $\mathbb{C}$ is simpler than $\mathbb{R}$, and allows more uniform approaches to many problems, like numerical/differential equations or the behaviour of functions and series. The term 'real' itself was coined to distinguish $\mathbb{R}$ from imaginary numbers. For example, the polynomial $x^3 - 15x - 4$ has a real zero, namely $x = 4$. Formula (9.5) elucidates that this root is generated by going through complex numbers:

$$x = \left(2 + \sqrt{-121}\right)^{1/3} + \left(2 - \sqrt{-121}\right)^{1/3}$$
$$= (2 + 11i)^{1/3} + (2 - 11i)^{1/3} = (2 + i) + (2 - i) = 4.$$

(The other two roots are $-2 \pm \sqrt{3}$, for the record.) Therefore complex numbers are present, even though not plainly, in 'real' situations, too. In this respect Hadamard [44, p.123] quipped " *The shortest path between two truths in the real domain passes through the complex domain.*" To see how real (no pun intended) this

---

[1] The dubious status of the number i saw its highest literary expression in *The confusions of young Törless* by Robert Musil (1880–1942).

© The Author(s), under exclusive license to Springer Nature Switzerland AG 2021
S. G. Chiossi, *Essential Mathematics for Undergraduates*,
https://doi.org/10.1007/978-3-030-87174-1_13

Fig. 13.1 The Argand–Gauß
plane (first described by
Wessel)

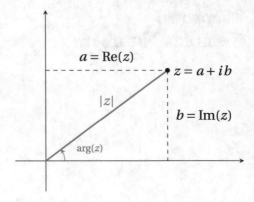

is, let's show that the product of the sums of two squares $(a^2 + b^2)(c^2 + d^2)$ is again
a sum of two squares:

$$(a^2 + b^2)(c^2 + d^2) = (a + bi)(a - bi)(c + di)(c - di)$$

$$= \underbrace{(a + bi)(c + di)}_{u+iv} \underbrace{(a + bi)(c - di)}_{u-iv} = u^2 + v^2.$$

For comparison, try proving this without complex numbers and without knowing
u, v in terms of a, b, c, d!

By the way, it was shown centuries later that a solution 'by radicals' of an
irreducible cubic must involve complex numbers, vindicating their necessity.

Definition (Hamilton 1833) A **complex number** z is a pair of real numbers $(a, b) \in$
$\mathbb{R}^2$.

The numbers $a =: \mathrm{Re}(z), b =: \mathrm{Im}(z)$ are called **real part** and **imaginary part**
of z.

It's customary to write z as $a + bi$, and the set of complex numbers (Fig. 13.1) is
indicated by

$$\mathbb{C} := \{ z = a + bi \mid a, b \in \mathbb{R} \}.$$

Definition The sum and the product of numbers $z_1 = a + bi, z_2 = c + di \in \mathbb{C}$ are

$$z_1 + z_2 := (a + b) + (c + d)i$$

$$z_1 \cdot z_2 := (ac - bd) + (ad + cb)i.$$

For any $b \in \mathbb{R}$, then, $(0 + bi)^2 = -b^2 + 0i$.

Example $(2 + i) + (2i - 7) = -5 + 3i;$ $(2 + i)(2i - 7) = -16 - 3i.$

The representation is unique, because $a+bi = a'+b'i \implies a-a' = (b-b')i \implies (a - a')^2 = -(b - b')^2 \in \mathbb{R} \implies a = a', b = b'$. Hence two complex numbers are equal (if and) only if they have the same real and imaginary parts.

In practice one can manipulate $z = a + bi$ as a polynomial in the variable i (the 'imaginary unit'), just remembering the rule

$$i^2 = -1.$$

Hence $(2 + i)(2i - 7) = 4i + 2i^2 - 14 - 7i = -16 - 3i$, and $(bi)^2 = -b^2 < 0$ for any $b \in \mathbb{R}^*$.

The reason (Cauchy, 1847) is that $\mathbb{R}[x]$ is a local ring with maximal ideal I generated by the irreducible polynomial $x^2 + 1$, and the quotient field is naturally isomorphic to $\mathbb{C}$

$$
\begin{array}{ccc}
\mathbb{R}[x]/I & \xrightarrow{\cong} & \mathbb{C} \\
[1] & \longmapsto & 1 \\
[x] & \longmapsto & i
\end{array}
$$

In fact, take a field extension $\mathbb{F}$ of $\mathbb{R}$ in which $x^2 + 1$ has roots, and let $i \in \mathbb{F}$ be one such root ($\mathbb{F}$ is called a *splitting field*). The subset $\mathbb{C} \subseteq \mathbb{F}$ of linear combinations $a + ib$, with $a, b \in \mathbb{R}$, is a subfield, and its structure doesn't depend on that of $\mathbb{F}$: if $\mathbb{F}'$ is a second extension containing a root i' of $x^2 + 1$, the corresponding set $\mathbb{C}'$ is isomorphic to $\mathbb{C}$.

Proposition $(\mathbb{C}, +, \cdot)$ *is a field.*

Proof The proof, left as exercise, boils down to checking that

- $+, \cdot$ are commutative and associative, $\cdot$ is distributive over $+$;
- $0 := 0 + 0i$ is the zero, $1 := 1 + 0i$ the unit;
- any $z = a + bi$ has opposite $-z := -a + (-b)i$;
- any $z = a + bi \in \mathbb{C} \setminus \{0\}$ has inverse $z^{-1} = \dfrac{a}{a^2 + b^2} - \dfrac{b}{a^2 + b^2}i$.

$\square$

Definition The **conjugate** of $z = a + bi$ is the complex number $\bar{z} := a - bi$:

$$\text{Re}(\bar{z}) = \text{Re}(z) \qquad \text{Im}(\bar{z}) = -\text{Im}(z).$$

The **modulus** of $z = a + bi$ is the non-negative real number

$$\rho := |z| := \sqrt{z\bar{z}} = \sqrt{a^2 + b^2} \geqslant 0.$$

Exercises 13.1 Prove

i) $|z| = |\bar{z}|$, $\qquad \text{Re}\, z = (z + \bar{z})/2$, $\qquad \text{Im}\, z = (z - \bar{z})/2i$

ii) $\overline{z_1 + z_2} = \overline{z_1} + \overline{z_2}$, $\overline{z_1 \cdot z_2} = \overline{z_1} \cdot \overline{z_2}$, $\overline{(\overline{z})} = z$

(conjugation is an involutive homomorphism)

iii) $|z_1 + z_2| \leqslant |z_1| + |z_2|$ (triangle inequality)

iv) $|z_1 \cdot z_2| = |z_1| \cdot |z_2|$

v) if $z \neq 0$ then $z^{-1} = \dfrac{\overline{z}}{|z|^2}$ and $\overline{z^{-1}} = (\overline{z})^{-1}$.

By identifying $a + 0i \in \mathbb{C}$ with $a \in \mathbb{R}$ (all operations are compatible!) we may think of real numbers as a subset of complex numbers:

$$\mathbb{R} \subset \mathbb{C}.$$

That's why $a + 0i \in \mathbb{C}$ is called real number, and $bi = 0 + bi \in \mathbb{C}$ is **imaginary**.

Remark 13.2 $\mathbb{C}$ is not an ordered field. Even worse, it doesn't admit order relations compatible with $(\mathbb{R}, \leqslant)$. As i^2 and i^4 are opposite, if one were declared positive the other would have to be negative. This would contradict that fact that the square of a non-zero imaginary number is negative ⚡. Or more simply, $i > 0 \implies 3i < 5i \implies 3i^2 < 5i^2 \implies -3 < -5$ ⚡ (similarly if $i < 0$).

Yet, by Zermelo's axiom 3.13 there exist (well!) orderings on $\mathbb{C}$. One such is the *lexicographic order* of the Cartesian product $\mathbb{R} \times \mathbb{R}$:

$$a + bi \leqslant_{\text{lex}} a' + ib' \iff a < a' \text{ or } (a = a' \text{ and } b \leqslant b'),$$

cf. Exercises 8.15. The point is that, $\mathbb{R}$ being ordered, that the lexicographic order implies $ni = n(0 + i) < (1 + 0i) = 1$, $\forall n$, which would turn $\mathbb{C} \cong \mathbb{R}^2$ into an undesirable non-Archimedean group.

The curious reader won't be disappointed to learn that the sequence of number sets $\mathbb{N} \subset \mathbb{Z} \subset \mathbb{Q} \subset \mathbb{R} \subset \mathbb{C}$ doesn't stop there. It continues with the *quaternions* $\mathbb{H}$, the *octonions* $\mathbb{O}$, and more .... The algebro-geometric theory of these *division algebras* is extremely rich: it explains why no Cartesian product $\mathbb{R}^n$ is a field except for the cases $n = 1, 2$, or that the spheres S^1, S^3 and S^7 are special among all other dimensions (☞ *spin geometry, algebraic topology*). The construction of division algebras is based on the duplication producing $\mathbb{C} = \mathbb{R} \oplus \mathbb{R}$ from $\mathbb{R}$. The general recipe, called *Cayley–Dickson process*, yields an increasing sequence $\{A_n\}$ of $\mathbb{R}$-algebras of real dimension 2^n, where $A_{n+1} = A_n \oplus A_n$. In fact

$$A_0 = \mathbb{R}, \quad A_1 = \mathbb{C}, \quad A_2 = \mathbb{H}, \quad A_3 = \mathbb{O}, \quad \ldots$$

Each step of the construction leaves a bill to foot: the loss of an algebraic property. $\mathbb{C}$ is no longer ordered, $\mathbb{H}$ is neither ordered not commutative, $\mathbb{O}$ isn't ordered, commutative, associative. (For the more inquisitive: the next iteration $\mathbb{O} \oplus \mathbb{O}$ isn't a normed division algebra: we enter the world of Jordan algebras...). What subsumes all this is the theory of ☞ *Clifford algebras*.

Beware that these aren't mere abstract artifices. Quaternions provide the perfect tool for understanding rotations in $\mathbb{R}^3$ (cf. p. 389), and as such are much in vogue in particle physics, video games and computer graphics, navigation and aerospace engineering (flight simulators). See [1] for their use in the study of surfaces. Octonions are the basic structure of ☞ *string-* and *$\mathcal{M}$-theory* in theoretical physics.

Exercises

i) Calculate modulus and conjugate of $2 - 5i, 3 - 2i, 4 - 3i, 1 + i, 1 - 3i, -3 - 3i$.
ii) Transform in 'Cartesian' form $a + bi$ the numbers:

$$(2 + 3i)^3 (2 + 2i)^{-1}, \qquad \frac{1+i}{i} - \frac{i}{i-1}, \qquad \frac{4+3i}{1+i\sqrt{3}}.$$

iii) Find a, b such that $(a^2 - 1) + (b^2 - 3)(a - 1)i$ is purely imaginary.
iv) Let z, w be complex numbers. Show $zw = 0 \iff z = 0 \lor w = 0$ ($\mathbb{C}$ is an integral domain).

13.1 Geometric Aspects

Since this is not an algebra textbook we refrain from digging deeper in the structure of $\mathbb{C}$. As a matter of fact the geometrically oriented viewpoint we'll adopt is better suited for an introduction to the subject.

Consider the Cartesian plane Oxy, identified with $\mathbb{R}^2 = \{(a, b) \mid a, b \in \mathbb{R}\}$. The map

$$\begin{array}{ccc} \mathbb{C} & \longrightarrow & \mathbb{R}^2 \\ a + bi & \mapsto & (a, b) \end{array}$$

is a bijection (exercise), and permits to identify numbers $z = a + bi \in \mathbb{C}$ with points on the plane of coordinates (a, b). (See Example 15.7.)

Thus the $x = \operatorname{Re} z$-axis is now called **real axis**, while the $y = \operatorname{Im} z$-axis is the **imaginary axis**. For example $\bar{z} = a - ib$ is mirror to $z = a + bi$ with respect to the real axis.

We are thinking of z as a vector $\vec{z}$ applied at the origin, and the modulus $|z| = \sqrt{a^2 + b^2}$ is the length $|\vec{z}|$, i.e. the distance between the points 0 and z. Under this vectorial interpretation the sum of complex numbers $z + w$ is the same thing as the sum $\vec{z} + \vec{w}$ of vectors, cf. Fig. 13.2.

Definition 13.3 The (principal) **argument** of $z \neq 0$ is the unique real number $\theta_0 =: \arg(z) \in [0, 2\pi)$ solution to

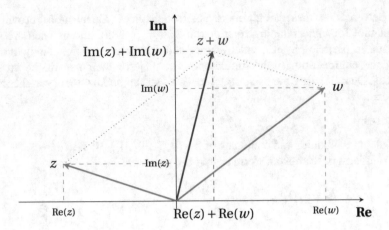

Fig. 13.2 Adding complex numbers

$$\cos \theta_0 = \frac{\operatorname{Re} z}{|z|}, \qquad \sin \theta_0 = \frac{\operatorname{Im} z}{|z|},$$

i.e. the angle between the real axis and the vector $\vec{z}$, see Fig. 13.1. The number $z = 0$ doesn't have argument.

Exercises Prove

i) if $|z_1 + z_2| = |z_1 - z_2|$ then $\arg(z_1) - \arg(z_2) = \dfrac{\pi}{2}$;

ii) if $\arg\left(\dfrac{z_1 + z_2}{z_1 - z_2}\right) = \dfrac{\pi}{2}$ then $|z_1| = |z_2|$.

iii) Fix $a, b \in \mathbb{C}$ and describe the loci of points z such that

$$\arg\left(\frac{z - a}{z - b}\right) = \text{constant} \qquad \text{or} \qquad \left|\frac{z - a}{z - b}\right| = \text{constant}.$$

Hint: [79, p.13].

iv) Let $z_1, z_2, z_3, z_4 \in \mathbb{C}$ be such that $\arg\left(\dfrac{z_3 - z_1}{z_3 - z_2}\right) = \arg\left(\dfrac{z_4 - z_1}{z_4 - z_2}\right)$. Prove that the four points belong on the same circle or straight line.

The quotient $\dfrac{z_3 - z_1}{z_3 - z_2} \bigg/ \dfrac{z_4 - z_1}{z_4 - z_2}$ is an invariant, called *cross-ratio* of the z_i in ☞ *projective geometry*, related to (17.6).

Remark 13.4 The argument's existence is proved inverting the complex exponential function

$$z \mapsto e^z := \sum_{n=0}^{\infty} \frac{z^n}{n!}$$

(which is the analytic prolongation of Theorem 12.3). For any $c, s \in \mathbb{R}$ satisfying $c^2 + s^2 = 1$ in fact, we have $\theta_0 = \log(c + si) \in [0, 2\pi)$ ($\bowtie$ complex analysis), which in turn allows to define the real trigonometric functions by

$$\cos x := \operatorname{Re}\left(e^{ix}\right), \quad \sin x := \operatorname{Im}\left(e^{ix}\right).$$

Putting $x = \pi$ gives a dazzling display of the five most important numbers in mathematics, namely the celebrated **Euler formula** $e^{i\pi} + 1 = 0$.

The use of (12.4) shows that every $z \in \mathbb{C} \setminus \{0\}$ possesses a **polar** or **exponential form**

$$z = \rho(\cos\theta + i\sin\theta) = \rho e^{i\theta}$$

as drawn in Fig. 12.8.

For instance $\bar{z} = \rho(\cos\theta - i\sin\theta) = \rho(\cos(-\theta) + i\sin(-\theta)) = \rho e^{-i\theta}$, meaning $\bar{z}$ lies on the same circle as z, forming a clockwise angle θ with the real axis.

Definition The (open) **unit disc** is the set

$$\mathbb{D} := \left\{ z \in \mathbb{C} \mid |z| < 1 \right\}$$

of points on the plane with distance from the origin less than one. Bounding this set is the **unit circle**

$$S^1 := \left\{ z \in \mathbb{C} \mid |z| = 1 \right\}.$$

More generally, we'll write $S^1(r) := \left\{ z \in \mathbb{C} \mid |z| = r \right\}$ for the circle centred at the origin of radius r.

Exercise Show that in polar coordinates the above objects read

$$\mathbb{D} = \left\{ \rho e^{i\theta} \mid 0 \leqslant \rho < 1, \theta \in [0, 2\pi) \right\}, \qquad S^1 = \left\{ e^{i\theta} \mid \theta \in [0, 2\pi) \right\}.$$

Therefore (cf. Fig. 12.9): $\mathbb{D} = \underbrace{[0, 1) \times [0, 2\pi)}_{\text{rectangle}}, \quad S^1 = \underbrace{\{1\} \times [0, 2\pi)}_{\text{segment}}.$

The Cartesian vs. polar correspondence of Corollary 12.6 reads

$$0 \neq z = \rho e^{i\theta} \longleftrightarrow (\rho = |z|, \theta_0 = \arg(z)). \tag{13.1}$$

See also Theorem 17.6.

Examples

i) $\dfrac{3}{2} - \dfrac{3\sqrt{2}}{2}i = 3\left(\dfrac{1}{2} - i\dfrac{\sqrt{2}}{2}\right) = 3\left(\cos\dfrac{\pi}{3} - i\sin\dfrac{\pi}{3}\right) = 3e^{i\pi/3}$;

ii) $i = \cos\dfrac{\pi}{2} + i\sin\dfrac{\pi}{2} = e^{i\pi/2}$;

iii) for all $k \in \mathbb{Z}$, $\quad i^k = \begin{cases} 1 & \text{if } k \equiv_4 0 \\ i & \text{if } k \equiv_4 1 \\ -1 & \text{if } k \equiv_4 2 \\ -i & \text{if } k \equiv_4 3 \end{cases}$. Therefore $i^{235} = i^{58\cdot4+3} = i^3 = -i$.

The exponential form gives a very simple mnemonic rule for multiplying complex numbers: writing $z_1 = \rho_1 e^{i\theta_1}$, $z_2 = \rho_2 e^{i\theta_2}$,

$$z_1 z_2 = \rho_1 e^{i\theta_1} \rho_2 e^{i\theta_2} = \rho_1 \rho_2 e^{i(\theta_1+\theta_2)}. \tag{13.2}$$

That is to say, $|z_1 z_2| = |z_1||z_2| = \rho_1\rho_2$, $\arg(z_1 z_2) = \arg(z_1) + \arg(z_2)$. As a consequence we obtain for free the addition formulas (12.7) for sine and cosine. Furthermore, using the exponential form is the best way by far for dividing complex numbers:

$$\frac{z}{w} = \frac{\rho_1 e^{i\theta_1}}{\rho_2 e^{i\theta_2}} = \frac{\rho_1}{\rho_2}e^{i(\theta_1-\theta_2)},$$

i.e. $\left|\dfrac{z}{w}\right| = \dfrac{|z|}{|w|}$, $\arg\left(\dfrac{z}{w}\right) = \arg(z) - \arg(w)$.

Corollary 13.5 (de Moivre) *For any* $n \in \mathbb{Z}, z \in \mathbb{C}^*$

$$\left|z^n\right| = |z|^n, \quad \arg\left(z^n\right) = n\arg(z).$$

Proof If $z = \rho e^{i\theta}$ then $z^n = \rho^n\left(e^{i\theta}\right)^n = \rho^n e^{in\theta}$. $\qquad\qquad\square$

Example Let's simplify the expression $\dfrac{3\left(\frac{\sqrt{3}}{2} + \frac{1}{2}i\right)^2}{i^{17}\left(\frac{\sqrt{2}}{2} + \frac{\sqrt{2}}{2}i\right)^2}$.

First of all write $\dfrac{\sqrt{3}}{2} + \dfrac{1}{2}i$ and $\dfrac{\sqrt{2}}{2} + \dfrac{\sqrt{2}}{2}i$ in exponential form:

$\left|\dfrac{\sqrt{3}}{2} + \dfrac{1}{2}i\right| = 1 \quad \arg\left(\dfrac{\sqrt{3}}{2} + \dfrac{1}{2}i\right) = \dfrac{\pi}{6} \implies \dfrac{\sqrt{3}}{2} + \dfrac{1}{2}i = e^{\pi i/6}$,

$\left|\dfrac{\sqrt{2}}{2} + \dfrac{\sqrt{2}}{2}i\right| = 1 \quad \arg\left(\dfrac{\sqrt{2}}{2} + \dfrac{\sqrt{2}}{2}i\right) = \dfrac{\pi}{4} \implies \dfrac{\sqrt{2}}{2} + \dfrac{\sqrt{2}}{2}i = e^{\pi i/4}$.

Since $i^{17} = i^{4\cdot4+1} = (i^4)^4 i = i$, we have

Fig. 13.3 Standard complex structure on $\mathbb{C}$

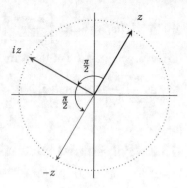

$$\frac{3\left(\frac{\sqrt{3}}{2} + \frac{1}{2}i\right)^2}{i^{17}\left(\frac{\sqrt{2}}{2} + \frac{\sqrt{2}}{2}i\right)^2} = \frac{3(e^{\pi i/6})^2}{i \cdot (e^{\pi i/4})^2} = \frac{3e^{\pi i/3}}{i \cdot i} = 3e^{\pi i/3 - \pi i} = 3e^{-2\pi i/3}.$$

Note: although the result is equal to $-3e^{\pi i/3}$, it's preferable not to use the exponential form with a negative coefficient in front, because $\rho \geqslant 0$ has the geometric meaning of a distance.

Remark 13.6 The exponential form also tells that complex numbers describe rotations and homotheties in the plane. Multiplying z by $w = re^{i\phi}$ means rescaling $|z|$ by the factor r and rotating it (anti-clockwise) by ϕ starting from $\arg(z)$.

Example: the product of $z = \rho(\cos\theta + i\sin\theta) = \rho e^{i\theta}$ by i corresponds (Fig. 13.3) to a $\pi/2$-rotation around the origin:

$$iz = \rho e^{i(\theta + \pi/2)} = \rho\big(\cos(\theta + \pi/2) + i\sin(\theta + \pi/2)\big) = \rho(-\sin\theta + i\cos\theta).$$

Hence the vector $\overrightarrow{iz}$ is perpendicular to $\vec{z}$ (which we knew already from p. 282). Two successive rotations act as a π-rotation, which is a reflection about the origin: $i^2 z = -z$. Put in different terms, multiplication by i is the **standard complex structure** of $\mathbb{C}$, cf. (15.4).

Exercises 13.7 Prove

i) $\cos(-ix) = \cosh x$, $i\sin(-ix) = \sinh x$, $\tanh x = i\tan(-ix)$, for all $x \in \mathbb{R}$;

ii) $i^{-i} = \sqrt{e\pi}$; [2]

---

[2] In 1864 the mathematician Benjamin Peirce, father of the known mathematician and philosopher Charles Peirce, during a class wrote this relation on the blackboard and told his students: *"Gentlemen, we have not the slightest idea what this equation means, but we may be sure that it means something very important."* The episode reminds us how much suspicion complex numbers raised even in the XIX century.

iii) if $|z|, |w| < 1$ then $\left| \dfrac{z - w}{1 - \overline{z}w} \right| < 1$;

iv) every expression of the form $e^{i\theta} \dfrac{z + z_0}{1 - \overline{z_0}z}$, where $\theta \in \mathbb{R}$, $z_0 \in \mathbb{C}$, can be written

as $\dfrac{az + b}{cz + d}$ for some $a, b, c, d \in \mathbb{R}$, and vice versa. Cf. (18.8).

13.2 Polynomial Factorisation

From the definition of product we know the nth power of $z = \rho e^{i\theta}$ equals $z^n = \rho^n e^{in\theta}$ (Corollary 13.5). But whereas in $\mathbb{R}$ n^{th} roots are borne by functions (and hence they are unique), in $\mathbb{C}$ the picture is completely different because complex roots do not define functions. They are interpreted as 'multi-valued functions' (functions on ☞ *Riemann surfaces*). The existence of complex roots is warranted by one of mathematics' pillars:

Theorem 13.8 (Fundamental Theorem of Algebra) *Every non-constant complex polynomial admits a root in* $\mathbb{C}$.

It applies in particular to real polynomials $\mathbb{R}[x] \subset \mathbb{C}[x]$. In the fewest possible words, the field $\mathbb{C}$ is algebraically closed. To show the strength of this condition we remind that

- no finite field $\mathbb{Z}_p = \{f_1, \ldots, f_p\}$ is algebraically closed, since the polynomial $\prod_{i=1}^{p}(x - f_i) + 1$ has no roots in it;
- $\mathbb{Q}$ isn't algebraically closed: $x^3 - 4 \in \mathbb{Q}[x]$ doesn't have rational roots (historically, this issue precipitated the need for the real numbers);
- neither $\mathbb{R}$ is algebraically closed, because $x^2 + 1 = 0$ is irreducible over the reals.

The book [28] puts the subject in a general perspective.

More explicitly, the fundamental theorem of algebra guarantees the algebraic equation

$$p(z) = \alpha_n z^n + \alpha_{n-1} z^{n-1} + \ldots + \alpha_1 z + \alpha_0 = 0$$

($\alpha_i \in \mathbb{C}$, $i = 0, \ldots, n$) always has a complex solution w_0. Hence $p(z) = (z - w_0)p'(z)$ factors as a product of a linear term and a polynomial of degree $n - 1$. As $p'(z) \in \mathbb{C}[z]$, it too has a root and can be factored. Iterating the argument by induction on the degree, after n iterations we reach the complete factorisation of $p(z)$. So every $p \in \mathbb{C}[z]$ of degree n can be decomposed as a product of n linear factors (with multiplicity):

$$p(z) = \alpha_n (z - w_0)^{r_0} (z - w_1)^{r_1} \cdots (z - w_k)^{r_k}$$

where $\sum_{j=0}^{k} r_j = n$.

Examples

i) $z^4 - 2 = (z^2 - \sqrt{2})(z^2 + \sqrt{2}) = (z - \sqrt[4]{2})(z + \sqrt[4]{2})(z - i\sqrt[4]{2})(z + i\sqrt[4]{2})$,

ii) $-2z^9 + 32z^6 - 128z^3 = -2z^3(z-2)^2 \left(z - (-1 + i\sqrt{3})\right)^2 \left(z - (-1 - i\sqrt{3})\right)^2$.

Corollary *Every complex number $a \neq 0$ has exactly n distinct complex n^{th} roots.*

Proof Write $a = \rho e^{i\theta}$ and define $\theta_k := \dfrac{\theta + 2k\pi}{n}$ with k integer. The n numbers spat out by **de Moivre's formula**

$$w_k = \sqrt[n]{\rho} e^{i\theta_k} = \sqrt[n]{\rho}\left(\cos\theta_k + i\sin\theta_k\right), \qquad k = 0, 1, \ldots, n-1,$$

solve $z^n = a$ by Corollary 13.5. Any two numbers w_i and w_j of the above form are distinct since $0 < |i - j| < n - 1 \implies |\theta_i - \theta_j| = \left|\dfrac{2\pi}{n}(i - j)\right| < 2\pi \implies \theta_i \neq \theta_j + 2\pi\mathbb{Z}$.

Since the polynomial's degree is n, the w_k are the nth roots of a. □

Now observe w_0 has argument $\theta_0 = \dfrac{\theta}{n}$. The roots w_k lie on $S^1(\sqrt[n]{\rho})$ and are equidistant, because they are constructed by adding each time the angle $\dfrac{2\pi}{n}$. Hence they are the vertices on a regular n-polygon inscribed in the circle of radius $\rho^{1/n}$, as in Fig. 13.4.

De Moivre's formula also says the $\sqrt[n]{z}$ are determined by: the modulus $\sqrt[n]{z\bar{z}} \in \mathbb{R}$ and the **roots of unity**

Fig. 13.4 The n-polygon of nth roots ($n = 5$)

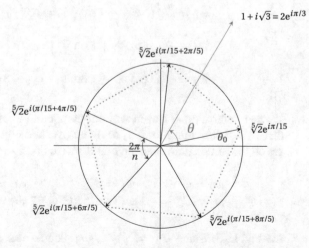

Fig. 13.5 The sixth roots of
$z = 1$

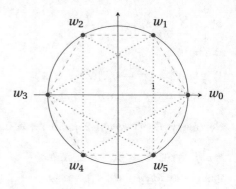

$$\sqrt[n]{1} = \cos\frac{2k\pi}{n} + i\sin\frac{2k\pi}{n}, \quad k = 1, \ldots, n-1.$$

Examples

i) The quotient $\dfrac{z^{n+1} - 1}{z - 1} = z^n + z^{n-1} + \ldots + z^2 + z + 1$ is called **cyclotomic polynomial** because its zeroes are the nth roots of unity (different from 1).

ii) Let's compute $\sqrt{-1}$. Notwithstanding we already know $i^2 = (-i)^2 = -1$, we may write $-1 = \cos\pi + i\sin\pi$, whose square roots are

$$w_0 = \cos\tfrac{\pi}{2} + i\sin\tfrac{\pi}{2} = i$$
$$w_1 = \cos\tfrac{\pi+2\pi}{2} + i\sin\tfrac{\pi+2\pi}{2} = -i$$

iii) We wish to compute $\sqrt[3]{3i}$. As $|3i| = 3$, $\arg(3i) = \pi/2$, the roots are

$$w_0 = \sqrt[3]{3}\left(\cos\tfrac{\pi}{6} + i\sin\tfrac{\pi}{6}\right) = \sqrt[3]{3}\left(\tfrac{\sqrt{3}}{2} + i\tfrac{1}{2}\right)$$
$$w_1 = \sqrt[3]{3}\left(\cos\tfrac{5\pi}{6} + i\sin\tfrac{5\pi}{6}\right) = \sqrt[3]{3}\left(-\tfrac{\sqrt{3}}{2} + i\tfrac{1}{2}\right)$$
$$w_2 = \sqrt[3]{3}\left(\cos\tfrac{3\pi}{2} + i\sin\tfrac{3\pi}{2}\right) = -\sqrt[3]{3}i$$

iv) We can determine $\sqrt[6]{1}$ even without computing anything explicitly: knowing that $1^6 = 1$, the first root is 1, of modulus 1 and argument $\theta_0 = 0$. The others are on the same circle at an angular distance of $2\pi/6 = \pi/3$ from one another:

$$w_0 = e^{i0} = 1 \quad w_1 = e^{i\pi/3} \qquad w_2 = e^{2i\pi/3}$$
$$w_3 = e^{i\pi} = -1 \quad w_4 = e^{-2i\pi/3} = \overline{w_2} \quad w_5 = e^{-i\pi/3} = \overline{w_1}$$

see Fig. 13.5. Since $\sqrt[6]{1} = \sqrt[3]{\sqrt[2]{1}}$, the picture also contains the square roots w_0, w_3 of 1, and the cubic roots w_0, w_2, w_4:

$$1 = w_0^2 = w_3^2, \qquad 1 = w_0^3 = w_2^3 = w_4^3, \qquad -1 = w_3^3 = w_1^3 = w_5^3.$$

The hexagon is determined by the two equilateral triangles with vertices at 1 and -1.

Corollary 13.9 *The non-real zeroes of a polynomial $p(z) \in \mathbb{R}[z]$ with real coefficients crop up in conjugate pairs, so with the same multiplicity:*

$$p(a) = 0 \iff p(\overline{a}) = 0, \ \forall a \in \mathbb{C}.$$

Proof Suppose $p(z) = \sum_{i=0}^{n} \alpha_i z^i \in \mathbb{R}[z]$ has a for a root. By Exercise 13.1 ii), $p(\overline{a}) = \overline{p(a)} = \overline{0} = 0$. □

Example Knowing that z_1, z_2 are solutions to $1 - z + z^2 = 0$, let's calculate $c :=$ $(z_1^4 - z_1^3 + 2z_1^2 + 1)^{2005} + (z_2^4 - z_2^3 + 2z_2^2 + 1)^{2005}$.

Dividing $z^4 - z^3 + 2z^2 + 1$ by $1 - z + z^2$ gives

$$z^4 - z^3 + 2z^2 + 1 = (z^2 + 1)(1 - z + z^2) + z$$

and so $z_i^4 - z_i^3 + 2z_i^2 + 1 = z_i$ for $i = 1, 2$. Since $1 - z + z^2$ has negative discriminant, $z_1 = \overline{z_2}$ are conjugate and therefore $c = z_1^{2005} + (z_2)^{2005} = 2\,\mathrm{Re}(z_1^{2005})$. To finish we write z_1 in polar form:

$$1 - z + z^2 = 0 \implies z_1 = \frac{1 + i\sqrt{3}}{2} = e^{i\pi/3}.$$

As $\frac{2005}{3} = 2 \cdot 334 + \frac{1}{3}$, we deduce $c = 2\,\mathrm{Re}(e^{2005\pi i/3}) = 2\,\mathrm{Re}(e^{\pi i/3}) = 1$.

Corollary 13.10 *If $a \in \mathbb{C} \setminus \mathbb{R}$ and $\deg p > 1$ then*

$$p(a) = 0 \iff p(z) = \left(z^2 - 2\,\mathrm{Re}(a)z + |a|^2\right) q(z)$$

for some $q(z) \in \mathbb{R}[z]$.

Proof It's enough to note $z^2 - 2\,\mathrm{Re}(a)z + |a|^2 = (z - a)(z - \overline{a})$. □

Example Let's factor $z^2 + 2z + 3$. Formula (9.2) tells us the (conjugate) roots read $z = \frac{-2 \pm \sqrt{4 - 12}}{2} = -1 \pm i\sqrt{2}$, so

$$z^2 + 2z + 3 = \left(z + 1 + i\sqrt{2}\right)\left(z + 1 - i\sqrt{2}\right).$$

The same can be seen, in an easier way perhaps, by 'completing the square': $z^2 + 2z + 3 = z^2 + 2z + 1 + 2 = (z + 1)^2 + \left(\sqrt{2}\right)^2$.

Corollary 13.11 *Any polynomial in $\mathbb{R}[x]$ of odd degree has a real root (and hence is reducible).*

Proof Exercise.   □

Corollary (Gauß) *The only irreducible polynomials $p(x) \in \mathbb{R}[x]$ over $\mathbb{R}$ are those of degree 1, or 2 if the discriminant Δ is negative.*

Proof Any p with $\deg p = 2h + 1$, $h > 0$, is reducible by Corollary 13.11.

In case p has degree $2k > 2$ its complex roots come in pairs $a, \overline{a}$, so p is reducible by Corollary 13.10, because $x^2 - 2\operatorname{Re}(a)x + |a|^2 \in \mathbb{R}[x]$.   □

Example The polynomial $p(x) = x^4 + 1 \in \mathbb{R}[x]$ is the sum of two squares ($p(x) > 0$ for all x), so it has no real roots. So we'll factor it over $\mathbb{C}$, finding its complex roots:

$$\sqrt[4]{-1} = w_k = e^{i\left(\frac{\pi}{4} + k\frac{\pi}{2}\right)} \quad k = 0, 1, 2, 3$$

where $w_2 = \overline{w_1}$, $w_3 = \overline{w_0}$. Then

$$
\begin{aligned}
x^4 + 1 &= (x - w_0)(x - w_3)(x - w_1)(x - w_2) \quad \text{[complex polynomials]} \\
&= (x - w_0)(x - \overline{w_0})(x - w_1)(x - \overline{w_1}) \\
&= \left(x^2 - 2\operatorname{Re}(w_0)x + |w_0|^2\right)\left(x^2 - 2\operatorname{Re}(w_1)x + |w_1|^2\right) \quad \text{[real (!) polynomials]} \\
&= (x^2 - \sqrt{2}x + 1)(x^2 + \sqrt{2}x + 1).
\end{aligned}
$$

This shows that the absence of roots doesn't imply irreducibility in $\mathbb{R}$.

A less amenable alternative is to impose $p(x) = (x^2 + ax + b)(x^2 + a'x + b')$ and find $a, b, a', b' \in \mathbb{R}$ (as p is monic we may limit ourselves to 4 parameters). Note the unique possible decomposition is '2+2', both irreducible (a degree-one factor would give a real root). Hence

$$(x^2 + ax + b)(x^2 + a'x + b') = x^4 + (a + a')x^3 + (b + b' + 2aa')x^2 + (ab' + a'b)x + bb'$$

implies

$$a' = -a, \qquad b + b' = a^2, \qquad a(b' - b) = 0, \qquad bb' = 1.$$

Since $a = 0$ is not admissible, we obtain $2b = 2b' = a^2 > 0$, $b^2 = 1$, whence $b = 1 = b'$, $a = \sqrt{2} = -a'$.

Exercises Solve

$$
\begin{aligned}
&z^2 + 3iz + 4 = 0, \quad z^2 + 2z + i = 0, \quad z|z| - 2z - 1 = 0, \quad |z|^2 z^2 = 1, \\
&z^4 = |z|^2 + 2, \qquad z + i\overline{z}^2 = -2i, \quad z^3\overline{z} + 3z^2 - 4 = 0, \quad z^2 + z\overline{z} = 1 + 2i.
\end{aligned}
$$

Remark The fundamental theorem of algebra is shocking in at least in two respects. First, it guarantees the existence of roots for any polynomial in $\mathbb{C}[z]$. But we shouldn't underestimate another aspect, namely that the number of solutions in $\mathbb{C}$ is *finite*. We are familiar with the fact that transcendental equations (think $\sin x = 0$) have an infinite number of roots. But even the algebraic equation $z^2 + 1 = 0$, when considered over the field $\mathbb{H}$ of *quaternions*, admits uncountably many solutions $bi + \sqrt{1 - b^2}j$ for any real $b \in [-1, 1]$.

Any proof of theorem 13.8 is based on arguments that transcend what we can offer here. With the help of calculus we may justify in two ways the existence of roots in $\mathbb{C}$, and the reader might choose the one that most appeals to them.

A) Consider a monic polynomial $p(x) \in \mathbb{R}[x]$. If $\deg p$ is odd, Corollary 13.11 tells us there is a (real) zero.

In case $\deg p$ is even, $\lim_{|x| \to \infty} p(x) = +\infty$. Therefore if there weren't any roots the minimum of p, say at the critical point α, would satisfy $p(\alpha) > 0$. Extend the polynomial to a complex variable $\mathbb{C} \ni \alpha + i\beta \longleftrightarrow (\alpha, \beta) \in \mathbb{R}^2$ and consider the map $|p| \colon \mathbb{R}^2 \to \mathbb{R}$. It is differentiable at (α, β) if and only if $p(\alpha, \beta) \neq 0$. So on the one hand $|p|$ must have minimum, on the other any critical point of $|p|$ is a saddle point. Hence at the minimum point p cannot be differentiable. ⨩

Example: $p(x) = x^2 + 1$ has one critical point $x = 0$ (corresponding to the parabola's vertex). As p is convex, it is a minimum, with $p(0) = 1 > 0$. Extending to two variables we obtain $|p|(\alpha, \beta) = \sqrt{(\alpha^2 - \beta^2 + 1)^2 + 4\alpha^2\beta^2}$, which has at 0 a saddle because $p(\epsilon, 0) = \epsilon^2 + 1 > |1 - \epsilon^2| = p(0, \epsilon)$ for any $\epsilon \neq 0$.

B) The second explanation follows in the footsteps of Gauß' original proof. We'll restrict to degree 2 since the generalisation is straightforward. Write $z^2 + rz + s = 0$ in polar coordinates (ρ, ϕ), so to obtain the intersection of two surfaces

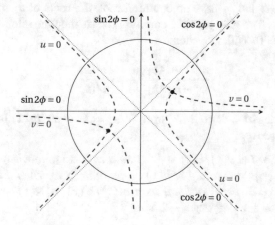

$$\begin{cases} u(\rho, \phi) := \rho^2 \cos 2\phi + r\rho \cos \phi + s = 0 \\ v(\rho, \phi) := \rho^2 \sin 2\phi + r\rho \sin \phi + s = 0 \end{cases}.$$

Fixing $\rho \gg 0$ (which means we are staying outside a large circle), the functions $\rho^{-2}u(\rho, \phi)$, $\rho^{-2}v(\rho, \phi)$ behave like $\cos 2\phi$, $\sin 2\phi$, so they assume alternating positive and negative values. The level curves $u(\rho, \phi) = 0 = v(\rho, \phi)$ are asymptotic to the (polar) straight lines $\cos 2\phi = 0 = \sin 2\phi$, so there must exist common points.

For higher degree, the number of asymptotes increases since $\cos n\phi = 0 = \sin n\phi$ represent two systems of n alternating lines through the origin each.

Eventually, we're ready to prove the partial-fraction decomposition promised in (9.1). We'll split it in two to simplify the statements and thus ease the reading.

Proposition 13.12 *Let $a(x)/b(x) \in \mathbb{R}(x)$ be a rational map with $a(x), b(x)$ coprime, and $\alpha \in \mathbb{R}$ a root of $b(x)$ of multiplicity m. Then there exist a unique constant $A \in \mathbb{R}$ and a unique rational map $a'(x)/b'(x) \in \mathbb{R}(x)$, with $a'(x), b'(x)$ coprime, such that $\deg b' < \deg b$ and*

$$\frac{a(x)}{b(x)} = \frac{A}{(x-\alpha)^m} + \frac{a'(x)}{b'(x)}.$$

Proof By hypothesis $b(x) = (x-\alpha)^m b_1(x)$, with $b_1(\alpha) \neq 0$. Then

$$\frac{a(x)}{b(x)} - \frac{A}{(x-\alpha)^m} = \frac{a(x) - Ab_1(x)}{(x-\alpha)^m b_1(x)},$$

and $\deg(a - Ab_1) < \deg b$. We seek A so that $a'(x)/b'(x)$ satisfies the required properties. Since a and b_1 have no common root, the roots of b_1 must differ from those of $a - Ab_1$. The only possible common root is then the number α solution to $a(\alpha) - Ab_1(\alpha) = 0$. With that choice

$$\frac{a(x) - Ab_1(x)}{(x-\alpha)^m b_1(x)} = \frac{a'(x)}{b'(x)},$$

where $\deg a' < \deg b' < \deg b$. Moreover, $b'(x) = (x-\alpha)^n b_1(x)$ for some $n < m$, and the claim is proved. $\square$

Proposition 13.13 *Let $a(x)/b(x) \in \mathbb{R}(x)$ be a rational map where $a(x), b(x)$ are coprime, and $\alpha + i\beta \notin \mathbb{R}$ a root of $b(x)$ of multiplicity m. Then there exist unique constants $B, C \in \mathbb{R}$ and a rational map $a'(x)/b'(x) \in \mathbb{R}(x)$, with $a'(x), b'(x)$ coprime, such that $\deg b' \leqslant \deg b - 2$ and*

$$\frac{a(x)}{b(x)} = \frac{Bx + C}{\left((x-\alpha)^2 + \beta^2\right)^m} + \frac{a'(x)}{b'(x)}.$$

Proof We know $b(x) = \left((x-\alpha)^2 + \beta^2\right)^m b_1(x)$, with $b_1(\alpha \pm i\beta) \neq 0$, by Corollary 13.10. We wish to find B, C so that

$$\frac{a(x)}{b(x)} - \frac{Bx + C}{\left((x - \alpha)^2 + \beta^2\right)^m} = \frac{a(x) - (Bx + C)b_1(x)}{\left((x - \alpha)^2 + \beta^2\right)^m b_1(x)}$$

satisfies the hypotheses. For that, let's impose $a(x) - (Bx + C)b_1(x)$ have roots $\alpha \pm i\beta$:

$$a(\alpha \pm i\beta) - (B(\alpha \pm i\beta) + C) b_1(\alpha \pm i\beta) = 0,$$

which, subtracting the two, implies $2i B\beta = \dfrac{a(\alpha + i\beta)}{b_1(\alpha + i\beta)} - \dfrac{a(\alpha - i\beta)}{b_1(\alpha - i\beta)}$. As the right-hand side is imaginary, and $\beta \neq 0$, we obtain a unique value $B \in \mathbb{R}$. Adding up the above relations produces a unique number C. Now it's easy to check these constants verify the claim.   □

13.3   Algebraic and Transcendental Numbers

This fascinating section may be viewed as a stepping stone towards ☞ *analytical number theory*.

Definition A complex number is called **algebraic** if it is the root of a polynomial with coefficients in $\mathbb{Q}$.

For instance, $\zeta = \sqrt{2} + \sqrt{3}$ is algebraic because $(\zeta^2 - 5)^2 - 24 = 0$.

Exercise Show that $\zeta \in \mathbb{C}$ is a zero of $p(x) \in \mathbb{Q}[x]$ if and only if it is a zero of some polynomial $\widetilde{p}(x) \in \mathbb{Z}[x]$.

Just like $2^{1/2} + 3^{1/2}$, every number obtained by a finite composition of sums, products and n^{th} roots is algebraic, and therefore algebraic numbers form a field $\overline{\mathbb{Q}}$, the algebraic closure of $\mathbb{Q}$. This fact doesn't mean any algebraic number can be represented in that way: the zeroes of $x^5 - x + 1$, albeit algebraic, cannot be expressed using arithmetic operations and extracting roots. (☞ *Galois theory*).

For any algebraic number a there's a unique monic polynomial $m_a(x) \in \mathbb{Q}[x]$ of lowest degree n (among those annihilated by a), called **minimal polynomial** of a. If so, a is algebraic of **degree** n.

Examples

- $\sqrt{5}$ is algebraic of degree 2, with minimal polynomial $x^2 - 5$.
- Irrational numbers $\sqrt[q]{r}$, with $r \in \mathbb{R}^+$, $q \in \mathbb{N}$ are algebraic: they solve $x^q - r = 0$.
- Suppose a is a real algebraic number. Then
$$\deg a = 1 \iff a \in \mathbb{Q}, \text{ since } \frac{p}{q} \text{ has minimal polynomial } x - p/q;$$

$\deg a = 2 \iff a = q_1 + q_2\sqrt{n}$ with $q_1, q_2 \in \mathbb{Q}$ and $n > 1$ square-free (a belongs to the field extension $\mathbb{Q}(\sqrt{n})$ and is called a quadratic irrational). These are the only numbers with periodic continuous fraction, cf. p. 173.

• Numbers of the form $\cos\theta$, $\sin\theta$ when $\theta \in \mathbb{Q}\pi$ are algebraic.

Proof If $\theta = m\pi/n$ with $m, n \in \mathbb{Z}$, then $z = \cos\theta + i\sin\theta$ is a $2n^{\text{th}}$ root of unity.   □

Examples:

$\cos(3\pi/7)$ is a solution to $8x^3 - 4x^2 - 4x + 1 = 0$;

$2\cos(\pi/5) = \varphi$ is the golden ratio of (10.1), root of $x^2 - x - 1 = 0$. See [36] for the relationship between π and φ.

• A complex number $a + bi$ is algebraic if and only if a, b are algebraic.

Non-algebraic numbers 'transcend' the reach of algebraic methods, to quote Euler who baptised them so. More precisely, let's consider polynomials with rational coefficients:

Definition 13.14 The real number $r \in \mathbb{R}$ is **transcendental (over** $\mathbb{Q}$) if no polynomial $q(x) \in \mathbb{Q}[x]$ (non-zero) has r as a root. Equivalently, the evaluation map $\mathbb{Q}[x] \longrightarrow \mathbb{R}$, $q(x) \longmapsto q(r)$ is one-to-one.

Examples

• π (proved by von Lindemann in 1882).
• e (proved by Hermite, 1873).
• e^a is transcendental for any $a \neq 0$ algebraic (*von Lindemann–Weierstraß theorem*). From that it follows that e is transcendental, and π alike: (by contradiction) $\pi \in \overline{\mathbb{Q}} \implies 2\pi i \in \overline{\mathbb{Q}}$ then $1 = e^{2\pi i} \notin \overline{\mathbb{Q}}$ $\mathcal{f}\mathcal{f}$.

• $\displaystyle\sum_{n=0}^{\infty} \beta^{2^n}$ with β algebraic, $0 < |\beta| < 1$.

• a^b is transcendental for algebraic numbers $a \neq 0, 1$ and b irrational (*Gelfond–Schneider theorem*, solution to Hilbert's 7th problem [55]).

Exercises Using the above facts, show

i) $\sin a$, $\cos a$, $\tan a$, $\log b$ are transcendental, for any algebraic $a \neq 0$, $b \neq 0, 1$.
   Example: e, $\sin 1$, $\log 2$.

ii) $e^\pi = (-1)^{-i}$, $i^i = e^{-\pi/2}$, $2^{\sqrt{2}}$, $\sqrt{2}^{\sqrt{2}}$ are transcendental.

iii) Is $\left(\sqrt{2}^{\sqrt{2}}\right)^{\sqrt{2}}$ algebraic? (*Hint*: compute it!)

Corollary *Algebraic numbers are countable, while transcendental numbers are more than countable.*

Proof The general degree-n algebraic equation

$$\sum_{i=0}^{n} a_i x^i = 0, \quad a_i \in \mathbb{Z},$$

depends on $n+1$ integers, so there are $\aleph_0 \cdot (n+1) = \aleph_0$ algebraic equations of degree n. Each one has n solutions, and therefore there exist $\aleph_0 \cdot n = \aleph_0$ algebraic numbers of degree n. In total, then, there are $\aleph_0 \cdot \aleph_0$ algebraic numbers: $\operatorname{card}(\overline{\mathbb{Q}}) = \aleph_0$.

Consequently there remain $|\mathbb{C} \setminus \overline{\mathbb{Q}}| = |\mathbb{C}| - \aleph_0 = \aleph_1 - \aleph_0 = \aleph_1$ non-algebraic numbers.   □

The above argument substantiates the fact that there are far more (uncountably more) transcendental numbers than algebraic numbers. It was used by Cantor (1874) to show that transcendental numbers do exist. But it is posterior to Liouville's discovery (1844) that $\sum_{j=1}^{\infty} 10^{-j!} = \frac{1}{10} + \frac{1}{100} + \frac{1}{1000000} + \cdots = 0.110001 \cdots$ is transcendental. This number is called Liouville constant, and belongs to a class of transcendental numbers named after him. Roughly, a Liouville number L is closer to a given rational more than any algebraic irrational:

$$\forall n > 0 \, \exists \, p, q \in \mathbb{Z}, q > 1 : \ 0 < \left| L - \frac{p}{q} \right| < \frac{1}{q^n}.$$

This is at first counter-intuitive if we remember the cardinality and density of the irrationals. It says that transcendental numbers possess better rational approximations than algebraic irrationals ($\mathbb{F}$ *Diophantine approximation*).

Examples (open Problems in $\mathbb{F}$ transcendental number theory)

- It isn't known whether $\pi \pm e, \pi e^{\pm 1}, \pi^{\pi}, e^e, \pi^e, \pi \sqrt{2}, e\pi^2$ are transcendental (nor if they are irrational, if it comes to that!).

  Yet, if $a, b \notin \overline{\mathbb{Q}}$ then at least one of $a + b$ and ab is transcendental: as $(x - a)(x - b) = x^2 - (a + b)x + ab$, if both $a + b, ab$ were algebraic also the roots a, b would be algebraic $\#$.

  On the other hand, $\pi + e^{\pi}, \pi e^{\pi}, e^{\pi \sqrt{n}}$ (for any $n \in \mathbb{N}$) have been proven transcendental.

- It's unknown whether the Euler–Mascheroni constant $\gamma := \lim_{n \to \infty} \left(\sum_{k=1}^{n} \frac{1}{k} - \log n \right)$

  (see p. 46) is algebraic or transcendental. Actually we don't even know if it's irrational.

- We know $\zeta(3) = \sum_{k=0}^{\infty} \frac{1}{k^3}$ is irrational (Apéry 1978), but nothing about it being

  transcendental (ζ is Riemann's zeta (3.1)). But no one knows whether $\zeta(2k + 1)$, $k > 1$, are irrational.

- The above theorems are generalised and subsumed by *Schanuel's conjecture*, aimed at measuring 'how' transcendental a set of complex numbers is. The conjecture posits that for linearly independent numbers $z_1, \ldots, z_n \in \mathbb{C}$ over $\mathbb{Q}$, the transcendence degree of the field $\mathbb{Q}(z_1, \ldots, z_n, e^{z_1}, \ldots, e^{z_n})$ over $\mathbb{Q}$ is no less than n.

A splendid monograph on these matters is [66].

Chapter 14
Enumerative Combinatorics

This is the odd-one-out chapter, in that it could fit in any place because its content relate to virtually everything else in the book. Its current placing reflects the emphasis that was put on functions. Apart from preparing for ☞ *discrete mathematics* in general, the techniques and ideas of *recurrence relations* turn out to be useful in view of ☞ *differential equations* and *complex analysis*. Once again we'll relinquish the complete discussion to specialised texts such as [29, 89], both winners of the Steele Prize for Mathematical Exposition. The intriguing lecture notes [86] were a source of inspiration.

Pre-requisites: Sect. 14.4 treats power series formally, as infinite sums. Group theory helps, but isn't necessary, for Sect. 14.1.

14.1 Permutations

Permutation groups were the sparkle that fired up and fuelled the development of abstract group theory. At the root of it all is Theorem 14.2, which states that every abstract group G can be realised as a subgroup of maps $G \to G$, thus reducing—theoretically—group theory to the study of permutations. It's here that ☞ *representation theory* originates, too.

Let $X = \{x_1, x_2, \ldots, x_{n-1}, x_n\}$ be a finite set of $n > 0$ distinct elements. As X has the same cardinality of the initial segment $[1, \ldots, n] \subset \mathbb{N}$, we identify each x_i with its index i and assume $X = \{1, 2, \ldots, n-1, n\}$.

Definition 14.1 Let X be a finite set with card $X = n$. Its bijections $\sigma : X \to X$ are called **permutations**, and are conventionally represented in this form

$$\sigma = \begin{pmatrix} 1 & 2 & \ldots & n \\ \sigma(1) & \sigma(2) & \ldots & \sigma(n) \end{pmatrix}.$$

For example, the bijection on $\{1, 2, 3, 4\}$ given by $\sigma(i) = i + 2 \pmod 4$ is represented by $\sigma = \left(\begin{smallmatrix} 1 & 2 & 3 & 4 \\ 3 & 4 & 1 & 2 \end{smallmatrix}\right)$.

Note that it suffices to demand a permutation be 1-1 (as X is finite).

Definition The permutations of a set X form the **symmetric group** of X

$$\mathfrak{S}(X) = \{\sigma : X \to X \text{ bijective}\}$$

under composition. If card $X = n$ we also write $\mathfrak{S}_n$. For example

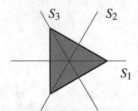

$\mathfrak{S}_2 \cong \mathbb{Z}_2$ is generated by $\left(\begin{smallmatrix} 1 & 2 \\ 1 & 2 \end{smallmatrix}\right)$ (identity) and $\left(\begin{smallmatrix} 1 & 2 \\ 2 & 1 \end{smallmatrix}\right)$ (swap);

$\mathfrak{S}_3 \cong D_3$ are the symmetries of an equilateral triangle, generated by the axial reflections $S_1 = \left(\begin{smallmatrix} 1 & 2 & 3 \\ 1 & 3 & 2 \end{smallmatrix}\right)$, $S_2 = \left(\begin{smallmatrix} 1 & 2 & 3 \\ 3 & 2 & 1 \end{smallmatrix}\right)$, $S_3 = \left(\begin{smallmatrix} 1 & 2 & 3 \\ 2 & 1 & 3 \end{smallmatrix}\right)$.

Exercise Write down the elements of $\mathfrak{S}_4$ and interpret them as symmetries of a square.

We remind, see (6.2), the factorial $n!$ of a natural number is defined recursively by $0! := 1$, $(n + 1)! := n!(n + 1)$, $\forall n \geqslant 0$. For instance $n!$ is the number of anagrams of an n-letter word, or the ways in which n of King Arthur's Knights can sit at the legendary round table.

Proposition If card $X = n$ then $|\mathfrak{S}_n| = n!$.

Proof (By induction) $\mathfrak{S}_0$ has one element only, the empty map $\varnothing \to \varnothing$, so $|\mathfrak{S}_0| = 1 = 0!$. Take a set $X = \{x_1, x_2, \ldots, x_n, x_{n+1}\}$ of cardinality $n+1$ and a permutation $\sigma \in \mathfrak{S}_{n+1}$. Let $Y \subseteq X$ be a subset of n elements. The restriction $\sigma\big|_Y$ belongs to $\mathfrak{S}_n$, so there are $n!$ ways to define $\sigma\big|_Y$. At the same time there exist $n + 1$ distinct subsets Y in X, given by $X \setminus \{x_i\}$ for $1 \leqslant i \leqslant n + 1$. Therefore we can define σ in $n!(n + 1) = (n + 1)!$ possible ways. $\square$

As $\mathfrak{S}_1$ is trivial and $\mathfrak{S}_2 \cong \mathbb{Z}_2$, one considers $\mathfrak{S}_n$ for $n \geqslant 3$. The following algebraic properties lie at the heart of the solvability of polynomial equations.

Proposition The group $\mathfrak{S}_n$, for $n \geqslant 3$,

i) is not Abelian (its centre is trivial: $Z(\mathfrak{S}_n) = \{\mathbb{1}\}$)
ii) is solvable $\iff n = 3, 4$.

Statement ii) manifests itself, in a non-trivial way ($\LARGE{\mathbb{\circledcirc}}$ Galois theory), in the formulas for solving polynomial equations of degree n:

  $n = 2$: formula (9.2) corresponds to $\mathfrak{S}_2 \cong \mathbb{Z}_2$ being Abelian;

  $n = 3$: formula (9.5), reducing a cubic to a quadratic equation, arises from a map $\mathfrak{S}_3 \longrightarrow \mathfrak{S}_2$ (the sign function);

$n = 4$: formula (9.6), whereby a cubic allows to solve a quartic, essentially depends on the group homomorphism $\mathfrak{S}_4/\mathfrak{S}_3 \cong D_2$, where the latter is the Klein group (cf. Example 14.3).

$n \geqslant 5$: corresponds to $\mathfrak{S}_n$ not being solvable.

The inverse of a permutation $\sigma = \left(\begin{smallmatrix} 1 & 2 & 3 & 4 & 5 & 6 & 7 \\ 2 & 4 & 6 & 3 & 5 & 7 & 1 \end{smallmatrix}\right)$ is found by swapping rows and rearranging terms: $\sigma^{-1} = \left(\begin{smallmatrix} 1 & 2 & 3 & 4 & 5 & 6 & 7 \\ 7 & 1 & 4 & 2 & 5 & 3 & 6 \end{smallmatrix}\right)$, and the product is defined by composition: $\sigma\tau := \tau \circ \sigma$

$$\underbrace{\left(\begin{smallmatrix} 1 & 2 & 3 & 4 & 5 & 6 & 7 \\ 2 & 4 & 6 & 3 & 5 & 7 & 1 \end{smallmatrix}\right)}_{\sigma} \underbrace{\left(\begin{smallmatrix} 1 & 2 & 3 & 4 & 5 & 6 & 7 \\ 3 & 5 & 7 & 1 & 2 & 4 & 6 \end{smallmatrix}\right)}_{\tau} = \left(\begin{smallmatrix} 1 & 2 & 3 & 4 & 5 & 6 & 7 \\ 5 & 1 & 4 & 7 & 2 & 6 & 3 \end{smallmatrix}\right)$$

Since σ fixes 5, we may simplify the expression and only write the elements on which it acts effectively: $\left(\begin{smallmatrix} 1 & 2 & 3 & 4 & 5 & 6 & 7 \\ 2 & 4 & 6 & 3 & 5 & 7 & 1 \end{smallmatrix}\right) = \left(\begin{smallmatrix} 1 & 2 & 3 & 4 & 6 & 7 \\ 2 & 4 & 6 & 3 & 7 & 1 \end{smallmatrix}\right)$ and dropping those that don't change. This idea justifies

Definition An *m*-cycle is a permutation $\sigma \in \mathfrak{S}_n$ that acts effectively on a set $\{x_1, \ldots, x_m\} \subseteq X$ of cardinality $m \leqslant n$ as follows:

$$\sigma(x_i) = x_{i+1}, \ i = 1, \ldots, m-1, \quad \sigma(x_m) = x_1.$$

The standard notation for a cycle is $(x_1 \ x_2 \ \ldots \ x_{m-1} \ x_m)$ (Fig. 14.1).

The previous σ is a 6-cycle: $\left(\begin{smallmatrix} 1 & 2 & 3 & 4 & 5 & 6 & 7 \\ 2 & 4 & 6 & 3 & 5 & 7 & 1 \end{smallmatrix}\right) = \left(\begin{smallmatrix} 1 & 2 & 3 & 4 & 6 & 7 \\ 2 & 4 & 6 & 3 & 7 & 1 \end{smallmatrix}\right)\left(\begin{smallmatrix} 5 \\ 5 \end{smallmatrix}\right) = (1\,2\,4\,3\,6\,7)(5)$.

The permutation $\tau = \left(\begin{smallmatrix} 1 & 2 & 3 & 4 & 5 & 6 & 7 \\ 3 & 5 & 7 & 1 & 2 & 4 & 6 \end{smallmatrix}\right)$ is no cycle: 2 cannot be reached from 4 (2 doesn't belong to the orbit of 4, in the jargon of Chap. 17).

Exercise Show the inverse cycle of $(1\ 2\ \cdots\ k{-}1\ k)$ is $(k\ k{-}1\ \cdots\ 2\ 1)$.

It can be proved that cycles are the constituents of any permutation:

Proposition *Every permutation can be written as a product of disjoint cycles, in an essentially unique way (up to order). It can also be decomposed in a product of 2-cycles (typically not uniquely).*

Example The permutation $\left(\begin{smallmatrix} 1 & 2 & 3 & 4 & 5 & 6 & 7 \\ 3 & 1 & 2 & 7 & 4 & 6 & 5 \end{smallmatrix}\right)$ equals $(1\,2\,3)(4\,7\,5)$ in term of disjoint cycles, or various products of 2-cycles: $(1\,3)(1\,2)(4\,7)(4\,5) = (2\,3)(1\,3)(4\,5)(4\,7)(5\,7)(4\,5)$.

Fig. 14.1 σ is an *m*-cycle

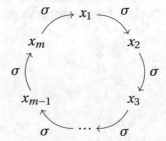

Here's the section's crowing theorem (revealing the link to Sect. 17.2):

Theorem 14.2 (Cayley 1854) *Any group* G *embeds in its permutation group* $\mathfrak{S}(G)$.

Proof Let $\rho(G) := \{\rho_g \in \mathfrak{S}(G) \mid \rho_g(h) = hg\} \subseteq \mathfrak{S}(G)$ denote the set of so-called right-translations. It is a subgroup because $\rho_g \rho_{g'} = \rho_{g'} \circ \rho_g = \rho_{g'g}$, and $\left(\rho_g\right)^{-1} = \rho_{g^{-1}}$.

Now look at the map $\rho \colon G \to \rho(G)$, $g \mapsto \rho_g$. It's clearly a surjective homomorphism, and the kernel

$$\operatorname{Ker}\rho = \{g \in G \mid \rho_g = \mathbb{1}\} = \{g \in G \mid hg = h,\ \forall h \in G\} = \{1\}$$

is trivial. (In the language of Definition 17.3, ρ is a faithful representation.) $\square$

Although Cayley's theorem appears to be rather elementary, we shouldn't forget that in the nineteenth century group theory was in its infancy. When Cayley introduced the objects we nowadays call groups, it wasn't at all clear that an abstract group was the same thing as a permutation group. Cayley's theorem merged the two notions.

Example 14.3 (Klein 1884) The group G generated by elements a, b such that $a^2 = b^2 = (ab)^2 = 1$ is the Klein group $D_2 \cong \mathbb{Z}_2 \oplus \mathbb{Z}_2$ (Abelian, not cyclic, with 4 elements). It has a linear representation on $\mathbb{R}^2$ that describes the symmetries of a rectangle: a, b are the axial symmetries, $c = ab$ the π-rotation. Viewed as 2×2 matrices,

$$1 = \begin{pmatrix} 1 & 0 \\ 0 & 1 \end{pmatrix}, \quad a = \begin{pmatrix} 1 & 0 \\ 0 & -1 \end{pmatrix}, \quad b = \begin{pmatrix} -1 & 0 \\ 0 & 1 \end{pmatrix}, \quad c = \begin{pmatrix} -1 & 0 \\ 0 & -1 \end{pmatrix}$$

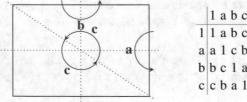

1 a b c		cycles:
1	1 a b c	$\rightsquigarrow \rho_1 = \mathbb{1}$
a	a 1 c b	$\rightsquigarrow \rho_a = (1\,a)(b\,c)$
b	b c 1 a	$\rightsquigarrow \rho_b = (1\,b)(a\,c)$
c	c b a 1	$\rightsquigarrow \rho_c = (1\,c)(a\,b)$

The *Yoneda lemma* in ☞ *category theory* conveys the same basic idea of Cayley's representation theorem.

14.2   The Twofold Way

Let's discuss a couple of classical enumerative problems concerning finite sets, a small morsel of Rota's *twelvefold way*. Generalising the idea of a permutation,

Definition Let $X = \{x_1, \ldots, x_n\} \approx \{1, 2, \ldots, n\}$ be a finite set and $k \leqslant n$ a natural number. We call k-**permutation** of n elements a one-to-one map $\{i_1, i_2, \ldots, i_k\} \hookrightarrow \{1, 2, \ldots, n\}$.

In practice a k-permutation is an ordered array $(x_{i_1}, x_{i_2}, \ldots, x_{i_k})$ of distinct elements $x_{i_j} \in X$ picked from n possibilities. Hence the alternative names partial permutation, or k-sequence without repetition.

Proposition *The number of k-permutations of n elements equals*

$$n^{\underline{k}} := \frac{n!}{(n-k)!} = \underbrace{n(n-1)(n-2) \cdots (n-k+1)}_{k \text{ factors}} = \prod_{j=1}^{k}(n+1-j), \qquad (14.1)$$

*called **falling factorial**.*

Proof Exercise.   □

Observe that if we took $k > n$ we would obtain a zero factor, reflecting the fact there are no injective maps $\{1, \ldots, n, \ldots, k\} \to \{1, \ldots, n\}$. This is a restatement of the Pigeonhole Principle 6.36.

Examples

  i) 30 students in the classroom can sit at tables for two in $30^{\underline{2}} = 30 \cdot 29 = 870$ different ways.
 ii) Ten people partake in a game with 4 prizes. Assuming that every person can gain only one prize, the number of possible winners' quadruples is $10^{\underline{4}} = 10 \cdot 9 \cdot 8 \cdot 7 = 5040$.
iii) The quantity of numbers made of 3 distinct digits equals $10^{\underline{3}} - 9^{\underline{2}} - 9^{\underline{1}} - 9^{\underline{0}} = 638$ (sequences of type $0ab$, $00b$ or 000 are excluded).

Exercises

  i) Prove $n^{\underline{2}} + n^{\underline{1}} = n^2$ for any $n \in \mathbb{N}^*$.
 ii) Solve equation $x^{\underline{3}} - x^3 = x^{\underline{2}} - 1$.

Allowing for duplicate choices gives rise to the notion of k-*permutation with repetition*:

Proposition *The number of k-permutations with repetition of n elements equals n^k.*

Proof Exercise.   □

Formulated in a slightly more general way, this is the content of the pompously named *principle of independent choices*: suppose we have to select n objects so that there are m_1 ways to choose the first object, m_2 ways to choose the second etc., and m_n ways to choose the last. The total number of possibilities is $\prod_{i=1}^{n} m_i = m_1 m_2 \cdots m_{n-1} m_n$.

Example 14.4 (Birthday Problem) The odds that two people in a group of k will be born on the same day equals $bday(k) := 365^{\underline{k}}/365^k$. (This assumes that birthdays are uniformly distributed over the year and no one was born on 29 February.)

In fact, if the individuals declare one by one their birthday, there are 365^k possibilities (principle of independent choices). Among those, the number necessary to prevent a coincidence is $365 \cdot 364 \cdot 363 \cdots (365 - k + 1) = 365^{\underline{k}}$. By the pigeonhole principle, the probability reaches 100% when $k = 366$. And here comes the (birthday) surprise: $bday(23) < 1/2$ says that just 23 people are enough to guarantee a 50% probability that at least two are born on the same day.

Exercises

   i) Tossing a coin three times, how many triples can be obtained?
  ii) In how many ways can we match the faces of two dice?
 iii) How many (meaningful) 4-digit numbers are there?

Let's consider now k-permutations where we don't distinguish reshuffled k-tuples.

Definition For any $k \leqslant n$ in $\mathbb{N}$, the numbers

$$\binom{n}{k} := \frac{n^{\underline{k}}}{k!} = \frac{n!}{k!(n-k)!}$$

are called **binomial coefficients** (a name invented by Stifel in 1544).

For instance $\binom{n}{0} = 1$, $\binom{n}{1} = n$, $\binom{n}{2} = \frac{n(n-1)}{2}$ for all n.
Notice the evident symmetry

$$\binom{n}{k} = \binom{n}{n-k}, \quad \forall k = 0, \ldots, n.$$

Binomial coefficients are described combinatorially as follows.

Definition Fix a finite set $X = \{x_1, \ldots, x_n\}$ and $k \leqslant n$. We call **k-combination** in X a subset $\{x_{i_1}, x_{i_2}, \ldots, x_{i_k}\} \subseteq X$ of cardinality k (note that the order is irrelevant).

Proposition *The binomial coefficient $\binom{n}{k}$ equals the number of k-combinations of n elements.*

Proof Since a k-combination is a k-permutation in which the ordering doesn't matter, the choices of k elements among n are given by k-permutations up to the $k!$ possible rearrangements (permutations): $n^{\underline{k}} = k! \binom{n}{k}$.   □

Examples

i) We can choose 2 aces in a standard deck of cards in 6 ways:

because there are $\binom{4}{2} = 6$ combinations of 2 elements chosen among 4. Analogously, there are $4 = \binom{4}{3}$ possible three-of-a-kind

and clearly $1 = \binom{4}{1}$ four-of-a-kind only. Poker champions know that the number of hands of each of the following type is:

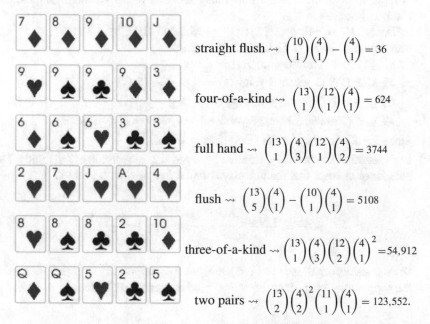

straight flush $\rightsquigarrow$ $\binom{10}{1}\binom{4}{1} - \binom{4}{1} = 36$

four-of-a-kind $\rightsquigarrow$ $\binom{13}{1}\binom{12}{1}\binom{4}{1} = 624$

full hand $\rightsquigarrow$ $\binom{13}{1}\binom{4}{3}\binom{12}{1}\binom{4}{2} = 3744$

flush $\rightsquigarrow$ $\binom{13}{5}\binom{4}{1} - \binom{10}{1}\binom{4}{1} = 5108$

three-of-a-kind $\rightsquigarrow$ $\binom{13}{1}\binom{4}{3}\binom{12}{2}\binom{4}{1}^2 = 54{,}912$

two pairs $\rightsquigarrow$ $\binom{13}{2}\binom{4}{2}^2\binom{11}{1}\binom{4}{1} = 123{,}552.$

ii) Consider an experiment where a unique action (like flipping a coin) is repeated identically, with possible outcomes labelled 'success' or 'failure' (heads or tails) and independent of one another (the odds are exactly the same at each toss, because the chance of heads is still 50% even if you've had 5000 tails in a row). This is called *Bernoulli trial*. The **binomial probability** function computes the chances of s successes in n attempts:

$$P(s, n) := \binom{n}{s} p^s (1 - p)^{n-s}$$

where $p \in [0, 1]$ are the odds of success of each trial (and $1 - p$ the odds of flop).

More concretely, suppose you are sitting a multiple-choice test: if there are 10 questions, each with 4 options, and you answer randomly, the odds of getting 6 correct answers (and passing) are approximately 4%:

$$P(6, 10) = \binom{10}{6} \left(\frac{1}{4}\right)^6 \left(\frac{3}{4}\right)^4 \approx 0.04.$$

Figuring out 7 or more answers sounds like 'mission: impossible': $P(7, 10) \approx 0.003$, $P(8, 10) \approx 0.0004$, $P(9, 10) \approx 0.00003$, $P(10, 10) < 0.000001$.

iii) Consider a set $S = B \cup G \cup Y$ of 6 black balls, 4 greens and 6 yellows (all different). We want to know how many sets of five balls contain at most one black and one green.

Define $K = \{R \in \mathscr{P}(S) \colon |R| = 5, |R \cap B| \leqslant 1, |R \cap G| \leqslant 1\}$ and decompose $K = K_1 \cup K_2 \cup K_3 \cup K_4$, where

$K_1 \subseteq Y$ (all yellows) $\Longrightarrow |K_1| = \binom{6}{5}$

$K_2 \subseteq Y \cup G$ (1 green, 4 yellows) $\Longrightarrow |K_2| = \binom{6}{4}\binom{6}{1}$

$K_3 \subseteq B \cup Y$ (1 black, 4 yellows) $\Longrightarrow |K_3| = \binom{6}{4}\binom{4}{1}$

$K_4 \subseteq S$ (1 green, 1 black, 3 yellows) $\Longrightarrow |K_4| = \binom{6}{1}\binom{4}{1}\binom{6}{3}$.

Since the K_j are disjoint, $|K| = \sum_i |K_i| = 636$.

iv) Let's return to the birthday problem 14.4 and try to approximate the odds. The number of chances that the birthdays of all k people are different is

$$\frac{365^{\underline{k}}}{} = \frac{\prod_{i=0}^{k-1}(365 - i)}{365^k} = \prod_{i=0}^{k-1}\left(1 - \frac{i}{365}\right).$$

If we consider small groups ($k \ll 365$) we may approximate each factor using the exponential map, since $1 - t \approx e^{-t}$ whenever $|t| \ll 1$. Hence

$$365^{\underline{k}} \approx \prod_{i=0}^{k-1} e^{-i/365} = e^{-(\sum_0^{k-1} i)/365} = \frac{1}{e^{\binom{k}{2}/365}}.$$

In the end $bday(k) \approx 1 - e^{-\binom{k}{2}/365}$.

A coarser estimate is found as follows. The odds any two people don't have the same birthday is $364/365$. If we treat the $\binom{k}{2}$ pairs with different birthday as independent events, we obtain about $\left(\frac{364}{365}\right)^{\binom{k}{2}}$ chances of k distinct birthdays, and so $bday(k) \approx 1 - \left(\frac{364}{365}\right)^{\binom{k}{2}}$.

Exercises Prove

i) $\binom{n+1}{k} = \binom{n}{k}\frac{n-k}{k+1}$, $\binom{n}{k}\binom{k}{m} = \binom{n}{m}\binom{n-m}{k-m}$.

ii) $\binom{n}{k} = \sum_{j=k}^{n}\binom{j-1}{k-1}$, $\binom{m+n}{k} = \sum_{j=0}^{k}\binom{m}{k-j}\binom{n}{j}$ (Vandermonde formula).

iii) A polygon with n edges has $\binom{n}{2} - n$ diagonals.

iv) An urn contains 20 distinct balls, of which 5 are white and 15 black. How many quadruples can be extracted so that each one contains at least one while ball? How many with exactly one while ball?

v) Consider n circles on the plane, none internal to any other. Determine the number of tangent lines to any two of the circles.

vi) Compute the number of cocktails of 4 ingredients a barman can mix if he has 30 different spirits, syrups and juices to choose from.

14.3 The Binomial Formula

The **Pascal triangle** is an infinite triangular-shaped array of binomial numbers $\binom{n}{k}$, arranged so that n represents the row number and k the column number, starting from zero (the apex):

The construction is recursive: for $0 < k < n$ the coefficient $\binom{n}{k}$ is the sum of the two numbers directly above it (right and left):

Lemma (Stifel) *For every* $1 \leqslant k \leqslant n$

$$\binom{n-1}{k-1} + \binom{n-1}{k} = \binom{n}{k}. \tag{14.2}$$

Proof Simply using the definition,

$$\binom{n-1}{k-1} + \binom{n-1}{k} = \frac{(n-1)!}{(k-1)!(n-k)!} + \frac{(n-1)!}{k!(n-1-k)!}$$

$$= \frac{k(n-1)!}{k!(n-k)!} + \frac{(n-1)!(n-k)}{k!(n-k)!}$$

$$= \frac{(k+n-k)(n-1)!}{k!(n-k)!} = \binom{n}{k}.$$

$\square$

Although it has the label Pascal, the triangular pattern was well known during the Renaissance by many, including Stifel, Tartaglia, Stevin and others.

Exercises Prove that

i) the sum of the terms on row n equals $\displaystyle\sum_{k=0}^{n} \binom{n}{k} = 2^n$;

ii) the alternating sum on row $n \geqslant 1$ vanishes: $\displaystyle\sum_{k=0}^{n} (-1)^k \binom{n}{k} = 0$;

iii) the sum of the terms in even position on row $n \geqslant 1$ is $\displaystyle\sum_{k=0}^{\lfloor n/2 \rfloor} \binom{n}{2k} = 2^{n-1}$.

What's the sum of the terms in odd position, then?

iv) the sum of the squares on row n is equal to $\displaystyle\sum_{k=0}^{n} \binom{n}{k}^2 = \binom{2n}{n}$.

This means that to choose n objects among $2n$ we may pick i objects from the first n and then $n - i$ from the remaining n. A solution is provided in Example 14.9 iv).

v) $\displaystyle\binom{2n}{n} - \binom{2n}{n+1} = \frac{1}{n+1}\binom{2n}{n}$ for $n \geqslant 1$.

Show that the coefficients $C_n := \dfrac{1}{n+1}\dbinom{2n}{n}$, called *Catalan numbers* (after Eugène Catalan), are integers. Explain why

$$C_{n+1} = C_0 C_n + C_1 C_{n-1} + \cdots + C_\alpha C_{n-\alpha} + \cdots + C_{n-1} C_1 + C_n C_0.$$

This fact is directly related to the *Catalan problem*, asking in how many ways C_n a convex n-polygon ($n \geqslant 3$) can be partitioned in triangles using non-intersecting diagonals. If we declare $C_1 = 0$, $C_2 = 1$, it's not difficult to see that $\{C_n\}_{n \geqslant 3}$ satisfies the aforementioned recurrence relation.

vi) Prove formula (14.2) by counting subsets of k elements, inside a set of cardinality n, that contain a chosen element z, and those without z.

The key point—and authentic *raison d'être*—of Pascal's triangle is that on any row n there appear the coefficients of the expansion of the binomial $(a + b)^n$, $n \in \mathbb{N}$. The rule is rather simple:

Theorem (Binomial Formula) *For every* $a, b \in \mathbb{R}, n \in \mathbb{N}$

$$(a + b)^n = \sum_{k=0}^{n} \binom{n}{k} a^k b^{n-k}. \tag{14.3}$$

Proof (By induction) When $n = 0$ we have $(a + b)^0 = 1 = \binom{0}{0} a^0 b^0$.

Supposing the formula holds for n, let's prove it for $n + 1$.

$$(a + b)^{n+1} = (a + b)(a + b)^n = (a + b) \sum_{k=0}^{n} \binom{n}{k} a^k b^{n-k}$$

$$= \sum_{k=0}^{n} \binom{n}{k} a^{k+1} b^{n-k} + \sum_{k=0}^{n} \binom{n}{k} a^k b^{n-k+1}$$

$$= \left[\binom{n}{n} a^{n+1} + \sum_{k=0}^{n-1} \binom{n}{k} a^{k+1} b^{n-k} \right]$$

$$+ \left[\sum_{k=0}^{n-1} \binom{n}{k+1} a^{k+1} b^{n-k} + \binom{n}{0} b^{n+1} \right]$$

$$= a^{n+1} + \sum_{k=0}^{n-1} \left[\binom{n}{k} + \binom{n}{k+1} \right] a^{k+1} b^{n-k} + b^{n+1}$$

$$\overset{(14.2)}{=} a^{n+1} + \sum_{k=0}^{n-1} \binom{n+1}{k+1} a^{k+1} b^{n-k} + b^{n+1}$$

$$= \left[\binom{n+1}{n+1} a^{n+1} b^0 + \sum_{k=0}^{n-1} \binom{n+1}{k+1} a^{k+1} b^{n-k} \right] + b^{n+1}$$

$$= \sum_{k=0}^{n} \binom{n+1}{k+1} a^{k+1} b^{n-k} + b^{n+1}, \quad \text{and reindexing with} \quad j = k+1$$

$$= \sum_{j=1}^{n+1} \binom{n+1}{j} a^j b^{n+1-j} + \binom{n+1}{0} a^0 b^{n+1}$$

$$= \sum_{j=0}^{n+1} \binom{n+1}{j} a^j b^{n+1-j}.$$

$\square$

The result isn't of easy attribution: it was probably first proved by al-Karajī in the Middle Ages.

Examples 14.5

i) Compute the sixth power of a complex number $a + ib$:

$$(a + ib)^6 = (a^6 - 15a^4 b^2 + 15a^2 b^4 - b^6) + i(6a^5 b - 20a^3 b^3 + 6ab^5).$$

Applying this relation to $\cos x + i \sin x = e^{ix}$, we obtain

$$\cos 6x = \cos^6 x - 15\cos^4 x \sin^2 x + 15\cos^2 x \sin^4 x - \sin^6 x$$

$$\sin 6x = 6\cos^5 x \sin x - 20\cos^3 x \sin^3 x + 6\cos x \sin^5 x.$$

The general formulas:

$$\cos(nx) = \sum_{k=0}^{\lfloor n/2 \rfloor} (-1)^k \binom{n}{2k} \cos^{n-2k}(x) \sin^{2k}(x),$$

$$\sin(nx) = \sum_{k=0}^{\lfloor (n-1)/2 \rfloor} (-1)^k \binom{n}{2k-1} \cos^{n-2k-1}(x) \sin^{2k+1}(x)$$

are proved by induction using the expansion of $e^{inx} = (\cos x + i \sin x)^n$.

ii) Expanding $(x + \epsilon)^n$ we may write

$$(x + \epsilon)^n - x^n = nx^{n-1}\epsilon + \sum_{j=2}^{n} \binom{n}{j} x^{n-j} \epsilon^j = nx^{n-1}\epsilon + o(\epsilon)$$

where $o(\epsilon)$ is a polynomial in ϵ of degree at least two. Dividing by ϵ and taking the limit as $\epsilon \to 0$ defines the derivative of the map $x \mapsto x^n$.

iii) We shall prove the inclusion-exclusion formula (6.5).

Proof Choose a in the union $\bigcup_{i=1}^{n} A_i$ and let $A_1, \ldots, A_t$ be the sets containing said element ($n \geqslant t > 0$). Since a is counted exactly once in the left-hand side of (6.5), we must show the same goes for the other side. On the right the contribution is non-zero only when every set in a term contains a, i.e. when all the sets are among the $A_1, \ldots, A_t$. For each one the contribution is ± 1, and hence it equals the number (with sign) of these sets in the term. Write $A_{ij} := A_i \cap A_j$, $A_{ijk} := A_i \cap A_j \cap A_k$ etc. for $1 \leqslant i < j < k < \cdots \leqslant t$. Therefore

$$
\left|\{A_i\}\right| - \left|\{A_{ij}\}\right| + \left|\{A_{ijk}\}\right| - \cdots + (-1)^{t+1} \left|\bigcap_{i=1}^{t} A_i\right|
$$

$$
= \binom{t}{1} - \binom{t}{2} + \binom{t}{3} - \cdots + (-1)^{t+1}\binom{t}{t} \overset{14.3}{=} \binom{t}{0} - (1-1)^t = 1.
$$

Thus a is counted only once in the right side of (6.5).   □

Exercises 14.6

i) Transform $(\sqrt{3} + \sqrt{2})^6$ into form $a + b\sqrt{6}$ for some $a, b \in \mathbb{Z}$.

ii) Prove that an integer $n \geqslant 2$ is prime if and only if all binomial numbers $\binom{n}{1}$, $\ldots, \binom{n}{n-1}$ are divisible by n. Deduce from this

   Fermat's little theorem: let p be prime and $a \in \mathbb{Z}^+$ such that $p \nmid a$. Then $a^{p-1} \equiv_p 1$.

iii) Find the number of non-negative integer solutions to the equation $x_1 + x_2 + \ldots + x_k = n$, for some $k \in \mathbb{N}, n \in \mathbb{Z}$.

iv) For $n, i \in \mathbb{N}$ let n_i be the largest integer such that $\binom{n_i}{i} < n$. Macaulay (1917) showed there exists a unique decomposition

$$
n = \binom{n_i}{i} + \binom{n_{i-1}}{i-1} + \ldots + \binom{n_j}{j},
$$

where $n_i > n_{i-1} > \ldots > n_j \geqslant 1$, called i^{th} binomial expansion of n.

   Example: $35 = \binom{8}{2} + \binom{7}{1}$, or $49 = \binom{7}{4} + \binom{5}{3} + \binom{3}{2} + \binom{1}{1}$.

   Explain with the help of examples why

$$
n > m \iff \begin{cases} n_i > m_i \\ n_{i-1} > m_{i-1} \\ \ldots \\ n_j > m_j \end{cases}
$$

(completing the sequence with 0 if necessary).

v) Recall (p. 31) that Bell numbers B_n count partitions (equivalence relations) on a set of n elements. Explain (or prove) the recursive formula $B_{n+1} = \sum_{k=0}^{n} \binom{n}{k} B_k$ that relates Bell numbers with binomials (this is an instance of a recurrence relation examined in Sect. 14.4).

(*Hint*: taking a partition of $n + 1$ elements and removing the set containing the first element leaves a partition of k elements, $0 \leqslant k \leqslant n$. But there are $\binom{n}{k}$ choices for the k elements, and B_k way to partition them.)

In spite of the fact that combinatorics is defined when n is a natural number, in the product formula of (14.1) no one stops us from taking $n \in \mathbb{R}$ and extend the theory to the reals. Thus $\binom{r}{k}$ makes sense for $r \in \mathbb{R}$ as well. The convention is

$$\binom{r}{k} := \begin{cases} \dfrac{r^{\underline{k}}}{k!} = \displaystyle\prod_{i=1}^{k} \dfrac{r+1-i}{i} & k > 0 \\ 1 & k = 0 \\ 0 & k < 0 \end{cases}.$$

For instance $\binom{4}{6} = \frac{4\cdot3\cdot2\cdot1\cdot0\cdot(-1)}{6!} = 0$, $\binom{\sqrt{2}}{3} = \sqrt{2}(\sqrt{2}-1)(\sqrt{2}-2)/6 = \frac{2}{\sqrt{6}} - 1$. Newton (1665) showed that formula (14.3) continues to hold, as a formal series:

(**binomial formula**) $(1+x)^r = \displaystyle\sum_{k=0}^{\infty} \binom{r}{k} x^k,$ when $|x| < 1$

for $r \in \mathbb{R}$. For example,

$r = \dfrac{1}{2}$: $\sqrt{1+x} = \displaystyle\sum_{k \geqslant 0} \binom{1/2}{k} x^k = 1 + \dfrac{1}{2}x - \dfrac{1}{8}x^2 + \dfrac{1}{16}x^3 - \dfrac{5}{128}x^4 + \cdots$

$r = -\dfrac{1}{2}$: $\dfrac{1}{\sqrt{1+x}} = \displaystyle\sum_{k \geqslant 0} \binom{-1/2}{k} x^k = 1 - \dfrac{1}{2}x + \dfrac{3}{8}x^2 - \dfrac{5}{16}x^3 + \dfrac{35}{128}x^4 - \cdots$

Exercises 14.7 Show

i) $\displaystyle\binom{-n}{k} = (-1)^k \binom{n+k-1}{k}$

ii) $\displaystyle\binom{n+k-1}{k}$ is the coefficient of x^n in $(1 + x + x^2 + \cdots + x^k)^n$.

iii) For any $k \geq 0$, $\binom{-1}{k} = (-1)^k$ and $\binom{-2}{k} = (-1)^k(k+1)$.

In general one proves [38, p. 15] that $\forall k \leq n$, $\binom{-n}{k} = \sum_{\substack{\ell+m=k \\ \ell,m \geq 0}} (-1)^\ell \binom{1-n}{m}$.

iv) $\binom{-n-1}{k-1} + \binom{-n-1}{k} = \binom{-n}{k}$, the 'negative' version of (14.2).

v) Let $bday(k)$ be the function of example (14.4). Deploy the inequality $1 - x \leq e^{-x}$, valid for any $x \in \mathbb{R}$, to show that $\log\big(bday(k)\big) \leq -\binom{k}{2}/365$.

As a final idiosyncrasy, we mention that binomials can take complex entries. For this one requires the *gamma function*

$$\Gamma : \mathbb{C} \setminus \{\mathbb{Z}^{\leq 0}\} \to \mathbb{C}, \quad \Gamma(t) = \int_0^\infty x^{t-1} e^{-x} \, dx.$$

In ☞ *complex analysis* one proves Γ is the *analytical prolongation* of the factorial: integrating by parts we obtain $\Gamma(t+1) = t\Gamma(t)$, whence $\Gamma(n+1) = n!$ for $n \in \mathbb{Z}^+$.

In general, for $r, s \in \mathbb{C}$ one defines $\binom{r}{s} := \dfrac{\Gamma(r+1)}{\Gamma(s+1)\Gamma(r-s+1)}$.

This is the inception of an incredibly exciting story, with ramifications in ☞ *analytic number theory* and *probability*. The paper [16] sheds light on the matter.

14.4 Generating Functions ☙

This section makes use of power series, but we'll ignore convergence issues completely (☞ *complex analysis*) and play with series by treating them as formal infinite sums.

Definition The **generating function** of a sequence $(a_k)_{k \in \mathbb{N}}$ is the formal power series $f(x) = \sum_{k=0}^\infty a_k x^k$.

Examples 14.8

i) $(1+x)^n = \sum_{k=0}^n \binom{n}{k} x^k$, $n \in \mathbb{N}$, generates the binomial coefficients: the sequence is finite, corresponding to a polynomial map.

ii) $\dfrac{1}{1-x} = \sum_{k=0}^\infty x^k$ generates the constant sequence $(1, 1, 1, 1, \ldots)$, corresponding to the geometric progression.

iii) $\dfrac{1}{1+x} = \sum_{k=0}^\infty (-1)^k x^k$ generates $(1, -1, 1, -1, 1, -1, \ldots)$.

$$
\begin{array}{ccccccccccccccc}
 & & & & & & & 1 & & & & & & & \\
 & & & & & & 1 & & 1 & & & & & & \\
 & & & & & 1 & & 2 & & 1 & & & & & \\
 & & & & 1 & & 3 & & 3 & & 1 & & & & \\
 & & & 1 & & 4 & & 6 & & 4 & & 1 & & & \\
 & & 1 & & 5 & & 10 & & 10 & & 5 & & 1 & & \\
 & 1 & & 6 & & 15 & & 20 & & 15 & & 6 & & 1 & \\
1 & & 7 & & 21 & & 35 & & 35 & & 21 & & 7 & & 1
\end{array}
$$

Fig. 14.2 Tip of Pascal's triangle

That's to say, $\binom{-1}{k} = (-1)^k$ for any $k \geqslant 0$, cf. Exercise 14.7 iii).

iv) $\dfrac{1}{1-x^2} = \dfrac{1/2}{1-x} + \dfrac{1/2}{1+x} = \displaystyle\sum_{j=0}^{\infty}\left(\dfrac{1}{2} + (-1)^j\dfrac{1}{2}\right)x^j = \sum_{k=0}^{\infty}x^{2k}$ generates

$(1, 0, 1, 0, 1, 0, \ldots)$

v) $\dfrac{1}{(1-x)^2} = \left(\displaystyle\sum_{j=0}^{\infty}x^j\right)^2 = \sum_{k=0}^{\infty}(k+1)x^k$ generates the positive integers.

From this, $\dfrac{1}{(1+x)^2}$ gives $(1, -2, 3, -4, \ldots)$, and we deduce $\binom{-2}{k} = (-1)^k(k+1)$, cf. Exercise 14.7 iii).

vi) $\dfrac{1}{(1-x)^3} = \displaystyle\sum_{k=0}^{\infty}\binom{k+2}{2}x^k$ generates the sequence of 'triangular numbers' $T_k = \binom{k+1}{2}$, $k \geqslant 1$, which show up along the third 'diagonal' of Pascal's triangle, Fig. 14.2).

vii) A relation indicating how complicated the interdependence between a sequence and a generating function can be is $\displaystyle\sum_{n=0}^{\infty}\phi(n)x^n = \prod_{m=1}^{\infty}\dfrac{1}{1-x^m}$, where ϕ is Euler's phi function. See Example 10.1 v) (☞ *modular forms*).

Exercises

i) Compute the sequences generated by the maps

$$
\dfrac{1}{1+\lambda x}, \quad \lambda \in \mathbb{R} \text{ fixed}; \quad e^x; \quad -\ln(1-x); \quad \sin x.
$$

ii) Prove that if $g(x)$ generates $(a_0, a_1, a_2, \ldots)$, then a primitive $\int g(x)$ generates
the sequence $(?, a_0, \frac{1}{2}a_1, \frac{1}{3}a_2, \ldots)$ (the first element is undetermined and can be anything).

iii) Verify what happens in ii) for $g(x) = e^x$ (reflecting the fact that e^x equals its own derivative).

iv) Use $e^x := \sum_{k=0}^{n} \frac{x^k}{k!}$ to show the homomorphic relation $e^{x+y} = e^x e^y$.

(*Hint:* $\dfrac{(x+y)^n}{n!} = \sum_{k=0}^{n} \left(\dfrac{1}{k!}x^k \right) \left(\dfrac{1}{(n-k)!} y^{n-k} \right)$, cf. Theorem 12.2.

When a generating function is rational we obtain a **linear recurrence sequence**:
$a_{n+1} = \sum_{j=0}^{n} q_j a_j$, where the $q_j \in \mathbb{R}$ are constants [38, 98].

Below are a few examples which serve as a nice warm-up for a serious undertaking of power series.

Examples 14.9

i) From $a_0 = 1$, $a_{n+1} = q a_n$ we infer $a_k = q^k$, giving the geometric progression.
The latter is generated by the rational map $(1 - qx)^{-1}$, by Example 14.8 ii);

ii) The Fibonacci sequence (see Examples 10.1) obeys the recurrence relation
$f_{n+2} = f_{n+1} + f_n$, $n \geqslant 0$. Indicating by f the corresponding generating function,

$$\sum_{k=2}^{\infty} f_k x^k = \sum_{k=2}^{\infty} f_{k-1} x^k + \sum_{k=2}^{\infty} f_{k-2} x^k$$

$$\sum_{k=0}^{\infty} f_k x^k - f_0 - f_1 x = \sum_{j=0}^{\infty} f_j x^{j+1} + \sum_{j=0}^{\infty} f_j x^{j+2}$$

$$f(x) - x = x f(x) + x^2 f(x)$$

and therefore $f(x) = \dfrac{x}{1 - x - x^2}$.

Now decompose f in partial fractions, remembering that the denominator has roots $\varphi = \frac{1+\sqrt{5}}{2}$ (golden ratio) and $\hat{\varphi} = -\varphi^{-1}$:

$$\frac{x}{1-x-x^2} = \frac{1}{\sqrt{5}} \left(\frac{1}{1-\varphi x} - \frac{1}{1-\hat{\varphi}x} \right) = \frac{1}{\sqrt{5}} \left(\sum_{i \geqslant 0} \varphi^i x^i - \sum_{i \geqslant 0} \hat{\varphi}^i x^i \right)$$

$$= \sum_{i \geqslant 0} \left(\frac{\varphi^i - \hat{\varphi}^i}{\sqrt{5}} \right) x^i.$$

Thus we recover formula (10.1).

iii) Let's find the generating function of the sequence $(k^2)_{k \geqslant 0}$. Start by writing k^2 as linear combination of binomials

$$k^2 = (k+1)(k-1) + 1 = (k+1)(k+2) - 3(k+1) + 1$$

$$= 2\binom{k+2}{2} - 3\binom{k+1}{1} + \binom{k}{0}.$$

Exploiting Examples 14.8 it follows that

$$\sum_{k=0}^{\infty} k^2 x^k = \frac{2}{(1-x)^3} - \frac{3}{(1-x)^2} + \frac{1}{1-x} = \frac{x(x+1)}{(1-x)^3}.$$

iv) Let's use generating functions to prove $\sum_{k=0}^{n} \binom{n}{k}^2 = \binom{2n}{n}$.

We have $(1+x)^n = \sum_{j=0}^{n} \binom{n}{j} x^j$, and $\left(1 + \dfrac{1}{x}\right)^n = \sum_{k=0}^{n} \binom{n}{k} x^{-k}$.

Therefore the sum $\sum_{k=0}^{n} \binom{n}{k}^2$ equals the constant term in the expansion of

$$(1+x)^n \left(1 + \frac{1}{x}\right)^n = \frac{(1+x)^{2n}}{x^n}.$$

Exercises 14.10

i) Show that the 'harmonic numbers' $h_n := \sum_{k=1}^{n} \dfrac{1}{k}$ are generated by

$$\frac{1}{1-x} \log \frac{1}{1-x}.$$

Aside: the *digamma function* $\psi_2(z) := \dfrac{d}{dz}(\log \Gamma(z))$ assumes, on natural numbers, the values $\psi_2(n+1) = h_n - \gamma$, where γ is the constant (3.2).

ii) Find the generating function of the sequence $a_m = \binom{m}{k_0}$, where k_0 is fixed.

iii) Prove the Fibonacci numbers f_n satisfy

$$\sum_{i=1}^{n} f_i^2 = f_n f_{n+1}, \quad \text{and} \quad \det\begin{pmatrix} f_n & f_{n+1} \\ f_{n+1} & f_{n+2} \end{pmatrix} = (-1)^n \quad \text{(Cassini identity)}.$$

Hint: use $\sum_{k=0}^{n} f_k = f_{n+2} - 1$, or the generating function.

iv) [91, p. 556] Consider the Catalan numbers $C_n := \dfrac{1}{n+1}\binom{2n}{n}$, see p. 324, and let $C(x)$ be their generating function.

a) Calculate $C(x)^2$ and show $\displaystyle\sum_{j=0}^{n} C_j C_{n-j}$ is the coefficient of x^n.

b) Prove $1 + xC(x)^2 = C(x)$, i.e. $C(x) = \dfrac{1 \pm \sqrt{1 - 4x}}{2x}$.

c) Conclude $C_n = \dfrac{1 \cdot 3 \cdot 5 \cdots (2n - 1)}{(n + 1)!} 2^n$.

Conversely, Lagrange's inversion formula for power series [29] (itself a smart application of the implicit function theorem) shows that the formula in b) gives back the expression of C_n.

v) Lagrange's inversion is a powerful general technique in the field. For instance, it provides straightaway the sequence generated by Lambert's highly implicit W function, defined as the (local) inverse to $z \mapsto ze^z$. Compute the first few terms in $W(xe^x) = x$ to convince yourself that $W(x) = \displaystyle\sum_{k=1}^{\infty} \dfrac{(-1)^{n-1}}{n!} x^n$ (sufficiently close to 0).

There are plenty of ways to write a sequence generated by a (suitable) function f (the Taylor series springs to mind). More intricate is the converse operation, as we saw in the previous examples. Actually if we are given a sequence $(a_k)_{k \in \mathbb{N}}$, it's not always possible to express $\displaystyle\sum_{k=0}^{\infty} a_k x^k$ in 'closed form'. When it happens the series is likely to converge to a function, for $|x|$ small enough. But 'how closed' is a matter of debate, and in that rests the usefulness of the notion.

Example 14.11 Suppose $g(x) = \displaystyle\sum_{j=0}^{\infty} \dfrac{x^j}{1 - x^j}$ expands in power series as $a_0 + a_1 x + a_2 x^2 + \dots$. The coefficient a_k is obtained summing from $j = 1$ to $j = k$ (this is enough), and equals the number of ways one can write x^k as $(x^j)^i$, for $1 \leqslant i \leqslant k$. Hence g is the generating function of the sequence $0, 1, 2, 2, 3, 2, 4, 1, 4, 3, 4, 2, \dots$, whose n^{th} term is the number of positive divisors of n (starting from $a_0 = 0$). The entry 2 represents prime numbers, entry 3 squares of primes. The '∞' in the series of g is purely symbolic: convergence is no issue because any coefficient can be computed by means of finite sums.

On the other hand, very often the convergence is guaranteed, which increases the chances of finding a closed form for the generating function. In case the sequence (a_k) is bounded, by comparison with the exponential series (where $a_k = 1$ for all k), the series

$$h(x) = \sum_{k=0}^{\infty} \frac{a_k}{k!} x^k$$

always converges, and h is called **exponential generating function** of (a_k).

Examples

a) e^x is the exponential generating function of $1, 1, 1, 1, \ldots$;

b) $(1+x)^r$ is the exponential generating function of $1, r, r(r-1), r(r-1)(r-2), \ldots$, the sequence of falling factorials $r^{\underline{k}}$;

c) expanding $\mathfrak{e}(x) := \dfrac{x}{e^x - 1}$ in the format $\displaystyle\sum_{k=0}^{\infty} \dfrac{\beta_k}{k!} x^k$ defines the Bernoulli numbers β_k. For example $\beta_0 = 1$, $\beta_1 = -1/2$, $\beta_2 = 1/6$, $\beta_3 = 0$, $\beta_4 = -1/30$, $\beta_5 = 0, \ldots$

It's not hard to show $\beta_{2k+1} = 0$, $k > 0$, since

$$1 + \sum_{k=2}^{\infty} \frac{\beta_k}{k!} x^k = \frac{x}{e^x - 1} + \frac{x}{2} = \frac{x/2}{\tanh(x/2)} \quad \text{is an even map.}$$

The Bernoulli numbers crop up in the Taylor expansion of tan, tanh, and hence in certain *characteristic classes* (the Todd genus is defined through $\mathfrak{e}(z)$, the L-genus by $\dfrac{\sqrt{z}}{\tanh \sqrt{z}}$ ☞ *complex geometry, algebraic topology*). The function $\dfrac{1}{\mathfrak{e}(-x)}$ is crucial in ☞ *Lie theory* because it governs the fundamental relationship (*BCH formula*) between a Lie group and its Lie algebra. What is more, the Laplace transform of $-\mathfrak{e}(x)$ is the *trigamma function* $\psi_3(z) := \frac{d^2}{dz^2} \log \Gamma(z) = \zeta(2)$.

d) $\displaystyle\sum_{k=0}^{\infty} \dfrac{B_k}{k!} x^k = e^{e^x - 1}$, where B_k are the Bell numbers (pp. 31 and 328). From there

descends the interesting expression $B_k = \dfrac{1}{e} \displaystyle\sum_{j=0}^{\infty} \dfrac{j^k}{j!}$.

Exercises

i) If g, h are the exponential generating functions of (a_n), (b_n), then the product map gh is the exponential generating function of the *convolution* (c_n), $c_n := \sum_{k=0}^{n} a_k b_{n-k}$.

ii) Use the series $e^x = \displaystyle\sum_{k=0}^{\infty} \dfrac{x^k}{k!}$ to show the Bernoulli numbers satisfy the recurrence relation $\displaystyle\sum_{k=0}^{n-1} \binom{n}{k} \beta_k = 0$.

For the sake of the present closing remarks a 'derangement' will indicate a permutation without fixed points. Let's call d_n the number of derangements of n objects: a simple thought experiment shows $d_1 = 0, d_2 = 1, d_3 = 2, d_4 = 9$.

Example ([86]) Shuffle two 52-card decks and place them next to each other, face up. Compare the two first cards and discard them if they are equal. If different, pass to the next ones below. Do the same comparing with the second pair, then repeat the operation until all pairs have been checked. The question is what are the odds of reaching the bottom of the decks without finding matching pairs.

It's enough to consider one deck only, which can be ordered in 52! possible ways. Among those, the number of shuffles not presenting matching pairs is d_{52}, so the answer to the question is $d_{52}/52!$. The next corollary will prove that this number is less than $1/2$, meaning that finding two equal cards is more likely than never finding equal pairs.

Proposition *The exponential generating function of the sequence (d_n) is $\dfrac{e^{-x}}{1-x}$.*

Proof Fix $0 \leqslant k \leqslant n$. The number of permutations of n elements with k fixed points equals $\binom{n}{k}d_{n-k}$ (where we force $d_0 = 1$, by decree). Hence

$$n! = d_n + nd_{n-1} + \binom{n}{2}d_{n-2} + \cdots + \binom{n}{n-2}d_2 + nd_1 + d_0 = \sum_{k=0}^{n}\binom{n}{n-k}d_k$$

$$\implies \quad 1 = \sum_{k=0}^{n}\frac{d_k}{k!}\frac{1}{(n-k)!}.$$

Let's call by h the function we're seeking. The sum on the right is the coefficient of x^n in
$$\left(\sum_{k=0}^{\infty}\frac{d_k}{k!}x^k\right)\left(\sum_{j=0}^{\infty}\frac{1}{j!}x^j\right) = h(x)e^x.$$
To finish we have to check (exercise) that the product of the two big brackets is equal to $1 + x + x^2 + x^3 + \cdots = (1-x)^{-1}$.
$\square$

Corollary 14.12 $d_n = n! \displaystyle\sum_{k=2}^{n}\frac{(-1)^k}{k!}.$

Proof Combine the previous result with the series $e^{-x} = \displaystyle\sum_{k=0}^{\infty}\frac{(-1)^k}{k!}x^k.$

An alternative argument goes as follows. Define

$$X_i = \{f \in \mathfrak{S}_n \mid f(i) = i\}, \quad X_{ij} = X_i \cap X_j, \quad X_{ijk} = X_i \cap X_j \cap X_k, \quad \ldots$$

with $i < j < k < \ldots$ Hence $|X_i| = (n-1)!$, $|X_{ij}| = (n-2)!$ etc. The number of 'non-derangements' of $n > 0$ objects is

$$n! - d_n = |X_1 \cup \cdots \cup X_n| = \binom{n}{1}(n-1)! - \binom{n}{2}(n-2)! + \binom{n}{3}(n-3)! - \cdots \mp 1$$

$$= n!\left(1 - \frac{1}{2!} + \frac{1}{3!} - \cdots \mp \frac{1}{n!}\right).$$

$\square$

For example, $\dfrac{1}{720}d_6 = \dfrac{1}{2} - \dfrac{1}{6} + \dfrac{1}{24} - \dfrac{1}{120} + \dfrac{1}{720} \implies d_6 = 265.$

Exercises Prove the recurrence relations:

i) $\displaystyle\sum_{k=2}^{n}(-1)^{k}n\underline{^{n-k}}=d_{n}=n\,d_{n-1}+(-1)^{n}$.

ii) $d_{n}=(n-1)(d_{n-1}+d_{n-2})$, using i).

iii) Prove ii) directly, looking at the effect of a permutation $f\in\mathfrak{S}_{n}$ on the last element n. *Hint*: distinguish the cases $f(f(n))=n$ and $f(f(n))\neq n$.

iv) Use Corollary 14.12 to convince yourself of the asymptotic approximation $d_{n}\approx\dfrac{1}{e}n!$ for very large n.

v) Show that the above alternative proof to Corollary 14.12 is consistent with (6.5) and Exercise 14.5 iii).

Part IV
Geometry Through Algebra

Chapter 15
Vector Spaces

This is the only chapter devoted specifically to linear algebra. We'll introduce just the fundamental algebraic concepts, given that we shall not accompany the subject with ☞ *analytic geometry*, and hence there's no real demand for exhaustiveness. Accordingly, proofs will be reduced to a minimum.

15.1 Lexicon and Basic Properties

Let's begin by reviewing the standard model of a vector space. For $n \in \mathbb{N}^*$ we write

$$\mathbb{R}^n := \left\{ \mathbf{u} := (u_1, u_2, \ldots, u_n) \mid u_i \in \mathbb{R}, \ \forall i = 1, \ldots, n \right\}$$

for the n-fold Cartesian product of the real line. To denote an element $\mathbf{u} \in \mathbb{R}^n$, called a **vector** of $\mathbb{R}^n$, sometimes the notation $\vec{u}$ is borrowed from physics. (We use boldface for better visibility.) The real number u_i is called i^{th} **coordinate**, or **component**, of the vector $\mathbf{u}$.

For every $\mathbf{u} = (u_1, \ldots, u_n), \mathbf{v} = (v_1, \ldots, v_n) \in \mathbb{R}^n$ and every $\lambda \in \mathbb{R}$ we set

$$\mathbf{u} + \mathbf{v} = (u_1, \ldots, u_n) + (v_1, \ldots, v_n) := (u_1 + v_1, \ldots, u_n + v_n)$$
$$\lambda \mathbf{u} = \lambda(u_1, \ldots, u_n) := (\lambda u_1, \ldots, \lambda u_n) \tag{15.1}$$

component-wise. The following definition is modelled on those operations.

Definition Fix a field $\mathbb{K}$ with unit 1. One calls $\mathbb{K}$-**vector space** a set V equipped with:

© The Author(s), under exclusive license to Springer Nature Switzerland AG 2021
S. G. Chiossi, *Essential Mathematics for Undergraduates*,
https://doi.org/10.1007/978-3-030-87174-1_15

i) a commutative and associative operation $+: V^2 \rightarrow V$ with unique neutral element $\mathbf{0}$ (**zero vector**) and unique opposites (making $(V, +)$ an Abelian group);

ii) a map $\mathbb{K} \times V \rightarrow V$, called **scalar multiplication** and denoted by juxtaposition,[1] such that:

$$\lambda(\mu\mathbf{u}) = (\lambda\mu)\mathbf{u}, \quad (\lambda + \mu)\mathbf{u} = \lambda\mathbf{u} + \mu\mathbf{u}, \quad 1\mathbf{u} = \mathbf{u}, \quad \lambda(\mathbf{u} + \mathbf{v}) = \lambda\mathbf{u} + \lambda\mathbf{v}$$

for every $\lambda, \mu \in \mathbb{K}, \mathbf{u}, \mathbf{v} \in V$.

With i) and ii) we can create **linear combinations** $\lambda_1\mathbf{u}_1 + \cdots + \lambda_s\mathbf{u}_s \in V$ of the vectors $\mathbf{u}_1, \ldots, \mathbf{u}_s$ with coefficients $\lambda_1, \ldots, \lambda_s \in \mathbb{K}$.

Almost invariably $\mathbb{K} = \mathbb{R}$, in which case V is said **real vector space**.

Examples

i) $\mathbb{R}^n$ with structure (15.1), for any $n \in \mathbb{N}^*$;
ii) real polynomials $p(x)$ (Chap. 9) are linear combinations of monomials x^k, with $k = 0, \ldots, \deg(p)$. Therefore $\mathbb{R}[x]$ is a real vector space.
iii) real-valued functions $\mathbb{R}^{\mathbb{R}}$, see Definition 10.2;
iv) linear maps, which will be presented in Sect. 15.2.

Definition A **(vector) subspace** W in a vector space V is a subset that is closed under linear combinations:

$$\mathbf{u}, \mathbf{v} \in W, \ \lambda, \mu \in \mathbb{K} \implies \lambda\mathbf{u} + \mu\mathbf{v} \in W.$$

Examples

i) $U_1 = \{(x, y) \in \mathbb{R}^2 \mid 3x + 2y = 0\}$ is a subspace of $\mathbb{R}^2$.
ii) $U_2 = \{(x, y) \in \mathbb{R}^2 \mid 3x + 2y = 6\}$ is not a subspace: $(0, 3), (2, 0) \in U_2$ but $(0, 3) + (2, 0) \notin U_2$ and $5(0, 3) \notin U_2$. The culprit is equation $3x + 2y = 6$: albeit linear, it is not homogeneous.
iii) $W = \{(x, x, xy, y, y) \in \mathbb{R}^5 \mid x, y \in \mathbb{R}\}$ isn't a vector subspace of $\mathbb{R}^5$ as not closed under addition: $(1, 1, 2, 2, 2) + (3, 3, 0, 0, 0) \notin W$. The problem is that $xy + ab \neq (x + a)(y + b)$ in the identity $(x, x, xy, y, y) + (a, a, ab, b, b) = (x + a, x + a, xy + ab, y + b, y + b)$.

    Nonetheless W contains the subspace $\{(x, x, 0, y, y) \in \mathbb{R}^5 \mid x, y \in \mathbb{R}\}$ obtained killing the problematic component. In the same way, W is contained in the subspace $\{(x, x, z, y, y) \in \mathbb{R}^5 \mid x, y, z \in \mathbb{R}\}$, where the third component is now arbitrary.

Exercise Prove that the (restrictions of the) operations make any subspace $W \subseteq V$ a vector space with the same zero vector $\mathbf{0}_W = \mathbf{0}_V$.

---

[1] Strictly speaking, calling this an 'operation' is incorrect. Yet, see Examples 4.12.

Examples Let V, V' be $\mathbb{K}$-vector spaces, and U, $W \subseteq V$ subspaces.

i) For any $\mathbf{u} \in V$, the set of its multiples $\mathbb{R}\mathbf{u} = \{\rho\mathbf{u} \mid \rho \in \mathbb{R}\}$ is a vector subspace;
ii) $U \cap W$ is a subspace of $\mathbb{R}^n$;
iii) $U \cup W$ is a subspace only if $U \subseteq W$ or $W \subseteq U$;
iv) the smallest subspace containing $U \cup W$ is the **sum** of U, W

$$U + W := \{\mathbf{u} + \mathbf{w} \mid \mathbf{u} \in U, \mathbf{w} \in W\}.$$

When $U \cap W = \{\mathbf{0}\}$, the sum is written $U \oplus W$ and called **direct sum**.
iv) $V' \times V$ is a $\mathbb{K}$-vector space with component-wise operations
$$(\mathbf{v}'_1, \mathbf{v}_1) + (\mathbf{v}'_2, \mathbf{v}_2) := (\mathbf{v}'_1 + \mathbf{v}'_2, \mathbf{v}_1 + \mathbf{v}_2)$$
$$\lambda(\mathbf{v}'_1, \mathbf{v}_1) := (\lambda\mathbf{v}'_1, \lambda\mathbf{v}_1)$$
for all $\mathbf{v}'_1, \mathbf{v}'_2 \in V'$, $\mathbf{v}_1, \mathbf{v}_2 \in V$, $\lambda \in \mathbb{K}$. In fact, $(V' \times \{\mathbf{0}_V\}) \oplus (\{\mathbf{0}_{V'}\} \times V) = V' \times V$.

In particular $\mathbb{K}^m \times \mathbb{K}^n$ is identified with $\mathbb{K}^{m+n}$, for every $m, n \in \mathbb{N}$, under

$$\big((\lambda_1, \ldots, \lambda_m), (\mu_1, \ldots, \mu_n)\big) \overset{1\text{-}1}{\longleftrightarrow} (\lambda_1, \ldots, \lambda_m, \mu_1, \ldots, \mu_n).$$

v) Fix a subspace $W \subseteq V$ and define the equivalence relation:

$$\mathbf{v}_1 \sim \mathbf{v}_2 \iff \mathbf{v}_1 - \mathbf{v}_2 \in W$$

on V. The quotient set V/W is a $\mathbb{K}$-vector space under

$$(\mathbf{v}_1 + W) + (\mathbf{v}_2 + W) := (\mathbf{v}_1 + \mathbf{v}_2) + W, \ \ \mu(\mathbf{v}_1 + W) := \mu\mathbf{v}_1 + W, \ \forall \mathbf{v}_1, \mathbf{v}_2 \in V, \mu \in \mathbb{K}.$$

Exercises Establish which of the following subsets of $\mathbb{R}[x]$ are vector subspaces:

i) $\{p(x) \in \mathbb{R}[x] \mid p \text{ is even}\}$
ii) $\{p(x) \in \mathbb{R}[x] \mid p \text{ is odd}\}$
iii) $\{p(x) \in \mathbb{R}[x] \mid p(1) = 1\}$
iv) $\{p(x) \in \mathbb{R}[x] \mid p''(x) - xp'(x) = x^2 p(x)\}$, where $p''(x), p'(x)$ are the polynomials corresponding to the first and second derivatives of the map p.

Definition Let $W \subseteq V$ be a subset. The **(linear) space generated**, or **spanned, by** W is the intersection of all subspaces U containing W

$$\mathscr{L}(W) := \bigcap_{W \subseteq U \subseteq V} U$$

Its elements are called **generators** of $\mathscr{L}(W)$. Note that $\mathscr{L}(\varnothing) = \{\mathbf{0}\}$.

Furthermore, the set W is said to be **linearly independent (LI)** if $\mathbf{v} \notin \mathscr{L}(W \setminus \{\mathbf{v}\})$ for all $\mathbf{v} \in W$, and **linearly dependent (LD)** otherwise.

Proposition *The space generated by W coincides with the set of linear combinations of elements of W*

$$\mathscr{L}(W) = \left\{ \sum\nolimits_{j=1}^{r} \lambda_j \mathbf{u}_j, \ \big| \ \lambda_j \in \mathbb{K}, \mathbf{u}_j \in W \right\}.$$

Proof Exercise.   □

Examples

  i) The vectors $\mathbf{u}_1 = (-1, 1, 0)$, $\mathbf{u}_2 = (-1, 0, 1) \in \mathbb{R}^3$ are LI. The typical element of $U = \mathscr{L}(\mathbf{u}_1, \mathbf{u}_2)$ is of the form $(-a - b, a, b) = a\mathbf{u}_1 + b\mathbf{u}_2$, that is $U = \left\{ (u_1, u_2, u_3) \in \mathbb{R}^3 \ \big| \ u_1 + u_2 + u_3 = 0 \right\}$.

     The latter is also generated by $\mathbf{u}_1 = (-1, 1, 0)$, $\mathbf{u}_3 = (0, 1, -1)$, $\mathbf{u}_4 = (0, -3, 3)$, $\mathbf{u}_5 = (0, -1, 1)$, since $c\mathbf{u}_1 + d\mathbf{u}_3 + f\mathbf{u}_4 + g\mathbf{u}_5 = (-c, c + d - 3f - g, -d + 3f + g)$ belongs to U for all c, d, f, g. In fact, $(-c) + (c + d - 3f - g) + (-d + 3f + g) = 0$. But $\mathbf{u}_1, \mathbf{u}_3, \mathbf{u}_4, \mathbf{u}_5$ are LD: $0\mathbf{u}_1 + 0\mathbf{u}_3 - \mathbf{u}_4 + 3\mathbf{u}_5 = \mathbf{0}$ implies $\mathbf{u}_4 \in \mathscr{L}(\mathbf{u}_5) \subseteq \mathscr{L}(\mathbf{u}_1, \mathbf{u}_3, \mathbf{u}_5)$.

 ii) The functions $\mathbf{u}_1(t) = \cos^2 t$, $\mathbf{u}_2(t) = \sin^2 t$, $\mathbf{u}_3(t) = 1 \in \mathbb{R}^{\mathbb{R}}$ are LD, because $\mathbf{u}_3(t) = \mathbf{u}_1(t) + \mathbf{u}_2(t)$ for every $t \in \mathbb{R}$.

iii) $\mathscr{L}(\mathbf{u}) = \mathbb{R}\mathbf{u}$ is called the **line** generated by $\mathbf{u} \in V$.

 iv) If $\mathbf{u}_1, \mathbf{u}_2 \in V$ are LI, $\mathscr{L}(\mathbf{u}_1, \mathbf{u}_2) = \mathbb{R}\mathbf{u}_1 + \mathbb{R}\mathbf{u}_2$ is the **2-plane** generated by $\mathbf{u}_1, \mathbf{u}_2$.

  v) In general, if $\mathbf{u}_1, \ldots, \mathbf{u}_r$ are LI, $\mathscr{L}(\mathbf{u}_1, \ldots, \mathbf{u}_r) = \mathbb{R}\mathbf{u}_1 + \ldots + \mathbb{R}\mathbf{u}_r$ is an **r-plane**. An $(n - 1)$-plane is called a **hyperplane**.

 vi) For $0 < k \leqslant n$ the space

$$\mathrm{Gr}_k(V^n) := \left\{ \mathscr{L}(\mathbf{u}_1, \ldots, \mathbf{u}_k) \subseteq V^n \ \big| \ \mathbf{u}_1, \ldots, \mathbf{u}_k \ \mathrm{LI} \right\} \tag{15.2}$$

is called **Graßmannian** of **k-planes** in V^n.

In particular, $\mathrm{Gr}_1(V^n)$ is the space of lines in V^n, $\mathrm{Gr}_{n-1}(V^n)$ the space of hyperplanes.

Exercises 15.1

  i) Prove that $\mathscr{L}(V) = V$, and that $U \subseteq W \implies \mathscr{L}(U) \subseteq \mathscr{L}(W)$.

 ii) Show $\{\mathbf{u}_1, \ldots, \mathbf{u}_r\}$ is LI if and only if

$$\sum_{j=1}^{r} \lambda_j \mathbf{u}_j = \mathbf{0} \implies \lambda_1 = \ldots = \lambda_r = 0.$$

In other words, the only linear combination equal to $\mathbf{0}$ is the trivial combination, the one with all zero coefficients.

iii) $\{\mathbf{u}_1, \ldots, \mathbf{u}_r\}$ is LD if there exist non-trivial linear combinations equal to $\mathbf{0}$.

 iv) If $\mathbf{0} \in \{\mathbf{u}_1, \ldots, \mathbf{u}_r\}$ then $\{\mathbf{u}_1, \ldots, \mathbf{u}_r\}$ is LD.

  v) Show that a set is LI if and only if every finite subset in it is LI.

 vi) The polynomials t^k, $k \in \mathbb{N}$, are LI in $\mathbb{R}[t]$.

In particular, two non-zero LD vectors $\mathbf{u}_1, \mathbf{u}_2$ are called **parallel**: $\mathbf{u}_1 = \alpha \mathbf{u}_2$ for some $\alpha \in \mathbb{R}^*$, with name borrowed from geometry.

Definition A **basis** of V is a maximal element in the collection of LI subsets of V, ordered by inclusion. (The existence of bases is guaranteed by Zorn's Lemma 3.12.)

Example In the previous example $\{\mathbf{u}_1, \mathbf{u}_2\}$, $\{\mathbf{u}_3, \mathbf{u}_5\}$ are bases of U, $\{\mathbf{u}_1, \mathbf{u}_4, \mathbf{u}_5\}$ isn't.

Theorem 15.2 *In a vector space V, if $W = \{\mathbf{w}_1, \ldots, \mathbf{w}_p\}$ is a generating set and $U = \{\mathbf{u}_1, \ldots, \mathbf{u}_m\}$ is LI, then $m \leqslant p$.*

Proof *(Sketch)* Consider the family $\mathscr{C} = \{(S \subseteq U, f : S \xrightarrow{1-1} W)\}$ where $(U \setminus S) \cup f(S)$ is LI, $U \cap W \subseteq S$ and $f(\mathbf{v}) = \mathbf{v}$ for all $\mathbf{v} \in U \cap W$.

Let's order $\mathscr{C}$ by saying $(S, f) \leqslant (T, g)$ if $S \subseteq T$ and f is a restriction of g. Observe that if there is a $\mathbf{v} \in U \setminus S$, then $W \not\subseteq \mathscr{L}\Big((U \setminus (S \cup \{\mathbf{v}\})) \cup f(S)\Big)$. Now one proves $\mathscr{C}$ is non-empty, that it has maximal elements $(\overline{S}, \overline{f})$, and one concludes $\overline{S} = U$. $\qquad\square$

Corollary *In a vector space $V \neq \{0\}$*

i) *every generating set contains at least one basis B;*
ii) *any basis B generates the entire space: $\mathscr{L}(B) = V$;*
iii) *all bases have the same cardinality.*

Definition The cardinality of a basis B is called the **dimension** of the vector space V:

$$\dim_{\mathbb{K}} V := \operatorname{card} B.$$

If $n := \dim_{\mathbb{K}} V < \infty$ one says V is **finite-dimensional** (written V^n), otherwise **infinite-dimensional**. (By convention $\dim\{0\} = 0$.)

Corollary $\dim \mathbb{R}^n = n$.

Proof The vectors $\mathbf{e}_1 = (1, 0, 0, \ldots, 0), \mathbf{e}_2 = (0, 1, 0, \ldots, 0), \ldots, \mathbf{e}_n = (0, \ldots, 0, 1) \in \mathbb{R}^n$ are LI and generate the whole space,

$$\mathbb{R}^n \ni \mathbf{u} = (u_1, u_2, \ldots, u_n) = u_1 \mathbf{e}_1 + u_2 \mathbf{e}_2 + \ldots + u_n \mathbf{e}_n \in \mathscr{L}(\mathbf{e}_1, \ldots, \mathbf{e}_n),$$

so they form a basis of $\mathbb{R}^n$, called **canonical basis**. $\qquad\square$

Sometimes, especially in a physics-related context, the canonical bases of $\mathbb{R}^2$ and $\mathbb{R}^3$ are indicated by $\mathbf{i} = (1, 0), \mathbf{j} = (0, 1)$, and $\mathbf{i} = (1, 0, 0), \mathbf{j} = (0, 1, 0), \mathbf{k} = (0, 0, 1)$.

It's a characteristic feature of the canonical basis that the components $u_1, u_2, \ldots, u_n$ of a vector of $\mathbb{R}^n$ coincide with the coefficients of the linear combination expressing the vector. This follows from $\pi_i(\mathbf{e}_j) = \delta_{ij}, i, j = 1, \ldots, n$, and is false for any other basis.

Corollary *Let $V \neq \{0\}$ be a vector space, U a subspace. Then*

$$\dim U = \max \{m \in \mathbb{N} \mid \exists\, m \text{ LI vectors in } U\} = \min \{p \in \mathbb{N} \mid \exists\, p \text{ generators in } U\}.$$

Proof Exercise.   □

Definition The number $\operatorname{codim} U := \dim V - \dim U$ is the **codimension** of the subspace $U \subseteq V$.

Exercise Let $U \subseteq V$ be a subspace. Prove $\dim U = r \iff U$ is an r-plane.
     Therefore U is a hyperplane $\iff \operatorname{codim} U = 1$.

Examples 15.3 Through these examples one can introduce affine geometry in parallel to linear algebra. Take $V = \mathbb{R}^n$, fix a point $\mathbf{x}_0$ and LI vectors $\mathbf{u}, \mathbf{v}$.

 i) The **affine line** $r = \{\mathbf{x}_0 + t\mathbf{v} \mid t \in \mathbb{R}\} = \mathbf{x}_0 + \mathscr{L}(\mathbf{v})$ isn't a vector subspace (unless $\mathbf{x}_0 = 0$). Geometrically, though, since r is a shift of the line $\mathscr{L}(\mathbf{v})$, its support is what we imagine when we visualise a straight line passing through the point $\mathbf{x}_0$ and with direction $\mathbf{v}$. In fact $t \mapsto \mathbf{x}_0 + t\mathbf{v}$ is a 1-1 correspondence between $\mathbb{R}$ and r.

 ii) The **affine plane** $\pi = \{\mathbf{x}_0 + t\mathbf{v} + s\mathbf{u} \mid s, t \in \mathbb{R}\} = \mathbf{x}_0 + \mathscr{L}(\mathbf{u}, \mathbf{v})$ isn't a subspace either (except if $\mathbf{x}_0 = 0$). As π is a shifted plane, its support is the geometric plane through $\mathbf{x}_0$ with direction $\mathscr{L}(\mathbf{u}, \mathbf{v})$.

 iii) Let $\mathbf{u}_1, \ldots, \mathbf{u}_{n-1}$ be LI. The **affine hyperplane** $\left\{ \mathbf{x}_0 + \sum_{j=1}^{n-1} \lambda_j \mathbf{u}_j \mid \lambda_j \in \mathbb{R} \right\}$ has direction given by the codimension-one space $\mathscr{L}(\mathbf{u}_1, \ldots, \mathbf{u}_{n-1})$.

Exercises 15.4

 i) Let U be a subspace in V. Show $\dim U \leqslant \dim V$, with equality holding if and only if $U = V$.

 ii) Let $\mathbb{R}[x]_d$ denote the vector space of real polynomials of degree d. Prove $\{1, x, x^2, \ldots, x^d\}$ are LI, and hence $\dim \mathbb{R}[x]_d = d + 1$.
     Given that the degree induces a stratification $\mathbb{R} = \mathbb{R}[x]_0 \subset \mathbb{R}[x]_1 \subset \cdots \subset \mathbb{R}[x]_k \subset \mathbb{R}[x]_{k+1} \subset \cdots$, conclude that $\mathbb{R}[x] = \bigcup_{d=0}^{\infty} \mathbb{R}[x]_d$ does not have finite dimension: $\dim \mathbb{R}[x] = \infty$, and *a fortiori* $\dim \mathbb{R}^{\mathbb{R}} = \infty$.

 iii) Show $\dim \operatorname{Gr}_k(\mathbb{R}^n) = \binom{n}{k}$.

Lemma (Basis Completion) *Let V^n be finite-dimensional and $U^k \subseteq V$ a subspace with basis $B = \{\mathbf{v}_1, \ldots, \mathbf{v}_k\}$. Then there exist $n - k$ vectors that complete B to give a basis of V.*

Proof Suppose $k < n$ (otherwise the claim is trivial). Take $\mathbf{v}_{k+1} \in V \setminus U^k$, so that $U^{k+1} := U^k + \mathscr{L}(\mathbf{v}_{k+1})$ has a basis $B \cup \{\mathbf{v}_{k+1}\}$. Iterating by induction we obtain a sequence $U^k \subseteq U^{k+1} \subseteq \ldots \subseteq U^{k+(n-k)} = U^n$. As the sequence of dimensions is increasing and bounded by $n < \infty$, the construction stops after $n - k$

steps ($U^n = V$), and eventually it rolls out a basis $B \cup \{\mathbf{v}_{k+1}, \mathbf{v}_{k+2}, \ldots, \mathbf{v}_{n-k}\}$ for V. $\qquad\qquad\square$

Theorem (Graßmann Formula[2]) *For any subspace U, W of a vector space V^n*

$$\dim(U + W) = \dim U + \dim W - \dim(U \cap W).$$

Proof Using the basis-completion lemma let $B_I = \{\mathbf{v}_1, \ldots, \mathbf{v}_t\}$, $B_U = B_I \cup \{\mathbf{v}_{t+1}, \ldots, \mathbf{v}_p\}$, $B_W = B_I \cup \{\mathbf{v}'_{t+1}, \ldots, \mathbf{v}'_s\}$ denote bases for $U \cap W, U, W$ respectively. The idea is that B_U, B_I, B_W will generate $U + W$, so let's take a generic linear combination

$$\mathbf{w} = \sum_{i=1}^{t} a_i \mathbf{v}_i + \sum_{j=1}^{p-t} b_j \mathbf{v}_{t+j} + \sum_{k=1}^{s-t} c_k \mathbf{v}'_{t+k}.$$

The first $t + p$ vectors belong in U, the others in W, hence the sum $U + W$ coincides with $\mathscr{L}(\mathbf{v}_1, \ldots, \mathbf{v}_t, \mathbf{v}_{t+1}, \ldots, \mathbf{v}_p, \mathbf{v}'_{t+1}, \ldots, \mathbf{v}'_s)$. Consequently we can produce $\mathbf{w}$ using $t + p - t + s - t = |B_U| + |B_W| - |B_I|$ vectors.

By contradiction, suppose $\mathbf{w} = \mathbf{0}$. Neither the b_j nor the c_k can all vanish (B_W, B_U are bases), so let's write

$$\underbrace{\sum_{j=1}^{p-t} b_j \mathbf{v}_{t+j}}_{\in U} = \underbrace{- \sum_{i=1}^{t} a_i \mathbf{v}_i - \sum_{k=1}^{s-t} c_k \mathbf{v}'_{t+k}}_{\in W}.$$

This non-null element belongs to $U \cap W$, which implies $U = U \cap W$ ⨏ (B_W is a basis). Therefore $B_U \cup B_I \cup B_W$ is a generating and LI set. $\qquad\square$

Corollary *For any subspaces U, W in V^n*

$$U \oplus W = V \iff \dim(U + W) = \dim U + \dim W.$$

Proof Immediate (exercise). $\qquad\qquad\square$

Remark 15.5 (Orientation) Take bases $(\mathbf{v}_i)$, $(\mathbf{u}_j)$ of V^n and consider the matrix $A = (a_{ij})$ of basis change: $\mathbf{u}_j = \sum_i a_{ij} \mathbf{v}_i$ (it's a standard linear-algebra fact that A is unique). As A is invertible, necessarily $\det A > 0$ or $\det A < 0$. The bases are said to have the *same orientation* when $\det A > 0$. An **orientation** on V^n is the choice of one of the two equivalence classes of bases with the same orientation, conventionally called 'positive' orientation (nothing to do with the determinant's sign).

---

[2] There are many incarnations of this formula in the sciences, all called 'Graßmann laws': in optics (theory of colour) for instance, and even one in linguistics!

If the space V is three-dimensional there is an extremely practical, alternative characterisation of orientation, see Remark 15.19.

15.2   Linear Maps ☕

The subject of this section is in other books disguised as the theory of matrices, and mostly applied to solving linear systems. The coffee-cup symbol is only there to signal an emphasis on the vector-space structure of linear maps and linear forms.

Let's begin with the unavoidable terminology.

Definition Let U, V be $\mathbb{K}$-vector spaces. A map $f : U \to V$ is $\mathbb{K}$-**linear** whenever

$$f(a\mathbf{u}_1 + b\mathbf{u}_2) = af(\mathbf{u}_1) + bf(\mathbf{u}_2)$$

for every $a, b \in \mathbb{K}$, $\mathbf{u}_1, \mathbf{u}_2 \in U$. The set of $\mathbb{K}$-linear maps $U \to V$ is indicated by

$$\mathrm{Hom}_{\mathbb{K}}(U, V).$$

A $\mathbb{K}$-linear bijection is a **(linear) isomorphism**. When $U = V$ one writes

$$\mathrm{End}_{\mathbb{K}} U := \mathrm{Hom}_{\mathbb{K}}(U, U),$$

called **endomorphisms**, and $\mathrm{Aut}_{\mathbb{K}}(U)$ for invertible endomorphisms (**automorphisms**).

When $V = \mathbb{K}$ a linear map $f : U \to \mathbb{K}$ is a **linear form** on U, and

$$U^{\vee} := \mathrm{Hom}(U, \mathbb{K})$$

is the **dual space** of V.

Proposition $\mathrm{Hom}_{\mathbb{K}}(U, V)$ *is a* $\mathbb{K}$-*vector space with the pointwise operations* $(f + g)(\mathbf{u}) := f(\mathbf{u}) + g(\mathbf{u})$, $(\mu f)(\mathbf{u}) := \mu f(\mathbf{u})$, *for any* $f, g \in \mathrm{Hom}_{\mathbb{K}}(U, V), \mathbf{u} \in V, \mu \in \mathbb{R}$.

Proof Exercise.   □

Exercise Prove

i) $f \in \mathrm{Hom}(V, V'), g \in \mathrm{Hom}(V', V'') \implies g \circ f \in \mathrm{Hom}(V, V'')$;

ii) $f \in \mathrm{Hom}(V, V')$ bijective $\implies f^{-1} \in \mathrm{Hom}(V', V)$;

iii) the first-derivative operator $y(x) \mapsto y'(x) := \dfrac{\mathrm{d}y}{\mathrm{d}x}$ is a linear map on (differentiable functions in) $\mathbb{R}^{\mathbb{R}}$. Conclude that the solutions to a homogeneous ODE with constant coefficients $a_n y^{(n)} + \ldots + a_1 y' + a_0 y = 0$ form a subspace of $\mathbb{R}^{\mathbb{R}}$.

Definition Let $f \in \mathrm{Hom}\,(V, V')$ be a linear map. The **kernel** or **nullspace** of f is

$$\mathrm{Ker}(f) := f^{-1}(\mathbf{0}_{V'})$$

(from the German *Kern*).

Examples

i) $f \colon \mathbb{R}^2 \to \mathbb{R}^2$, $f(x, y) = (2x + y, 4x + 2y)$ has nullspace $\mathrm{Ker}(f) = \{(x, -2x) \in \mathbb{R}^2 \mid x \in \mathbb{R}\} = \mathscr{L}(1, -2)$, the line of equation $2x + y = 0$.

ii) The axial projection $p \colon \mathbb{R}^3 \to \mathbb{R}^3$, $p(x, y, z) = z$ has as kernel the Oxy plane $\{z = 0\} = \mathscr{L}\{\mathbf{i}, \mathbf{j}\}$.

iii) The kernel of the operator $y(x) \mapsto \dfrac{dy}{dx}$ is the space of constant maps $\mathbb{R}$.

Exercises 15.6 Take $f \in \mathrm{Hom}\,(V, V')$ and show

i) $\mathrm{Ker}(f)$, $\mathrm{Im}(f)$ are subspaces, as is $\mathrm{Coker}(f) := V'/\mathrm{Im}(f)$, called **cokernel**;
ii) $\mathrm{Ker}(f) = \{\mathbf{0}_V\} \iff f$ is 1-1;
iii) $\mathrm{Coker}(f) = \{\mathbf{0}_{V'}\} \iff f$ is onto;
iv) $f(\mathbf{v}_1) = f(\mathbf{v}_2) \iff \mathbf{v}_1 - \mathbf{v}_2 \in \mathrm{Ker}\,f$, $\forall \mathbf{v}_1, \mathbf{v}_2 \in V$.

Next is a lovely characterisation for kernels of linear forms. As a side effect, it shows that nullspaces come in spades.

Theorem *Let V be a $\mathbb{K}$-vector space and π a subspace. Then*

$$\pi \text{ is a hyperplane} \iff \exists\, f \in V^{\vee} \setminus \{0\} \text{ such that } \pi = \mathrm{Ker}\,f.$$

Proof ($\Longleftarrow$) Since $f \neq 0$ there exists an $\mathbf{a} \in V$ such that $f(\mathbf{a}) \neq 0$. Define $\mathbf{e} := f(\mathbf{a})^{-1}\mathbf{a}$. As $f(\mathbf{e}) = 1$, every vector $\mathbf{v}$ is a sum $\mathbf{v} = f(\mathbf{v})\mathbf{e} + \mathbf{v}'$ for some $\mathbf{v}'$, in a unique way. Notice that $f(\mathbf{v}') = 0$, so $V = \mathscr{L}(\mathbf{e}) \oplus \mathrm{Ker}\,f$. But this implies $\dim \mathrm{Ker}\,f = \dim V - 1$.

($\Longrightarrow$) From $\mathrm{codim}\,\pi = 1$ we know there is a vector $\mathbf{e} \neq \mathbf{0}$ that doesn't belong to π, and so $V = \pi \oplus \mathscr{L}(\mathbf{e})$. Then any vector $\mathbf{v} \in V$ decomposes uniquely as $\mathbf{v} = \mathbf{v}' + \lambda\mathbf{e}$ with $\mathbf{v}' \in \pi, \lambda \in \mathbb{K}$. Define $f \colon V \to \mathbb{K}$ by $f(\mathbf{v}) = \lambda$, which is clearly linear. Now observe $\mathbf{v} \in \pi \iff \lambda = 0 \iff x \in \mathrm{Ker}\,f$. □

Theorem *Let V be a vector space with basis $\mathbf{v}_1, \ldots, \mathbf{v}_n$, V' another vector space and $\mathbf{v}'_1, \ldots, \mathbf{v}'_n$ vectors in V'. There exists a unique map $f \in \mathrm{Hom}\,(V, V')$ such that $f(\mathbf{v}_j) = \mathbf{v}'_j, j = 1, \ldots, n$.*

Proof Set $f\left(\sum_{j=1}^{n} a_j \mathbf{v}_j\right) = \sum_{j=1}^{n} a_j f(\mathbf{v}_j)$ for all $a_j \in \mathbb{R}$. This warrants linearity. The uniqueness is automatic because $\mathbf{v}_j$ is a basis. □

The consequence is that we may always assume, without any loss in generality, that a finite-dimensional real vector space is $\mathbb{R}^n$ for some n, up to isomorphisms.

Corollary *Every $\mathbb{R}$-vector space of dimension $n < \infty$ is isomorphic to $\mathbb{R}^n$.*

Proof Let $\mathbf{u}_1, \ldots, \mathbf{u}_n$ be a basis. The bijection $\mathbf{e}_j \mapsto \mathbf{u}_j$, $j = 1, \ldots, n$, extends linearly to an isomorphism $\mathbb{R}^n \to V$, $\sum_{j=1}^{n} \lambda_j \mathbf{e}_j \mapsto \sum_{j=1}^{n} \lambda_j \mathbf{u}_j$.   □

The isomorphism is not canonical (choice-free) because there is no special basis in $\mathbb{R}^n$ (the so-called canonical basis $(\mathbf{e_i})$ is preferred only because it's components are simple); nor can it be unique, for it depends on the choice of one among infinitely many bases.

Example 15.7 The map $\mathbb{R}^2 \to \mathbb{C}$ associating $\mathbf{e_1}, \mathbf{e_2}$ with the complex numbers $1, i$ extends to an isomorphism:

$$(u_1, u_2) = u_1 \mathbf{e_1} + u_2 \mathbf{e_2} \mapsto u_1 + iu_2$$

Then $\mathbb{C}$ is an $\mathbb{R}$-vector space of dimension 2, which is why it's common to identify $\mathbb{R}^2 \cong \mathbb{C}$.

The vectors $\mathbf{v}_1 = 2 + 3i$, $\mathbf{v}_2 = -\frac{1}{\sqrt{5}} + 2i \in \mathbb{C}$ are not parallel so they form a basis, and $(u_1, u_2) \mapsto u_1 \mathbf{v}_1 + u_2 \mathbf{v}_2$ is another isomorphism $\mathbb{R}^2 \cong \mathbb{C}$.

Exercises Let V', V be finite-dimensional $\mathbb{K}$-vector spaces. Prove $\dim (V' \times V) = \dim V' + \dim V$.

Conclude $\mathbb{C}^m$ is a real vector space of dimension $2m$, isomorphic to $\mathbb{R}^{2m}$.

Definition Let V, V' be $\mathbb{K}$-vector spaces, $f \in \mathrm{Hom}\,(V, V')$ a linear map. The **rank** of f is the dimension of the range

$$\mathrm{rk}(f) := \dim_{\mathbb{K}} \mathrm{Im}(f).$$

The **nullity** of f is the dimension $\dim_{\mathbb{K}} \mathrm{Ker}(f)$ of the nullspace.

The following result is the linear instance of an isomorphism theorem in universal algebra (see also exercise v) p. 351).

Theorem (Rank-Nullity) *Let V, V' be vector spaces with $\dim V < \infty$, and $f \in \mathrm{Hom}\,(V, V')$ a linear map. Then*

$$\dim \mathrm{Ker}(f) + \mathrm{rk}(f) = \dim V.$$

Proof By Exercise 15.6 f induces a canonical quotient map

$$V/\mathrm{Ker}(f) \to \mathrm{Im}(f), \quad \mathbf{v} + \mathrm{Ker}(f) \mapsto f(\mathbf{v}),$$

which becomes a $\mathbb{K}$-isomorphism in view of Eq. (3.7).   □

The rank-nullity theorem can be formulated in terms of the *index*

$$\mathrm{ind}(f) := \dim \mathrm{Ker}\, f - \dim \mathrm{Coker}\, f$$

of the linear mapping $f: V \to V'$. Intuitively $\dim \operatorname{Ker} f$ is the number of independent solutions $\mathbf{x}$ to equation $f\mathbf{x} = 0$, while $\dim \operatorname{Coker} f$ is the number of independent constraints (relations) that must be imposed on $\mathbf{y}$ so that $f(\mathbf{x}) = \mathbf{y}$ has solutions. The theorem (when $\dim V' < \infty$) is equivalent to $\operatorname{ind}(f) = \dim V - \dim V'$. The same effect occurs in a much deeper result: the *Atiyah–Singer index theorem* states that the index of certain differential operators (analytical notions) can be extracted from the geometry of the spaces involved (topological notions), without studying the operator in detail.

Exercises Let $f \in \operatorname{Hom}(V^m, V^n)$ be a linear map between finite-dimensional spaces. Deduce

 i) f is onto $\iff \operatorname{rk}(f) = n \implies m \geqslant n$;
 ii) f is 1-1 $\implies m \leqslant n$;
 iii) if $m = n$, then $\dim \operatorname{Ker}(f) = 0 \iff \operatorname{rk}(f) = n$.
 iv) If $W \subseteq V$ is a subspace, then $\dim(V/W) = \dim V - \dim W$.

Statement iii) is reminiscent of Exercise 3.6 ii).

Take a basis $(\mathbf{v}_1, \ldots, \mathbf{v}_n)$ of V^n, so that any $\mathbf{u} \in V$ has the unique representation $\mathbf{u} = \sum_{j=1}^{n} \lambda_j(\mathbf{u})\mathbf{v}_j$. The coordinates $\lambda_j(\mathbf{u})$ of $\mathbf{u}$ in this basis can be viewed as elements of the dual space $V^\vee = \operatorname{Hom}(V, \mathbb{R})$:

$$\lambda_j : V \to \mathbb{R}, \ \mathbf{u} \mapsto \lambda_j(\mathbf{u}).$$

In this way $\lambda_j(\mathbf{v}_i) = \delta_{ij}$.

Proposition *The above elements* $(\lambda_1, \ldots, \lambda_n)$ *form a basis for* $V^\vee = \operatorname{Hom}(V, \mathbb{R})$, *called* **dual basis** *to* $(\mathbf{v}_1, \ldots, \mathbf{v}_n)$.

Proof Let $\sum_{j=1}^{n} \alpha_j \lambda_j = 0$ be the zero linear form. Then $0 = 0(\mathbf{v}_i) = \sum_{j=1}^{n} \alpha_j \lambda_j(\mathbf{v}_i) = \alpha_i$ for all i, making the λ_j linearly independent.

Take $f \in V^\vee$, and let's say $f(\mathbf{v}_i) = b_i \in \mathbb{R}$. Then the linear map $g = f - \sum_{j=1}^{n} b_j \lambda_j$ maps $\mathbf{v}_i$ to 0, and therefore $g \equiv 0$ and $\mathscr{L}(\lambda_1, \ldots, \lambda_n) = V^\vee$. $\square$

Corollary 15.8 *If* $\dim V < \infty$ *then* $V \cong V^\vee$, *and so* $\dim V = \dim V^\vee$.

Proof Let $(\mathbf{v}_1, \ldots, \mathbf{v}_n)$ be a basis (and $\dim V = n$). By the previous proposition the correspondence $\mathbf{v}_i \mapsto \lambda_i$ extends to an isomorphism (not canonical). $\square$

The dimensional constraint is essential. In case $\dim V = \infty$, V and $V^\vee$ can't be isomorphic, as we proved in Exercise 7.28.

Definition The **dual linear map** to $G \in \operatorname{Hom}(W, U)$ is a mapping $G^\vee \in \operatorname{Hom}(U^\vee, W^\vee)$ defined by right-composition: $G^\vee(f) := f \circ G$, for any $f \in U^\vee$.

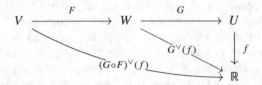

Exercises Take $F \in \mathrm{Hom}\,(V, W)$, $G \in \mathrm{Hom}\,(W, U)$. Show

i) $(G \circ F)^\vee = F^\vee \circ G^\vee$ and $(\mathbb{1}_V)^\vee = \mathbb{1}_{V^\vee}$. Dualisation $^\vee : V \mapsto V^\vee$, $F \mapsto F^\vee$ is therefore an endofunctor on the category $\mathbb{K}$-**Vect**.

ii) There exists an isomorphism $\mathrm{Hom}\,(V, W) \cong \mathrm{Hom}\,(W^\vee, V^\vee)$ (non-canonical), in case $\dim V$, $\dim W$ are finite.

Proposition *Take $F \in \mathrm{Hom}\,(V, W)$, its dual map $F^\vee \in \mathrm{Hom}\,(W^\vee, V^\vee)$, and $f \in W^\vee$, $g \in V^\vee$. Then*

$$f \in \mathrm{Ker}\,F^\vee \iff \mathrm{Im}\,F \subseteq \mathrm{Ker}\,f, \quad \text{and} \quad g \in \mathrm{Im}\,F^\vee \iff \mathrm{Ker}\,F \subseteq \mathrm{Ker}\,g.$$

In the finite-dimensional case $\mathrm{rk}(F) = \mathrm{rk}(F^\vee)$. *Put equivalently, F is 1-1/onto if and only if $F^\vee$ is respectively 1-1/onto.*

Proof Exercise.   □

Eventually, we'll give in to the temptation and consider the **bidual** space $V^{\vee\vee} := (V^\vee)^\vee$. Fix $\mathbf{v} \in V$ and define the map

$$\mathrm{ev}_\mathbf{v} : V^\vee \longrightarrow \mathbb{K}, \ \mathrm{ev}_\mathbf{v}(f) := f(\mathbf{v}). \tag{15.3}$$

It's easy to see $\mathrm{ev}_\mathbf{v}$ is linear, i.e. $\mathrm{ev}_\mathbf{v} \in V^{\vee\vee}$. Furthermore,

Theorem *The correspondence*

$$\mathrm{ev} : V \longrightarrow V^{\vee\vee}, \ \mathrm{ev}(\mathbf{v}) := \mathrm{ev}_\mathbf{v}$$

is linear and one-to-one. Moreover, ev is an isomorphism if and only if $\dim V < \infty$.

Proof The linearity is obvious. The injectivity is proved completing $\{\mathbf{v}\}$ to a basis of V in case $\dim V < \infty$, or using Zorn's Lemma 3.12 for infinite dimension.   □

Exercise Does the definition of ev involve any currying? (Cf. Remark 7.18.)

Corollary *Assuming $\dim V, W < \infty$, there exists a natural isomorphism $F \mapsto F^{\vee\vee}$, for every $F \in \mathrm{Hom}\,(V, W)$.*

Naturality refers to the non-dependence on choices (like bases): in ☞ *category theory* ev is a *natural transformation* between the functor $\mathbb{1}$ and the bidual functor $^{\vee\vee}$ of $\mathbb{K}$-**Vect**.

Definition 15.9 Suppose $\dim V < \infty$ and let $W \subseteq V$ be a subspace. The **annihilator (space)** of W is the subspace of linear forms killing W:

$$W^o := \{L \in V^\vee \mid L(W) = 0\} = \{L \in V^\vee \mid W \subseteq \operatorname{Ker} L\}.$$

The two extreme cases are $\{0\}^o = V^\vee$ and $V^o = \{\text{zero map}\}$.

Exercises Let $U, W \subseteq V$ be vector subspaces. Prove

i) $U \subseteq W \Longrightarrow W^o \subseteq U^o$
ii) $(U^o)^o = U$
iii) $(U \cap W)^o = U^o + W^o$ and $(U + W)^o = U^o \cap W^o$
iv) $\dim W = \operatorname{codim} W^o$.

The next exercises, albeit simple, form the embryo of ☞ *homological algebra*.

Exercises 15.10 Consider a family $\{V_i\}_{i \in I}$ of finite-dimensional $\mathbb{K}$-vector spaces and homomorphisms $f_i \in \operatorname{Hom}_{\mathbb{K}}(V_i, V_{i-1})$ for every $i \in I$. Take the diagram

$$\cdots \xrightarrow{f_{i+3}} V_{i+2} \xrightarrow{f_{i+2}} V_{i+1} \xrightarrow{f_{i+1}} V_i \xrightarrow{f_i} V_{i-1} \xrightarrow{f_{i-1}} V_{i-2} \xrightarrow{f_{i-2}} \cdots$$

and suppose $\operatorname{Im} f_{i+1} \subseteq \operatorname{Ker} f_i$ for every i. In other words, all composites $f_{i+1} \circ f_i = 0$ are zero. (Such a configuration is called a *complex* in $\mathbb{K}$-**Vect**.) We say the diagram is an **exact sequence** if $\operatorname{Ker} f_i = \operatorname{Im} f_{i+1}$ for all i.

Now, for any mapping $f \in \operatorname{Hom}_{\mathbb{K}}(V, V')$ prove that

i) the sequence $\{0\} \xrightarrow{0} \operatorname{Ker} f \xhookrightarrow{\iota} V \xrightarrow{f} V' \xtwoheadrightarrow{\pi} \operatorname{Coker} f \xrightarrow{0} \{0\}$

is exact (ι, π are the canonical inclusion and projection);

ii) $\{0\} \xrightarrow{0} V \xrightarrow{f} V'$ is exact $\iff f$ is 1-1;

iii) $V \xrightarrow{f} V' \xrightarrow{0} \{0\}$ is exact $\iff f$ is onto;

iv) $\{0\} \xrightarrow{0} V \xrightarrow{f} V' \xrightarrow{0} \{0\}$ is exact $\iff f$ is an isomorphism;

v) $W \xhookrightarrow{i} V \xtwoheadrightarrow{\pi} V/W$ is exact, for every subspace $W \subseteq V$.

15.3   Complexification ✋

Definition Let V be an $\mathbb{R}$-vector space of finite dimension n. Its **canonical complexification** is the $\mathbb{C}$-vector space $V_{\mathbb{C}} = V \times V$ endowed with operations

$$(\mathbf{u}, \mathbf{v}) + (\mathbf{w}, \mathbf{y}) := (\mathbf{u} + \mathbf{w}, \mathbf{v} + \mathbf{y}), \qquad (a + bi)(\mathbf{u}, \mathbf{v}) := (a\mathbf{u} - b\mathbf{v}, b\mathbf{u} + a\mathbf{v})$$

for all $\mathbf{u}, \mathbf{v}, \mathbf{w}, \mathbf{y} \in V, a + bi \in \mathbb{C}$.

We set $V_{\mathbb{R}} := \{(\mathbf{u}, \mathbf{v}) \in V_{\mathbb{C}} \mid \mathbf{v} = \mathbf{0}\}$, whose elements are called **real vectors**. The linear map $V_{\mathbb{R}} \to V, (\mathbf{u}, \mathbf{0}) \mapsto \mathbf{u}$ is manifestly bijective.

Proposition $V_{\mathbb{R}}$ *is an $\mathbb{R}$-vector space isomorphic to V. It is not a $\mathbb{C}$-subspace of $V_{\mathbb{C}}$.*

Proof Exercise.   □

The complexification of the vector space $\mathbb{R}$ is $\mathbb{C}$ (with basis $1, i$), so we can identify the pair $(u, v) \in \mathbb{R} \times \mathbb{R}$ with the number $u + iv \in \mathbb{C}$. In this way the multiplication by i in the canonical complexification $\mathbb{C}$

$$i(u, v) = (-v, u)$$

corresponds exactly to the product

$$i(u + iv) = -v + iu$$

in the vector space $\mathbb{C}$. Furthermore

$$\mathbb{C} = \mathbb{R} + i\mathbb{R}. \tag{15.4}$$

The analogue of multiplying by i in higher dimensions is described in the following

Definition The mapping $J: V_{\mathbb{C}} \to V_{\mathbb{C}}$ defined by $J(\mathbf{u}, \mathbf{v}) = (-\mathbf{v}, \mathbf{u})$ is called **standard complex structure** on $V_{\mathbb{C}}$.

Exercises Prove

i) J is $\mathbb{C}$-linear;
ii) J is an anti-involution, i.e. $J^2 = -\mathbb{1}_{V_{\mathbb{C}}}$;
iii) J fixes the real space $V_{\mathbb{R}}$;
iv) $V_{\mathbb{C}} = V_{\mathbb{R}} \oplus J(V_{\mathbb{R}})$. In other words any 'complex vector' can be written in a unique way as $(\mathbf{u}, \mathbf{v}) = (\mathbf{u}, \mathbf{0}) + J(\mathbf{v}, \mathbf{0})$.

Corollary $\dim_{\mathbb{C}} V_{\mathbb{C}} = \dim_{\mathbb{R}} V = \dfrac{1}{2} \dim_{\mathbb{R}} V_{\mathbb{C}}$.

Proof Let $(\mathbf{e}_i)_{i=1}^n$ be a basis of V^n. Then

353

$$(\mathbf{u}, \mathbf{v}) = \left(\sum_{j=1}^{n} u_j \mathbf{e}_j, \sum_{j=1}^{n} v_j \mathbf{e}_j \right) = \sum_{j=1}^{n} u_j (\mathbf{e}_j, \mathbf{0}) + \sum_{j=1}^{n} v_j J(\mathbf{e}_j, \mathbf{0})$$

$$= \sum_{j=1}^{n} u_j (\mathbf{e}_j, \mathbf{0}) + \sum_{j=1}^{n} i v_j (\mathbf{e}_j, \mathbf{0}) = \sum_{j=1}^{n} (u_j + i v_j)(\mathbf{e}_j, \mathbf{0}),$$

showing the $(\mathbf{e}_j, \mathbf{0})$ generate $V_{\mathbb{C}}$. But $V \cong V_{\mathbb{R}}$, so they are LI as well.   □

In a more general context:

Definition Let V be a vector space over $\mathbb{R}$. A **complexification** is a $\mathbb{C}$-vector space $V_{\mathbb{C}}$ together with an $\mathbb{R}$-linear, injective map $i \colon V \to V_{\mathbb{C}}$ whose image $i(V) \cong V$ satisfies $V_{\mathbb{C}} = V \oplus iV$. The subspace V is called a **real form** of $V_{\mathbb{C}}$.

The construction is universal, because the complexification is isomorphic to the tensor product $V \otimes_{\mathbb{R}} \mathbb{C}$ in the category $\mathbb{R}$-**Vect**. It also implies any two complexifications of V are $\mathbb{C}$-isomorphic (in a unique way), so we may as well consider the canonical case: $V \oplus iV = V \otimes_{\mathbb{R}} \mathbb{C}$, $V_{\mathbb{R}} \cong V$.

Definition Let V be an $\mathbb{R}$-vector space. An endomorphism $J \colon V \to V$ such that $J^2 = -\mathbb{1}_V$ is said to be a **complex structure** on V.

A complex structure J is an automorphism since $J^{-1} = -J$. It defines a canonical complexification $V_{\mathbb{C}}$ by extending the scalars $\mathbb{R} \rightsquigarrow \mathbb{C}$ so that $J(\mathbf{v}) = i\mathbf{v}$:

$$(a + bi)\mathbf{v} = a\mathbf{v} + bJ(\mathbf{v}).$$

Corollary *Let V be an $\mathbb{R}$-vector space with complex structure J. Then there exist 1-1 correspondences between*

   i) $\mathbb{R}$-linear maps on V and $\mathbb{C}$-linear maps on $V_{\mathbb{C}}$ that commute with J:

$$\mathrm{End}_{\mathbb{R}}(V) \longleftrightarrow \left\{ f \in \mathrm{End}_{\mathbb{C}}(V_{\mathbb{C}}) \,\middle|\, f \circ J = J \circ f \right\};$$

   ii) $\mathbb{R}$-anti-linear maps on V and $\mathbb{C}$-linear maps on $V_{\mathbb{C}}$ that anti-commute with J.

Proof Exercise.   □

The real forms of a space $V_{\mathbb{C}}$ are far from unique, and in fact there are infinitely many of them.

Example (See p. 399 for the notation) Complex matrices $\mathrm{End}\,(\mathbb{C}^n)$ are complexifications of real matrices $\mathrm{End}\,(\mathbb{R}^n)$. Said differently, real matrices are a real form of complex matrices. Let's define $M^* := \overline{M}^{\mathsf{T}}$ for a square $n \times n$ matrix M with complex entries. **Skew-Hermitian** matrices

$$\mathfrak{u}(n) := \left\{ M \in \mathrm{End}\,(\mathbb{C}^n) \,\middle|\, M^* = -M \right\}$$

are another real form, $\mathfrak{u}(n) \otimes \mathbb{C} = \text{End}\,(\mathbb{C}^n)$, just like **Hermitian** matrices $i\mathfrak{u}(n) = \{M \in \text{End}\,(\mathbb{C}^n) \mid M^* = M\}$. The decomposition

$$\text{End}\,(\mathbb{C}^n) = \mathfrak{u}(n) \oplus i\mathfrak{u}(n)$$
$$M = \frac{M - M^*}{2} + \frac{M + M^*}{2} \tag{15.5}$$

is the complex version of the analogue sum skew-symmetric + symmetric for real matrices: $\text{End}\,(\mathbb{R}^n) = \bigwedge^2 \mathbb{R}^n \oplus \bigodot^2 \mathbb{R}^n$.

Definition An $\mathbb{R}$-linear map $f \in \text{Hom}_{\mathbb{R}}(V, V')$ between real vector spaces V, V' defines a (unique) $\mathbb{C}$-linear extension $f_{\mathbb{C}} \in \text{Hom}_{\mathbb{C}}(V_{\mathbb{C}}, V'_{\mathbb{C}})$

$$f_{\mathbb{C}}(\mathbf{v} + i\mathbf{u}) = f(\mathbf{v}) + if(\mathbf{u}), \qquad \mathbf{v}, \mathbf{u} \in V.$$

Exercises Let $g \in \text{Hom}\,(V, V')$, $f \in \text{Hom}\,(V', V'')$ be arbitrary linear maps. Prove

i) $f_{\mathbb{C}} \circ g_{\mathbb{C}} = (f \circ g)_{\mathbb{C}}$ and $(\mathbb{1}_V)_{\mathbb{C}} = \mathbb{1}_{V_{\mathbb{C}}}$;
ii) $(f_{\mathbb{C}})^{-1} = (f^{-1})_{\mathbb{C}}$ (if invertible);
iii) $\text{Ker}(f_{\mathbb{C}}) = (\text{Ker}\, f)_{\mathbb{C}}$ and $\text{Im}(f_{\mathbb{C}}) = (\text{Im}\, f)_{\mathbb{C}}$.

(In this way the complexification is an (exact) functor $- \otimes_{\mathbb{R}} \mathbb{C}$ from $\mathbb{R} - \mathsf{Vect}$ to $\mathbb{C} - \mathsf{Vect}$.)

In particular, a complex structure J on V defines a complex structure $J_{\mathbb{C}}$ on $V_{\mathbb{C}}$, whose eigenspaces are

$$V^{1,0} := \{\mathbf{v} \in V_{\mathbb{C}} \mid J_{\mathbb{C}}\mathbf{v} = i\mathbf{v}\} \quad \text{(called \textbf{holomorphic vectors})}$$
$$V^{0,1} := \{\mathbf{v} \in V_{\mathbb{C}} \mid J_{\mathbb{C}}\mathbf{v} = -i\mathbf{v}\} \quad \text{(called \textbf{anti-holomorphic vectors})}$$

These spaces are conjugated: $V^{1,0} \cong \overline{V^{0,1}}$, and therefore

$$V_{\mathbb{C}} \cong V^{1,0} \oplus iV^{1,0} \cong V^{1,0} \oplus \overline{V^{1,0}}.$$

Proposition *The projections*

$$\pi^{1,0} \colon V_{\mathbb{C}} \longrightarrow V^{1,0} \qquad\qquad \pi^{0,1} \colon V_{\mathbb{C}} \longrightarrow V^{0,1}$$
$$\mathbf{v} \longmapsto \tfrac{1}{2}(\mathbf{v} - iJ_{\mathbb{C}}\mathbf{v}) \qquad\qquad \mathbf{v} \longmapsto \tfrac{1}{2}(\mathbf{v} + iJ_{\mathbb{C}}\mathbf{v})$$

define, by restriction, isomorphisms $\pi^{1,0}\big|_V \colon V \to V^{1,0}$ *($\mathbb{C}$-linear) and* $\pi^{0,1}\big|_V \colon V \to V^{0,1}$ *($\mathbb{C}$-anti-linear).*

Proof Exercise.   □

This result explains that the complexification $(V_{\mathbb{C}}, J_{\mathbb{C}})$ contains a copy of (V, J) and one of $(V, -J)$.

Proposition *Let V be a real vector space and suppose $V_{\mathbb{C}} = W \oplus \overline{W}$ for some subspace W. There exists a unique complex structure $J \in \mathrm{End}\, V$ such that $W = V^{1,0}$ (and $\overline{W} = V^{0,1}$).*

Proof The uniqueness descends from $V^{1,0}$ being the graph of $-J$ if we use the isomorphism $V_{\mathbb{C}} = V \oplus V$.

Let $J_{\mathbb{C}} \in \mathrm{End}_{\mathbb{C}}\, V_{\mathbb{C}}$ denote the unique mapping with eigenspaces $W, \overline{W}$ for eigenvalues $i, -i$. Then $J_{\mathbb{C}}$ is an anti-involution that commutes with the operation of conjugation. Its restriction $\mathrm{Re}\, J_{\mathbb{C}} = J \circ \pi^{1,0}$ to $W \cong V$ is a complex structure $J \in \mathrm{End}\, V$.   □

In the end a complex structure $J \in \mathrm{End}\, V$ corresponds to a choice of subspace $W \subseteq V_{\mathbb{C}}$ such that $V_{\mathbb{C}} = W \oplus \overline{W}$. For instance the standard complex structure on $\mathbb{C}$ boils down to selecting the real axis $\{z \in \mathbb{C} \mid \mathrm{Im}(z) = 0\} \cong \mathbb{R}$ as the 'real numbers'.

Remark 15.11 The holomorphic and anti-holomorphic spaces crop up in ☞ *complex geometry*. The space of tangent vectors (to a real, differentiable $2n$-manifold), with canonical complex structure J, has a real basis $\partial_{x_1}, \ldots, \partial_{x_n}, \partial_{y_1}, \ldots, \partial_{y_n}$. The vectors

$$\partial_{z_j} := \frac{1}{2}\left(\partial_{x_j} - i\partial_{y_j}\right) \qquad \partial_{\overline{z_j}} := \frac{1}{2}\left(\partial_{x_j} + i\partial_{y_j}\right), \qquad j = 1, \ldots, n$$

form conjugate bases of holomorphic and anti-holomorphic vectors for $J_{\mathbb{C}}\colon \mathbb{C}^{2n} \to \mathbb{C}^{2n}$, because $\pi^{1,0}(\partial_{z_j}) = \partial_{x_j}$ and $\pi^{0,1}(\partial_{z_j}) = \partial_{y_j}$. The notation comes from the identification between (tangent) vectors in $\mathbb{R}^{2n}$ and derivations on $\mathbb{R}^{2n}$. With these conventions the *Cauchy–Riemann equations* $\partial_{\overline{z_j}} f = 0$, $j = 1, \ldots, n$, characterise holomorphic maps $f\colon \mathbb{C}^n \to \mathbb{C}$ as those whose differential $\mathrm{d}f\colon (\mathbb{C}^n, J) \to \mathbb{C}$ is $\mathbb{C}$-linear (i.e., J commutes with $\mathrm{d} = \partial + \overline{\partial}$).

15.4 Bilinear and Quadratic Forms ☕

Definition Let V be a finite-dimensional $\mathbb{R}$-vector space. A **bilinear form** on V is an element in the dual space $(V \times V)^{\vee}$, that's to say a map $\phi\colon V \times V \to \mathbb{R}$ such that

$$\phi(a\mathbf{u} + b\mathbf{v}, c\mathbf{w} + d\mathbf{y}) = ac\,\phi(\mathbf{u}, \mathbf{w}) + bc\,\phi(\mathbf{v}, \mathbf{w}) + ad\,\phi(\mathbf{u}, \mathbf{y}) + bd\,\phi(\mathbf{v}, \mathbf{w})$$

for all $\mathbf{u}, \mathbf{v}, \mathbf{w}, \mathbf{y} \in V$, $a, b, c, d \in \mathbb{R}$.

A bilinear form ϕ is called

- **symmetric** if $\phi(\mathbf{u}, \mathbf{w}) = \phi(\mathbf{w}, \mathbf{u})$ for every $\mathbf{u}, \mathbf{w} \in V$;
- **skew-symmetric** if $\phi(\mathbf{u}, \mathbf{w}) = -\phi(\mathbf{w}, \mathbf{u})$ for all $\mathbf{u}, \mathbf{w} \in V$.

The **kernel** of ϕ is the subspace

$$\operatorname{Ker}\phi := \{\mathbf{u} \in V \mid \phi(\mathbf{u}, \mathbf{y}) = 0 \ \forall \mathbf{y} \in V\} =: V^o,$$

and the bilinear form is **singular** if $\dim \operatorname{Ker}\phi > 0$, **non-singular** if $\operatorname{Ker}\phi = \{\mathbf{0}\}$.

Examples Consider on $\mathbb{R}^2$ the following bilinear forms:

i) $\phi\big((x_1, y_1), (x_2, y_2)\big) = x_1 x_2 - y_1 y_2$ is symmetric, and its kernel is trivial because $\phi = 0 \implies x_1 = 0$ (taking $x_2 = 0, y_2 = 1$) and $y_1 = 0$ (for $x_2 = 1, y_2 = 0$);

ii) $\phi_3\big((x_1, y_1), (x_2, y_2)\big) = x_1 x_2$ symmetric and singular, $\operatorname{Ker}\phi_3 = \mathscr{L}(0, 1) \cup \mathscr{L}(1, 0)$;

iii) $\phi_1\big((x_1, y_1), (x_2, y_2)\big) = \det \begin{pmatrix} x_1 & y_1 \\ x_2 & y_2 \end{pmatrix}$ is skew and non-singular: the vanishing of the determinant implies the first column is proportional to every vector, in particular to $\mathbf{0}$;

iv) $\phi_2\big((x_1, y_1), (x_2, y_2)\big) = x_1 y_1 - x_2 y_2$ is skew-symmetric and singular: $\operatorname{Ker}\phi_2 = \operatorname{Ker}\phi_3$.

Exercises Consider a bilinear form ψ on V.

i) Prove ψ can be written in a unique fashion as a sum of a symmetric bilinear form ψ' plus a skew-symmetric bilinear form ψ''. (Aside: what does this say about matrices?)

ii) Let $\alpha \colon \mathbb{K}^n \to (\mathbb{K}^n)^\vee$ be the mapping $\alpha(x) = \psi_x \colon \mathbb{K}^n \to \mathbb{K}$, $\psi_x(y) = \psi(x, y)$. Prove $\operatorname{Ker}\psi = \operatorname{Ker}\alpha$ (which justifies the use of the name 'kernel').

iii) Explain the relationship between $\operatorname{Ker}\psi$ and the annihilator (Definition 15.9) of a linear subspace.

   Assume ψ is non-singular. Using the above finding, show that $\dim V = \dim U + \dim U^o$ for any subspace $U \subseteq V$. If ψ is further symmetric, or skew-symmetric, then $(U^o)^o = U$.

Even if ϕ is globally non-singular, its restriction to some subspace $U \subseteq V$ might be singular. In view of the above exercise we may set $U^o = \operatorname{Ker}\phi\big|_{U \times U}$.

Examples Retaining the notation of the previous examples,

i) ϕ is non-singular, but when restricted to $U = \mathscr{L}(1, 1)$ it becomes singular: $U^o = U$.

ii) ϕ_3 is singular, but on $U = \mathscr{L}(1, 0)$ it is non-singular, because $U^o = \{(0, 0)\}$.

Lemma *For any subspace* $U \subseteq V$, $\phi\big|_{U \times U}$ *is non-singular if and only if* $V = U \oplus U^o$.

Proof Since $\operatorname{Ker}\phi\big|_{U \times U} = U \cap U^o$, if we assume $V = U \oplus U^o$ then ϕ is non-singular on U.

   Vice versa, if ϕ is non-singular on U then $U \cap U^o = \{\mathbf{0}\}$. To show $V = U + U^o$ it suffices to note $\mathbf{u} \mapsto \phi(\mathbf{u}, \cdot)$ is an isomorphism $U \to U^\vee$. Hence for any $\mathbf{v} \in V$ there exists a $\mathbf{u} \in U$ such that $\phi(\mathbf{u}, \cdot) = \phi(\mathbf{v}, \cdot)$ on U, and therefore $\mathbf{v} - \mathbf{u} \in U^o$. $\square$

The lemma is false when $\dim U = \infty$. For example, take $V = \ell^2(\mathbb{R}) = \{(a_n)_{n \in \mathbb{N}} \mid \|a\|_{L^2} < \infty\}$, the space of square-summable real sequences (see (19.4)), and let U be the subspace of sequences that are null almost everywhere ($a_n \neq 0$ for a finite number of indices n). The standard L^2-product guarantees $U^o = \{0\}$, because the previous lemma's map $\mathbf{u} \mapsto \phi(\mathbf{u}, \cdot)$ is 1-1 but not onto (☞ *functional analysis*).

For the purposes of the next result, we say a basis $(\mathbf{v}_i)$ of V diagonalises a linear form $f : V \to \mathbb{R}$ if $f(\mathbf{v}_i) = \beta_i \mathbf{v}_i$ for some coefficients β_i (the matrix associated to f in that basis is diagonal).

Proposition 15.12 *Let V be a $\mathbb{K}$-vector space of finite dimension n, with $\operatorname{char} \mathbb{K} \neq 2$. For any symmetric bilinear form ϕ on V there exists a basis of V that diagonalises ϕ.*

Proof (by induction on n] If $\phi \equiv 0$ or $n = 1$ the claim is obvious, so suppose the result true for spaces of dimension $< n$ and take $\phi \not\equiv 0$. There exists $\mathbf{v}_1 \in V$ such that $\phi(\mathbf{v}_1, \mathbf{v}_1) \neq 0$ (as $\operatorname{char} \neq 2$). Therefore ϕ is non-singular on $U = \mathscr{L}(\mathbf{v}_1)$, and by the lemma $V = U \oplus U^o$. By the induction hypothesis U^o has a basis B diagonalising $\phi\big|_{U^o \times U^o}$. Then $B \cup \{\mathbf{v}_1\}$ diagonalises ϕ. $\qquad\square$

Until the section's end we shall deal with symmetric bilinear forms.

Definition Let ϕ be a symmetric bilinear form on a finite-dimensional $\mathbb{R}$-vector space V^n. The **quadratic form** associated to ϕ is the mapping $\Phi : V \to \mathbb{R}$,

$$\Phi(\mathbf{u}) = \phi(\mathbf{u}, \mathbf{u}).$$

Consequently Φ is homogeneous of degree 2: $\Phi(\lambda \mathbf{u}) = \lambda^2 \Phi(\mathbf{u})$, for $\lambda \in \mathbb{R}, \mathbf{u} \in V$. Cf. Exercise 9.3 iii).

Example Pick a basis $(\mathbf{v}_1, \ldots, \mathbf{v}_n)$ for V and set $\mathbf{x}^T = (x_1, \ldots, x_n)$ in this basis. Define the symmetric matrix $A = (a_{ij}) = (\phi(\mathbf{v}_i, \mathbf{v}_j))$. Then

$$\phi(\mathbf{u}, \mathbf{w}) = \mathbf{u}^T A \mathbf{w},$$

with associated quadratic form

$$\Phi(\mathbf{x}) = \mathbf{x}^T A \mathbf{x} = \sum_{i=1}^n a_{ii} x_i^2 + 2 \sum_{i<j} a_{ij} x_i x_j.$$

The next proposition is a kind of Taylor formula for symmetric bilinear forms:

Proposition *The quadratic form Φ determines its symmetric bilinear form ϕ completely, by means of the **polarisation formula**:*

$$\phi(\mathbf{u}, \mathbf{v}) = \frac{1}{2}\big(\Phi(\mathbf{u} + \mathbf{v}) - \Phi(\mathbf{u}) - \Phi(\mathbf{v})\big), \qquad \mathbf{u}, \mathbf{v} \in V.$$

Proof Exercise.   □

From now on, then, we shall not distinguish between ϕ and the associated Φ.

The (vector) space of symmetric bilinear forms on V (or of quadratic forms) is denoted by

$$\odot^2(V) = \left\{ \phi \in (V \times V)^\vee \mid \phi(\mathbf{u}, \mathbf{v}) = \phi(\mathbf{v}, \mathbf{u}), \ \mathbf{u}, \mathbf{v} \in V \right\}$$
$$\cong \left\{ \Phi \in V^\vee \mid \Phi(\lambda \mathbf{v}) = \lambda^2 \Phi(\mathbf{v}), \ \lambda \in \mathbb{R}, \mathbf{v} \in V \right\}.$$

Polarisation has an important application to the description of linear isometries:

Proposition 15.13 *Let ϕ be a symmetric bilinear form on V associated with the quadratic form Φ. Then a function $f: V \to V$ such that $f(\mathbf{0}) = \mathbf{0}$ and $\Phi\big(f(\mathbf{u}) - f(\mathbf{v})\big) = \Phi(\mathbf{u} - \mathbf{v})$, $\mathbf{v}, \mathbf{u} \in V$ satisfies:*

i) $\phi(\mathbf{u}, \mathbf{v}) = \phi\big(f(\mathbf{u}), f(\mathbf{v})\big)$ for any $\mathbf{v}, \mathbf{u} \in V$;
ii) f is linear.

Proof Using the polarisation formula part i) is immediate.

So, $\Phi\big(f(\mathbf{u} + \mathbf{v}) - f(\mathbf{u}) - f(\mathbf{v})\big) = 0$, i.e. $f(\mathbf{u} + \mathbf{v}) = f(\mathbf{u}) + f(\mathbf{v})$. Still by polarisation, $\Phi\big(f(a\mathbf{u}) - af(\mathbf{u})\big) = \phi(a\mathbf{u}, a\mathbf{u}) + a^2\phi(\mathbf{u}, \mathbf{u}) + 2a\phi(a\mathbf{u}, \mathbf{u}) = 0$, implying $f(a\mathbf{u}) = af(\mathbf{u})$. Therefore f commutes with the addition and the multiplication by scalars, i.e. it's linear.   □

Theorem (Representation of Linear Maps) *Let ϕ be a non-singular bilinear form on a finite-dimensional space V. For any $f \in V^\vee$ there exist a unique vector $\mathbf{f} \in V$ such that*

$$f(\mathbf{u}) = \phi(\mathbf{f}, \mathbf{u}) \qquad \text{for every } \mathbf{u} \in V.$$

Proof (Existence) Let $\{\mathbf{v}_j\}$ be an basis de V such that $\phi(\mathbf{v}_i, \mathbf{v}_j) = \delta_{ij}$ (later this will be called an ON basis). Set $\mathbf{f} = \sum_j f(\mathbf{v}_j)\mathbf{v}_j$. For any $\mathbf{v}_i$, thus, $\phi(\mathbf{f}, \mathbf{v}_i) = \sum_j f(\mathbf{v}_j)\phi(\mathbf{v}_j, \mathbf{v}_i) = f(\mathbf{v}_i)$. By linearity the claim follows.

(Uniqueness) If $\mathbf{g}$ were another representing vector then $\phi(\mathbf{g}, \mathbf{u}) = f(\mathbf{u}) = \phi(\mathbf{f}, \mathbf{u})$. Hence $\phi(\mathbf{f} - \mathbf{g}, \mathbf{u}) = 0$ for all $\mathbf{u}$, and so $\mathbf{f} - \mathbf{g} \in \text{Ker } \phi = \{\mathbf{0}\}$.   □

The representation theorem establishes the explicit **duality** between V and $V^\vee$: the correspondence

$$\mathbf{f} \longleftrightarrow f = \phi(\mathbf{f}, \cdot)$$

tells the vector $\mathbf{f}$ represents f by means of ϕ. The generalisation of this duality to Hilbert spaces goes under the name of *Riesz's representation theorem* (☞ *functional analysis*).

15.5 Euclidean Spaces

Definition A bilinear form ϕ (or the quadratic form Φ) is **positive definite** if non-singular and $\phi(\mathbf{u}, \mathbf{u}) \geqslant 0$ for all $\mathbf{u} \in V$, i.e.

$$\Phi(\mathbf{u}) > 0 \qquad \forall \mathbf{u} \neq \mathbf{0}.$$

It's called **indefinite** if there exist $\mathbf{v}, \mathbf{w} \in V$ such that $\Phi(\mathbf{v}) < 0 < \Phi(\mathbf{w})$.

Proposition *Let* $\phi \in \odot^2(V)$ *be positive definite. For any* $\mathbf{u}, \mathbf{v} \in V$,

$$\phi(\mathbf{u}, \mathbf{v})^2 \leqslant \Phi(\mathbf{u})\Phi(\mathbf{v}) \quad \text{(\textbf{Cauchy–Schwarz inequality})}$$

$$\sqrt{\Phi(\mathbf{u} + \mathbf{v})} \leqslant \sqrt{\Phi(\mathbf{u})} + \sqrt{\Phi(\mathbf{v})} \quad \text{(\textbf{Minkowski inequality}, or \textbf{triangle inequality})}$$

Proof (Cauchy–Schwarz) Obvious if $\mathbf{v} = \mathbf{0}$, so take $\mathbf{v} \neq \mathbf{0}$. Take any $a \in \mathbb{R}$, so

$$0 \leqslant \Phi(\mathbf{u} + a\mathbf{v}) = \Phi(\mathbf{u}) + 2a\phi(\mathbf{u}, \mathbf{v}) + a^2\Phi(\mathbf{v}).$$

This quadratic polynomial in the variable a is non-negative for all a, whence its discriminant must be non-negative: $\Delta/4 = \phi(\mathbf{u}, \mathbf{v})^2 - \Phi(\mathbf{u})\Phi(\mathbf{v}) \leqslant 0$.

(Minkowski) The condition we seek to prove, once we square, is $\Phi(\mathbf{u} + \mathbf{v}) \leqslant \Phi(\mathbf{u}) + \Phi(\mathbf{v}) + 2\sqrt{\Phi(\mathbf{u})\Phi(\mathbf{v})}$, i.e. (polarising) $\phi(\mathbf{u}, \mathbf{v}) \leqslant \sqrt{\Phi(\mathbf{u})\Phi(\mathbf{v})}$. $\qquad \square$

Exercise Explain what the above triangle inequality has to do with earlier namesakes: Proposition 7.6 and Exercise 13.1 iii). (We'll go there in a sec...)

Definition 15.14 Let V be an $\mathbb{R}$-vector space. A symmetric and positive definite bilinear form ϕ is called **inner product**, and the pair (V, ϕ) is called **Euclidean**, or **inner-product space**.

The non-negative real number

$$\|\mathbf{u}\| := \sqrt{\Phi(\mathbf{u})} \tag{15.6}$$

is the **norm** of $\mathbf{u}$. A vector of norm 1 is a **unit vector**, and the set of unit vectors in V is the **unit sphere**

$$S^{n-1} := \left\{ \mathbf{u} \in V^n \mid \|\mathbf{u}\| = 1 \right\}.$$

Example On $V = \mathbb{R}$ ($n = 1$) we may choose the absolute value as norm of a real number $\mathbf{u} \in \mathbb{R}$. For $V = \mathbb{C}$ ($n = 2$) we can take the modulus of a complex number $\mathbf{u} \in \mathbb{C}$. In either case the respective ordinary multiplication is an inner product.

The unit sphere in $\mathbb{R}$ is the set $\{\pm 1\}$, whilst in $\mathbb{R}^2 \cong \mathbb{C}$ it is the unit circle $S^1 = \{e^{i\theta} \mid \theta \in \mathbb{R}\}$.

Given that any norm defines in a natural way a distance (Chap. 19),

Corollary *Every Euclidean vector space* (V, ϕ) *is a metric space with distance*

$$d(\mathbf{u}, \mathbf{v}) := \|\mathbf{u} - \mathbf{v}\| = \sqrt{\Phi(\mathbf{u} - \mathbf{v})}.$$

Example Consider real functions $t \mapsto f(t)$ defined on an interval $I \subseteq \mathbb{R}$ and such that $\left| f(t) \right|^2$ is Lebesgue integrable. They form a vector space $L^2(I)$ that becomes Euclidean when endowed with the inner product $(f, g)_{L^2} := \int_I f(t) g(t) dt$ and norm

$$\|f\|_{L^2} := \left(\int_I \left| f(t) \right|^2 dt \right)^{1/2}. \tag{15.7}$$

The vectors of this space are the $\mathbf{L^2}$ **functions**, and are the central players in the theory of *Fourier series* (☞ *functional analysis, harmonic analysis, ODEs*) and a great deal of physical applications (*vibrations, acoustics, optics, electromagnetism, quantum mechanics . . .*).

Example The standard inner product on $V = \mathbb{R}^n$

$$\mathbf{u} \cdot \mathbf{v} := \sum_{j=1}^{n} u_j v_j \tag{15.8}$$

is called **dot product**, and $\|\mathbf{u}\| := \left(\sum_{j=1}^{n} u_j^2 \right)^{1/2}$ is the **standard norm** of $\mathbb{R}^n$. We'll refer to $\mathbb{R}^n$ with the above dot product as the **(standard) Euclidean n-space** $\mathbb{R}^n_{\text{Eucl}}$.

There is an analogue special inner product on $\mathbb{C}^n$ as well, see (16.2).

Proposition 15.15 *For any* $\mathbf{u}, \mathbf{v}, \mathbf{w} \in \mathbb{R}^n, \lambda \in \mathbb{R}$

i) $\|\mathbf{u} \cdot \mathbf{v}\| = \|\mathbf{u}\| \|\mathbf{v}\| \iff \mathbf{u}, \mathbf{v}$ *are LD;*
ii) $\mathbf{u} \cdot \mathbf{v} = \|\mathbf{u}\| \|\mathbf{v}\| \cos \theta$, *where* θ *is the angle between the geometric vectors* $\vec{\mathbf{u}}, \vec{\mathbf{v}}$.

Proof Exercise.   □

As a matter of fact, using ii) above one can *define* the angle θ between non-zero vectors, setting $\cos \theta := \dfrac{\mathbf{u} \cdot \mathbf{v}}{\|\mathbf{u}\| \|\mathbf{v}\|}$.

Definition Let (V, ϕ) be a Euclidean space. Two vectors $\mathbf{u}, \mathbf{v} \in V$ are **orthogonal**, written $\mathbf{u} \perp \mathbf{v}$, if $\phi(\mathbf{u}, \mathbf{v}) = 0$. The **orthogonal space to** $U \subseteq V$ is the subspace

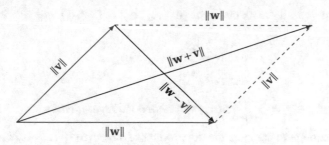

Fig. 15.1 Triangle inequality & parallelogram rule

$$U^{\perp} = \left\{ \mathbf{v} \in V \mid \phi(\mathbf{u}, \mathbf{v}) = 0 \ \forall \mathbf{u} \in U \right\}.$$

Exercises Referring to Fig. 15.1, prove that

i) two orthogonal vectors $\mathbf{u}, \mathbf{v} \neq \mathbf{0}$ are LI;
ii) $\mathbf{u} \perp \mathbf{v} \implies \|\mathbf{u} + \mathbf{v}\|^2 = \|\mathbf{u}\|^2 + \|\mathbf{v}\|^2;$ (*Pythagoras theorem*)
iii) $\|\mathbf{w} + \mathbf{v}\|^2 + \|\mathbf{w} - \mathbf{v}\|^2 = 2\|\mathbf{w}\|^2 + 2\|\mathbf{v}\|^2;$ (*parallelogram rule*)
iv) $\|\mathbf{w} + \mathbf{v}\|^2 = \|\mathbf{w} - \mathbf{v}\|^2 \iff \mathbf{w} \perp \mathbf{v}$ (a parallelogram's diagonals have the same length if and only if the parallelogram is a rectangle.)

Examples

i) The canonical basis $(\mathbf{e}_j)_{j=1,\dots,n}$ of $\mathbb{R}^n$ is **orthonormal** (ON) for the dot product: the basis vectors are pairwise-orthogonal, unit vectors:

$$\mathbf{e}_i \cdot \mathbf{e}_j = \delta_{ij}.$$

ii) On $V = \mathbb{R}^2$: $\mathscr{L}(\mathbf{i}) = \left\{ (x, 0) \mid x \in \mathbb{R} \right\} = \mathbf{j}^{\perp}$ is the x-axis (of Cartesian equation $y = 0$), whereas $\mathbf{i}^{\perp} = \left\{ (0, y) \mid y \in \mathbb{R} \right\} = \mathscr{L}(\mathbf{j})$ is the y-axis (equation $x = 0$).
iii) On $V = \mathbb{R}^3$: $\mathbf{k}^{\perp} = \left\{ (x, y, 0) \mid x, y \in \mathbb{R} \right\} = \mathscr{L}(\mathbf{i}, \mathbf{j})$ is the Oxy plane ($z = 0$).
iv) On $V = \mathbb{R}^3$: the space orthogonal to a non-zero vector $\mathbf{n} = (a, b, c)$ is the (hyper)plane

$$\mathbf{n}^{\perp} = \left\{ \mathbf{x} \in \mathbb{R}^3 \mid \mathbf{x} \cdot \mathbf{n} = 0 \right\} = \left\{ (x, y, z) \in \mathbb{R}^3 \mid ax + by + cz = 0 \right\}.$$

If $a \neq 0$, for example, the vectors $\mathbf{v}_1 = (-c, 0, a)$, $\mathbf{v}_2 = (-b, a, 0)$ are LI, and orthogonal to $\mathbf{n}$. Hence $\mathbf{n}^{\perp} = \mathscr{L}(\mathbf{v}_1, \mathbf{v}_2)$.

Exercises Let $U, W \subseteq (V, \phi)$ be subspaces. Prove

i) $W \cap W^{\perp} = \{\mathbf{0}\}$, $(U + W)^{\perp} = U^{\perp} + W^{\perp}$, $W \subseteq (W^{\perp})^{\perp};$
ii) $\dim V < \infty \implies V = W \oplus W^{\perp}$ and $W = (W^{\perp})^{\perp}$.

Proposition *Relative to an ON basis* $\{\mathbf{u}_1, \ldots, \mathbf{u}_n\}$ *of* (V^n, ϕ), *any vector* $\mathbf{v}$ *splits as*

$$\mathbf{v} = \sum_{j=1}^{n} \phi(\mathbf{u}_j, \mathbf{v}) \mathbf{u}_j.$$

In particular: $\|\mathbf{v}\|^2 = \sum_{j=1}^{n} \phi(\mathbf{u}_j, \mathbf{v})^2$, *called **Parseval identity**.*

Proof It's enough to compute the dot product of $\mathbf{v} = v_1 \mathbf{u}_1 + \ldots + v_n \mathbf{u}_n$ with $\mathbf{u}_j$ to obtain v_j, $j = 1, \ldots, n$.   □

The proposition is a manifestation of a more general situation:

Lemma *Let* $\mathbf{v}$ *and* $\mathbf{u} \neq \mathbf{0}$ *be vectors. Then* $\mathbf{v} = \dfrac{\phi(\mathbf{u}, \mathbf{v})}{\phi(\mathbf{u}, \mathbf{u})} \mathbf{u} + \mathbf{v}^{\perp}$ *for some vector* $\mathbf{v}^{\perp} \perp \mathbf{u}$.

Proof Define $\mathbf{v}^{\perp} = \mathbf{v} - \dfrac{\phi(\mathbf{u}, \mathbf{v})}{\phi(\mathbf{u}, \mathbf{u})} \mathbf{u}$. Immediately, then, $\phi(\mathbf{v}^{\perp}, \mathbf{u}) = \phi(\mathbf{u}, \mathbf{v}) -$
$\dfrac{\phi(\mathbf{u}, \mathbf{v})}{\phi(\mathbf{u}, \mathbf{u})} \phi(\mathbf{u}, \mathbf{u}) = 0$.   □

Definition Let $\mathbf{u} \neq \mathbf{0}$. The component $\pi_{\mathbf{u}}(\mathbf{v}) := \dfrac{\phi(\mathbf{u}, \mathbf{v})}{\phi(\mathbf{u}, \mathbf{u})} \mathbf{u}$ of the vector $\mathbf{v}$ parallel to $\mathbf{u}$ in the decomposition

$$\mathbf{v} = \pi_{\mathbf{u}}(\mathbf{v}) + \mathbf{v}^{\perp}$$

is called **orthogonal projection of v along u**.

Now take linearly independent vectors $\mathbf{v}_1, \ldots, \mathbf{v}_k$ and compute the following:

$$\mathbf{u}_1 = \mathbf{v}_1$$
$$\mathbf{u}_2 = \mathbf{v}_2 - \pi_{\mathbf{u}_1}(\mathbf{v}_2)$$
$$\mathbf{u}_3 = \mathbf{v}_3 - \pi_{\mathbf{u}_1}(\mathbf{v}_3) - \pi_{\mathbf{u}_2}(\mathbf{v}_3)$$

$$\vdots$$

$$\mathbf{u}_k = \mathbf{v}_k - \sum_{j=1}^{k-1} \pi_{\mathbf{u}_j}(\mathbf{v}_k).$$

Exercise Show this recipe produces pairwise-orthogonal vectors $\{\mathbf{u}_\beta\}$ that span the same subspace as the $\{\mathbf{v}_\alpha\}$: $\mathscr{L}(\mathbf{u}_1, \ldots, \mathbf{u}_k) = \mathscr{L}(\mathbf{v}_1, \ldots, \mathbf{v}_k)$.

The process taking LI vectors $\{\mathbf{v}_\alpha\}$ and returning an orthogonal collection $\{\mathbf{u}_\beta\}$ is called **Gram–Schimidt orthogonalisation**. The numbers $\dfrac{\phi(\mathbf{u}_j, \mathbf{v}_k)}{\phi(\mathbf{u}_j, \mathbf{u}_j)}$ are the **Fourier coefficients** of $\mathbf{v}_k$ in the basis $\{\mathbf{u}_\beta\}$.

Example Take the vector space $\mathscr{C}^0([0, 2\pi], \mathbb{R}) = \{f : [0, 2\pi] \to \mathbb{R} \text{ continuous}\}$. The functions

$$1, \ \sin x, \ \sin 2x, \ldots, \ \sin nx, \ \ldots, \ \cos x, \ \cos 2x, \ \ldots, \ \cos nx, \ldots \qquad n \in \mathbb{N}^*$$

are orthogonal vectors for the L^2 inner product, cf. (15.7). The classical Fourier series of a function $f \in \mathscr{C}^0([0, 2\pi], \mathbb{R})$ is

$$a_0 + \sum_{n>0} b_n \cos nx + \sum_{n>0} c_n \sin nx$$

where $a_0 = \dfrac{1}{2\pi} \displaystyle\int_0^{2\pi} f(x)\mathrm{d}x, \ b_n = \dfrac{1}{\pi} \displaystyle\int_0^{2\pi} f(x) \cos(nx)\mathrm{d}x, \ c_n = \dfrac{1}{\pi} \displaystyle\int_0^{2\pi} f(x) \sin(nx)\mathrm{d}x$ are the Fourier coefficients defined above.

Exercise Prove that $\displaystyle\sum_{n=1}^{\infty} \frac{1}{(2n-1)^2} = \frac{\pi^2}{8}$ is the Fourier series of $f(x) = |x|$ evaluated at $x = 0$, and $\displaystyle\sum_{n=1}^{\infty} \frac{1}{n^2} = \frac{\pi^2}{6}$ is that of $f(x) = x^2$ at $x = \pi$.

Definition Let $V_1, \ldots, V_k$ be subspaces of a Euclidean space (V, ϕ) whose sum is precisely V. If the V_j are pairwise orthogonal we write

$$V = \bigoplus_{j=i}^{k} V_j := V_1 \oplus V_2 \oplus \cdots \oplus V_{k-1} \oplus V_k,$$

called **orthogonal decomposition**, or **orthogonal splitting**, of V.

So in an orthogonal decomposition $V_i \cap V_j = \{0\}$ if $i \neq j$. In case $\dim V < \infty$, any non-trivial subspace $W \subset V$ induces an orthogonal decomposition $V = W \oplus W^\perp$, and every vector then splits as $\mathbf{v} = \mathbf{w} + \mathbf{w}^\perp$, $\mathbf{w} \in W$, $\mathbf{w}^\perp \in W^\perp$ in a unique manner. Generalising Definition 15.5, we have

Definition Let V be a finite-dimensional Euclidean space and $W \subseteq V$ a subspace. An **orthogonal projection** on W is a map $\pi_W : W \oplus W^\perp \to W$

$$\pi_W(\mathbf{w} + \mathbf{w}^\perp) := \mathbf{w}.$$

Exercises Prove that

i) π_W is linear and onto; $\operatorname{Ker} \pi_W = W^\perp$ and $(\pi_W)^2 = \pi_W$ (orthogonal projections are idempotent).

ii) If $V = \bigoplus_{j=1}^{k} V_j$ then any vector decomposes as $\mathbf{v} = \pi_{V_1}(\mathbf{v}) + \ldots \pi_{V_k}(\mathbf{v})$ in a unique way.

Remarks 15.1 Proposition 15.12 is the finite-dimensional version of the *spectral theorem* (Sect. 16.2), by which every quadratic form $\Phi \in \odot^2 V^n$ (possibly singular) can be diagonalised over $\mathbb{C}$

$$\Phi(x_1, \ldots, x_n) = \sum_{i=1}^{m} x_i^2, \qquad m \leqslant n,$$

and over $\mathbb{R}$

$$\Phi(x_1, \ldots, x_n) = \sum_{i=1}^{n_+} x_i^2 - \sum_{j=1}^{n_-} x_j^2, \qquad n_+ + n_- \leqslant n.$$

Sylvester's law of inertia asserts that $n_\pm$ (inertia indices) are invariants of Φ. Geometrically it means $n_\pm$ is the maximal dimension of a subspace on which Φ is positive/negative definite. The pair (n_+, n_-) is called **signature** of Φ.

Exercises Let $\Phi \in \odot^2 \mathbb{R}^n$ be a quadratic form. Show

i) $n_+ + n_- = n \iff \Phi$ is non-singular (whence $\dim \operatorname{Ker} \Phi = n - n_+ - n_-$ is the nullity of Φ);
ii) $n_+ = n$ (i.e. the signature is $(n, 0)$) $\iff \Phi$ is positive definite;
iii) $n_- = n \iff -\Phi$ is positive definite (Φ is negative definite).
iv) Let $W, Z \subseteq \mathbb{R}^n$ be subspaces such that $\Phi\big|_W > 0$ and $\Phi\big|_Z \leqslant 0$. Prove $W \cap Z = \{\mathbf{0}\}$, and in particular that $\dim W + \dim Z \leqslant n$.

While we've got out hands in physics, let's take a (bilinear, symmetric and) non-singular form with arbitrary signature $(n - n_-, n_-)$, possibly indefinite ($n_- > 0$). This relaxes notion Definition 15.14, and defines a **pseudo-Euclidean** inner product. A vital example is **Minkowski space** $\mathbb{R}^{3,1}$, namely spacetime $\mathbb{R}^3 \times \mathbb{R}$ equipped with the **Lorenz product**

$$(x, y, z, t) \cdot (x', y', z', t') := xx' + yy' + zz' - tt'.$$

The pseudo-Euclidean signature $(3, 1)$ determines the causal structure of the universe we allegedly live in. It forces the existence of vectors of zero length (*light-like*) and negative length (*time-like*), beside the usual vectors (*space-like*) (☞ *Riemannian geometry*).

15.6 Cross Product

This short section is dedicated to a special operation of a three-dimensional vector space. Here, just for convenience, we'll use the standard Euclidean space $\mathbb{R}^3$. Apart from being relevant *per se* (physics, anyone?), it provides a good warm-up and the ground material for the study of ☞ *multilinear algebra*.

Definition The map $\times : \mathbb{R}^3 \times \mathbb{R}^3 \longrightarrow \mathbb{R}^3$

$$(\mathbf{u}, \mathbf{v}) \mapsto \mathbf{u} \times \mathbf{v} := (u_2 v_3 - u_3 v_2, \ u_3 v_1 - u_1 v_3, \ u_1 v_2 - u_2 v_1),$$

for $\mathbf{u} = (u_1, u_2, u_3)$, $\mathbf{v} = (v_1, v_2, v_3)$, is called **cross product**.

The cross product can be characterised by a number of properties,[3] which are included in the list below:

Proposition *For any* $\mathbf{u}, \mathbf{v}, \mathbf{w} \in \mathbb{R}^3, \lambda \in \mathbb{R}$

 i) $(\mathbf{u} + \mathbf{v}) \times \mathbf{w} = \mathbf{u} \times \mathbf{w} + \mathbf{v} \times \mathbf{w}$, $\quad (\lambda\mathbf{u}) \times \mathbf{w} = \lambda(\mathbf{u} \times \mathbf{w})$; $\quad$ *($\mathbb{R}$-linearity)*

 ii) $\mathbf{u} \times \mathbf{v} = -\mathbf{v} \times \mathbf{u}$ $\quad$ *(i.e.,* $\mathbf{u} \times \mathbf{u} = 0$); $\qquad\qquad$ *(skew-symmetry)*

 iii) $\mathbf{u} \times \mathbf{v} = 0 \iff \mathbf{u}, \mathbf{v}$ *are LD;*

 iv) $\mathbf{u} \times \mathbf{v} \perp \mathbf{u}, \mathbf{v}$, *that's to say:* $(\mathbf{u} \times \mathbf{v}) \cdot \mathbf{u} = 0 = (\mathbf{u} \times \mathbf{v}) \cdot \mathbf{v}$;

 v) *if* $\mathbf{u}, \mathbf{v}$ *are LI then* $\{\mathbf{u}, \mathbf{v}, \mathbf{u} \times \mathbf{v}\}$ *is a basis of* $\mathbb{R}^3$;

 vi) $(\mathbf{u} \times \mathbf{v}) \times \mathbf{w} = (\mathbf{u} \cdot \mathbf{w})\mathbf{v} - (\mathbf{v} \cdot \mathbf{w})\mathbf{u}$;

 vii) $(\mathbf{u} \times \mathbf{v}) \times \mathbf{w} + (\mathbf{v} \times \mathbf{w}) \times \mathbf{u} + (\mathbf{w} \times \mathbf{u}) \times \mathbf{v} = 0$ $\qquad$ *(Jacobi identity)*

Proof Exercise. $\qquad\qquad\qquad\qquad\qquad\qquad\qquad\qquad\qquad\qquad\qquad\qquad$ □

The Jacobi's identity is the measure of the non-associativity of the cross product, for instance $\mathbf{i} \times (\mathbf{i} \times \mathbf{j}) \neq (\mathbf{i} \times \mathbf{i}) \times \mathbf{j}$. Another point is that $\times$ doesn't have a multiplicative unit. It's called a 'product' simply because it distributes over sums.

Borrowing the determinant symbol one symbolically writes

$$\mathbf{u} \times \mathbf{v} = \begin{vmatrix} \mathbf{i} & \mathbf{j} & \mathbf{k} \\ u_1 & u_2 & u_3 \\ v_1 & v_2 & v_3 \end{vmatrix}.$$

Albeit formally nonsense, the notation is useful to bear in mind the above properties (provided we know the properties of determinants, that is). For example, $\mathbf{u} \times \mathbf{v} = \mathbf{0} \iff \operatorname{rk} \left(\begin{smallmatrix} u_1 & u_2 & u_3 \\ v_1 & v_2 & v_3 \end{smallmatrix} \right) \leqslant 1$. Another notation is $\mathbf{u} \wedge \mathbf{v}$ (here $\wedge$ is not the logical connective, but an operation called *wedge product* that allows for higher-dimensional generalisations).

---

[3] We won't prove this fact, but for the record they are i), ii) and vii).

Examples 15.16

i) the canonical basis of $\mathbb{R}^3$ satisfies

$$\mathbf{i} \times \mathbf{j} = \mathbf{k}, \quad \mathbf{j} \times \mathbf{k} = \mathbf{i}, \quad \mathbf{k} \times \mathbf{i} = \mathbf{j}.$$

   More generally $\mathbf{e}_i \times \mathbf{e}_{i+1} = \mathbf{e}_{i+2}$, $\forall i \in \{1, 2, 3\}$ mod. 3, for any ON basis $(\mathbf{e}_j)$.

ii) The equation for an affine line $\mathbf{x} = \mathbf{x}_0 + t\mathbf{v}$, see Example 15.3 i), can be put in the form $\mathbf{x} \times \mathbf{v} = \mathbf{q}$, where $\mathbf{q} = \mathbf{x}_0 \times \mathbf{v}$ is a fixed vector orthogonal to $\mathbf{v}$.

iii) Let $\pi_1 \colon \mathbf{x} \cdot \mathbf{n}_1 = d_1$, $\pi_2 \colon \mathbf{x} \cdot \mathbf{n}_2 = d_2$ be non-parallel planes in $\mathbb{R}^3$: $\mathbf{n}_1 \nparallel \mathbf{n}_2$. The intersection $\pi_1 \cap \pi_2$ defines a linear system with infinitely many solutions depending on 1 free parameter t. Hence they determine a line of equation $\mathbf{x} = \mathbf{x}_0 + t(\mathbf{n}_1 \times \mathbf{n}_2)$, for any $\mathbf{x}_0 \in \pi_1 \cap \pi_2$.

Lemma *Let $\mathbf{u}, \mathbf{v}$ be vectors in $\mathbb{R}^3$ forming an angle θ. Then*

$$\|\mathbf{u} \times \mathbf{v}\|^2 = \|\mathbf{u}\|^2 \|\mathbf{v}\|^2 - (\mathbf{u} \cdot \mathbf{v})^2, \quad \text{that is,} \quad \|\mathbf{u} \times \mathbf{v}\| = \|\mathbf{u}\| \, \|\mathbf{v}\| \sin \theta.$$

Proof Write $\mathbf{u} = (u_i)$, $\mathbf{v} = (v_i)$, $i = 1, 2, 3$, so

$$
\begin{aligned}
\|\mathbf{u} \times \mathbf{v}\|^2 &= (u_2 v_3 - u_3 v_2)^2 + (u_3 v_1 - u_1 v_3)^2 + (u_1 v_2 - u_2 v_1)^2 \\
&= (u_1^2 + u_2^2 + u_3^2)(v_1^2 + v_2^2 + v_3^2) - (u_1 v_1)^2 - (u_2 v_2)^2 - (u_3 v_3)^2 \\
&\quad - 2(u_2 v_2 u_3 v_3 + u_1 v_1 u_3 v_3 + u_1 v_1 u_2 v_2) \\
&= (u_1^2 + u_2^2 + u_3^2)(v_1^2 + v_2^2 + v_3^2) - (u_1 v_1 + u_2 v_2 + u_3 v_3)^2 \\
&= \|\mathbf{u}\|^2 \|\mathbf{v}\|^2 - (\mathbf{u} \cdot \mathbf{v})^2
\end{aligned}
$$

By Proposition 15.15 ii) the claim follows.   □

Corollary 15.17 $\|\mathbf{u} \times \mathbf{v}\|$ *equals the area of the parallelogram with adjacent sides* $\mathbf{u}, \mathbf{v}$.

Proof Exercise.   □

Exercises

i) Fix $\mathbf{u}, \mathbf{v} \in \mathbb{R}^3$ and determine all vectors $\mathbf{x} \in \mathbb{R}^3$ such that $\mathbf{u} \times \mathbf{x} = \mathbf{u} \times \mathbf{v}$. *Hint*: use Corollary 15.17 and the figure below.

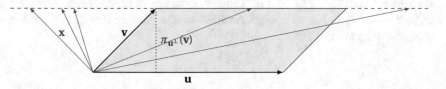

ii) Show the Jacobi identity can be interpreted geometrically as the existence of a triangle's orthocentre (the intersection of the three altitudes).[4]

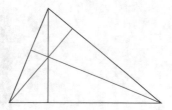

Remark 15.18 (☞ Multilinear Algebra) Let's define the vector space $\bigwedge^2 \mathbb{R}^3$ by prescribing that its elements ('bivectors') are formal objects of type $\mathbf{v} \wedge \mathbf{w}$, where $\mathbf{v}, \mathbf{w} \in \mathbb{R}^3$. Since $\mathbf{i}, \mathbf{j}, \mathbf{k}$ span $\mathbb{R}^3$, then $\mathbf{i} \wedge \mathbf{j}, \ \mathbf{j} \wedge \mathbf{k}, \ \mathbf{k} \wedge \mathbf{i}$ is a basis of $\bigwedge^2 \mathbb{R}^3$. So by mapping $\mathbf{k} \mapsto \mathbf{i} \wedge \mathbf{j}$ (and permuting cyclically) we obtain $\left(\bigwedge^2 \mathbb{R}^3, +, \wedge \right)$ is an anti-commutative, non-associative $\mathbb{R}$-algebra of dimension 3, in agreement with Example 15.16 i).

Let $\mathfrak{so}(3)$ indicate the space of real 3×3 skew-symmetric matrices, and call $E_{ij} \in \mathfrak{so}(3)$ the matrix with entries: 1 in position (i, j), -1 in position (j, i) and 0 elsewhere. Fix an ON basis $(\mathbf{e}_\alpha)$ of $\mathbb{R}^3$. Then the map $\mathbf{e}_i \wedge \mathbf{e}_j \mapsto E_{ij}$ extends linearly to an isomorphism $\bigwedge^2 \mathbb{R}^3 \cong \mathfrak{so}(3)$. Under this identification, the cross product (or $\wedge$) corresponds to the commutator of matrices $XY - YX$ (☞ *Lie theory*).

Exercises Let $\mathbf{u} = (u_i), \mathbf{v} = (v_i), \mathbf{w} = (w_i) \in \mathbb{R}^3$ be arbitrary vectors. Show

i) $(\mathbf{u} \times \mathbf{v}) \cdot \mathbf{w} = \det \begin{pmatrix} u_1 & u_2 & u_3 \\ v_1 & v_2 & v_3 \\ w_1 & w_2 & w_3 \end{pmatrix}$ (called **mixed/triple product**);

ii) $(\mathbf{u} \times \mathbf{v}) \cdot \mathbf{w} = (\mathbf{v} \times \mathbf{w}) \cdot \mathbf{u} = (\mathbf{w} \times \mathbf{u}) \cdot \mathbf{v}$;

iii) $\|(\mathbf{u} \times \mathbf{v}) \cdot \mathbf{w}\|$ equals the volume of the parallelepiped of edges $\mathbf{u}, \mathbf{v}, \mathbf{w}$;

iv) $(\mathbf{u} \times \mathbf{v}) \cdot \mathbf{w} = 0$ if and only if $\mathbf{u}, \mathbf{v}, \mathbf{w}$ are LD.

Remark 15.19 As regards orientations (Remark 15.5), a basis $\{\mathbf{u}, \mathbf{v}, \mathbf{w}\}$ of $\mathbb{R}^3$ is said to be **positive** or **right-handed** when $(\mathbf{u} \times \mathbf{v}) \cdot \mathbf{w} > 0$, and **negative** or **left-handed** in case $(\mathbf{u} \times \mathbf{v}) \cdot \mathbf{w} < 0$.

The commonly called **right-hand rule**, for instance, states that $\{\mathbf{i}, \mathbf{j}, \mathbf{k}\}$, or index finger-middle finger-thumb in Fig. 15.2, is positive.

The different notions of handedness gives rise to the concept of *chirality* we see in hyperplane reflections (☞ *theoretical physics, representation theory, probability*).

---

[4] Kudos and thanks to Letterio Gatto for pointing this out.

Fig. 15.2 Andrea orienting
his basis positively (Photo
courtesy of the author)

15.7 Detour into Physics ☕

The little gem [99] enlightens us as to how the importance of physics in mathematics
(and vice versa) cannot be overestimated. Heeding Wigner's advice we'll take a
break and show how the dot and cross products of $\mathbb{R}^3$ allow to formalise in a simple
and effective way the kinematics of a point-particle (and its application to Kepler's
laws) and classical electromagnetism.

- *Point-particle kinematics.*

The trajectory of a point-particle moving in space is described by a rectifiable
curve $\mathbf{r}\colon I \subseteq \mathbb{R} \to \mathbb{R}^3, t \mapsto \mathbf{r}(t)$. Its velocity is the vector field $\mathbf{v}(t) := \mathbf{r}'(t) =$
$d\mathbf{r}/dt$ with norm (speed) $v(t) := \|\mathbf{r}'(t)\|$. Let's indicate with $s = s(t)$ the arc-length
function, that is the parameter giving unit speed $v(s) = 1$, so that $\mathbf{r}(s) \perp \mathbf{v}(s)$ and
$\mathbf{r}(s) /\!/ \mathbf{v}'(s)$ for any s. Since $v(t) = ds/dt$, the velocity $\mathbf{v}(t) = v(t)\mathbf{v}(s)$ is entirely
tangential to the curve. The acceleration

$$\mathbf{a}(t) := \mathbf{v}'(t) = v'(t)\mathbf{v}(s) + v(t)^2\kappa(t)\mathbf{r}(s)$$

on the other hand, is partly tangential and partly centripetal. The coefficient in the
normal direction $\kappa(t) = \dfrac{\mathbf{v}(t) \times \mathbf{a}(t)}{v(t)^3}$ is the *curvature* of $\mathbf{r}$, and equals $\dfrac{1}{v(t)}\theta'(t)$ in
terms of the tangent vector's turning angle $\theta(t)$.

- *Planetary motion.*

The astronomer Kepler discovered the celebrated three laws (Fig. 15.3) empiri-
cally whilst attempting to complete and correct the Copernican theory:

   I. [*Law of elliptic orbits*] planets describe elliptical orbits, with the sun at one
      focus;

Fig. 15.3 Kepler's laws

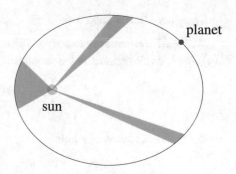

II. [*Area law*] the segment joining a planet to the sun sweeps out equal areas over equal time intervals;

III. [*Period law*] the square of a planet's period of revolution is directly proportional to the cube of the orbit's semi-major axis.

We'll sketch their proof. Call $\mathbf{r} = \mathbf{r}(t)$ the position vector of the planet (with mass m) with respect to the sun (mass $M \gg m$) at time t, and write G for the universal gravitational constant. *Newton's law* $\mathbf{F} = -\dfrac{GmM}{\|\mathbf{r}\|^3}\mathbf{r}$ guarantees the acceleration is radial, i.e. $\mathbf{0} = \mathbf{r} \times \mathbf{a} = (\mathbf{r} \times \mathbf{v})'$, with the consequence that $\mathbf{r} \times \mathbf{v} = \mathbf{c}$ is a constant vector (non-zero since the orbit isn't a straight line). Hence $\mathbf{r} \cdot \mathbf{c} = 0$ implies $\mathbf{r} \in \mathbf{c}^\perp$, which for starters tells the orbit lies on a plane.

Introduce polar coordinates $r := \|\mathbf{r}\|$ and θ on the fixed plane $\mathbf{c}^\perp$, and define the ON frame system $\mathbf{u}_r = \mathbf{r}/r$, $\mathbf{u}_\theta = \mathbf{r}^\perp/r$. Since $\mathbf{v}(t) := \mathbf{r}'(r(t), \theta(t)) = r'\mathbf{u}_r + r\theta'\mathbf{u}_\theta$, we deduce that $\mathbf{c} = r\mathbf{u}_r \times \mathbf{v}(t) = r^2\theta'\mathbf{u}_r \times \mathbf{u}_\theta$ has length $c := \|\mathbf{c}\| = |r^2\theta'|$. By *Green's theorem* the variation of the area $A(t)$ swept by $\mathbf{r}$ equals $|A'(t)| = c/2$. This settles II.

Since $(\mathbf{v} \times \mathbf{c})' = \mathbf{a} \times \mathbf{c} = (GM\mathbf{u}_r)'$, integrating we obtain $\mathbf{v} \times \mathbf{c} = GM\mathbf{u}_r + \mathbf{e}$ for some constant vector $\mathbf{e}$. Now, for simplicity call $e := \|\mathbf{e}\|$, and let ϕ denote the angle between $\mathbf{e}$ and $\mathbf{r}$. Exploiting the mixed product $\mathbf{r} \cdot \mathbf{v} \times \mathbf{c}$ we find equation

$$r = \frac{ed}{e\cos\phi + 1},$$

which describes a conic with eccentricity e, and where $d := c^2/GMe$. Because the orbit is closed we must be looking at an ellipse ($e < 1$) with focus in the sun. This concludes I.

Finally, let a, b be the lengths of the ellipse's semi-axes ($2a$ is then the distance from aphelion to perihelion). Elementary geometric considerations tell $b^2 = a^2(1 - e^2) = aed$, and also $T = \dfrac{2\pi ab}{c}$. Therefore $T^2 = \dfrac{4\pi^2 a^4(1 - e^2)}{GMa(1 - e^2)} = \dfrac{4\pi^2}{GM}a^3$, proving III.

- *Classical electromagnetism.*

We recall preliminarily that if we have a $\mathscr{C}^1$-vector field $\mathbf{F}(\mathbf{r}) = \big(F_1(\mathbf{r}), F_2(\mathbf{r}), F_3(\mathbf{r})\big)$ in $\mathbb{R}^3$, its *divergence* is the function

$$\operatorname{div} \mathbf{F} := \frac{\partial F_1}{\partial x} + \frac{\partial F_2}{\partial y} + \frac{\partial F_3}{\partial z} =: \nabla \cdot \mathbf{F}$$

whilst its *curl* is the vector field

$$\operatorname{curl} \mathbf{F} := \left(\frac{\partial F_2}{\partial z} - \frac{\partial F_3}{\partial y}, \frac{\partial F_3}{\partial x} - \frac{\partial F_1}{\partial z}, \frac{\partial F_1}{\partial y} - \frac{\partial F_2}{\partial x} \right) =: \nabla \times \mathbf{F},$$

where $\nabla = \left(\frac{\partial}{\partial x}, \frac{\partial}{\partial y}, \frac{\partial}{\partial z} \right)$ is thought of as a vector of differential operators. Observe

$$\nabla \times \nabla f = \mathbf{0}, \qquad \nabla \cdot \nabla \times \mathbf{F} = 0,$$

which is reminiscent of formal triple products. Then the operators fit in an exact sequence (cf. Exercises 15.10)

$$\mathscr{C}^1(\mathbb{R}^3, \mathbb{R}) \xrightarrow{\ \operatorname{grad}\ } \mathscr{C}^1(\mathbb{R}^3, \mathbb{R}^3) \xrightarrow{\ \operatorname{curl}\ } \mathscr{C}^1(\mathbb{R}^3, \mathbb{R}^3) \xrightarrow{\ \operatorname{div}\ } \mathscr{C}^1(\mathbb{R}^3, \mathbb{R})$$

called *de Rham complex* (☞ *differential geometry, algebraic topology*).

A charge q at $\mathbf{r} = (x, y, z)$ in $\mathbb{R}^3$, moving with velocity $\mathbf{v}$, is subject to the Lorentz force $q(\mathbf{E} + \mu \mathbf{v} \times \mathbf{H})$, where $\mathbf{E}$ is the electric field, $\mathbf{H}$ the (intensity of the) magnetic field and μ a physical constant (the magnetic permeability of the material. To simplify we'll consider a material that is linear, isotropic, non-dispersive, homogeneous, non-magnetic nor polarisable).

Suppose initially $\mathbf{E} = \mathbf{E}(\mathbf{r}), \mathbf{H} = \mathbf{H}(\mathbf{r})$ do not depend on time t (static case). *Coulomb's law* says $\mathbf{E}(\mathbf{r}) = \dfrac{q}{4\pi\varepsilon} \dfrac{\mathbf{r}}{\|\mathbf{r}\|^3}$, where ε is constant (permittivity). At any point in space, bar the origin where there is a 'singularity', the field is conservative with scalar potential $\phi(\mathbf{r}) = -\dfrac{q}{4\pi\varepsilon r}$. Hence $\mathbf{E}$ is irrotational and solenoidal. The flux through any regular surface $S = \partial R$ enclosing a bounded region $R \subseteq \mathbb{R}^3$ that contains q is $\displaystyle\int_S \mathbf{E} \cdot d\mathbf{S} = \dfrac{q}{\varepsilon}$. When the charge $q = \displaystyle\int_R \rho\, d\mathbf{r}$ is distributed with density ρ, the *divergence theorem* imposes the *Gauß law*

$$\operatorname{div} \mathbf{E} = \frac{\rho}{\varepsilon}. \tag{15.9}$$

Under physically reasonable hypotheses on ρ, one can prove $\mathbf{E}$ is conservative everywhere, so

$$\text{curl } \mathbf{E} = 0. \tag{15.10}$$

In an electrodynamical context magnetic fields are generated by currents, i.e. moving charges. The current flowing through a straight wire is governed by *Ampère's force law*, $I = \oint_C \mathbf{H} \cdot d\mathbf{r}$ being the line integral along a closed curve C going around the wire. More generally, in presence of a density of current $\mathbf{J}$ (replacing the wire), *Stokes' theorem* tells

$$\text{curl } \mathbf{H} = \mathbf{J}. \tag{15.11}$$

The field $\mathbf{H}$ has vector potential $\mathbf{A}(\mathbf{r}) = \dfrac{I}{4\pi} \displaystyle\int \dfrac{\mathbf{J}(s)}{\|\mathbf{r}\|}$, so curl $\mathbf{A} = \mathbf{H}$ implies

$$\text{div } \mathbf{H} = 0. \tag{15.12}$$

Relations (15.9)–(15.12) are the fundamental equations governing static electromagnetic fields. The divergence equations speak about sources/sinks of electric and magnetic force. Formula (15.9) warrants the existence of positive electric charges (positrons) and negative ones (electrons). Hence the field lines of $\mathbf{E}$ can be closed or open.

Equation (15.12) explains there are no solitary magnetic charges, and the field lines of $\mathbf{H}$ must be closed, between a magnetic dipole. Those created by two parallel wires carrying currents of equal intensity and flowing in the same direction are, on a plane orthogonal to the wires, *Cassini ovals* with foci the intersection points with the wires. The same is true for the equipotential lines of $\mathbf{E}$ created by two parallel wires with uniform, identical charge. That said, the existence of magnetic 'monopoles' is highly expected, also (not only) for reasons of formal symmetry between $\mathbf{H}$ and $\mathbf{E}$. The mathematical theory of monopoles and generalisations (after Nahm, Atiyah, Hitchin, Donaldson, Taubes, Manton among others) is the subject of very intense current research work (☞ *gauge theory*).

When $\mathbf{E}(\mathbf{r}, t)$, $\mathbf{H}(\mathbf{r}, t)$ depend on the variable t as well (dynamical case), the two potentials combine as $\mathbf{E} = -\nabla\phi(\mathbf{r}) - \dfrac{\partial \mathbf{A}}{\partial t}$, and Eqs. (15.11), (15.10) are replaced by the *Maxwell–Ampère law* and *Faraday's induction law*:

$$\text{curl } \mathbf{E} = -\mu \frac{\partial \mathbf{H}}{\partial t}, \qquad \text{curl } \mathbf{H} = \mathbf{J} + \varepsilon \frac{\partial \mathbf{E}}{\partial t}. \tag{15.13}$$

Definition The four equations in (15.9), (15.12), (15.13) are called *Maxwell equations* for the classical electromagnetic field, and are invariant under the Poincaré group $O(3, 1) \ltimes \mathbb{R}^4$.

It's know that the Gauß and Faraday laws imply the *continuity equation* $\text{div } \mathbf{J} + \dfrac{\partial \rho}{\partial t} = 0$. The constants ε and μ in vacuum are defined in such a way to obey

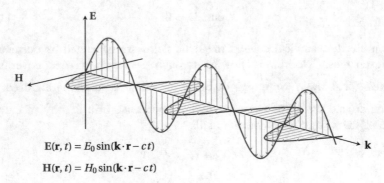

$$E(\mathbf{r}, t) = E_0 \sin(\mathbf{k} \cdot \mathbf{r} - ct)$$

$$H(\mathbf{r}, t) = H_0 \sin(\mathbf{k} \cdot \mathbf{r} - ct)$$

Fig. 15.4 The electromagnetic field $\mathbf{E} + i\mathbf{H}$

the constraint $\varepsilon\mu = c^{-2}$, and thus Maxwell's equations contain the information regarding the speed c of propagation of electromagnetic waves.

In regions with no charges nor currents (such as vacuum), the curl equations give rise to the electromagnetic *wave equation*:

$$\Box(\mathbf{E} + i\mathbf{H}) = 0,$$

where $\Box = \nabla^2 - \dfrac{1}{c^2}\dfrac{\partial^2}{\partial t^2}$ is the Laplace–Beltrami operator on Minkowski spacetime $\mathbb{R}^{3,1}$, and $\nabla^2 = \nabla \cdot \nabla = \dfrac{\partial}{\partial x^2} + \dfrac{\partial}{\partial y^2} + \dfrac{\partial}{\partial z^2}$ is the Laplacian on $\mathbb{R}^3$, see Fig. 15.4 (☞ *Riemannian geometry*).

An impeccable approach—both mathematically and physically—to what we've raced through here, but still accessible and hands-on, can be found in [3, ch.9].

Chapter 16
Orthogonal Operators ☕

This chapter focuses on the structure of linear transformations of vector spaces that preserve the inner product, a.k.a. linear isometries.

There is no major conceptual difficulty, but the reader should be familiar with groups and the linear algebra of matrices.

For the sake of completeness, we recall that invertible matrices form the **general linear group**

$$GL(n, \mathbb{R}) := \{ M \in \text{End } \mathbb{R}^n \mid \det M \neq 0 \}.$$

The subset of matrices with determinant equal to one make up the **special linear group** $SL(n, \mathbb{R}) := \{ M \in GL(n, \mathbb{R}) \mid \det M = 1 \}$. More matrix groups can be found on p. 399.

16.1 Orthogonal Groups

Definition 16.1 A square matrix $A \in \mathbb{R}^{n \times n}$ is said **orthogonal** if it preserves the standard dot product:

$$A\mathbf{v} \cdot A\mathbf{w} = \mathbf{v} \cdot \mathbf{w} \quad \text{for all } \mathbf{v}, \mathbf{w} \in \mathbb{R}^n$$

The set of orthogonal matrices is the **orthogonal group**

$$O(n) := \{ A \in \text{End } (R^n) \mid A\mathbf{v} \cdot A\mathbf{w} = \mathbf{v} \cdot \mathbf{w} \text{ for any } \mathbf{v}, \mathbf{w} \in \mathbb{R}^n \}.$$

Exercises 16.2 Prove the following are equivalent:

i) $A \in O(n)$
ii) $AA^T = A^T A = \mathbb{1}_n$

© The Author(s), under exclusive license to Springer Nature Switzerland AG 2021
S. G. Chiossi, *Essential Mathematics for Undergraduates*,
https://doi.org/10.1007/978-3-030-87174-1_16

iii) the matrix columns $\{A^j\}$ form an ON basis of $\mathbb{R}^n$, as do the rows $\{A_j\}$
iv) A is invertible, with $A^{-1} = A^T$
 v) $\|A\mathbf{v}\| = \|\mathbf{v}\|$ for all $\mathbf{v} \in \mathbb{R}^n$.

Statement v) is telling us that an orthogonal matrix preserves the Euclidean norm. In the language of metric spaces this fact ensures every orthogonal matrix is a **linear isometry** (Sects. 18.1.2 and 19.2) of the normed vector space $\mathbb{R}^n_{\mathrm{Eucl}} := (\mathbb{R}^n, \|\cdot\|_{\mathrm{Eucl}})$.

Example We wish to fill in the ? entries in the matrix $P = (p_{ij}) = \begin{pmatrix} -2/3 & 2/3 & ? \\ 2/3 & ? & ? \\ ? & ? & ? \end{pmatrix}$

so that it becomes orthogonal.

Rows and columns must be ON vectors. Hence the row vector $P_1 = (-\frac{2}{3}, \frac{2}{3}, p_{13})$ has unit norm if $p_{13} = \pm\sqrt{1 - (-\frac{2}{3})^2 - (\frac{2}{3})^2} = \pm\frac{1}{3}$. Let's choose the $+$ sign. In the same argument on column P^1 we pick $p_{31} = \frac{1}{3}$. Now let's find $p_{22} = x$, $p_{23} = y$ so that

$$\begin{cases} 1 = \|P_2\|^2 = (\frac{2}{3})^2 + x^2 + y^2 \\ 0 = P_1 \cdot P_2 = -\frac{4}{9} + \frac{2}{3}x + \frac{1}{3}y \end{cases}.$$

Take $x = \frac{1}{3}$, $y = \frac{2}{3}$, resulting in $\begin{pmatrix} -2/3 & 2/3 & 1/3 \\ 2/3 & 1/3 & 2/3 \\ 1/3 & ? & ? \end{pmatrix}$. The coefficients that are still missing are found from P^2, P^3, P_3 having to be ON to the remaining vectors, whence $p_{23} = \frac{2}{3} = -p_{33}$. Putting together this solution plus the other choices gives back 8 possibilities:

$$\begin{pmatrix} -2/3 & 2/3 & -1/3 \\ 2/3 & 1/3 & -2/3 \\ \pm1/3 & 2/3 & \pm2/3 \end{pmatrix}, \begin{pmatrix} -2/3 & 2/3 & 1/3 \\ 2/3 & 1/3 & 2/3 \\ \pm1/3 & \pm2/3 & \mp2/3 \end{pmatrix}, \begin{pmatrix} -2/3 & 2/3 & 1/3 \\ 2/3 & 11/15 & 2/15 \\ \pm1/3 & \mp2/15 & 14/15 \end{pmatrix}, \begin{pmatrix} -2/3 & 2/3 & 1/3 \\ 2/3 & 11/15 & 2/15 \\ \pm1/3 & \mp2/15 & \mp14/15 \end{pmatrix},$$

and the transposes.

Another important feature of orthogonal matrices is that their spectrum is bounded:

Corollary 16.3 *Let $A \in \mathrm{O}(n)$. Then*

 i) *every eigenvalue is a unit complex number $e^{i\theta}$, for some $\theta \in \mathbb{R}$.*
 ii) $\det A = \pm 1$.

Proof If $\lambda \in \mathbb{C}$ is an eigenvalue, there exists an eigenvector $\mathbf{0} \neq \mathbf{v}$ such that $\|\mathbf{v}\| = \|A\mathbf{v}\| = \|\lambda\mathbf{v}\| = |\lambda|\|\mathbf{v}\|$, so $|\lambda| = 1$.

As for the determinant, it is given by the product $\prod_j e^{i\theta_j} = e^{i\sum_j \theta_j}$ of the eigenvalues and is real, so it must be equal to ± 1. $\qquad\square$

Exercise Prove $A \in \mathrm{O}(n) \implies \det A = \pm 1$ in an alternative way. Show the real eigenvalues can only be ± 1, then use Exercise 16.2 ii).

It's hard to overestimate the importance of the group $O(n)$ in geometry. We'll see in Sect. 18.1.2 that it is the core constituent of the isometry group of $\mathbb{R}^n$, and it resurfaces under many guises throughout the text.

Definition 16.4 The set

$$SO(n) := O(n) \cap SL(n, \mathbb{R}) = \left\{ A \in \mathbb{R}^{n \times n} \mid \|A\mathbf{v}\| = \|\mathbf{v}\|, \ \mathbf{v} \in \mathbb{R}^n; \ \det A = 1 \right\}$$

is called **special orthogonal group** of order n.

Let's analyse the cases $n \leqslant 3$. We discard immediately $n = 1$ since $GL(1, \mathbb{R}) \cong \mathbb{R}^*$, $O(1) = \{\pm 1\}$ and thus $SO(1) = \{1\}$ is trivial.

Lemma 16.5 $A \in SO(2) \iff A = \begin{pmatrix} \cos\alpha & -\sin\alpha \\ \sin\alpha & \cos\alpha \end{pmatrix}$ *for some $\alpha \in \mathbb{R}$.*

Proof Let $\begin{pmatrix} p & q \\ r & s \end{pmatrix}$ be orthogonal, so that its columns are ON. The only vectors that are orthogonal to (p, r) are $\pm(-r, p)$, so the column (q, s) must be one of those. This gives us two possibilities: $\begin{pmatrix} p & -r \\ r & p \end{pmatrix}$ or $\begin{pmatrix} p & r \\ r & -p \end{pmatrix}$. Because $p^2 + r^2 = 1$, the first instance belongs in $SO(2)$, while the second has determinant -1. To finish just recall that $p^2 + r^2 = 1$ implies $p = \cos\alpha, q = \sin\alpha$ for some angle α. $\square$

If we look at columns, the operator $\begin{pmatrix} \cos\alpha & -\sin\alpha \\ \sin\alpha & \cos\alpha \end{pmatrix}$ maps $\begin{pmatrix} 1 \\ 0 \end{pmatrix}$ to $\begin{pmatrix} 0 \\ 1 \end{pmatrix}$ and $\begin{pmatrix} 0 \\ 1 \end{pmatrix}$ to $\begin{pmatrix} -1 \\ 0 \end{pmatrix}$, meaning it is a rotation in $\mathbb{R}^2$ around the origin by angle α. Identifying $\mathbb{R}^2 \cong \mathbb{C}$, it maps the ordered basis $\{1, i\}$ to $\{i, -1\}$. To all effects, we have a 1-1 correspondence

$$SO(2) \cong S^1, \qquad \begin{pmatrix} \cos\alpha & -\sin\alpha \\ \sin\alpha & \cos\alpha \end{pmatrix} \longleftrightarrow e^{i\alpha} \qquad (16.1)$$

cf. Remark 13.4. This agrees with the fact that, excluding $\pm 1_2$, the characteristic polynomial $t^2 - 2(\cos\alpha)t + 1$ gives eigenvalues $e^{\pm i\alpha} \notin \mathbb{R}$, and a non-trivial rotation does not have fixed points in $\mathbb{C}^*$.

Elements in $O(2)$ of the form $\begin{pmatrix} \cos\alpha & \sin\alpha \\ \sin\alpha & -\cos\alpha \end{pmatrix}$ correspond to reflections, cf. (18.1).

Exercise Show $SO(2)$ is an Abelian group.

With an eye to (16.1), what does the product $e^{i\alpha}e^{i\beta} = e^{i(\alpha+\beta)}$ tell us about the trigonometric functions?

The structure of the three-dimensional group $SO(3)$ is not much more complicated. The main differences are:

Exercise Show $SO(3)$ is not Abelian. Which matrices $A \in SO(3)$ commute with every other matrix in the group?

Proposition *Any matrix $R \in SO(3)$, $R \neq 1$, has eigenvalue 1 with multiplicity 1.*

Proof Call $\lambda_1, \lambda_2, \lambda_3$ the eigenvalues. Since the characteristic polynomial is cubic, there is a real eigenvalue, say λ_1, and $\lambda_2 = \overline{\lambda_3}$. We also know $\lambda_1 = \pm 1$ and $|\lambda_2| = 1$ from Corollary 16.3. But $1 = \det R = \lambda_1\lambda_2\lambda_3 = \lambda_1|\lambda_2|^2$, so $\lambda_1 = 1$.

If the multiplicity of λ_1 were larger than one then $1 = \lambda_1 = \lambda_2 = \lambda_3$, implying $R = \mathbb{1}$. ⚡

An alternative proof worth mentioning is based on the following observation:

$$|R - \mathbb{1}| = \left|R - R^{\mathrm{T}}R\right| = \left|(\mathbb{1} - R^{\mathrm{T}})R\right| = \left|(\mathbb{1} - R)^{\mathrm{T}}\right| = |\mathbb{1} - R| = -|R - \mathbb{1}|$$

so that $|R - \mathbb{1}| = 0$. This argument, like the lemma, can be adapted to $\mathrm{SO}(2k + 1)$, $k \in \mathbb{N}$. The crucial point is that the last equality holds because the order of the matrix is odd.

Consequently,

Corollary *For any matrix $R \in \mathrm{SO}(3)$ there is an ON basis of $\mathbb{R}^3$ in which*

$$R = \begin{pmatrix} 1 & 0 & 0 \\ 0 & \cos\theta & -\sin\theta \\ 0 & \sin\theta & \cos\theta \end{pmatrix}$$

for some angle $\theta \in \mathbb{R}$.   □

Proof Let's say $\lambda_1 = 1$ to fix ideas. The freedom in choosing λ_2, λ_3 is limited: if they are real, $\lambda_2 = \lambda_3 = \pm 1$, and

$$R = \mathbb{1} \qquad \text{or} \qquad R = \begin{pmatrix} 1 & 0 & 0 \\ 0 & -1 & 0 \\ 0 & 0 & -1 \end{pmatrix}.$$

If complex, $\lambda_2 = e^{i\theta}$, $\lambda_3 = e^{-i\theta}$, i.e. $R = \begin{pmatrix} 1 & 0 & 0 \\ 0 & \cos\theta & -\sin\theta \\ 0 & \sin\theta & \cos\theta \end{pmatrix}$, $\theta \notin \mathbb{Z}\pi$.   □

By this result $\mathrm{SO}(3)$ is indeed the group of rotations in $\mathbb{R}^3$. The choice of $\lambda_1 = 1$ as first eigenvalue is arbitrary. Position apart, it's the block structure that is preserved. In fact if we permute the basis elements we can achieve three basic forms:

$$\begin{pmatrix} \cos\theta & -\sin\theta & 0 \\ \sin\theta & \cos\theta & 0 \\ 0 & 0 & 1 \end{pmatrix} \qquad \begin{pmatrix} \cos\theta & 0 & -\sin\theta \\ 0 & 1 & 0 \\ \sin\theta & 0 & \cos\theta \end{pmatrix} \qquad \begin{pmatrix} 1 & 0 & 0 \\ 0 & \cos\theta & -\sin\theta \\ 0 & \sin\theta & \cos\theta \end{pmatrix}$$

that respectively correspond to rotations about the z-, y- and x-axis.

The notion of orthogonality in $\mathbb{R}^n$ has a kindred concept in $\mathbb{C}^n$.

Definition Let W be a vector space over the complex numbers $\mathbb{C}$. A **Hermitian** (inner) **product** is a non-negative map $\langle \cdot, \cdot \rangle \colon W \times W \to [0, +\infty)$ satisfying

- $\langle \mathbf{w}, \mathbf{v} \rangle = \overline{\langle \mathbf{v}, \mathbf{w} \rangle}$ (*conjugate-symmetry*)
- $\langle \mathbf{v} + \mathbf{v}', \mathbf{w} \rangle = \langle \mathbf{v}, \mathbf{w} \rangle + \langle \mathbf{v}', \mathbf{w} \rangle,$ $\qquad \langle \alpha\mathbf{v}, \mathbf{w} \rangle = \alpha\langle \mathbf{v}, \mathbf{w} \rangle$
- $\langle \mathbf{v}, \mathbf{v} \rangle = 0 \iff \mathbf{v} = \mathbf{0}$ (*positive definiteness*)

for all $\mathbf{w}, \mathbf{v}, \mathbf{v}' \in W, \alpha \in \mathbb{C}$. The pair $\left(W, \langle \cdot, \cdot \rangle\right)$ is called **Hermitian** vector space.

In particular, $\langle \cdot, \cdot \rangle$ is $\mathbb{C}$-linear in the first variable but $\mathbb{C}$-antilinear in the second: $\langle \mathbf{v}, \beta \mathbf{w} \rangle = \overline{\beta} \langle \mathbf{v}, \mathbf{w} \rangle$. In analogy to the Euclidean situation we can associate with a Hermitian inner product a norm $\|\mathbf{v}\|_{\text{Herm}} := \sqrt{\langle \mathbf{v}, \mathbf{v} \rangle}$. The sovereign example is $\mathbb{C}^n$ with the standard Hermitian product

$$\langle z, w \rangle := \sum_{i=1}^{n} z_i \overline{w_i}. \tag{16.2}$$

The associated norm is the Euclidean norm $\sum_{i=1}^{n} |z_i|^2$ after we identify $\mathbb{C}^n = \mathbb{R}^{2n}$ (☞ *complex geometry*). Vectors $v_1, v_2, \ldots \in W$ will be called unitary when $\langle v_i, v_j \rangle = \delta_{ij}$, in analogy to ON vectors. Definitions 16.1 and 16.4 also have corresponding objects in this context:

Definition The groups

$$\mathrm{U}(n) := \left\{ M \in \mathbb{C}^{n \times n} \mid M \overline{M}^{\mathrm{T}} = 1 \right\}$$

$$\mathrm{SU}(n) := \mathrm{U}(n) \cap \mathrm{SL}(n, \mathbb{C})$$

are called **unitary group** and **special unitary group**.

In particular

$$\mathrm{U}(1) \cong \left\{ e^{i\theta} \in \mathbb{C} \mid \theta \in \mathbb{R} \right\}$$

gets identified to the multiplicative group S^1 of unit complex numbers 1, and hence with $\mathrm{SO}(2)$ as well.

Exercises Prove that

  i) the eigenvalues of a matrix in $\mathrm{U}(n)$ have modulus 1;
 ii) $\det \colon \mathrm{U}(n) \to \mathrm{U}(1)$ is a group homomorphism.
 ii) $T^n := \left\{ \operatorname{diag}(e^{i\theta_1}, \ldots, e^{i\theta_n}) \mid e^{i\theta_j} \in \mathrm{U}(1), j = 1, \ldots, n \right\}$ is a subgroup of $\mathrm{U}(n)$, called **n-torus**, isomorphic to $\mathrm{U}(1)^n$.

And since (16.2) is the complex analogue of the dot product, in the same manner

$$\mathrm{U}(n) = \left\{ Q \in \mathrm{GL}(n, \mathbb{C}) \mid \langle Q\mathbf{v}, Q\mathbf{w} \rangle = \langle \mathbf{v}, \mathbf{w} \rangle, \ \forall \mathbf{v}, \mathbf{w} \in \mathbb{C}^n \right\}$$

are isometries of $\left(\mathbb{C}^n, \langle \cdot, \cdot \rangle\right)$, among which are the complex rotations $\mathrm{SU}(n)$.

Example 16.6 Show that for $n = 2$

$$SU(2) = \left\{ \begin{pmatrix} z & w \\ -\overline{w} & \overline{z} \end{pmatrix} \in \text{End } \mathbb{C}^2 \ \middle| \ |z|^2 + |w|^2 = 1 \right\}.$$

In this way SU(2) is determined by pairs $(z = a + bi, w = c + di) \in \mathbb{C}^2 \cong \mathbb{R}^4$ subject to the constraint $1 = |z|^2 + |w|^2 = a^2 + b^2 + c^2 + d^2$. The conclusion is that SU(2) can be viewed as the sphere S^3 of unit vectors (a, b, c, d) in $\mathbb{R}^4$.

16.2   The Spectral Theorem ☕

This section address the interaction between eigenvalues and bilinear forms, and proves one of most useful theorems in mathematics. We shall fix the ground field $\mathbb{K} = \mathbb{C}$ and consider Hermitian vector spaces $\left(W, \langle \cdot, \cdot \rangle\right)$ of finite dimension n. We remind that by identifying $W = \mathbb{C}^n$ we can take the standard Hermitian product $\langle \mathbf{z}, \mathbf{w} \rangle := \mathbf{z}^{\mathrm{T}} \overline{\mathbf{w}}$, and $\mathbf{v}$ is a unit vector when $\|\mathbf{v}\|_{\text{Herm}} := \sqrt{\langle \mathbf{v}, \mathbf{v} \rangle} = 1$.

It's an old habit to call maps $f : \left(W, \langle \cdot, \cdot \rangle\right) \to \left(W, \langle \cdot, \cdot \rangle\right)$ operators, probably because of their employ in ☞ operator theory.

Proposition *Let f be a linear operator on the Hermitian vector space $\left(W, \langle \cdot, \cdot \rangle\right)$ with $\dim W = n < \infty$. There exists a unique linear operator $f^* : W \to W$ such that*

$$\langle f\mathbf{v}, \mathbf{w} \rangle = \langle \mathbf{v}, f^*\mathbf{w} \rangle, \qquad \forall \mathbf{v}, \mathbf{w} \in W. \tag{16.3}$$

Proof (Existence) Since $n < \infty$ we fix a unitary basis $(\mathbf{v}_i)$ and a matrix M for f in that basis. Defining

$$M^* := \overline{M}^{\mathrm{T}}$$

we obtain $\langle M\mathbf{v}, \mathbf{w} \rangle = (M\mathbf{v})^{\mathrm{T}}\overline{\mathbf{w}} = \left(\mathbf{v}^{\mathrm{T}}M^{\mathrm{T}}\right)\overline{\mathbf{w}} = \mathbf{v}^{\mathrm{T}}\left(M^{\mathrm{T}}\overline{\mathbf{w}}\right) = \mathbf{v}^{\mathrm{T}}\overline{M^*\mathbf{w}} = \langle \mathbf{v}, M^*\mathbf{w} \rangle$. The operator f^* associated with M^* satisfies (16.3).

(Uniqueness) If f^*, g^* fulfil the requirements then $\langle \mathbf{v}, f^*\mathbf{w} - g^*\mathbf{w} \rangle = 0$ for every $\mathbf{v}, \mathbf{w}$. In particular $\| f^*\mathbf{w} - g^*\mathbf{w} \| = 0$, meaning $f^* = g^*$. $\square$

Exercise For any functions f, g and $\lambda \in \mathbb{C}$, show that the map $* : f \mapsto f^*$

i) is a $\mathbb{C}$-antilinear involution: $(f^*)^* = f$, $(f + g)^* = f^* + g^*$, $(\lambda f)^* = \overline{\lambda} f^*$,
ii) satisfies: $(fg)^* = g^* f^*$, $\mathbb{1}^* = \mathbb{1}$.

Definition Take $f \in \text{End}_\mathbb{C} W$ with matrix M and suppose $\dim W < \infty$. The operator $f^* \in \text{End}_\mathbb{C} W$ satisfying (16.3) is the **adjoint** of f, and its matrix M^* is the **adjoint** of M.

Furthermore, f and M are called

f	Condition	M	Condition
Self-adjoint	$f = f^*$	Hermitian	$M = M^*$
Skew-adjoint	$f = -f^*$	Skew-Hermitian	$M = -M^*$
Unitary	$f^* = f^{-1}$	Unitary	$M^* = M^{-1}$
Normal	$ff^* = f^*f$	Normal	$MM^* = M^*M$

Exercises 16.7 Let M be a complex $n \times n$ matrix. Prove

i) M real unitary $\iff$ M orthogonal.
ii) M Hermitian or skew-Hermitian or unitary $\implies$ M normal.
iii) M is unitary if and only if its columns form a unitary basis.
iv) Decompose a complex matrix $Z = (z_{ij})$ in the sum of real matrices

$$Z = S + iA$$
$$z_{ij} = \operatorname{Re} z_{ij} + i \operatorname{Im} z_{ij}.$$

Show: Z Hermitian $\implies$ S symmetric and A skew-symmetric.
Conclude that: real Hermitian $\iff$ symmetric.

Spectral Theorem Let $(W, \langle \cdot, \cdot \rangle)$ be a Hermitian space of finite dimension. An operator $f: W \to W$ is normal (it commutes with its adjoint) if and only if there exists a unitary basis of W made by eigenvectors of f.

The equivalent matrix version states:

Theorem 16.8 (Spectral Theorem) $M \in \operatorname{End}(\mathbb{C}^n)$ *is diagonalisable by unitary matrices*

$$M = QDQ^{-1}, \qquad Q \in \mathrm{U}(n), \quad D \text{ diagonal}$$

if and only if M is normal.

Proof ($\implies$) To prove the necessary condition one uses the properties of transposition and the fact that diagonal matrices lie in the centre of $\operatorname{End}(\mathbb{C}^n)$. If $M = QDQ^{-1}$ with $Q^{-1} = \overline{Q}^{\mathsf{T}}$ and D is diagonal, then

$$MM^* = QD \underbrace{Q^{-1}\overline{Q^{-1}}^{\mathsf{T}}}_{\mathbb{1}} \overline{D}\,\overline{Q}^{\mathsf{T}} = QD\overline{D}Q^{-1} = Q\overline{D}DQ^{-1}$$

$$\tag{16.4}$$

$$= \overline{Q^{-1}}^{\mathsf{T}}\overline{D}^{\mathsf{T}} DQ^{-1} = \overline{Q^{-1}}^{\mathsf{T}}\overline{D}^{\mathsf{T}} \underbrace{\overline{Q}^{\mathsf{T}} Q}_{\mathbb{1}} DQ^{-1} = (\overline{QDQ^{-1}})^{\mathsf{T}} QDQ^{-1} = M^*M.$$

($\impliedby$) More interesting is the converse, which we'll prove in the more flexible 'operator' notation by using a cascade of short lemmas.

Lemma 16.9 *If $f \in \operatorname{End}_{\mathbb{C}} W$ is normal,*

(1) $\langle f\mathbf{v}, f\mathbf{w} \rangle = \langle f^*\mathbf{v}, f^*\mathbf{w} \rangle$ *for all $\mathbf{v}, \mathbf{w} \in V$.*
(2) $\operatorname{Ker} f = \operatorname{Ker} f^*$.
(3) $\operatorname{Ker}(f - \lambda\mathbb{1}) = \operatorname{Ker}(f^* - \overline{\lambda}\mathbb{1})$.

Proof

(1) $\langle f\mathbf{v}, f\mathbf{w}\rangle = \langle \mathbf{v}, f^*f\mathbf{w}\rangle = \langle \mathbf{v}, ff^*\mathbf{w}\rangle = \langle \mathbf{v}, (f^*)^*f^*\mathbf{w}\rangle = \langle f^*\mathbf{v}, f^*\mathbf{w}\rangle$.

(2) By part (1), $\|f\mathbf{v}\| = \|f^*\mathbf{v}\|$, so $f\mathbf{v} = \mathbf{0} \iff 0 = \|f\mathbf{v}\| = \|f^*\mathbf{v}\| \iff f^*\mathbf{v} = \mathbf{0}$.

(3) The identity $\mathbb{1}$ commutes with every operator, so $(f - \lambda\mathbb{1})^* = f^* - \bar{\lambda}\mathbb{1}$. The claim now follows from (2).
$\square$

Exercise Prove: $\|f\mathbf{v}\| = \|f^*\mathbf{v}\| \iff f$ is normal $\iff f^*$ is normal.

Finally to the key point:

Lemma 16.10 *If $f \in \mathrm{End}_{\mathbb{C}} W$ is normal, there exists a basis of W made of eigenvectors of f.*

Proof Let $\sigma(f) = \{\lambda_1, \ldots, \lambda_k\}$ be the spectrum, with the λ_i distinct, and call $W_{\lambda_i} = \{\mathbf{w} \in W \mid f\mathbf{w} = \lambda_i\mathbf{w}\} \subseteq W$ the eigenspaces. Set $U = \bigoplus_{i=1}^{k} W_{\lambda_i}$. We claim $U = W$. For that, consider

$$U^{\perp} = \{\mathbf{v} \in W \mid \langle \mathbf{v}, \mathbf{w}\rangle = 0 \text{ for every } \mathbf{w} \text{ such that there exists } i \text{ with } f\mathbf{w} = \lambda_i\mathbf{w}\}.$$

Take $\mathbf{v} \in U^{\perp}$ and $\mathbf{w} \in W_i$, so that $\langle f\mathbf{v}, \mathbf{w}\rangle = \langle \mathbf{v}, f^*\mathbf{w}\rangle = \langle \mathbf{v}, \bar{\lambda}_i\mathbf{w}\rangle = \lambda_i\langle \mathbf{v}, \mathbf{w}\rangle = 0$. Then $U^{\perp}$ is f-invariant, and the restriction $f\big|_{U^{\perp}} : U^{\perp} \to U^{\perp}$ is a linear operator.

Suppose, by contradiction. $U \neq W$. Assume in other words that $U^{\perp}$ were non-empty, so there would be an eigenvector $\mathbf{v} \in U^{\perp}$ of $f\big|_{U^{\perp}}$ ($\mathbb{C}$ is algebraically closed). This would force $f\big|_{U^{\perp}}\mathbf{v} = \lambda\mathbf{v}$, and therefore $\lambda = \lambda_i$ for some i. But then $\mathbf{v} \in U^{\perp} \cap U = \{\mathbf{0}\}$ *≸*.
$\square$

Lemma 16.11 *If f is normal and $\lambda \neq \mu$ are eigenvalues, then W_λ and W_μ are orthogonal spaces (with respect to $\langle \cdot, \cdot\rangle$).*

Proof Take $\mathbf{v} \in V_\lambda$ and $\mathbf{w} \in V_\mu$. Then

$$\lambda\langle \mathbf{v}, \mathbf{w}\rangle = \langle \lambda\mathbf{v}, \mathbf{w}\rangle = \langle f\mathbf{v}, \mathbf{w}\rangle = \langle \mathbf{v}, f^*\mathbf{w}\rangle = \langle \mathbf{v}, \bar{\mu}\mathbf{w}\rangle = \mu\langle \mathbf{v}, \mathbf{w}\rangle.$$

We conclude $(\lambda - \mu)\langle \mathbf{v}, \mathbf{w}\rangle = 0 \implies \langle \mathbf{v}, \mathbf{w}\rangle = 0$.
$\square$

At last we can finish the proof of the spectral theorem. Choose in every eigenspace of f a unitary basis. The union of these bases, by Lemma 16.10, is a basis of W. It is further unitary by Lemma 16.11.
$\square$

The first consequences are these special results:

Corollary *Let $\left(W, \langle \cdot, \cdot\rangle\right)$ be a finite-dimensional Hermitian vector space and $f \in \mathrm{End}_{\mathbb{C}} W$. Then*

  i) *f is self-adjoint $\iff f$ is normal and its spectrum $\sigma(f)$ is real.*
 ii) *f is skew-adjoint $\iff f$ is normal and $\sigma(f) \subseteq i\mathbb{R}$.*
iii) *f is unitary $\iff f$ is normal and $\sigma(f) \subseteq \mathrm{U}(1)$.*

Proof Exercise.   □

Exercises Let $(W, \langle \cdot, \cdot \rangle)$ be a Hermitian vector space with dim $W < \infty$, and take a normal endomorphism $f: W \to W$. Show that

i) if $U \subseteq W$ is f-invariant then also $U^\perp$ is f-invariant;
ii) the restriction $f\big|_U : U \to U$ is normal.

Before we discuss the real version of Theorem 16.8 let's observe the following.

Proposition *The spectrum $\sigma(S)$ of a real symmetric matrix S is real.*

Proof Suppose $\lambda \in \mathbb{C}$ is a root of the characteristic polynomial $p_S(t) \in \mathbb{R}[t]$ of S. If we view S as a complex matrix, it has a complex λ-eigenvector $\mathbf{v}$. Moreover, $\overline{S} = S = S^T = S^*$, so

$$\lambda \overline{\mathbf{v}}^T \cdot \mathbf{v} = \overline{\mathbf{v}}^T \cdot (\lambda \mathbf{v}) = \overline{\mathbf{v}}^T \cdot (S\mathbf{v}) = S^* \overline{\mathbf{v}}^T \cdot \mathbf{v} = (\overline{S\mathbf{v}})^T \cdot \mathbf{v} = \overline{\lambda} \overline{\mathbf{v}}^T \cdot \mathbf{v}.$$

But this forces $\lambda = \overline{\lambda}$ (as $\overline{\mathbf{v}}^T \cdot \mathbf{v} = \|\mathbf{v}\|^2 > 0$).   □

The next exercise cements the relationship between symmetric and orthogonal matrices.

Exercises

i) Consider $A = \begin{pmatrix} -1 & 1 & 1 \\ 1 & -1 & 1 \\ 1 & 1 & -1 \end{pmatrix}$ and its eigenvectors $\mathbf{v}_1 = (1, 1, 1)$, $\mathbf{v}_2 = (1, -1, 0)$
    and $\mathbf{v}_3 = (1, 0, -1)$ in $V = \mathbb{R}^3$, with eigenvalues 0 (simple) and -3 (double).
    Verify that $\mathbf{v}_1 \perp \mathbf{v}_2, \mathbf{v}_3$ but $\mathbf{v}_2 \cdot \mathbf{v}_3 \neq 0$.
    Show there exists a $\mathbf{v}_3' \in V_{-3}$ orthogonal to $\mathbf{v}_2$.
    Normalise $\mathbf{v}_1, \mathbf{v}_2, \mathbf{v}_3'$ to obtain an orthogonal matrix P that diagonalises A.
    Is it possible to choose $\mathbf{v}_3'$ so that det $P = 1$?
ii) Adapt the proof of Lemma 16.11 to show that the eigenspaces of a real symmetric matrix are orthogonal.

The relationship between symmetric and orthogonal matrices is fleshed out by the

Theorem (Real Spectral Theorem) $A \in \mathrm{End}\,(\mathbb{R}^n)$ *is diagonalisable by orthogonal matrices*

$$A = PDP^{-1}, \qquad P \in O(n), \quad D \text{ diagonal}$$

if and only if A is symmetric.

Proof ($\Longrightarrow$) Suppose $A = PDP^{-1}$ with $P = (P^{-1})^T$ and D diagonal. Then $A^T = (PDP^{-1})^T = (P^{-1})^T D^T P^T = PDP^{-1} = A$. (This is just a reformulation of (16.4).)

($\Longleftarrow$) Conversely, assume A symmetric. Seen as complex matrix, A is Hermitian, and by the spectral theorem it is diagonalisable with real spectrum. The geometric multiplicity of an eigenvalue λ doesn't change if we compute it over $\mathbb{R}$ or $\mathbb{C}$ (it only depends on $\mathrm{rk}(A - \lambda\mathbb{1})$, not on the field). Since the geometric and algebraic multiplicities are equal, A is diagonalisable over $\mathbb{R}$. To finish it's enough to note that the standard Hermitian product, restricted to $\mathbb{R}^n \times \mathbb{R}^n$, is the dot product (cf. Exercises 16.7). □

We end the chapter with a few considerations on positive definite, symmetric matrices. For conciseness we'll indicate the space of such by

$$\odot^+(n) := \{S \in \mathrm{End}\,(\mathbb{R}^n) \mid S = S^{\mathrm{T}}, \ S\mathbf{v} > 0, \mathbf{v} \neq \mathbf{0}\}.$$

Exercises Prove that

i) If A is invertible, then $A^{\mathrm{T}}A$ is symmetric and positive definite;
ii) if S is symmetric and positive definite then so are S^2, S^{-1}.

Proposition *For any matrix $T \in \odot^+(n)$ there exists a unique matrix $S \in \odot^+(n)$ such that $S^2 = T$ (in which case one writes $S = \sqrt{T}$).*

Proof We know $D = P^{-1}TP$ is diagonal, with $\sigma(D) = \sigma(T) = (\lambda_1, \ldots, \lambda_k)$ (in this order). Define $\sqrt{D}$ to be the diagonal matrix with ordered entries $\sqrt{\lambda_1}, \ldots,$ $\sqrt{\lambda_k}$. (The notation suggests $\sqrt{D}$ is the unique matrix whose square is D.) Then $S^2 = (P^{-1}\sqrt{D}P)^2 = T$. Regarding uniqueness, if there exists $C \in \odot^+(n)$ with $C^2 = T = S^2$ then C, B must be equal on each eigenspace (the λ-eigenspace of C is the λ^2-eigenspace of T). But C, S are diagonalisable, so they coincide. Therefore $\sqrt{T} = P^{-1}\sqrt{D}P$. □

Using the spectral theorem we can now show that symmetric matrices and orthogonal matrices are the constituents of any invertible real matrix:

Theorem (Polar Decomposition) *For any $A \in \mathrm{GL}(n, \mathbb{R})$ there exist unique matrices $P \in \mathrm{O}(n)$ and $S \in \odot^+(n)$ such that $A = PS$. In other words*

$$\mathrm{GL}(n, \mathbb{R}) = \mathrm{O}(n) \times \odot^+(n).$$

Proof (Uniqueness) Assume $A = PS$ for certain matrices $P \in \mathrm{O}(n)$, $S \in \odot^+(n)$. Then $A^{\mathrm{T}}A = SP^{\mathrm{T}}PS = S^2$. Consequently $S = \sqrt{A^{\mathrm{T}}A}$ exists and is unique, and $P = AS^{-1}$ alike.

(Existence) Take $S = \sqrt{A^{\mathrm{T}}A}$ (symmetric, positive definite) and define $P = AS^{-1}$. The computation $P^{\mathrm{T}}P = (S^{-1})^{\mathrm{T}}A^{\mathrm{T}}AS^{-1} = S^{-1}S^2S^{-1} = \mathbb{1}$ demonstrates P is orthogonal. □

There exists a polar decomposition for complex matrices, the simplest example being 1×1 matrices: $\mathrm{GL}(1, \mathbb{C}) = \mathrm{U}(1) \times \mathbb{R}^+$. This is but the polar/exponential form of non-zero complex numbers $\rho e^{i\theta}$, as described in Theorem 17.6. More generally we have $\mathrm{GL}(n, \mathbb{C}) = \mathrm{U}(n) \times \mathrm{Herm}^+(n)$, where $\mathrm{Herm}^+(n)$ denotes the vector space of Hermitian matrices with positive diagonal entries. We have seen something similar in (15.5), which ☞ *Lie theory* says can be viewed as the infinitesimal counterpart to the polar decomposition.

There's no shortage of decomposition formulas for real square matrices M. Some examples:

- *LU decomposition*: $M = LU$ where L, U are triangular (lower and upper triangular, respectively). This is just the matrix version of the Gauß elimination of variables in linear systems. Its generalisation is called *Bruhat decomposition*.
- *PT decomposition*: $M = PT$ with P orthogonal, T upper triangular. It's another name for the well-known Gram–Schmidt process, see p. 362. Its generalisation is the *Iwasawa decomposition*.
- *Singular-value decomposition (SVD)*: $M = PDR$, where P, R are orthogonal, D is diagonal and non-negative. The SVD is intimately related to diagonalisation, and its generalisation is called *Cartan decomposition*.

It's also worth mentioning the *Cholesky decomposition*: $M \in \odot^{+}(n)$ can be written as TT^{T} for some lower triangular T with positive diagonal. Its relevance resides in its efficiency for solving linear systems by numerical methods or, in general, *Monte Carlo methods* (☞ *numerical analysis*). The aforementioned generalisations belong to the study of ☞ *algebraic groups* and *Lie groups*.

We have spent some time attempting to clarify the relationship between orthogonal matrices and positive definite, symmetric matrices. Even more pivotal is the love affair between $\mathrm{O}(n)$ and skew-symmetric matrices

$$\mathfrak{so}(n) := \left\{ A \in \mathrm{End}\,(\mathbb{R}^n) \mid A = -A^{\mathrm{T}} \right\}.$$

In ☞ *Lie theory*, in fact, one proves that the infinitesimal variation ('derivative') of an orthogonal matrix gives rise to a skew-symmetric matrix. Vice versa, the exponential (an 'integral' of sorts) of a skew matrix returns an orthogonal matrix:

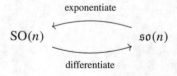

$$\mathrm{SO}(n) \underset{\text{differentiate}}{\overset{\text{exponentiate}}{\rightleftarrows}} \mathfrak{so}(n)$$

Chapter 17
Actions and Representations ☕

Pre-requisites: familiarity with the algebra of groups and matrices.

Although it is possible to discuss actions of groups G on sets X, the theory's most spectacular manifestation occurs in geometric contexts, where X is a topological manifold (real, of class $\mathscr{C}^r$, complex etc.) and G is a group of transformations in the same category (homeo-, diffeo-, biholo-morphisms etc.). Core references for the theory of representations are [26, 34, 51].

17.1 Group Actions

Definition 17.1 A group G **acts on a set** $X \neq \varnothing$ if there exists a mapping

$$\rho: G \times X \longrightarrow X$$
$$(g, x) \longmapsto \rho(g, x)$$

called **action**, such that

$$\rho(e, x) = x, \qquad \rho\big(g_1, \rho(g_2, x)\big) = \rho(g_1 g_2, x) \qquad (17.1)$$

for any $x \in X$, $g_1, g_2 \in G$, where e is the group's neutral element.

Notation: the symbol $G \curvearrowright X$ is employed to indicate an action without writing it explicitly. The shorthand notation we prefer is multiplicative-like: $gx := \rho(g, x)$. In this way requisites (17.1) become the more intuitive: $ex = x$, $g_1(g_2 x) = (g_1 g_2)x$.

Example Let X be a finite set of cardinality card $X = n$. The symmetric group $G = \mathfrak{S}_n$ acts on X by permuting elements, see Definition 14.1:

© The Author(s), under exclusive license to Springer Nature Switzerland AG 2021
S. G. Chiossi, *Essential Mathematics for Undergraduates*,
https://doi.org/10.1007/978-3-030-87174-1_17

$$\rho_g = \begin{pmatrix} 1 & 2 & \dots & n \\ g(1) & g(2) & \dots & g(n) \end{pmatrix}.$$

Using the exponential law (7.17) we may view an action as a function $G \to X^X, g \mapsto \rho_g$ satisfying (17.1), or a family of maps $\{\rho_g : X \to X\}_{g \in G}$

$$\rho_g(x) := \rho(g, x),$$

parametrised by the group G, and called **translations**. The name comes from the following situation: take $X = \mathbb{R}^n$ and consider the translations $T_r : x \mapsto x + r$ for $r \in \mathbb{R}^n$. These are parametrised by vectors $r \in \mathbb{R}^n$, so $r \mapsto T_r$ defines an action of $\mathbb{R}^n$ (seen as group of transformations) on $\mathbb{R}^n$ (seen as space).

Taking in account

$$\rho_e = \mathbb{1}_X, \qquad \rho_{g_1 g_2} = \rho_{g_1} \circ \rho_{g_2}, \qquad \rho_g^{-1} = \rho_{g^{-1}},$$

we have

Corollary *An action $\rho : G \curvearrowright X$ is a homomorphism from G to the symmetric group $\mathfrak{S}(X)$ of bijections of X.*

Taking $X = G$, *Cayley's Theorem* 14.2 says translations $\{\rho_g\}_{g \in G}$ form a group isomorphic to G itself, and hence any abstract group G becomes a subgroup of $\mathfrak{S}(G)$.

A standard example is a group of matrices $G \subseteq \mathrm{End}\,(\mathbb{R}^n)$ acting by matrix multiplication, where the action is called **standard**:

$$G \times \mathbb{R}^n \longrightarrow \mathbb{R}^n, \qquad (M, \mathbf{v}) \mapsto M\mathbf{v}.$$

Examples 17.2 A grand source of examples comes from topology: if X is a topological space, every subgroup $G \subseteq \mathrm{Homeo}(X)$ of homeomorphisms acts on X by $\rho_g(x) = g(x)$. For instance

(a) the translations $\{T_r : x \mapsto x + r \mid r \in \mathbb{R}^n\}$ act on $\mathbb{R}^n$;
(b) the rotations $O(n)$ and $SO(n)$ act on $\mathbb{R}^n$;
(c) $U(n) \curvearrowright \mathbb{C}^n$ and $SU(n) \curvearrowright \mathbb{C}^n$;
(d) $\{\rho_\theta : z \mapsto e^{i\theta} z \mid \theta \in \mathbb{R}\} \cong U(1)$ acts on $\mathbb{C}$, on the unit disc $\mathbb{D} = \{z \in \mathbb{C} \mid |z| < 1\}$ and also on itself, thought of as the circle $\partial \mathbb{D}$ bounding the disc.

It's known that any abstract group G can be seen as a transformation group acting on a suitable topological space $X = K(G, n)$ (☞ algebraic topology).

Definition 17.3 An action $G \curvearrowright X$ is called **faithful** (or **effective**) if $g \mapsto \rho_g$ is a 1-1 homomorphism. In other words no transformation $g \in G \setminus \{e\}$ fixes every points of X.

An element $x \in X$ is a fixed point of $g \in G$ if $gx = x$ (written $x \in \text{Fix}(g)$), and a **fixed point** of the action if it belongs to

$$\text{Fix}(G) := \bigcap_{g \in G} \text{Fix}(g).$$

An action $G \curvearrowright X$ is **free** (of fixed points) if no $g \in G \setminus \{e\}$ has fixed points:

$$\exists x \in X \colon \rho_g x = x \implies \rho_g = \mathbb{1}_X.$$

Equivalently, under a free action each ρ_g, except ρ_e, transforms every point.

Exercise Show that a free action is faithful. Consider Examples 17.2 and decide which actions are faithful and which free.

Definition Let $G \curvearrowright X$ be an action, and introduce the equivalence relation

$$x \sim y \iff \exists g \in G \colon x = gy.$$

The cosets are denoted Gx, or $\mathcal{O}_x$ when G is clear, and are called **G-orbits**. The quotient X/G, whose elements are the various orbits, is the **orbit space**.

Theorem 2.8 ensures $X = \bigsqcup_{x \in X} \mathcal{O}_x$ is a disjoint union, because distinct orbits do not meet: $\mathcal{O}_x \cap \mathcal{O}_y \neq \emptyset \iff \mathcal{O}_x = \mathcal{O}_y$, and $\mathcal{O}_x \cap \mathcal{O}_y = \emptyset \iff x \notin \mathcal{O}_y$. The partition into orbits induces the quotient map $X \to X/G$, $x \mapsto \mathcal{O}_x$.

Examples 17.4

i) The additive group $G = \mathbb{R}$ acts on $\mathbb{R}^2$ under $G \times \mathbb{R}^2 \to \mathbb{R}^2$, $(t, (x, y)) \mapsto (x + ty, y)$. The orbits of a point (x, y) are of two kinds:
   if $y \neq 0$ then $\mathcal{O}_{(x,y)} = \{p + \lambda(1, 0) \mid \lambda \in \mathbb{R}\}$ is the horizontal line through p;
   if $y = 0$ then $\mathcal{O}_{(x,0)} = \{(x, 0)\}$ is a point on the x-axis.

ii) The group of rotations in $\mathbb{R}^3$ around the z-axis

$$G = \left\{ \begin{pmatrix} \cos\alpha & \sin\alpha & 0 \\ -\sin\alpha & \cos\alpha & 0 \\ 0 & 0 & 1 \end{pmatrix}, \ \alpha \in \mathbb{R} \right\} \cong SO(2)$$

acts on the sphere $S^2 = \left\{ \begin{pmatrix} x_1 \\ x_2 \\ x_3 \end{pmatrix} \mid x_1^2 + x_2^2 + x_3^2 = 1 \right\} \subseteq \mathbb{R}^3$ in the standard way. The action is faithful but not free, since the points $(0, 0, \pm 1)$ are fixed by every element of G. The typical orbit is a parallel $\mathcal{O}_m = \{x_1^2 + x_2^2 = 1 - m^2, \ x_3 = m \in (-1, 1)\}$ obtained slicing the sphere with a horizontal plane. There are two 'exceptional' orbits corresponding to $m = \pm 1$, if we think of the two poles as 'singular' parallels of null radius. As $S^2 =$

$$\bigsqcup_{m\in[-1,1]} \mathcal{O}_m,$$ we may describe the orbit space by the closed interval $[-1,1]$. The projection $\mu\colon S^2 \to [-1,1]$, $\mu(\mathcal{O}_m) = m$ is a simple instance of a *moment map* (☞ *symplectic geometry*).

iii) The standard action of $G = \mathrm{GL}(n,\mathbb{R})$ on $\mathbb{R}^n$ has two types of orbit: $\mathcal{O}_x = \mathbb{R}^n \setminus \{0\}$ for $x \neq 0$, and $\mathcal{O}_0 = \{0\}$. The action, being linear, induces an action of $\mathrm{GL}(n,\mathbb{R})$ on the projective space $\mathbb{RP}^n$. And since the centre $Z(\mathrm{GL}(n,\mathbb{R})) = \{\lambda \mathbb{1}\}_{\lambda\in\mathbb{R}^*} \cong \mathbb{R}^*$ acts trivially, we obtain an action of $\mathrm{PGL}(n,\mathbb{R}) := \mathrm{GL}(n,\mathbb{R})/\mathbb{R}^*$ on $\mathbb{RP}^n$.

Definition The set X is said **homogeneous** under the action of G when, for any $x,y \in X$, there exists a $g \in G$ such that $y = gx$. An action $G \curvearrowright X$ is **transitive** if there is only one orbit: $\exists x \in X$ such that $X = \mathcal{O}_x$.

Proposition $G \curvearrowright X$ *is transitive if and only if X is homogeneous under G.*

Proof Exercise.   □

Definition Choose a point $x \in X$. The subgroup $G_x = \{g \in G \mid gx = x\}$ is called **isotropy**, or **stabiliser**, of x.

Exercises Show that

i) isotropy subgroups of points in the same orbit are conjugate: $G_{gx} = gG_xg^{-1}$ for all $x \in X, g \in G$.
ii) $G \curvearrowright X$ is free $\iff G_x = \{e\}$ for all $x \in X$.
iii) $x \in \mathrm{Fix}(G) \iff G_x = G \iff \mathcal{O}_x = \{x\}$.

Examples

i) Every group G acts on itself by multiplication $g \mapsto \rho_g\colon h \mapsto gh$. This action is transitive and free.

   If $H \subseteq G$ is a subgroup, the quotient set $G/H = \{gH \mid g \in G\}$ is homogeneous under the action

$$(g', gH) \longmapsto (g'g)H,$$

and the isotropy of the identity coset $H \in G/H$ is the subgroup H itself. Consequently any subgroup in a group is the isotropy group, at some point, of an action. This leads to show that *every G-homogeneous space is a quotient G/H*, for some isotropy subgroup $H \subseteq G$ (see the next proposition).

Exercise Suppose H is normal in G, meaning $gHg^{-1} = H, \forall g \in G$. Show that the multiplication $(g_1H)(g_2H) := (g_1g_2)H$ makes G/H a group, and $H = \mathrm{Ker}(G \to G/H)$.

ii) A group may act on itself $G \curvearrowright G$ in a number of ways. A relevant one is by conjugation: $g \mapsto \rho_g : h \mapsto ghg^{-1}$. The kernel of this action is the group's centre

$$Z(G) := \{y \in G \mid xy = yx \ \forall x \in X\}.$$

Because the orbit of $x \in G$ is its conjugacy class, the isotropy at x is the centraliser $Z_G(x) := \{y \in G \mid xy = yx\}$, so $\text{Fix}(G) = \bigcap_{x \in G} Z_G(x) = Z(G)$.

iii) Let G, N be groups and $\rho \colon G \curvearrowright N$ an action. The **semi-direct product** $G \ltimes N$ is the group $G \times N$ with multiplication $(g, n)(h, m) = (gh, n\rho_g(m))$. The obvious maps $G \xrightarrow{j} G \times \{e\}$ and $N \xrightarrow{i} \{e\} \times N$ show that $G \cong j(G)$ is a subgroup of $G \ltimes N$ and $N \cong i(N)$ is normal:

$$N \hookrightarrow^{i} G \ltimes N \xrightarrow{\text{pr}_2} G$$
$$\xleftarrow{j}$$

Examples are the **affine group** $\text{Aff}(\mathbb{R}^n) := \text{GL}(n, \mathbb{R}) \ltimes \mathbb{R}^n$, the **Euclidean group** $E(n) := O(n) \ltimes \mathbb{R}^n$, or the *Poincaré group* $O(3, 1) \ltimes \mathbb{R}^{3,1}$ from relativity.

Now, since the semi-direct construction produces a group, interesting things happen when we let this group acts on a space, cf. p.395.

Our next result generalises *Lagrange's theorem* from group theory.

Proposition *For any point $x \in X$, an action $G \curvearrowright X$ establishes a 1-1 correspondence $G/G_x \approx \mathcal{O}_x$.*

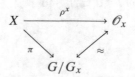

Proof Define the 'orbit' map $\rho^x \colon G \to \mathcal{O}_x, g \mapsto \rho^x(g) := gx$. First, ρ^x is onto by definition of orbit. Since $g_1 x = g_2 x \iff g_2^{-1} g_1 \in G_x \iff g_2 \in g_1 G_x \iff G_x = \{e\}$, its kernel is $\text{Ker} \rho^x = G_x$. By (3.7), then, the induced map $G/G_x \longrightarrow \mathcal{O}_x, gG_x \longmapsto gx$ becomes bijective. $\square$

Observe that the relationship between the action ρ_g and the orbit maps ρ^x involves the exponential law (Proposition 7.17).

Examples 17.5

i) $G = \text{SO}(3)$ acts on $S^2 = \{(x_1\ x_2\ x_3)^\mathsf{T} \mid x_1^2 + x_2^2 + x_3^2 = 1\} \subseteq \mathbb{R}^3$ transitively. The isotropy of the north pole $N = (0, 0, 1)$, and the south pole $(0, 0, -1)$ alike, is the group $\text{SO}(2)$ of Example 17.4 ii). By the previous proposition

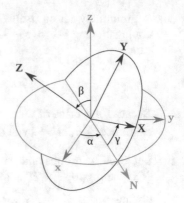

$SO(3)/SO(2) = \mathcal{O}_N = SO(3)N$. But every point p on the sphere can be moved to any other point q by a rotation R with axis $p \times q$ and angle $2 \arccos(p \cdot q)$. Said equivalently, the action is transitive and S^2 is a homogeneous space under $SO(3)$.

In terms of rotations $R_{\mathbf{v}}(\phi)$ about an axis $\mathbf{v}$ by angle ϕ, we have

$$R = R_z(\gamma) \circ R_x(\beta) \circ R_z(\alpha)$$

where β, γ, α are the *Euler angles* of classical mechanics (yaw, pitch and roll, depending on the convention; Fig. 17.1).

Explicitly

$$R = \begin{pmatrix} \cos\gamma & -\sin\gamma & 0 \\ \sin\gamma & \cos\gamma & 0 \\ 0 & 0 & 1 \end{pmatrix} \begin{pmatrix} 1 & 0 & 0 \\ 0 & \cos\beta & -\sin\beta \\ 0 & \sin\beta & \cos\beta \end{pmatrix} \begin{pmatrix} \cos\alpha & -\sin\alpha & 0 \\ \sin\alpha & \cos\alpha & 0 \\ 0 & 0 & 1 \end{pmatrix} \quad (17.2)$$

Summing up, then,

$$\frac{SO(3)}{SO(2)} \cong S^2.$$

ii) Consider the unit disc $\mathbb{D} := \{z \in \mathbb{C} \mid |z| < 1\}$ in the complex plane and the action of $\mathrm{Aut}\,(\mathbb{D}) = \{f : \mathbb{D} \to \mathbb{D} \text{ biholomorphism}\}$. *Schwarz's lemma* implies the stabiliser of the origin is $\mathrm{Aut}\,(\mathbb{D})_0 = \{f \in \mathrm{Aut}\,(\mathbb{D}) \mid f(z) = e^{i\theta}z\}$. From there one can prove that the group of biholomorphisms is the **Möbius group**

$$\mathcal{M} := \left\{ f(z) = e^{i\theta} \frac{z + z_0}{1 - \overline{z_0}z} \;\middle|\; \theta \in \mathbb{R}, z_0 \in \mathbb{D} \right\}$$

cf. (18.8). Then $\mathbb{D} \cong \mathcal{M}/S^1$. (☞ *Riemann surfaces*)

iii) The group

$$\mathrm{SU}(1,1) := \left\{ \begin{pmatrix} a & b \\ \overline{b} & \overline{a} \end{pmatrix} : |a|^2 - |b|^2 = 1 \right\} \subset \mathrm{End}\, \mathbb{C}^2$$

acts on $\mathbb{D}$ by the rational map $z \mapsto \dfrac{az + b}{\overline{b}z + \overline{a}}$. Since the matrix $-\mathbb{1}$ acts like $\mathbb{1}$ we consider $\mathbb{Z}_2 = \{\pm\mathbb{1}\}$ and $\mathrm{PSU}(1,1) := \mathrm{SU}(1,1)/\mathbb{Z}_2$. This acts transitively and $\mathrm{PSU}(1,1)_0 \cong S^1$, implying $\mathbb{D} \cong \mathrm{PSU}(1,1)/S^1$. (☞ *Riemann surfaces*)

iv) The group $\mathrm{SL}(2, \mathbb{R})$ acts on the upper half-plane $\mathscr{H}^+ := \{z \in \mathbb{C} \mid \mathrm{Im}\, z > 0\}$ by $\begin{pmatrix} a & b \\ c & d \end{pmatrix} z = \dfrac{az + b}{cz + d}$. Said action is not free because $\begin{pmatrix} 1 & 0 \\ 0 & 1 \end{pmatrix} z = \begin{pmatrix} -1 & 0 \\ 0 & -1 \end{pmatrix} z$ for all $z \in \mathbb{D}$. Nonetheless, identifying the two matrices $\pm\mathbb{1}$ we obtain the group

$$\mathrm{PSL}(2, \mathbb{R}) := \frac{\mathrm{SL}(2, \mathbb{R})}{\mathbb{Z}_2}$$

which acts freely and transitively on $\mathscr{H}^+$. From there it can be shown that $\mathrm{Aut}\,(\mathscr{H}^+) = \mathrm{PSL}(2, \mathbb{R})$, and then $\mathscr{H}^+ \cong \mathrm{PSL}(2, \mathbb{R})/\mathrm{U}(1)$. (☞ *Riemann surfaces*)

v) The group $\mathrm{U}(1)$ acts on the sphere $S^3 = \{(z, w) \in \mathbb{C}^2 \mid |z|^2 + |w|^2 = 1\}$ under $e^{i\theta}(z, w) := (e^{i\theta}z, e^{i\theta}w)$. Writing $S^2 = \{(z, x) \in \mathbb{C} \times \mathbb{R} \mid |z|^2 + x^2 = 1\}$ gives the *Hopf fibration*

$$
\begin{array}{ccc}
& & S^3 \longleftarrow \mathbb{C}^2 \setminus \{0\} \\
S^3 \xrightarrow{H} S^2 & & {\scriptstyle H} \downarrow {\scriptstyle S^1} \qquad \downarrow {\scriptstyle \mathbb{C}^*} \\
(z, w) \longmapsto (2z\overline{w}, |z|^2 - |w|^2) & & S^2 \xrightarrow{\;\cong\;} \mathbb{CP}^1
\end{array}
$$

which can be viewed as the restriction of the quotient map defining the complex projective line. The bottom map in the commutative diagram is the stereographic identification (18.12). The typical fibre $H^{-1}(p)$ is isomorphic to S^1, and then

$$S^3 \Big/ S^1 \cong S^2.$$

Fibres over distinct points are entwined circles (☞ *principal fibre bundles, K-theory*).

vi) $\mathrm{SO}(n + 1) \curvearrowright S^n \subseteq \mathbb{R}^{n+1}$ transitively. Looking at the isotropy at the sphere's north pole $(0, \ldots, 0, 1)$

$$SO(n+1)_{(0,\ldots,0,1)} = \left\{ \left(\begin{array}{c|c} \tilde{} & \begin{array}{c} 0 \\ \vdots \\ 0 \end{array} \\ \hline 0 \ldots 0 & 1 \end{array} \right) \right\} \cong SO(n),$$

we may describe the sphere as the homogeneous space

$$\frac{SO(n+1)}{SO(n)} \cong S^n.$$

Analogously, $SU(n+1)$ acts transitively on $S^{2n+1} \subseteq \mathbb{C}^{n+1}$, and so $\frac{SU(n+1)}{SU(n)} \cong S^{2n+1}$.

vii) Let $l = \mathbb{C}\mathbf{u}$ be the line through $\mathbf{u} \in \mathbb{C}^{n+1} \setminus \{0\}$. Since $l \cap S^{2n+1} = \{\lambda\mathbf{u} \in \mathbb{C}^{n+1} \mid |\lambda| = |\mathbf{u}|^{-1}\} \cong S^1$, and every line through $\mathbf{0}$ is determined by a point $\mathbf{u}$, we have $l_1 \neq l_2 \iff (l_1 \cap S^{2n+1}) \cap (l_2 \cap S^{2n+1}) \neq \varnothing$. Therefore $\mathbb{CP}^n = S^{2n+1}/\sim$ where $\mathbf{x} \sim \mathbf{y}$ whenever $\mathbf{x}, \mathbf{y} \in S^{2n+1}$ are collinear. Eventually

$$\mathbb{CP}^n = S^{2n+1}\Big/S^1 .$$

(When $n = 1$ we recover example v).)

viii) Finally, consider the action $G \times G \curvearrowright G$ given by $(g_1, g_2)g = g_1 g g_2^{-1}$. The orbit of e is G, and its isotropy is the diagonal $\Delta(G) := \{(g, g) \mid g \in G\} \cong G$. This gives rise to a bijection $G \cong (G \times G)/\Delta(G)$, by which any group becomes a homogeneous space.

Exercises

i) Prove that the Möbius group $\mathcal{M}$ acts transitively on $\mathbb{D}$, and $\mathcal{M} \supset \mathrm{Aut}\,(\mathbb{D})_0$. Conclude $\mathcal{M} = \mathrm{Aut}\,(\mathbb{D})$.

ii) What is the relationship between $PSU(1, 1)$ and $\mathcal{M}$?

iii) Determine the orbits of the action $\mathbb{R} \curvearrowright \mathbb{R}^2$, $t(x, y) = (tx, t^{-1}y)$.

17.2 Representations

Fix a group G and a $\mathbb{K}$-vector space V of finite dimension n (we'll have in mind $\mathbb{K} = \mathbb{R}$ and $V \cong \mathbb{R}^n$ practically always). Call $GL(V)$ the automorphism group of V, which is isomorphic to $GL(n, \mathbb{R})$ after choosing a basis for V.

Definition A homomorphism $\rho\colon G \to GL(V)$ is a **(linear) G-representation on V**

$$\rho(g)\rho(h) = \rho(gh), \quad \forall g, h \in G.$$

The space V itself is often called representation.

Remark An action $\rho: G \curvearrowright X$ induces a representation $g \mapsto \rho(g)$ of G on $\mathrm{Aut}\,(X)$. Conversely any representation $\rho: G \to \mathrm{GL}(V)$ defines an action $G \curvearrowright V$, by declaring $g\mathbf{v} = \rho(g)\mathbf{v}$. Hence the theory of group representations runs parallel, and is equivalent, to the theory of actions.

Underneath this is the fact that an action is a functor $\mathsf{G} \to \mathsf{Set}$, and a representation is a functor $\mathsf{G} \to \mathsf{Vect}$. The correspondence actions–representations is an equivalence of categories between $\mathsf{G\text{-}Set}$ and $\mathsf{G\text{-}Rep}$.

Examples

i) The representation $g \mapsto \rho(g) = \mathbb{1}_V$ is called trivial (the action doesn't do anything).

ii) The **standard representation** of a matrix group $G \subseteq \mathrm{End}\,(\mathbb{R}^n)$ is a representation $\rho: G \to \mathrm{GL}(n, \mathbb{R})$ given by matrix-vector multiplication. If $g = \begin{pmatrix} g_{11} & \cdots & g_{1n} \\ \vdots & \ddots & \vdots \\ g_{n1} & \cdots & g_{nn} \end{pmatrix}$, the map $\rho_g: \mathbb{R}^n \to \mathbb{R}^n$ is defined by

$$\rho_g \begin{pmatrix} x_1 \\ \vdots \\ x_n \end{pmatrix} = \begin{pmatrix} g_{11} & \cdots & g_{1n} \\ \vdots & \ddots & \vdots \\ g_{n1} & \cdots & g_{nn} \end{pmatrix} \begin{pmatrix} x_1 \\ \vdots \\ x_n \end{pmatrix}.$$

It corresponds to the standard action $G \curvearrowright \mathbb{R}^n$.

iii) The representation $\mathbb{C} \to \mathrm{End}\,\mathbb{R}^2$

$$a + bi \mapsto \begin{pmatrix} a & b \\ -b & a \end{pmatrix} \tag{17.3}$$

is faithful. It can be interpreted in terms of polar coordinates 12.6:

Theorem 17.6 (Polar Decomposition) $\mathrm{GL}(1, \mathbb{C}) \cong \mathbb{R}^+ \, \mathrm{SO}(2)$.

Proof Identify $\mathrm{GL}(1, \mathbb{C}) = \mathbb{C}^*$ and represent a non-zero complex number $a+bi$ by the matrix $\begin{pmatrix} a & b \\ -b & a \end{pmatrix}$. This correspondence is an isomorphism, and the generators $1, i$ of $\mathbb{C}$ correspond to $\mathbb{1}_2$, $\begin{pmatrix} 0 & 1 \\ -1 & 0 \end{pmatrix}$. Moreover $|a + bi| = \det \begin{pmatrix} a & b \\ -b & a \end{pmatrix}$ under the identification, so that the determinant represents the modulus. Therefore the polar form $z = \rho e^{i\theta}$ is the required the polar decomposition, after we identify $\theta \longleftrightarrow e^{i\theta}$ with (16.1). $\qquad \square$

iv) One can extend (17.3) to pairs of complex numbers (z, w), and obtain a faithful representation

$$\mathbb{C}^2 \;\rightarrow\; \text{End } \mathbb{C}^2$$
$$(z, w) \mapsto \begin{pmatrix} z & w \\ -\overline{w} & \overline{z} \end{pmatrix} \tag{17.4}$$

Under it, $\|(z, w)\|^2 = \|z\|^2 + \|w\|^2 = \det\left(\begin{smallmatrix} z & w \\ -\overline{w} & \overline{z} \end{smallmatrix}\right)$. A special basis of these matrices are the elements

$$\mathbb{1} = \begin{pmatrix} 1 & 0 \\ 0 & 1 \end{pmatrix}, \quad \mathbb{I} := \begin{pmatrix} i & 0 \\ 0 & -i \end{pmatrix}, \quad \mathbb{J} := \begin{pmatrix} 0 & 1 \\ -1 & 0 \end{pmatrix}, \quad \mathbb{K} := \begin{pmatrix} 0 & i \\ i & 0 \end{pmatrix}. \tag{17.5}$$

Note that $-i\mathbb{I}$, $-i\mathbb{J}$, $-i\mathbb{K}$ are traceless matrices, known in ☞ *quantum mechanics* as **Pauli matrices**.[1] Examining this algebraic structure of $\mathbb{C}^2$ (related to ☞ *quaternions*), one can prove

- the group SU(2) of Example 16.6 is isomorphic to the unit sphere S^3 in $\mathbb{C}^2 = \mathbb{R}^4$;
- the group SO(3) is homeomorphic to the projective space $\mathbb{RP}^3$.

v) Take the standard representation of $SL(2, \mathbb{R})$ on $\mathbb{R}^2$

$$g = \begin{pmatrix} a & b \\ c & d \end{pmatrix} \mapsto \rho_g \begin{pmatrix} x \\ y \end{pmatrix} = \begin{pmatrix} a & b \\ c & d \end{pmatrix}\begin{pmatrix} x \\ y \end{pmatrix} = \begin{pmatrix} ax + by \\ cx + dy \end{pmatrix}.$$

The matrix $H = \begin{pmatrix} 0 & -1 \\ 1 & 0 \end{pmatrix}$ generates a subgroup isomorphic to $U(1) \subset SL(2, \mathbb{R})$ by (17.3). The complexification $SL(2, \mathbb{R}) \otimes_{\mathbb{R}} \mathbb{C} = SL(2, \mathbb{C})$ has a special basis

$$H = -i\mathbb{J}, \quad X = \frac{\mathbb{K} - i\mathbb{1}}{2}, \quad Y = \frac{-\mathbb{K} - i\mathbb{1}}{2},$$

cf. (17.5), satisfying the relations $HX - XH = X$, $HY - YH = -Y$, $XY - YX = H$. Again, this is pivotal in ☞ *Lie theory*.

vi) Consider the vector space of homogeneous polynomials of degree n in two variables:

$$\mathbb{R}[x, y]_n := \left\{ p(x, y) = \sum_{i=0}^{n} p_i x^i y^{n-i} \;\middle|\; p_i \in \mathbb{R} \right\} \subset \mathbb{R}[x, y].$$

The representation $\rho\colon SL(2, \mathbb{R}) \curvearrowright \mathbb{R}^2$ seen above induces a representation $\widetilde{\rho}$ on $\mathbb{R}[x, y]_n$

$$\widetilde{\rho}_g\left(p(x, y)\right) := p\left(dx - by, ay - cx\right).$$

---

[1] Not accidentally, $\mathbb{I}$, $\mathbb{J}$, $\mathbb{K} \in \mathfrak{su}(2)$ are unit quaternions, cf. Remark 13.2.

The previous examples all involve some representation via matrices. It's worth bearing in mind that not every group G can be described by matrices. The examples typically involve groups of operators that act on infinite-dimensional vector spaces.

17.3   The Erlangen Programme

Virtually every mathematical object involves, in one way or another, some form of spatial representation in our minds. Think of the arrow symbol for a map $X \to Y$, or an element x in a set Z: if we replace them by the words 'transformation' or 'point in a space', the evocative power of geometry makes concrete what looks abstract.

The present section's contents constitute the very heart of the scientific revolution kick-started by the celebrated *Erlangen Programme*, whose repercussions in geometry, algebra, topology, dynamical systems, differential equations, physics and many, many more, are still being understood [11, 53]. The name refers to the inaugural dissertation Felix Klein presented as he assumed the Chair of Mathematics at the University of Erlangen,[2] Germany, in 1872. Klein put forward the idea that geometry should be considered the study of the actions of **transformation groups** on spaces, thought of as the **symmetries** of the space, and of their **invariants**.

Let's explain with examples of transformation groups G and their actions.

Affine geometry   let V be a $\mathbb{K}$-vector space. The **affine group**

$$\mathrm{Aff}(V) := \mathrm{GL}(V) \ltimes V$$

acts transitively on V under $\mathbf{x} \mapsto A\mathbf{x} + \mathbf{b}$, where $A \in \mathrm{GL}(V)$ and $\mathbf{b} \in V$. Affine geometry is thus the study of the invariants of V under this action.

Exercise Show that an affine transformation of $\mathbb{R}^2$ maps straight lines to straight lines, the mid-point of a segment to the mid-point of the image segment, and ellipses to ellipses. Hence these are instances of (real) affine invariants.

Euclidean geometry   on $\mathbb{R}^n_{\mathrm{Eucl}}$ take the orthogonal group $\mathrm{O}(n)$ and the translations $\mathbb{R}^n$. Angles and the Euclidean distance $d_{\mathrm{Eucl}}(x, y) := \|x - y\|$ are invariants for the transitive action of the **Euclidean group**

$$\mathrm{E}(n) := \mathrm{O}(n) \ltimes \mathbb{R}^n.$$

More details can be found in Sect. 18.1.2.

Projective geometry   let V be a $\mathbb{K}$-vector space. A linear representation $G \to \mathrm{GL}(V)$ given by an action $(g, \mathbf{v}) \mapsto g\mathbf{v}$ induces a projective representation $G \to \mathrm{PGL}(V)$, $(g, [\mathbf{v}]) \mapsto [g\mathbf{v}]$, where

---

[2] Today called *Friedrich-Alexander University Erlangen-Nüremberg*.

$$PGL(V) := GL(\mathbb{P}(V)) = \frac{GL(V)}{\mathbb{K}^*}$$

is called **projective linear group**. Projective subspaces in $\mathbb{P}(V)$ of a given dimension, or the rank of quadrics, are projectively invariant notions.

Another remarkable example is the **cross-ratio** of 4 distinct, collinear points A, B, C, D:

$$[ABCD] := \frac{\overline{DA} \cdot \overline{CB}}{\overline{CA} \cdot \overline{DB}}. \tag{17.6}$$

Proposition *Consider a pencil of four lines in the plane passing through O and intersecting two transversals r, s at A, B, C, D and A', B', C', D' respectively (Fig. 17.2). Then $[ABCD] = [A'B'C'D']$.*

Proof Call h the common altitude of the triangles with bases along r and opposite vertex O. The area of $\overset{\triangle}{ADO}$ is equal to $\frac{1}{2}\overline{AD} \cdot h = \frac{1}{2}\overline{AO} \cdot \overline{DO} \cos A\hat{O}D$, and similarly for $\overset{\triangle}{CBO}, \overset{\triangle}{CAO}, \overset{\triangle}{DBO}$ and the corresponding primed triangles. Substituting, the cross-ratio reads

$$[ABCD] = \frac{\cos A\hat{O}D \cdot \cos B\hat{O}C}{\cos A\hat{O}C \cdot \cos B\hat{O}D},$$

which only depends on the angles at O, and not on the line r. This is telling that the central projection $\pi_O : r \to s$ mapping $A \mapsto A'$ etc. is a projective transformation, i.e. an element of $PGL(\mathbb{R}^2)$. □

A glance at Fig. 17.2 brings up the notion of perspective, a type of projective transformation. Perspective (and thus projective geometry with it) was born during the Italian Renaissance. Among the many foundational artists of that time, the painter Piero della Francesca (1416–1492) is renowned for writing mathematical

Fig. 17.2 The projectively invariant cross-ratio

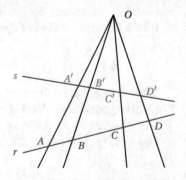

textbooks on the subject. His painting 'The Flagellation of Christ' is geometrically so perfect that he was named '*the best geometer of his time*'.

Conformal geometry   let $\Phi \in \odot^2 V^n$ be a quadratic form on V with bilinear form ϕ, and define the **conformal group**

$$\mathrm{Conf}(V, \Phi) := \big\{ f \in \mathrm{PGL}(V) \mid \Phi(f\mathbf{v}) = f(\Phi(\mathbf{v})) \big\}.$$

This group acts on $\mathbb{P}(V)$, but not transitively: the quadric $\mathcal{Q} = \big\{ [\mathbf{v}] \mid \Phi(\mathbf{v}) = 0 \big\}$ coincides with the orbit of any of its points: $\mathcal{Q} = \mathcal{O}_{[\mathbf{v}]}$, $\forall [\mathbf{v}] \in \mathcal{Q}$. Conformal geometry studies the invariants of a non-singular quadric $\mathcal{Q} \subseteq \mathbb{P}(V)$ under $\mathrm{Conf}(V, \Phi)$. Note that for any $[\mathbf{v}], [\mathbf{w}] \in \mathbb{P}(V) \setminus \mathcal{Q}$ the quantity

$$\gamma(\mathbf{v}, \mathbf{w}) := \frac{\phi(\mathbf{v}, \mathbf{w})^2}{\Phi(\mathbf{v})\Phi(\mathbf{w})}$$

is a conformal invariant.

Example: when $V = \mathbb{R}^{n+1}$ and $\Phi(\mathbf{x}) = x_0^2 - x_1^2 - \ldots - x_n^2$ we have $\mathcal{Q} = \mathbb{R}\mathbb{P}^1$ for $n = 2$, and $\mathcal{Q} = \mathbb{C}\mathbb{P}^1$ for $n = 3$. Hence in dimension 2 or 3 conformal geometry is the geometry of a projective line.

Spherical geometry   (retaining notations) take $V = \mathbb{R}^{n+1}$ and $\Phi(\mathbf{x}) = x_0^2 + x_1^2 + \ldots + x_n^2$. As Φ is positive on $V \setminus \{\mathbf{0}\}$, the space $\mathbb{R}\mathbb{P}^n$ is homogeneous. The distance $d([\mathbf{v}], [\mathbf{w}]) = 2 \arccos \sqrt{\gamma(\mathbf{v}, \mathbf{w})}$ is invariant under projective transformations. When $n = 2$: viewing $\mathbb{R}\mathbb{P}^2 = S^2/\mathbb{Z}_2$ as a sphere with antipodal points identified, $d([\mathbf{v}], [\mathbf{w}])$ measures the length of the equatorial arc joining $\mathbf{v}$ and $\mathbf{w}$ on S^2. Through distinct points there is a unique equator, and the intersection of 2 equators is unique (an antipodal pair, hence one point in $\mathbb{R}\mathbb{P}^2$). Therefore spherical equators behave like straight lines in $\mathbb{R}^2$, safe for the fact that there are no parallel equators on S^2. All of this will be spelt out in Sect. 18.2.

Hyperbolic geometry   define $\mathcal{H}^n := \big\{ [\mathbf{x}] \in \mathbb{R}\mathbb{P}^n \mid \Phi(\mathbf{x}) = x_0^2 - x_1^2 - \ldots - x_n^2 > 0 \big\}$. Supposing $x_0 \neq 0$, we may take inhomogeneous coordinates $y_i = x_i/x_0$, so that

$$\mathcal{H}^n \cong \big\{ (y_1, \ldots, y_n) \in \mathbb{R}^n \mid y_1^2 + \ldots + y_n^2 < 1 \big\}$$

is described as the interior of the unit ball. Let's examine the cases $n = 1, 2$:

- *The hyperbolic line*

$$\mathcal{H}^1 := \big\{ [x_0, x_1] \in \mathbb{R}\mathbb{P}^2 \mid x_0^2 - x_1^2 > 0 \big\} \cong \big\{ \lambda = \tfrac{x_1}{x_0} \in \mathbb{R} \mid |\lambda| < 1 \big\} = (-1, 1).$$

Recall that $\tanh(x) = \frac{e^{2x}-1}{e^{2x}+1}$ is a bijection $(-1, 1) \to \mathbb{R}$ with inverse $\tanh^{-1}(x) = \frac{1}{2} \log \frac{1+x}{1-x}$ (cf. Exercises 12.5). The **hyperbolic distance** of $\lambda, \mu \in \mathcal{H}^1$ is

$$d_{\mathcal{H}}(\lambda, \mu) := \left|\tanh^{-1}\lambda - \tanh^{-1}\mu\right| = \tanh^{-1}\left|\frac{\mu - \lambda}{1 - \lambda\mu}\right|.$$

Using the cross-ratio $[-1, \lambda, \mu, 1] = \frac{(1+\mu)(\lambda-1)}{(\lambda+1)(\mu-1)}$ we can write

$$d_{\mathcal{H}}(\lambda, \mu) = \frac{1}{2}\log\frac{(1+\mu)(1-\lambda)}{(1-\mu)(1+\lambda)}. \tag{17.7}$$

Exercise Check this by verifying that $\tanh\left(\frac{1}{2}\log[-1, \lambda, \mu, 1]\right) = \frac{\mu-\lambda}{1-\mu\lambda}$.

When λ or μ tend to the boundary ± 1 of the interval, $d_{\mathcal{H}} \longrightarrow \infty$.

- *The hyperbolic plane*

$$\mathcal{H}^2 := \left\{[x_0, x_1, x_2] \in \mathbb{RP}^3 \mid x_0^2 - x_1^2 - x_2^2 > 0\right\} \cong \left\{(\lambda, \mu) \in \mathbb{R}^2 \mid \lambda^2 + \mu^2 < 1\right\}$$

is the unit disc $\mathbb{D} \subseteq \mathbb{R}^2$. (An equivalent choice would be the region enclosed by any conic in $\mathbb{RP}^2$.) The distance on the hyperbolic plane is still given by formula (17.7), because two points determine a line (cf. (18.5)).

A line on the hyperbolic plane is the intersection of a projective line with $\mathcal{H}^2$, hence a **chord**. And we declare disjoint chords to be parallel (like ordinary lines in $\mathbb{R}^2$). Therefore (Fig. 17.3) if we pick a point $P \in \mathcal{H}^2$ outside a chord r, there's a whole pencil of chords disjoint from r passing through P:

$$P \notin r \implies \exists \text{ infinitely many lines } p_i : P \in p_i \wedge p_i /\!/ r.$$

The hyperbolic plane, thus, is a geometric model in which the parallel postulate fails (Sect. 18.3).

The hierarchy of geometries is describable through a hierarchy of symmetry groups G and invariants. For example

Fig. 17.3 The Beltrami disc $\mathcal{H}^2$

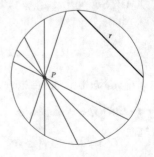

- lengths, angles and areas are preserved by the Euclidean group E(n), whereas the more general projective transformations PGL(n, $\mathbb{R}$) preserve only incidence structures (intersections) and the cross-ratio;
- since Aff(V) $\subsetneq$ PGL(V), it's no surprise that the affine notion of parallelism is meaningless in the projective world. At the same time, a projectively invariant notion may *a priori* not be relevant from an affine point of view.

The relationships between the various geometries can be re-established at the level of the underlying symmetry groups. If we add symmetries we end up with a theory possessing more concepts and theorems. This in turn narrows its scope, because it will speak about special situations. Conversely, shrinking the group of symmetries gives us more freedom to cast a wider net. The ensuing theory will be larger, for it will consist of deeper, more general, results. Look for instance at the following table, to which Klein's vision applies:

Setting	G	Examples of invariants
Set theory	$\mathfrak{S}$	Cardinality
Linear algebra	GL	Dimension, spectrum (determinant, trace,...)
Topology	Homeo	Dimension, cardinal invariants
Hermitian geometry	U	Complex structures
Conformal geometry	Conf	Angles
Differential geometry	Diff	Atlases
Symplectic geometry	Sp	Symplectic capacities
Projective geometry	PGL	Collinearity, cross-ratio, degree
Affine geometry	Aff	Mid-points, ellipses
Riemannian geometry	Isom	Lengths

In the light of all this we also present a short list of groups (☞ *topological groups*) that are mentioned throughout the text.

Definition ($\mathbb{K} = \mathbb{R}, \mathbb{C}$)	Name	$\dim_{\mathbb{R}}$
$\mathrm{GL}(n, \mathbb{K}) := \left\{ M \in \mathrm{End}\, \mathbb{K}^n \mid \det M \neq 0 \right\}$	General linear group	$(\dim_{\mathbb{R}} \mathbb{K})^2$
$\mathrm{SL}(n, \mathbb{K}) := \left\{ M \in \mathrm{GL}(n, \mathbb{K}) \mid \det M = 1 \right\}$	Special linear group*	$(\dim_{\mathbb{R}} \mathbb{K})^2 - 1$
$\mathrm{O}(n) := \left\{ M \in \mathrm{GL}(n, \mathbb{R}) \mid M M^{\mathsf{T}} = \mathbb{1} \right\}$	Orthogonal group*	$\binom{n}{2}$
$\mathrm{SO}(n) := \mathrm{O}(n) \cap \mathrm{SL}(n, \mathbb{R})$	Special orthogonal group*	$\binom{n}{2}$
$\mathrm{Sp}(2n, \mathbb{R}) := \left\{ M \in \mathrm{SL}(2n, \mathbb{R}) \mid M^{\mathsf{T}} J M = J \right\}$	Real symplectic group*	$n(2n + 1)$
$\mathrm{U}(n) := \left\{ M \in \mathrm{GL}(n, \mathbb{C}) \mid M \overline{M}^{\mathsf{T}} = \mathbb{1} \right\}$	Unitary group	n^2
$\mathrm{SU}(n) := \mathrm{U}(n) \cap \mathrm{SL}(n, \mathbb{C})$	Special unitary group*	$n^2 - 1$
$\mathrm{Sp}(n) := \mathrm{U}(2n) \cap \mathrm{Sp}(2n, \mathbb{C})$	(Unitary) quaternionic group*	$n(2n + 1)$

where $J := \begin{pmatrix} 0 & \mathbb{1}_n \\ -\mathbb{1}_n & 0 \end{pmatrix} \in \mathrm{End}\, \mathbb{R}^{2n}$. Entries marked with * are called **classical Lie groups**.

Exercise Show that the intersection of any two of $Sp(2n, \mathbb{R})$, $SO(2n)$, $GL(n, \mathbb{C})$ always gives $U(n)$. This observation is the beginning of ☞ *Hermitian geometry*.

Historically, the most profound reshaping of geometry successive to Klein took place with Grothendieck in the twentieth century. He formulated geometry purely in terms of rings of objects similar to numbers, which can be multiplied in a commutative way. This set-up concretises the ultimate partnership of geometry and algebra, after almost three millennia of tormented romance. Grothendieck's contributions are so numerous and momentous they cannot be listed. He created algebraic *K-theory* and *topos theory*, and invented the notions of *scheme, stack, Abelian category, derived category, motive*, plus various forms of *cohomology*. His work touched on and revolutionised ☞ *functional analysis, algebraic geometry, commutative algebra, number theory, algebraic topology, category theory, logic, representation theory*, . . . you name it.

Chapter 18
Elementary Plane Geometry

The present subject matter is usually part of a separate lecture course. Early exposure to it is desirable because it is complementary to the rest of the book's contents. It picks up from where Sect. 17.3 stopped, as its ideal prosecution. Here the idea is to provide glimpses of elementary Euclidean and non-Euclidean geometry, for which reason the proofs will be few and laconic. Axiomatic aspects are emphasised,[1] with focus on the notion of isometry. Other manuals (e.g. [2, 9, 83, 90]) should be consulted for a comprehensive treatment.

Prerequisites: Section 18.1 is fully accessible. The others require the notion of group and some analytical geometry (familiarity with coordinates and equations in $\mathbb{R}^2$, $\mathbb{R}^3$).

Retrospectively, the discovery of non-Euclidean geometries, and the ensuing deflagration of the relative theories, represented traumatic events. First, they allowed to recognise that it is possible to conceive a notion of space other than Euclidean space. The consequences in the epistemology of science and the philosophical repercussions turned out to be even greater than the Copernican revolution. The attempts to understand what space really *is* led to the demolition of the concept of 'absolute space', one 'synthetic a priori' conceived by the philosopher Immanuel Kant (1724–1804) [65]. But it wasn't the issue of parallel lines that brought about the modern idea of space (and time). More or less simultaneously to the introduction of non-Euclidean geometries, questions over the concept of space in relation to curvature made their appearance. Riemann's pioneering 1854 lecture in Göttingen stressed the locally Euclidean structure of space. After the formulation of general relativity by Einstein, which essentially replaced the Newtonian concept of gravity as a force with the notion of curvature of space, Riemann's work secured a general scientific meaning and changed for good our opinion about space.

---

[1] The author, by admission, is a raving prescriptivist.

© The Author(s), under exclusive license to Springer Nature Switzerland AG 2021   401
S. G. Chiossi, *Essential Mathematics for Undergraduates*,
https://doi.org/10.1007/978-3-030-87174-1_18

18.1   The Euclidean Plane

The tale of Euclidean geometry might be said to start in the VI century BC
with Thales' Theorem 18.5 (which brings into geometrical being the notion of
proportionality) and Pythagoras' Theorem 18.2 (which gives a simple construction
of irrational lengths). Around 300 BC Euclid's masterpiece appeared, *The Elements*
(στοιχεῖα). The latter's importance lies in the fact that for the very first time in
mathematics someone adopted the axiomatic/deductive viewpoint. Six of the 13
books treat plane geometry and contain the following axioms, conventionally called
postulates:

  I. There is a unique straight line passing through any two distinct points.
 II. Any segment can be extended indefinitely to form a straight line.
III. Choosing a point and a distance (positive number), it's possible to construct a
     circle centred at that point with radius equal to the given distance.
 IV. All right angles are congruent.
  V. If two straight lines intercept a third straight line so that the sum of the internal
     angles on one side is less than two right angles, the two lines must meet on that
     side.

It's notorious that the fifth axiom is equivalent to the more famous version called
parallel postulate (Playfair, 1795):

  V$^{/\!/}$. For any straight line l and any point P a unique straight line $l^{/\!/}$ can be drawn
through P that never meets l ($l^{/\!/}$ is parallel to l).

The most relevant geometries (geometric theories) include I–IV among their axioms.
A geometry is called **Euclidean** if V$^{/\!/}$ is true, **non-Euclidean** if false. The existence
of both types of models shows the V postulate independence from the other four. A
model (in the sense of Sect. 4.3) of non-Euclidean geometry where there exists no
parallel line $l^{/\!/}$ is called **spherical** geometry, while we call the model a **hyperbolic**
geometry if there exist infinitely many such $l^{/\!/}$. The latter was discovered by Gauß,
Lobachevskiĭ, Bolyai, Beltrami and others. Far-reaching generalisations were then
developed by Riemann, who founded the subject we know as ☞ *Riemannian
geometry*, crucial for Einstein to formulate the theory of general relativity. Today
non-Euclidean geometry is essential to the physical theories attempting to unify
gravity with the other fundamental forces. Axiom I reminds us that to construct a
straight line in practice we use a laser, and photons travel along geodesic trajectories
in spacetime.

Further modern axiomatisations, consistent with Euclid's postulates, are due to
Hilbert, Veblen, MacLane and Tarski. See [30] for a comparison of these set-ups.
In this chapter we'll take an informal set of premises proposed by George Birkhoff
in 1932 [10]. Birkhoff was the first to construct the foundations of geometry on the
real numbers. This enabled him to introduce a small number of axioms, in contrast
to the 20 or so in Hilbert's system.

The primitive concepts, apart from $\mathbb{R}$, are **points**, whose set $\mathscr{E}$ is called **plane**, and certain special subsets of $\mathscr{E}$ called **lines**. We'll assume we can associate to every pair of points a number, called the 'distance' between them, and to every pair of lines another number, called the 'angle' between the lines. Let's take a look Birkhoff's 5 postulates in some detail.

Bı (ruler). For any line l there exists a bijection $f : l \to \mathbb{R}$ measuring distances.

The existence of the function f allows to label each point on l with a real number unambiguously, and thus call l 'real line'. We may place 0 anywhere on l, but distances on the line should coincide with distances on $\mathbb{R}$:

$$\text{dist}(A, B) := |f(A) - f(B)|.$$

The postulate also defines the idea of 'betweenness' of two points A, B on a line l (which Euclid doesn't have), by introducing

$$\textbf{segments } \overline{AB} := \{C \in l \mid f(A) \leqslant f(C) \leqslant f(B)\} \text{ and}$$
$$\textbf{rays } \overrightarrow{AB} := \overline{AB} \cup \{D \in l \mid f(B) \leqslant f(D)\},$$

so that $l = \overrightarrow{AB} \cup \overrightarrow{BA}$, $\overline{AB} = \overrightarrow{AB} \cap \overrightarrow{BA}$. Once such a correspondence has sunk in, it becomes unnecessary to refer to f again, so $\overline{AB}$ can indicate both a segment and its length.

The second axiom strengthens Euclidean postulate II:

Bıı (two-point line). There is exactly one line passing through two distinct points A, B.

We'll indicate this line by AB given it's unique.

An **angle** $A\hat{O}B$ is defined as a pair of rays $(\overrightarrow{OA}, \overrightarrow{OB})$ with the same origin O. The notation is shortened to $\hat{O}$ when the rays are clear.

Bııı (angle measure). There exists a 1-1 correspondence g between the set of rays with common origin and $[0, 2\pi)$, which measures angles in radians.

In the Cartesian plane we place O at $(0, 0)$. In that way $g^{-1} : \theta \mapsto (\cos\theta, \sin\theta)$ identifies rays emanating from the origin O with points on the unit circle centred there, and

$$A\hat{O}B := \left| g(\overrightarrow{OA}) - g(\overrightarrow{OB}) \right|.$$

The third postulate actually contains another statement about continuity, so to guarantee that the rays extending a triangle's edges intercept the opposite side in a continuous way.

A half-turn is an angle whose defining rays are the two halves of a line.

Bɪᵥ (angle gauge). Every half-turn corresponds to π radians.

Let's agree to indicate by $\overset{\triangle}{ABC}$ a triangle:[2] that would be a configuration of three non-collinear **vertices** A, B, C and three **edges** $\overline{BC}, \overline{CA}, \overline{AB}$ of lengths a, b, c, in which the angles $\alpha = \hat{A}, \beta = \hat{B}, \gamma = \hat{C}$ are all supposed to measure less than π.
Hence the notion of proportional triangles arises:

Definition Two triangles are said **similar**, $\overset{\triangle}{ABC} \sim \overset{\triangle}{A'B'C'}$, if

$$\begin{cases} \alpha = \alpha', \ \ \beta = \beta', \ \ \gamma = \gamma' \\ a = ka', \ b = kb', \ c = kc' \end{cases}$$

for some real number $k > 0$ (stretching factor), and **congruent** when $k = 1$.

Exercise Prove that similarity is an equivalence relation among triangles.

The final Birkhoff postulate warrants similarity if we assume only three of the above six equations:

Bᵥ (similarity). If $\alpha = \alpha'$, $b = kb'$ and $c = kc'$ then $\overset{\triangle}{ABC} \sim \overset{\triangle}{A'B'C'}$.

In school we called this the SAS criterion (side-angle-side). Notice that similarity doesn't follow by taking $\alpha = \alpha'$ and $a/a' = c/c'$ (the angle must be the one between the two edges).

Exercise The 'pentagram' below is drawn prolonging the edges of a regular pentagon. Determine the angles of the triangles. If the inner pentagon has side length 1, what's the length of the star's edge? (*Hint*: (10.1).)

Remark There exist in geometry objects that are self-similar, meaning that at any magnification scale there is a smaller piece that is similar to the whole object.

---

[2] To be pedantic, a triangle should be the 1-skeleton of the convex hull of 3 points on the plane, but we did admit at the beginning we'd be a little sloppy.

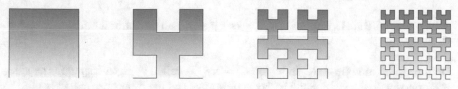

Fig. 18.1 Peano's space-filling curve

Fig. 18.2 A winter fractal

Fig. 18.3 The Euclidean excess $\pi - (\alpha + \beta + \gamma)$ is null

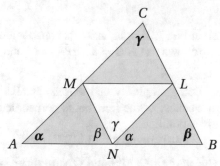

Motivated by Cantor's counterintuitive result that the segment $I = [0, 1]$ has the same cardinality as the square I^2, Peano defined a *space-filling curve* $\gamma : I \twoheadrightarrow I^2$ which is (continuous and) *onto* (Fig. 18.1), and thus shattered the accepted idea of dimension (☞ *dimension theory*).

Self-similarity is a typical property of *fractals*, so called because their dimension is fractional. Examples include the renowned Mandelbrot set, or the Koch snowflake (Fig. 18.2).

Generalisations of self-similarity are instrumental in stock market movements, cellular automata and ☞ *chaos theory*, to name but a few.

Theorem 18.1 (Euclidean Excess) *The sum of the angles of a triangle $\overset{\triangle}{ABC}$ equals $\alpha + \beta + \gamma = \pi$ (Fig. 18.3).*

Proof Let L, M, N denote the mid-points of $\overline{BC}, \overline{CA}, \overline{AB}$ (which exist by **B**ı). The coloured triangles are similar to $\overset{\triangle}{ABC}$ by **B**v, so in particular $\alpha = B\hat{N}L$, $\beta = A\hat{N}M$. The inner triangle's edges are half as long as the edges of $\overset{\triangle}{ABC}$, whence

$A\overset{\triangle}{B}C\sim L\overset{\triangle}{M}N$. But $\mathbf{B_{III}}$ implies $A\hat{N}B = \alpha + \beta + \gamma$, and the latter is equal to π under $\mathbf{B_{IV}}$. $\qquad\qquad\qquad\qquad\qquad\qquad\qquad\qquad\qquad\qquad\qquad\qquad\qquad\square$

Definition A triangle $A\overset{\triangle}{B}C$ in which $\hat{C} = \pi/2$ is called a **right** triangle. The side $\overline{AB}$ opposite to the right angle is the **hypotenuse** and the others are called **legs**.

The next theorem is so well-known it needs no introduction.

Theorem 18.2 (Pythagoras) *If $\hat{C} = \pi/2$ then $a^2 + b^2 = c^2$.*

Proof Here are two classical arguments based on the notion of area.

1) In Fig. 18.4 left, the proof boils down to comparing

$$\underbrace{(a+b)^2}_{\text{big external square}} = \underbrace{c^2 + 4 \cdot \frac{1}{2}ab}_{\text{white + green regions}} .$$

2) In Fig. 18.4 right, $A\overset{\triangle}{B}C$ is constructed by similarity with magnifying factor c. It follows that $\overline{HB} = a^2$, $\overline{HA} = b^2$ and hence $c^2 = \overline{AB} = \overline{AH} + \overline{HB} = a^2 + b^2$. $\qquad\qquad\qquad\qquad\qquad\qquad\qquad\qquad\qquad\qquad\qquad\qquad\qquad\square$

Using similarity and axioms $\mathbf{B_I}$, $\mathbf{B_{III}}$ it is possible to prove the existence and uniqueness of the line that is perpendicular to a chosen line and passes through a given point.

Furthermore (cf. (7.6)),

Corollary 18.3 (Triangle Inequality) *In any triangle, $c \leqslant a + b$.*

Proof Remember the altitude is a segment perpendicular to an edge or its prolongation, traced from the opposing vertex. The edge is called base of the altitude, and the point where altitude meets base is the altitude's foot.

Let $C' \in AB$ be the foot of the altitude from C. Then $\overline{AC}, \overline{CB}$ are hypotenuses and $b > \overline{AC'}, \overline{C'B} < a$. As A, B, C' are collinear, then $c \leqslant \overline{AC'} + \overline{C'B} < a + b$. $\qquad\qquad\qquad\qquad\qquad\qquad\qquad\qquad\qquad\qquad\qquad\qquad\qquad\square$

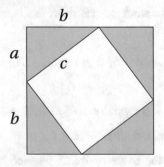
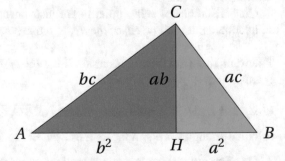

Fig. 18.4 Pythagoras' theorem

Note that $c = a + b$ corresponds to a degenerate triangle, where $C \in AB$. Hence the strict inequality is interpreted as the condition for the existence for triangles, cf. Fig. 15.1.

Remark (Pythagorean Numerology) We'd like to determine what are the possible lengths of a right triangle's edges if we want them to be all integers.

A **Pythagorean triple** is an ordered set $(a, b, c) \in \mathbb{Z}^3$ such that $a^2 + b^2 = c^2$. We assume $a, b, c > 0$ and that the numbers are coprime, to avoid an inflation of proportional triples. In particular a, b cannot be both even (exercise). Moreover, if they were both odd then c^2 would be divisible by 4 whereas $a^2 + b^2$ just by 2. Hence we may seek b even and a, c odd.

Proposition *Every coprime Pythagorean triple with b even is of the form*

$$a = p^2 - q^2, \qquad b = 2pq, \qquad c = p^2 + q^2$$

where $p > q > 0$ are coprime and not both odd. Conversely, any such choice produces a coprime Pythagorean triple.

Proof The last claim is clear since $(p^2 - q^2)^2 + (2pq)^2 = (p^2 + q^2)^2$.

By assuming the prime r divides a and c, moreover, first of all $r \neq 2$ because a, c are odd. Secondly, r would divide $a + c = 2p^2$ and $a - c = 2q^2$, i.e. $r = p = q$ ⚡. Let's rewrite the Pythagorean equation as $\left(\dfrac{b}{2}\right)^2 = \left(\dfrac{c + a}{2}\right)\left(\dfrac{c - a}{2}\right)$, where every fraction is an integer by hypothesis. If the prime $r \geq 2$ divides $b/2$ then r^2 divides $(c + a)/2$ or $(c - a)/2$ (otherwise r would divide the sum, but the triple is coprime). Factoring $b/2$ in primes shows there exist $p, q > 0$ such that $\dfrac{c + a}{2} = p^2$, $\dfrac{c - a}{2} = q^2$, and $p > q$ cannot have common factors by the above argument. $\square$

Exercise Let a, b, c, p, q as in the proposition. Exploiting Exercise 12.9 ii) verify that $a = c \cos \theta$, $b = c \sin \theta$, where $\tan(\theta/2) = p/q$. Conclude there exist infinitely many points (x, y) on the circle $x^2 + y^2 = 1$ with rational coordinates.

Here's a partial list of triples with $b = 2pq$. If q is odd we can take p even and vice versa:

q	p	triples $(a, 2pq, c)$
1	2, 4, 6	(3, 4, 5), (15, 8, 17), (35, 12, 37)
2	3, 5, 7	(5, 12, 13), (21, 20, 29), (45, 28, 53)
3	4, 6, 8	(7, 24, 25), (27, 36, 45), (55, 48, 73)
4	5, 7, 9	(9, 40, 41), (33, 56, 65), (65, 72, 97)

Already for $q = 1$ there is an infinite number of coprime triples, of the form $(p^2 - 1, 2p, p^2 + 1)$.

Regarding the natural generalisation of the problem to arbitrary integer powers (☞ *algebraic number theory*), we have

Theorem (Wiles 1995) *Equation $a^n + b^n = c^n$ doesn't have positive integer solutions a, b, c for any integer $n > 2$.*

The above is known in the literature as **Fermat's last theorem** for historical reasons. The mathematical community agrees on not accepting the notorious one-liner of Fermat, who in 1637 bragged "*... cuius rei demonstrationem mirabilem sane detexi. Hanc marginis exiguitas non caperet.*" [... but I found an outstanding proof of this fact. The narrowness of this page margin does not contain it.]

18.1.1 Parallelism

We return to the central crux of Euclidean geometry, the notion of parallelism, and stick to the plane $\mathcal{E}$ for simplicity.

Definition Two lines l, m on the plane $\mathcal{E}$ are **parallel**, written $l /\!/ m$, if either they are disjoint or they coincide.

Exercise Prove that $/\!/$ is an equivalence relation on the set of lines on the plane.

Theorem 18.4 *Let l be a line and P a point. There exists a unique line $l^{/\!/}$ parallel to l passing through P.*

Proof (Existence) We can assume $P \notin l$ (otherwise $l^{/\!/}$ must be l). We know there exist a line k perpendicular to l through P, and a line m perpendicular to k through P, by axiom $\mathbf{B}_{\text{III}}$. Since they are distinct ($m \ni P \notin l$), one proves l, m are disjoint because they are both orthogonal to k.

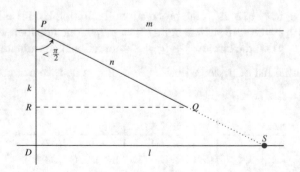

(Uniqueness) Let n be another line through P, so that $l \cap n \neq \emptyset$. Suppose n forms an angle less than $\pi/2$ with k at P, and pick $Q \in n$ different from P. Let R denote the foot of the perpendicular line to k through Q, and $D \in l$ the

Fig. 18.5 The geometric dawn of the Hellenic theory of proportions

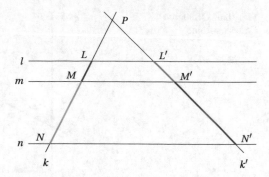

foot of the perpendicular drawn from P. Then there exists a point $S \in l$ such that $\overline{PR}/\overline{PD} = \overline{RQ}/\overline{DS}$. (Note: we still can't conclude $S \in PQ = n$.)

The triangles $\stackrel{\triangle}{PDS}$, $\stackrel{\triangle}{PRQ}$ are similar by **B**v. Hence $D\hat{P}S = R\hat{P}Q$ and $PS = n$ by **B**ɪɪɪ. Therefore $S \in n$, so n cannot be parallel to l. $\qquad\square$

The uniqueness statement is the parallel postulate $V^{/\!/}$!

Let's consider now all possible lines that are parallel to some line l: in the picture above there will be one for each point on m.

Proposition *A line k intersects two lines m, n forming equal angles if and only if $m \,/\!/\, n$.*

Proof

($\Longrightarrow$) We already know this is true if the angle is right. By contradiction, if m, n weren't parallel their intersection would be the vertex V of a triangle whose angles add up to $\alpha + (\pi - \alpha) + \hat{V} > \pi$ ⚡.

($\Longleftarrow$) Suppose k intersects m at point P. Draw a line m' meeting k at P with angle α. From what we've seen $l \,/\!/\, m'$ and $m \,/\!/\, l$, so $m \,/\!/\, m'$. Since m, m' have a common point, $m = m'$.

$\qquad\square$

This shows the parallel postulate is strongly related to Theorem 18.1.

A line k meeting each element in a family of parallel lines at one point, with the same angle, is called transversal to the family (Fig. 18.5).

Theorem 18.5 (Thales' Intercept Theorem) *Let l, m, n be parallel lines, and k, k' two transversals. Then the lengths of the segments cut out on the transversals are proportional:*

$$\frac{\overline{LM}}{\overline{L'M'}} = \frac{\overline{MN}}{\overline{M'N'}}.$$

Proof In case $k \cap k' = \{P\}$ we invoke the previous proposition, since triangles have three equal angles and hence are similar:

Fig. 18.6 Orthogonal
Cartesian frame

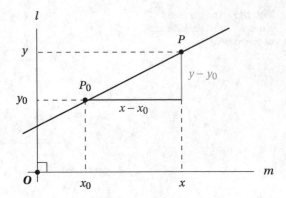

$$\overset{\triangle}{PLL'} \sim \overset{\triangle}{PMM'} \sim \overset{\triangle}{PNN'}.$$

Consequently $\dfrac{\overline{PL}}{\overline{P'L'}} = \dfrac{\overline{MP}}{\overline{M'P'}} = \dfrac{\overline{PN}}{\overline{P'N'}}$, and then $\dfrac{\overline{MP} - \overline{PL}}{\overline{M'P} - \overline{PL'}} = \dfrac{\overline{PN} - \overline{MP}}{\overline{PN'} - \overline{M'P}}$.
The case $k \,/\!/\, k'$ is left as exercise (you will need a third line $k'' = LN'$ cutting m at some point M''). $\square$

Thales' theorem enables us to construct parallel networks, that is, define a **Cartesian frame system**. We are interested in **orthogonal** systems, for simplicity, although no-one vetoes considering the generic 'oblique' case. Draw two lines $l \perp m$ intersecting at $l \cap m = \{O\}$, called the frame system's **origin**. Take a family of perpendicular lines to l parametrised by the distances $y \in \mathbb{R}$ (positive or negative) between O and the foot of the perpendicular ($y = y_0$). Similarly, for each $x \in \mathbb{R}$ we have a line through a point with distance x to O ($x = x_0$). What we get is a system of two families of parallel lines, where each line is orthogonal to every member of the other family (Fig. 18.6).

The parameters x, y are called **Cartesian coordinates** of point $P = (x, y)$, and the lines l, m, of equation $x = 0$, $y = 0$ are the system's **axes**. Eventually

Theorem 18.6 *There is a one-to-one correspondence between the (geometric) plane $\mathscr{E}$ and the (algebraic) Cartesian product $\mathbb{R}^2$*

$$P \longleftrightarrow (x, y).$$

The theorem (a reformulation of the identification between the affine plane and $\mathscr{E}$) is usually the starting point of analytic geometry courses, and the work we've done thus far becomes a prequel in that story. The next step would be to show that the 'lines' of axioms $\mathbf{B_I}$– $\mathbf{B_V}$ coincide with the familiar straight lines from Cartesian geometry. We shall leave it here.

Remark 18.7 (Projective Axiomatics) To highlight the role played by the axioms that define a geometry we briefly present an axiomatic system of different nature (☞ *projective geometry*).

Let X be a set (the space of points) and $L \subseteq \mathscr{P}(X)$ a subset (the lines). An **abstract projective plane** is an incidence structure (X, L) satisfying

$P_I.$ $\forall x, y \in X \; \exists! \; l \in L$ such that $x, y \in l$;
$P_{II}.$ $\forall l \neq l' \in L \; l \cap l' = \{x\}$ is a point in X;
$P_{III}.$ There exist 3 non-collinear points in X;
$P_{IV}.$ Any line $l \in L$ contains at least 3 points.

A low-budget model for these axioms is the *Fano plane* $X = \mathrm{PG}_2(2)$, i.e. the (unique) finite projective plane of order 2. It is made of

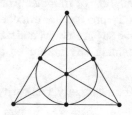

$|X| = 7$ points •

$|L| = 7$ lines: the edges and the medians of the triangle, plus the circle.

Each line contains exactly 3 points, and dually, exactly 3 lines meet at any point.

The points have homogeneous coordinates

$$\underbrace{\begin{pmatrix}1\\0\\0\end{pmatrix}, \begin{pmatrix}0\\1\\0\end{pmatrix}, \begin{pmatrix}0\\0\\1\end{pmatrix}}_{\substack{\text{triangle's}\\\text{vertices}}}, \underbrace{\begin{pmatrix}1\\1\\0\end{pmatrix}, \begin{pmatrix}1\\0\\1\end{pmatrix}, \begin{pmatrix}0\\1\\1\end{pmatrix}}_{\substack{\text{edge}\\\text{mid-points}}}, \underbrace{\begin{pmatrix}1\\1\\1\end{pmatrix}}_{\substack{\text{centre}\\\text{point}}}.$$

Three points A_i, A_j, A_k on one line satisfy $A_i + A_j + A_k \equiv_2 \mathbf{0}$ (☞ *finite geometry, graph theory*). The Fano matroid is intimately related to the non-associative algebra $\mathbb{O}$ of ☞ *octonions*, in that it encodes the only higher-dimensional analogue of the cross product (Sect. 15.6).

Exercise Check whether B_I–B_V hold on the Fano plane.

18.1.2 Euclidean Isometries ☕

Recall the length

$$\overline{PQ} := d_{\mathrm{Eucl}}(P, Q) = \sqrt{(x_P - x_Q)^2 + (y_P - y_Q)^2}$$

defines a distance on $\mathscr{E} \cong \mathbb{R}^2$, see Chap. 19.

Definition A mapping $\phi \colon \mathscr{E} \to \mathscr{E}$ is an **isometry** whenever

$$d\big(\phi(P), \phi(Q)\big) = d(P, Q) \quad \text{for all } P, Q \in \mathscr{E}.$$

That's to say, when it preserves distances.

Examples

i) Rotations R_θ about a point O by angle θ. By construction the unique fixed point of a rotation is O, the so-called centre.
ii) Axial reflections S_l about a mirror line l. Fixed points are the points on l.
iii) Translations $T_v : x \mapsto x + v$ by v, which haven't got fixed points (unless $v = 0$).

An isometry is completely determined by the images of three non-collinear points (a triangle), because two edges are enough to fix a basis for $\mathbb{R}^2 \cong \mathscr{E}$. Moreover, an isometry preserves triangles, i.e. it maps a triangle to a congruent triangle.

Rotations and translations preserve the orientation of triangles (roughly, the cyclic order of vertices). Reflection on the contrary change the orientation from clockwise to counter-clockwise and vice versa. One says rotations and translations are **even** and reflections are **odd**, implicitly defining the notion of parity for isometries.

Exercise Show that the composite of isometries is an isometry, and parity behaves as in the table:

$\circ$	even	odd
even	even	odd
odd	odd	even

under composition.

Example Let v be an oriented segment and l a line containing v. The composite $ST_v = T_v \circ S_l = S_l \circ T_v$ is an isometry of different type from the previous three: it is odd and doesn't fix any point. It's called a *glide reflection*.

With the exception of the identity (which is even and fixes every point), there exist 4 basic isometries, classified by the parity and the number of fixed points:

$\phi \neq \mathbb{1}$	R_θ	S_l	T_v	ST_v
parity	even	odd	even	odd
fixed points	O	l	$\varnothing$	$\varnothing$

and it can be proved that

Theorem 18.8 *Every isometry ϕ of the plane $\mathscr{E}$ is obtained composing rotations, translations and reflections.*

Exercises Show that

i) composing two rotations or two reflections gives a rotation, a reflection or a translation.
ii) Every rotation, and every translation, can be seen as the composite of appropriate reflections.

iii) Any isometry is bijective. For instance:

$(R_\theta)^{-1} = R_{-\theta}$: reversing a clockwise turn gives an counter-clockwise turn;

$S_l^{-1} = S_l$: reflections are involutive;

$T_v^{-1} = T_{-v}$: reversing a right shift produces a left shift.

Theorem (Classification by Fixed Points) *Let* $\phi\colon \mathcal{E} \to \mathcal{E}$ *be an isometry.*

i) If ϕ has 3 non-collinear fixed points, then $\phi = \mathbb{1}$.

ii) If it fixes just one line l, then $\phi = S_l$ is an axial reflection about l.

iii) If it has a unique fixed point, then $\phi = R_\theta$ is a rotation by $\theta \notin 2\pi\mathbb{Z}$.

iv) If ϕ has no fixed points, it is a glide reflection or a translation.

Proof Exercise.   □

Corollary (Three Reflections) *Every isometry ϕ of the Euclidean plane is the composite of 3 reflections at most (Fig. 18.7).*

Proof (Sketch) Take non-collinear points $A, B, C \in \mathcal{E}$. If A is not a fixed point of ϕ, we reflect in the axis l of $A, \phi(A)$ (the line orthogonal to $\overline{A\phi(A)}$ through the mid-point, so $S_l(A) = \phi(A)$). Then we set $B' = S_l(B), C' = S_l(C)$. If $B' \neq \phi(B)$ we do the analogous thing, reflecting in their axis m (note $\phi(A)$ doesn't move). If $C'' = S_m(C') \neq \phi(C)$ we reflect once more in the axis n of $C'', \phi(C)$ (neither $\phi(A)$ nor $\phi(B)$ are displaced).

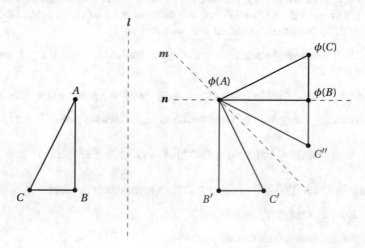

Fig. 18.7 Mirrors, mirrors, on the wall...

In summary, the three mirrors have this effect:

$$A \mapsto S_n(S_m(S_l(A))) = S_n(S_m(\phi(A))) = \phi(A)$$

$$B \mapsto S_n(S_m(S_l(B))) = S_n(S_m(B')) = S_n(\phi(B)) = \phi(B)$$

$$C \mapsto S_n(S_m(S_l(C))) = S_n(S_m(C')) = S_n(C'') = \phi(C).$$

Hence $S_n \circ S_m \circ S_l$ acts as ϕ on A, B, C. But we saw that an isometry is determined by a generic configuration of 3 points, resulting in $S_n \circ S_m \circ S_l = \phi$.   □

Contemplating further reflections doesn't add more possibilities in the plane, because the extra symmetries can be rearranged to cancel out. Besides, notice that composing any two reflections produces a rotation.

At last, the salient concept of the chapter:

Definition The **isometry group** of $\mathscr{E}$ is denoted Isom($\mathscr{E}$).

Let $O \in \mathscr{E}$ be a point and call $O' = \phi(O)$, for $\phi \in \text{Isom}(\mathscr{E})$. The composite $T_{OO'}^{-1} \circ \phi$ fixes O, so any isometry can be written as composite of a translation (element of $T_{\mathscr{E}}$) and an element in the subgroup

$$O(\mathscr{E}) := \{\phi \in \text{Isom}(\mathscr{E}) \mid \phi(O) = O\},$$

i.e. a rotation, a reflection, or $\mathbb{1}_{\mathscr{E}}$.

Let's consider for a moment a slightly more general situation: take $A \in O(n)$, $b \in \mathbb{R}^n$ and set

$$\phi(A, b)\mathbf{x} := A\mathbf{x} + b.$$

As $\phi(A, 0) = A$ is orthogonal, whilst $\phi(0, b) = T_b$ is the translation $\mathbf{x} \mapsto \mathbf{x} + b$, it follows that $O(n)$ and $T_{\mathscr{E}}$ (translations) are subgroups of $E(n)$. Because of that the element $\phi(A, b)$ is sometimes called *roto-translation*.

Lemma 18.9 *The transformation $\phi(A, b)$ is a Euclidean isometry, for any $A \in O(n), b \in \mathbb{R}^n$.*

Proof $\|\phi(A, b)\mathbf{x} - \phi(A, b)\mathbf{y}\| = \|A\mathbf{x} - b - A\mathbf{y} + b\| = \|A(\mathbf{x} - \mathbf{y})\| = \|\mathbf{x} - \mathbf{y}\|$.   □

Definition 18.10 The **group of rigid motions** or **Euclidean group** is

$$E(n) := \left\{\phi(A, b) \in \mathbb{R}^{n \times n} \mid A \in O(n), b \in \mathbb{R}^{n \times 1}\right\}.$$

As we saw in Chap. 17, $E(n) = O(n) \ltimes \mathbb{R}^n$ is a semi-direct product.

Lemma *$E(n)$ is a subgroup of $GL(n, \mathbb{R})$.*

Proof Composing two transformations gives

$$\phi(A, b)\big(\phi(A', b')\mathbf{x}\big) = AA'\mathbf{x} + Ab' + b = \phi(AA', Ab' + b)\mathbf{x},$$

so $\phi(A, b) \circ \phi(A', b') = \phi(AA', Ab' + b) \in E(n)$.

Since $\phi(\mathbb{1}, 0) = \mathbb{1}$, immediately $\phi(A, b)^{-1} = \phi(A^{-1}, -A^{-1}b) \in E(n)$.   □

On the other hand every isometry of $\mathbb{R}^n$ is of the form $\phi(A, b)$, since:

Theorem 18.11 *Every Euclidean isometry is a rigid motion, and conversely:*

$$E(n) = \text{Isom}\left(\mathbb{R}^n_{Eucl}\right).$$

Proof By Lemma 18.9 a roto-translation is isometric, so $\text{Isom}\left(\mathbb{R}^n_{Eucl}\right) \supseteq E(n)$. Vice versa, if f is an isometry then also $k\colon \mathbf{x} \mapsto f(\mathbf{x}) - f(\mathbf{0})$ is an isometry, and $k(\mathbf{0}) = \mathbf{0}$. Lemma 15.13 says k is then linear, say represented by a matrix A. Hence $f(\mathbf{x}) = f(\mathbf{0}) + A\mathbf{x} = \phi(A, f(\mathbf{0}))\mathbf{x}$ for any $\mathbf{x}$, and $\text{Isom}\left(\mathbb{R}^n_{Eucl}\right) \subseteq E(n)$. □

Returning to two dimensions, Lemma 16.5 proves $O(\mathscr{E}) \cong O(2)$ is generated by

$$R_\theta = \begin{pmatrix} \cos\theta & \sin\theta \\ -\sin\theta & \cos\theta \end{pmatrix}, \qquad S_\theta = \begin{pmatrix} \cos\theta & \sin\theta \\ \sin\theta & -\cos\theta \end{pmatrix}, \tag{18.1}$$

where S_θ is the reflection whose axis form an angle $\theta/2$.

Exercises Show that $R_\theta \circ R_\phi = R_{\theta+\phi}$ and $S_\theta \circ S_\phi = R_{\theta-\phi}$.

Conclude that rotations form a subgroup of $O(\mathscr{E})$, whilst the composition of two reflections about non-parallel axes is a rotation (about the axes' intersection point).

Let's use the notation

$$SO^+(\mathscr{E}) := \left\{ R_\theta \mid \theta \in [0, 2\pi) \right\} \cong SO(2), \qquad SO^-(\mathscr{E}) := \left\{ S_\theta \mid \theta \in [0, 2\pi) \right\}$$

so $O(\mathscr{E}) = SO^+(\mathscr{E}) \sqcup SO^-(\mathscr{E})$. It reminds us that the determinant's sign $\det R_\theta = 1$, $\det S_\theta = -1$ reflects the parity (even / odd). Reformulating Theorem 18.11,

Theorem *The isometry group of the plane $\mathscr{E}$ is the semi-direct product*

$$\text{Isom}(\mathscr{E}) = O(\mathscr{E}) \ltimes T_\mathscr{E} \cong O(2) \ltimes \mathbb{R}^2 = E(2).$$

To close the section we have a few exercises that translate in complex terms the classification of plane isometries.

Exercises Identify $\mathscr{E} \cong \mathbb{C}$ and fix $a, b \in \mathbb{C}$.

i) Show that the automorphism $f(z) = az + b$ is a
    translation $\iff a = 1$,
    rotation $\iff a \in U(1) \setminus \{1\}$, $b = 0$.
   What kind of automorphism does one get for $a \in \mathbb{R} \setminus \{1\}$, $b = 0$?
ii) Show that the automorphism $\tilde{f}(z) = a\overline{z} + b$ is a
    reflection $\iff a\overline{b} + b = 0$,
    isometry $\iff |a| = 1$.
   What does one get for $|a| \neq 1$?

iii) Let u, v, w be the lengths of the edges of a triangle $\overset{\triangle}{ABC}$. Prove $\overset{\triangle}{ABC}$ is equilateral if and only if $u + v\omega + w\omega^2 = 0$, where $\omega = e^{\pm 2\pi i/3}$ is a cubic root of unity.

18.2 The Sphere ☕

The objective of the section is to present a model of geometry on the 2-sphere. Sometimes spherical geometry is called 'elliptic' to remind of the ubiquitous division (cf. Sect. 18.4) in properties of parabolic type (here, those of the plane $\mathscr{E}$), elliptic type (the sphere is an ellipsoid!) and hyperbolic type (Sect. 18.3). We'll highlight the analogies and differences between spherical isometries and triangles, and plane (Euclidean) ones.

18.2.1 Spherical Coordinates

Fix Cartesian coordinates on $\mathbb{R}^3$ with origin $O = (0, 0, 0)$ and consider the punctured space $\mathbb{R}^3 \setminus \{O\}$. This set can be parametrised using **spherical coordinates**: the **longitude** $\theta \in [0, 2\pi)$, the **colatitude** $\phi \in [0, \pi]$ and the **distance from the origin** $\rho > 0$. Fixing $\rho = r$, any point P on the sphere

$$S^2(r) := \{ P \mid d_{\mathscr{E}}(P, O) = r \} = \{ (x, y, z) \in \mathbb{R}^3 \mid x^2 + y^2 + z^2 = r^2 \}$$

of radius $r > 0$ and centred at O is described by two parameters θ, ϕ (the latter's suggestive names give away their roots in early sea navigation). Spherical coordinates identify $\mathbb{R}^3 \setminus \{O\}$ with $S^2 \times \mathbb{R}^+$, under the bijection

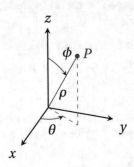

$$\begin{cases} x(\rho, \theta, \phi) = \rho \cos\theta \sin\phi, \\ y(\rho, \theta, \phi) = \rho \sin\theta \sin\phi, \\ z(\rho, \theta, \phi) = \rho \cos\phi, \end{cases} \qquad (18.2)$$

and stratify, so to speak, the punctured space into concentric spheres of radius ρ, a 3D version of Fig. 12.9. The levels of the *foliation* are thus described by the parameters θ, ϕ.

In analogy to the parallel grid of coordinate lines on the plane, on $S^2(r)$ too there are two special families of circles (Fig. 18.8):

- **parallels**, of parametric equations $\rho = r$, $\phi = \phi_0$, are 'horizontal' slices

$$\begin{cases} x^2 + y^2 + z^2 = r^2 \\ z = c \end{cases} \iff \begin{cases} x(\theta) = \sqrt{r^2 - c^2} \cos\theta \\ y(\theta) = \sqrt{r^2 - c^2} \sin\theta \\ z(\theta) = c \in [-r, r] \end{cases}$$

(the orbits of the action described in Example 17.4 ii));
- **meridians**, of equations $\rho = r$, $\theta = \theta_0$, are intersections with 'vertical' planes through the z-axis (axial sections)

$$\begin{cases} x^2 + y^2 + z^2 = r^2 \\ ax - by = 0 \end{cases} \iff \begin{cases} x(\phi) = br \sin\phi \\ y(\phi) = ar \sin\phi \\ z(\phi) = r \cos\phi \end{cases}.$$

Meridians play a leading role, because they are the 'lines' of spherical geometry, see Sect. 18.2.2.

Let's study an important map, especially for ☞ *topology* and *differential geometry*. Consider the unit sphere $S^2 = S^2(1)$ with north pole $N = (0, 0, 1)$, and a point $P \in S^2 \setminus \{N\}$. Take the intersection $\pi_N(P)$ of the line NP with the equatorial plane $\{(x, y, 0) \in \mathbb{R}^3 \mid x, y \in \mathbb{R}\} \cong \mathbb{R}^2$:

Fig. 18.8 Coordinate lines on the 2-sphere

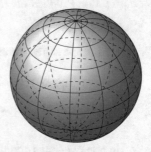

Definition 18.12 The **stereographic projection** of the sphere S^2 is the map

$$\pi_N : S^2 \setminus \{N\} \longrightarrow \mathbb{R}^3$$
$$(x, y, z) \longmapsto \left(\frac{x}{1-z}, \frac{y}{1-z}, 0 \right)$$

Exercises 18.13

i) Sketch a picture of the stereographic projection, taking inspiration from Fig. 7.2.

ii) Prove that π_N is bijective, with inverse

$$\pi_N^{-1}(u, v) = \left(\frac{2u}{u^2 + v^2 + 1}, \frac{2v}{u^2 + v^2 + 1}, \frac{u^2 + v^2 - 1}{u^2 + v^2 + 1} \right).$$

iii) Show π_N maps meridians to lines through the origin lying on $z = 0$, and sends parallels to circles centred at the origin of the plane.

Extending the stereographic projection, ☞ *projective geometry* allows to identify the sphere S^2 with the one-point compactification $\mathbb{CP}^1$ of $\mathbb{C}$: first we identify $\mathbb{R}^2 = \mathbb{C}$ under $(x, y) \longleftrightarrow x + iy$, then write $\pi_N(x, y, z) = \dfrac{x + iy}{1-z}$, Thus we can extend (18.12) to a 1-1 correspondence

$$S^2 \to \mathbb{CP}^1 = \mathbb{C} \cup \{\infty\}$$

by imposing $\pi_N(0, 0, 1) = \infty$. In the 'projective' interpretation of this correspondence ∞ is the point of $\mathbb{C}$ at infinity, i.e. the intersection between the plane $z = 0$ and any line $\{z = 1, \, ax + by = 0\}$ (geometric limit of PN as $P \to N$).

Definition The sphere S^2, identified with $\mathbb{CP}^1$ under stereographic projection, is called **Riemann sphere**.

Then the stereographic projection of S^2 really is the two-dimensional version of construction Proposition 7.14, that identifies $\mathbb{RP}^1 \cong S^1$.

Exercises Prove that

i) the inverse stereographic projection can be described through

$$\pi_N^{-1} : \mathbb{CP}^1 \longrightarrow S^2 \subseteq \mathbb{R}^3$$
$$[z, w] \longmapsto \left(\frac{z\overline{w} + w\overline{z}}{|z|^2 + |w|^2}, \frac{z\overline{w} - w\overline{z}}{|z|^2 + |w|^2}, \frac{|z|^2 - |w|^2}{|z|^2 + |w|^2} \right),$$

in homogeneous coordinates.

ii) $(\pi_N \circ \pi_N^{-1})([z, w]) = \dfrac{z}{w}.$

iii) Let $S = (0, 0, -1)$ denote the south pole of S^2. Show that the sets $U_N = \pi_N(S^2 \setminus \{N\})$, $U_S = \pi_N(S^2 \setminus \{S\})$ are a covering (cf. Definition 3.19) of $\mathbb{CP}^1$ (called an **atlas** in ☞ *differential geometry*). Describe the intersection $U_N \cap U_S$.

iv) Generalise (18.12) by taking the unit m-sphere $S^m := \{x \in \mathbb{R}^{m+1} \mid \|x\| = 1\}$ with north pole $N = (0, \ldots, 0, 1)$ and equatorial plane $\{(x_1, \ldots, x_m, 0) \in \mathbb{R}^{m+1}\} \cong \mathbb{R}^m$. The function $\pi_N \colon S^m \setminus \{N\} \to \mathbb{R}^{m+1}$, $\pi_N(x_1, \ldots, x_{m+1}) = \dfrac{1}{1 - x_{m+1}}(x_1, \ldots, x_m, 0)$ is the stereographic projection mapping $P \in S^m$ to the intersection of PN with the hyperplane $\mathbb{R}^m$.

Any point $P \in S^2$ specifies a unique line OP through the origin, so we have a function

$$(\theta, \phi) \mapsto r_{\theta, \phi}.$$

Since the antipodal point $-P$ lies on OP, we mat restrict to half a sphere, and choose a smaller parameter space:

$$(\theta, \phi) \in [0, \pi) \times [0, \pi) = \underbrace{(0, \pi) \times (0, \pi)}_{\approx \mathbb{R}^2} \sqcup \underbrace{\{0\} \times (0, \pi)}_{\approx \mathbb{R}} \sqcup \underbrace{\{(0, 0)\}}_{\{\infty\}}$$

$$\underbrace{\qquad\qquad\qquad\qquad\qquad}_{\mathbb{RP}^1 \,\cong\, S^1,\ \text{see Proposition 7.14}}$$

- lines $r_{\theta, \phi}$ with $\theta \in (0, \pi)$ are associated to points $P, -P$ on the hemispheres

$$S^+ := \left\{(x, y, z) \in S^2 \mid y > 0\right\} \approx \mathbb{R}^2, \qquad S^- := \left\{(x, y, z) \in S^2 \mid y < 0\right\} \approx \mathbb{R}^2$$

- lines $r_{0, \phi}$ with $\phi \in (0, \pi)$ are associated to $\pm Q$ on the half-circles

$$S^1 \setminus \{N, S\} = \{y = 0\} \cap \left\{x^2 + z^2 = 1,\ |z| \neq 1\right\} \approx \mathbb{R}$$

- the z-axis (line $r_{0,0}$) is associated to the poles $N, S = -N$.

To sum up, we've set up a 1-1 correspondence between pairs of antipodal points and the set $\mathscr{R}$ of lines through the origin

$$\begin{aligned} \mathscr{R} &\longrightarrow \mathbb{R}^2 \sqcup \mathbb{R} \sqcup \{\infty\} \\ r_{\theta, \phi} &\longmapsto P \\ r_{0, \phi} &\longmapsto \quad\ Q \\ r_{0, 0} &\longmapsto \qquad\quad N \end{aligned}$$

In ☞ *projective geometry* $\mathscr{R}$ is called **projective plane** $\mathbb{RP}^2$. It was shown above that we may look at it as the quotient of the sphere under the equivalence relation

$-P \sim P$. In terms of Sect. 17.1 one says the antipodal map $-\mathbb{1} \in \mathrm{Aut}\,(S^2)$ generates a free action of the subgroup $\mathbb{Z}_2 = \{\,\pm\,\mathbb{1}\}$ on the sphere. Hence

$$\mathbb{RP}^2 = S^2 \big/ \mathbb{Z}_2 \,. \tag{18.3}$$

The disjoint unions $\mathbb{RP}^2 \cong \mathbb{R}^2 \sqcup \mathbb{R} \sqcup \{\infty\}$ and $\mathbb{RP}^1 \cong \mathbb{R} \sqcup \{\infty\}$ are instances of a general phenomenon known as *CW-decomposition* (☞ *algebraic topology*).

18.2.2 Geometry on the Sphere

The positive number $K_{S^2(r)} = \dfrac{1}{r^2}$ is called **curvature** of the sphere $S^2(r)$. An asymptotic argument allows to treat the plane $\mathscr{E}$ as a sphere with radius $r = \infty$, for which the curvature is zero. The first shock caused by non-Euclidean geometry is that the theory of Sect. 18.1 is but a limiting case of elliptic geometry (☞ *Riemannian geometry*).

The crucial point is understanding what axiomatically defined lines are on the sphere.

Definition A spherical line, called **great circle** or **equator**, is the intersection between the sphere and a plane passing through the centre.

In suitable coordinates (that is, up to rotations) lines on S^2 are meridians.

Corollary *In spherical geometry there are no parallel lines.*

Proof Two planes π, τ through the sphere's centre determine great circles $\gamma_1 = \pi \cap S^2$, $\gamma_2 = \tau \cap S^2$. Their intersection $\gamma_1 \cap \gamma_2 = \pi \cap \tau \cap S^2$ equals the intersection of the line $r = \pi \cap \tau$ with the sphere. But as r goes through the centre, it punctures the surface in two opposite points $r \cap S^2 = \{u, v = -u\}$.   □

In the proof, the points u, v belong to infinitely many great circles, which are intersections of the pencil of planes with axis r. Therefore Birkhoff's axiom $\mathbf{B}_{\mathrm{II}}$ is false. That said, it fails only because of antipodal pairs of points, because if we take P and $Q \neq -P$ there is a unique equator through them, namely the intersection of the sphere with the (unique) plane $\mathscr{L}(PO, QO)$ through P, Q, O. Draw a picture to convince yourself this is true.

Definition 18.14 Take points $w = (w_1, w_2)$, $z = (z_1, z_2) \in \mathbb{C}^2$ and their standard Hermitian product $\langle z, w \rangle := \overline{w_1}z_1 + \overline{w_2}z_2$, cf. (16.2). The **Fubini–Study distance** of $[w], [z] \in \mathbb{CP}^1 = S^2$ is

$$d_{FS}\big([w], [z]\big) := 2 \arccos \frac{|\langle w, z \rangle|}{\langle w, w \rangle \langle z, z \rangle}.$$

The right side can be interpreted as cross-ratio of 4 suitable points, or as transition probability between pure quantic states [18].

On the sphere of radius r the distance between two points $P = [w]$, $Q = [z] \in S^2(r)$ is but a variation of the angular distance:

$$d_{S^2(r)}(P, Q) = r \, d_{FS}([w], [z]) = r \, P\hat{O}Q, \tag{18.4}$$

where $P\hat{O}Q$ is the acute angle ($\leqslant \pi$) formed by the radial segments $\overline{OP}, \overline{OQ}$.

Axiom $\mathbf{B_I}$ is false because lines are bounded: $d_{S^2}(P, Q) \leqslant \pi$, $\forall P, Q \in S^2$. The same argument also says $\mathbf{B_{III}}$ holds, since from any point there depart equatorial arcs of any length.

Two equators divide the surface in 4 regions called lunes. Each lune determines a unique dihedral angle (the aperture), which equals the plane angle formed by the tangent lines to the great circles. Therefore axiom $\mathbf{B_{IV}}$ is valid.

Exercise

i) Use the stereographic projection to verify which Birkhoff axioms transfer from the plane to the sphere.
ii) Identify antipodal points on the sphere, obtaining the projective plane $\mathbb{RP}^2$ as explained in (18.3). Decide which Birkhoff axioms still hold, and find counterexamples for the false ones.

Definition Let $P_1, P_2, P_3 \in S^2$ be non-collinear (not belonging to one equator). A **spherical triangle** is the portion of sphere bounded by equatorial arcs joining the vertices P_1, P_2, P_3.

Spherical trigonometry is in many respects similar to Euclidean trigonometry, *mutatis mutandis* (see the exercises at the end of the chapter). However, on the sphere there is a novelty, which doesn't hold on the plane: on $S^2(r)$ axiom $\mathbf{B_V}$ is typically false:

Proposition *The lengths a, b, c of a spherical triangle's edges are determined by the angles α, β, γ. Equivalently: similar spherical triangles are always congruent.*

Proof Exercise.   □

We shall not define the concept of area of a region, a delicate matter that requires caution and different instruments. Relying on intuition we might be convinced to believe

Theorem 18.15 (Girard) *The area of a triangle $\overset{\triangle}{ABC}$ on $S^2(r)$ with internal angles α, β, γ equals*

$$\mathscr{A}(\overset{\triangle}{ABC}) = r^2(\alpha + \beta + \gamma - \pi).$$

Proof First, $\overset{\triangle}{ABC}$ is the intersection of three lunes. For example, the lune opening between antipodal points $\pm B$ has meridian lines passing through A and C, and similarly the other two. Knowing the sphere's total area $4\pi r^2$, it follows that the area of a lune with aperture θ is $2r^2\theta$, and therefore

$$2\pi r^2 = 2\alpha r^2 + 2\beta r^2 + 2\gamma r^2$$

$$- 3\mathscr{A}(A\overset{\triangle}{B}C) + \mathscr{A}(A\overset{\triangle}{B}C),$$

whence the claim.   □

Regarding Theorem 18.1, therefore, in the spherical world we have:

Corollary 18.16 (Spherical Excess) *The angles of a spherical triangle add up to no less than π.*

Exercise Prove that there exist equilateral right triangles on S^2, i.e. with all right angles. (*Hint*: divide the sphere in 8 equal parts.)

Exercise What happens in Theorem 18.15 when we take the limit $r \longrightarrow \infty$? The answer is not obvious since α, β, γ secretly depend on r. (*Hint*: think of the stereographic projection.)

Finally, let's describe the sphere's isometries (see also Examples 17.5).

Lemma *Let $\phi \in Isom(S^2, d_{FS})$ be an isometry with three fixed points A, B, C not on the same great circle (not 'collinear'). Then $\phi = \mathbb{1}$.*

Proof If O is the centre, A, B, C determine LI vectors $A - O, B - O, C - O \in \mathbb{R}^3$. A vector $P - O$ is determined by the dot products with that basis. Hence any $P \in S^2$ is determined by its three spherical distances from A, B, C, which means the mutual distances of A, B, C. If ϕ fixes the latter, then it fixes P too.   □

Exercise Why is it necessary to assume above the points do not lie on an equator?

Theorem 18.17 $Isom(S^2, d_{FS}) \cong O(3)$.

Proof Take $A = (1, 0, 0), B = (0, 1, 0), C = (0, 0, 1)$ and an isometry ϕ. Then $\phi(A), \phi(B), \phi(C)$ are ON vectors in $\mathbb{R}^3$, and there exists a matrix $g \in O(3)$ such that $\phi(A) = g(A), \phi(B) = g(B), \phi(C) = g(C)$. In this way $\phi \circ g^{-1}$ has three fixed points and must be the identity, so $\phi = g \in O(3)$. This proves one inclusion. The other is automatic because orthogonal transformations are isometric (for d_{FS}) almost by definition.   □

As in Euclidean geometry, one could continue this train of thought by considering rotations and various types of reflections on the sphere. Reflections in points and great circles, for instance, can be shown to be conjugate to

$$\text{either} \quad \begin{pmatrix} 1 & 0 & 0 \\ 0 & -1 & 0 \\ 0 & 0 & -1 \end{pmatrix} \in SO(3), \quad \text{or} \quad \begin{pmatrix} 1 & 0 & 0 \\ 0 & 1 & 0 \\ 0 & 0 & -1 \end{pmatrix} \in SO^-(3).$$

(This covers all cases, as the determinant makes $O(3) = SO(3) \sqcup SO^-(3)$ a partition). Moreover, any isometry is the product of 3 reflections of this kind, at most.

18.3 The Hyperbolic Plane ☕

Section 18.2 is about a geometry with positive curvature. Here we wish to address the opposite situation, that is non-Euclidean models characterised by negative curvature. In this respect the object that most resembles the sphere $S^2(r)$ is

Definition (Beltrami 1868) The **pseudosphere** $\Psi^2(r)$ is the surface in $\mathbb{R}^3$ with constant curvature $K_{\Psi^2(r)} = -\dfrac{1}{r^2} < 0$.

Example The pseudosphere can be realised in the following way. In the Oxz plane consider the parametric curve

$$x(t) = \sin t, \quad z(t) = \cos t + \log \tan \frac{t}{2},$$

called **tractrix** (Fig. 18.9). As a geometric locus, it is the plane curve through $(1, 0)$ for which the tangent segment to the vertical axis has constant length (here, 1). Prosaically, it's the path of a stubborn dog x wanting to walk rightwards as he pulls (or is dragged by) his master z on a fixed leash.

Another parametrisation of this *pursuit curve* reads

$$x(s) = e^{-s}, \quad z(s) = \text{sgn}(s) \int_0^s \sqrt{1 - e^{-2u}} \, du, \quad s \in \mathbb{R}.$$

Its revolution around its asymptote $x = 0$ generates the surface $\Psi^2(1)$, which may be parametrised by $(x(t) \cos \phi, x(t) \sin \phi, z(t))$, where $(t, \phi) \in [0, +\infty) \times [0, 2\pi)$.

Fig. 18.9 Tractrix

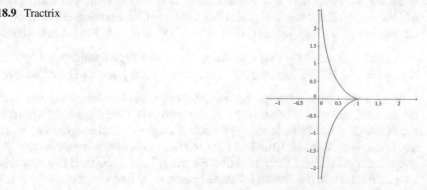

The negative curvature is responsible for the sharp boundary arising when rotating the cusp point, and for the unbounded aspect corresponding to the asymptote (☞ *Riemannian geometry*)

There exist other abstract models of hyperbolic geometry in two dimensions: each one is called **hyperbolic plane** because it is hyperbolically equivalent to the pseudosphere. Beside Ψ^2, the better known ones are

- the Poincaré disc $\mathbb{D}$, see Definition 18.18
- the Beltrami–Klein disc (Fig. 17.3)
- the (Poincaré) upper half-plane $\mathcal{H}^+$ (Definition 18.19)
- the upper hyperboloid Hyp^+ (Definition 18.23).

Despite being rather simple to describe (essentially because all of these surfaces can be immersed in $\mathbb{R}^3$), a detailed analysis requires the tools of ☞ *Riemannian geometry*. An outstanding reference on these topics, and much more, is [83]. We shall content ourselves with summarising some properties, evoking one model or another depending on the convenience.

Definition 18.18 The **Poincaré disc** is the interior of the unit circle in $\mathcal{E} \cong \mathbb{R}^2$:

$$\mathbb{D} := \left\{ (x, y) \in \mathbb{R}^2 \mid x^2 + y^2 < 1 \right\}.$$

We adopted the same symbol employed for the open unit disc in $\mathbb{C}$, and we'll indicate by $S^1 = \partial\mathbb{D}$ the boundary. A **hyperbolic line** is a circular arc in $\mathbb{D}$ that meets S^1 at right angles. Diameters are hyperbolic lines, seen as portions of circles with infinite radius.

Lemma *Hyperbolic lines come in two flavours:*

i) diameters d: $ax + by = 0$ or
ii) circular arcs γ: $x^2 + y^2 + 2dx + 2ey + 1 = 0$, with $d^2 + e^2 > 1$

for some $a, b, d, e \in \mathbb{R}$ and any $(x, y) \in \mathbb{D}$.

Proof A diameter is a line through the origin O, which has equation i); and vice versa. Regarding ii), the curve γ is a circle centred at $C = (-d, -e)$ with radius $r = (d^2 + e^2 - 1)^{1/2}$. Referring to Fig. 18.10 call B one of the intersection points of γ with the boundary $\partial\mathbb{D}$. Observe that $\overline{OC} = (d^2 + e^2)^{1/2}$, $\overline{OB} = 1$. The existence condition of the radius (hence of γ) says precisely $O\overset{\triangle}{B}C$ is a right triangle with hypotenuse OC, hence the leg BC has length r. Therefore S^1 and γ meet perpendicularly.   □

It is known that in any hyperbolic plane (with suitable distance function) axioms $\mathbf{B_I}$–$\mathbf{B_{IV}}$ hold, and we shall prove only the first two in $\mathbb{D}$. First though, let's remark that the failure of postulate $\mathbf{B_V}$ is conspicuous. Although we shall not prove it, looking at any of the Figs. 17.3, 18.10 or 18.11 it's not hard to convince ourselves that given a line l and a point $P \notin l$ there are infinitely many lines through P that don't meet l, i.e. parallel to it. In the Poincaré disc, for instance, to find arcs disjoint from a given

Fig. 18.10 Hyperbolic lines
in the disc

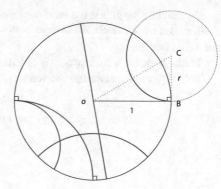

Fig. 18.11 Lines in the
upper half-plane: red lines are
parallel

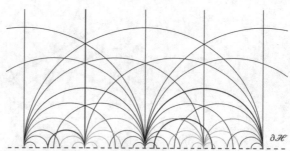

arc or diameter we have 2 coefficients to play around with, meaning that the system
of equations is generically overdetermined.

Proposition *For any $P, Q \in \mathbb{D}$ there exists a unique hyperbolic line $l \ni P, Q$.*
($\models_{\mathbb{D}} \boldsymbol{B}_{II}$)

Proof Suppose $P = (x_1, y_1)$, $Q = (x_2, y_2)$ lie on the arc $x^2 + y^2 + 2dx + 2ey + 1 =$
0 and consider the linear system $\begin{pmatrix} x_1 & y_1 \\ x_2 & y_2 \end{pmatrix} \begin{pmatrix} d \\ e \end{pmatrix} = \begin{pmatrix} k_1 \\ k_2 \end{pmatrix}$ where k_1, k_2 are constants.
Then we have two cases. If P, Q, O are aligned, clearly P, Q belong to a diameter.
If P, Q aren't collinear with O then $P - O, Q - O$ are LI vectors, the system's
determinant is non-zero and the unique solution satisfies $d^2 + e^2 > 1$. □

Definition The **hyperbolic angle** formed by two lines l, m through a point A is the
Euclidean angle formed by the tangents.

For example: let l, m be diameters through the centre O, and suppose $l \cap S^1 =$
$\{P_0, P_3\}$ and $m \cap S^1 = \{P_1, P_2\}$. Then

$$\cos^2\left(\tfrac{1}{2} P_2 \hat{O} P_3\right) = [P_0 P_1 P_2 P_3]$$

is the cross-ratio of the four points, cf. Definition 18.14.

Completely similar is the case of generic lines (red). The cross-ratio of the Q_i is determined by the angle at the intersection of the two arcs (any angle, since all have the same cosine).

That, albeit far from a proof, is at least leading us to believe in $\mathbf{B_{III}}$, $\mathbf{B_{IV}}$.

The **hyperbolic distance** between $z_1, z_2 \in \mathbb{D} = \{z \in \mathbb{C} \mid |z| < 1\}$ is

$$d_{\mathcal{H}}(z_1, z_2) := 2 \tanh^{-1} \left| \frac{z_1 - z_2}{1 - \overline{z_1} z_2} \right|. \tag{18.5}$$

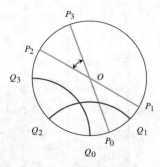

It will be proved in Proposition 19.3 that $d_{\mathcal{H}}$ truly is a distance.

For example, if $z_1 = 0$ is the origin O

$$d_{\mathcal{H}}(z, 0) = \frac{1}{2} \log \frac{1 + |z|}{1 - |z|}.$$

This map establishes a 1-1 correspondence between $\mathbb{R}$ and the diametral segment $(-1, 1) \times \{0\} = \mathbb{D} \cap \{\text{Im}(z) = 0\}$. Exercise 12.5 ii) actually shows it's an isometry, and therefore axiom $\mathbf{B_I}$ holds.

Exercise Interpret the hyperbolic distance's radial symmetry $d_{\mathcal{H}}(z, 0) = d_{\mathcal{H}}(|z|, 0)$, $\forall z \in \mathbb{D}$ as invariance under rotations $z \mapsto e^{i\theta} z$, $\theta \in \mathbb{R}$.

Convince yourself (or prove)

$$\lim_{P \to \partial \mathbb{D}} d_{\mathcal{H}}(O, P) = \lim_{|z| \to 1} d_{\mathcal{H}}(0, z) = \infty,$$

meaning that the distance of any $z \in \mathbb{D}$ to the boundary is infinite.

In order to describe the hyperbolic plane's isometries we'll use the apt third model we mentioned:

Definition 18.19 The (Poincaré) **upper half-plane**

$$\mathcal{H}^+ := \{z \in \mathbb{C} \mid \text{Im } z > 0\}$$

is a 2-dimensional hyperbolic model in which lines have equation $|z - x_0| = c^2$, for $x_0 \in \mathbb{R}$ and $c \in \mathbb{R}^+ \cup \{\infty\}$. In a down-to-earth fashion, these are

- half-circles with centre x_0 on the real axis (if $c \in \mathbb{R}^+$)
- vertical rays $z = x_0$ (corresponding to $c = \infty$).

Exercise 18.20 Can you see why the above lines intercept $\partial \mathcal{H}^+$ orthogonally? (*Hint:* imagine cutting the boundary $\partial \mathbb{D}$ in Fig. 18.10, flattening it out and stretching it to become a line.)

The corresponding distance is

$$d_{\mathcal{H}^+}(z, z') = \cosh^{-1}\left(1 + \frac{|z - z'|^2}{2(\operatorname{Im} z)(\operatorname{Im} z')}\right). \tag{18.6}$$

(The space $(\mathcal{H}^+, d_{\mathcal{H}^+})$ is also known as **Siegel half-plane**.)

Examples The following are isometries of $(\mathcal{H}^+, d_{\mathcal{H}^+})$:

i) horizontal translations: $T_v(x, y) = (x + v, y)$, $v \in \mathbb{R}$;
ii) dilations (homotheties): $D_\lambda(z) = \lambda^2 z$, $\lambda > 0$;
iii) reflections:

- the circular inversion about the unit circle $R_1(z) = \dfrac{z}{|z|^2}$; hence inversions about any circle centred at $\partial \mathcal{H}^+$ (by translating the former);
- the reflection $R_\infty(z) = -\bar{z}$ in the y-axis (which is a circle centred at ∞), and hence reflections in any vertical ray;
- the involution $R(z) = -\dfrac{1}{z} = (R_1 \circ R_\infty)z$;

iv) rotations R_θ: these can be thought of as composites of two reflections about lines (circles or straight lines) at an angle $\theta/2$.

Exercises Prove that

i) $D_\lambda \circ T_v \circ D_{\lambda^{-1}} = T_{\lambda v}$.
ii) All above examples preserve the orientation of the plane.
iii) $\mathcal{H}^+$ is a homogeneous space under translations and dilations.
(*Hint:* $T_{y_1 - \frac{y_2 x_1}{y_1}} \circ D_{y_2/y_1}$ maps (x_1, y_1) to (x_2, y_2).)
iv) Revisit Exercise 18.20 in the light of the above examples.

The isometries T_v, D_λ, R correspond to the standard action $z \mapsto \dfrac{az + b}{cz + d}$ of $\begin{pmatrix} a & b \\ c & d \end{pmatrix} \in \mathrm{PSL}(2, \mathbb{R})$ on $\mathcal{H}^+$, so we may write

$$T_v = \begin{pmatrix} 1 & v \\ 0 & 1 \end{pmatrix}, \qquad D_\lambda = \begin{pmatrix} \lambda & 0 \\ 0 & 1/\lambda \end{pmatrix}, \qquad R = \begin{pmatrix} 0 & 1 \\ -1 & 0 \end{pmatrix}.$$

Proposition *The group* $\mathrm{PSL}(2,\mathbb{R})$ *is generated by* T_v, D_λ, R *for* $v \in \mathbb{R}$, $\lambda \in \mathbb{R}^+$.

Proof As $\begin{pmatrix} a & b \\ c & d \end{pmatrix} = \begin{pmatrix} 1 & -ac^{-1} \\ 0 & 1 \end{pmatrix} \begin{pmatrix} 0 & 1 \\ -1 & 0 \end{pmatrix} \begin{pmatrix} 1 & -ca^{-1} \\ 0 & 1 \end{pmatrix} \begin{pmatrix} c^{-1} & 0 \\ 0 & c \end{pmatrix}$, the generic element of $\mathrm{PSL}(2,\mathbb{R})$ can be written as $T_{-a/c} \circ R \circ T_{-c/a} \circ D_{c^{-1}}$. □

Theorem $\mathrm{Isom}(\mathcal{H}^+, d_{\mathcal{H}^+}) = \mathrm{PSL}(2,\mathbb{R})$.

(See Examples 17.5.)

Let's focus on fixed points.

Exercises Let $f \in \mathrm{PSL}(2,\mathbb{R})$ be the isometry $f(z) = \dfrac{az+b}{cz+d}$. Prove

i) $\mathrm{Im}\, f(z) = \dfrac{\mathrm{Im}\, z}{|cz+d|^2}$;

ii) $z \in \partial\mathcal{H}^+ = \{z \in \mathbb{C} \mid \mathrm{Im}\, z = 0\}$ is a real fixed point if and only if $a + d \geqslant 2$.

By the exercise, isometries $\phi \in \mathrm{PSL}(2,\mathbb{R})$ can be distinguished algebraically by the trace:

Type	Trace	Fixed points	Jordan normal form
Elliptic	$\mathrm{tr}\,\phi > 2$	1 (and 1, conjugated, in $\mathcal{H}^-$)	D_λ
Parabolic	$\mathrm{tr}\,\phi = 2$	1 real, in $\partial\mathcal{H}^+$	T_v
Hyperbolic	$\mathrm{tr}\,\phi < 2$	2 distinct, in $\partial\mathcal{H}^+$	R_θ

Of particular interest are the discrete subgroups $\Gamma \subseteq \mathrm{PSL}(2,\mathbb{R})$, such as the modular group $\mathrm{PSL}(2,\mathbb{Z})$ (☞ *Fuchsian groups, automorphic functions*).

The second crucial point is

Theorem 18.21 *The spaces* $(\mathcal{H}^+, d_{\mathcal{H}^+})$ *and* $(\mathbb{D}, d_{\mathcal{H}})$ *are isometric.*

Proof The **Cayley transform**

$$C : \mathcal{H}^+ \to \mathbb{D}, \ z \mapsto i\frac{z-i}{z+i} \tag{18.7}$$

is the isometry we seek. In real coordinates it reads $C(x,y) = \left(\dfrac{x^2+y^2-1}{x^2+(y-1)^2}, \dfrac{2xy}{x^2+(y-1)^2} \right)$. □

In summary, we may work in any model indifferently. A **hyperbolic triangle** is given by three vertices in $\mathbb{D}$ joined by 'geodesic' segments, meaning line segments measured with distance $d_{\mathcal{H}}$. From that one proves the analogue of Theorem 18.15:

Theorem *The area of a hyperbolic triangle equals* $\mathscr{A}(\overset{\triangle}{ABC}) = \pi - \alpha + \beta + \gamma$.

And then, cf. Theorem 18.1,

Corollary 18.22 (Hyperbolic Excess) *The sum of the angles of a hyperbolic triangle is less than π.*

Finally, here is the last model we promised, which somewhat justifies the omnipresent adjective 'hyperbolic'.

Definition 18.23 The **upper hyperboloid**

$$\text{Hyp}^+ = \{(x, y, z) \in \mathbb{R}^3 \mid z^2 = 1 + x^2 + y^2, z > 0\}$$

is a model of the hyperbolic plane whose lines are the intersections of Hyp^+ with planes through the origin (hyperbolas).

The model is strikingly analogous to the sphere for at least two reasons. First, let's embed Hyp^+ in the pseudo-Euclidean space $\mathbb{R}^{2,1}$ with indefinite Lorenz product $(x, y, z) \cdot_L (x', y', z') := -xx' - yy' + zz'$ (see Remarks 15.1). Then the upper hyperboloid is indeed one half of the unit sphere of $\mathbb{R}^{2,1}$.

Second, the geometry is entirely governed by a stereographic projection $\pi_S \colon \text{Hyp}^+ \to \mathbb{D} \subseteq \mathbb{R}^2$ exiting the south pole $S = (0, 0, -1)$. If $P = (x, y, z) \in \text{Hyp}^+$ then the Euclidean line PS intersects $\{z = 0\}$ at $\left(\dfrac{x}{1+z}, \dfrac{y}{1+z}, 0\right)$:

$$\pi_S^{-1} \colon \mathbb{D} \longrightarrow \text{Hyp}^+$$
$$(u, v) \longmapsto \left(\frac{2u}{1 - u^2 - v^2}, \frac{2v}{1 - u^2 - v^2}, \frac{1 + u^2 + v^2}{1 - u^2 - v^2}\right)$$

Exercise Sketch a picture for the hyperbolic projection, and show that π_S^{-1} maps lines in $\mathbb{D}$ to lines in Hyp^+.

Taking the Poincaré distance, π_S becomes an isometry, and it can be proved that

$$\text{Isom}(\text{Hyp}^+) \left\{ M = (m_{ij}) \in \text{End } \mathbb{R}^3 \mid Mx \cdot_L My = x \cdot_L y, \ m_{33} > 0 \right\},$$

the pseudo-Euclidean version of $O(3)$, called $O^+(2, 1)$.

Summarising,

Theorem 18.24 *The following spaces are isometric models of two-dimensional hyperbolic geometry:*

$$\Psi^2 \cong \mathscr{H}^+ \cong \mathbb{D} \cong \text{Hyp}^+.$$

*Isometries form the **Möbius group***

$$\left\{ z \mapsto e^{i\theta} \frac{z + z_0}{1 - \overline{z_0} z} \;\middle|\; \theta \in \mathbb{R}, z_0 \in \mathbb{D} \right\} \cong \mathrm{PSL}(2, \mathbb{R}) \cong \mathrm{PSU}(1, 1) \cong \mathrm{O}^+(2, 1).$$

$$(18.8)$$

Needless to say, the same pattern is replicated in higher dimensions. The n-dimensional analogue of $\mathscr{H}^+$ is the space of matrices $\{X + iY \mid X \in \odot(n), Y \in \odot^+(n)\}$, called *Siegel half-space*, where $\odot(n)$ denotes real symmetric matrices $\odot^+(n)$ the subset of positive definite ones.

Perhaps more relevant is that $\mathscr{H}^+$ is an instance of a *Siegel domain* $\{(z, u) \in \mathbb{C} \times \mathbb{C}^n \mid \mathrm{Im}(z) - \|u\|_{\mathrm{Herm}} > 0\}$. This reduces to $\mathscr{H}^+$ for $u = 0$. A Siegel domain is biholomorphic to the ball $B(0, 1) \subset \mathbb{C}^n$ under $z \mapsto \left(\dfrac{z - i}{z + i}, \dfrac{2u_1}{z + i}, \ldots, \dfrac{2u_n}{z + i} \right)$, which involves the Cayley transform (18.7) and a form of stereographic projection (☞ *complex geometry, Lie theory*).

18.4 Trichotomy ☕

The synopsis below reviews the differences between the 3 geometries from the axiomatic point of view:

	B$_\mathrm{I}$	B$_\mathrm{II}$	B$_\mathrm{III}$	B$_\mathrm{IV}$	B$_\mathrm{V}$	n
$\mathscr{E}$	✓	✓	✓	✓	✓	1
S^2	✗	✗	✓	✓	✗✗	0
Ψ^2	✓	✓	✓	✓	✗✗	∞

Tolerable failures are marked with ✗ (as we saw, any such can be amended by passing to the quotient $\mathbb{RP}^2$, which still is a geometry of spherical type). On the contrary, ✗✗ denotes an ingrained problem, in-built in the geometry itself. Models are distinguished, and characterised, by the number n of parallels to a given line that can be drawn from a point outside the line.

The next objective is to swiftly lay out the trigonometry in $\Sigma = \mathscr{E}, S^2, \mathbb{D}$, and highlight the 'absolute' aspects, those that do not depend on the model under exam.

The notation is the same as always: in triangle $\overset{\triangle}{ABC}$ we have $\alpha = \hat{A}, \beta = \hat{B}, \gamma = \hat{C}$ and $\overline{BC} = a, \overline{CA} = b, \overline{AB} = c$.

- *Euclidean trigonometry*

Let's start with a few preparatory exercises

Exercises On $\mathscr{E}$, prove that

   i) as a consequence of **B$_\mathrm{III}$–B$_\mathrm{IV}$**, opposite angles are equal. Hence for $P \in l$ there exists a unique line $l^\perp$ perpendicular to l through P;

ii) if $\alpha = \alpha'$ and $\beta = \beta'$ then $\overset{\triangle}{ABC} \sim \overset{\triangle}{A'B'C'}$;

iii) if $(a, b, c) = k(a', b', c')$, $k > 0$, then $\overset{\triangle}{ABC} \sim \overset{\triangle}{A'B'C'}$;

iv) $\alpha = \beta \iff a = b$. Deduce that all equilateral triangles are similar.

Pythagoras's theorem lets itself generalise to arbitrary triangles $\overset{\triangle}{ABC}$, as follows:

Proposition (Cosine Law) *In the above conventions,* $c^2 = a^2 + b^2 - 2ab \cos \gamma$.

Proof Let $H \in BC$ be the foot of altitude AH drawn from A, and set $h = \overline{AH}$. By Theorem 18.2 we have $c^2 = h^2 + (a - b\cos\gamma)^2$ on $\overset{\triangle}{AHB}$ and $b^2 = h^2 + (b \sin\gamma)^2$ on $\overset{\triangle}{AHC}$. $\qquad\square$

Similarly,

Corollary (Sine Law) $\dfrac{\sin\alpha}{a} = \dfrac{\sin\beta}{b} = \dfrac{\sin\gamma}{c}$.

Proof It suffices to note that identity $\sin^2\gamma = 1 - \cos^2\gamma = 4a^2b^2 - (a^2+b^2-c^2)^2$ is symmetric in a, b, c. To finish, remember $(\sin\gamma)/c > 0$.

An alternative argument is to use the altitude $\overline{AA'}$, whence $b\sin\gamma = h = c\sin\beta$. The rest follows by symmetry. $\qquad\square$

- *Spherical trigonometry*

Let $\mathbf{v}_1, \mathbf{v}_2, \mathbf{v}_3 \in S^2$ be non-collinear vectors, identified with points on the sphere $P_i \longleftrightarrow \mathbf{v}_i = P_i - O$. As $\mathbf{v}_1 \cdot (\mathbf{v}_2 \times \mathbf{v}_3) \neq 0$, they form an ON basis of $\mathbb{R}^3$. Let V denote the matrix with those vectors as columns. Because of (18.4) the edges' lengths equal the angles

$$\theta_1 = P_2\hat{O}P_3, \qquad \theta_2 = P_3\hat{O}P_1, \qquad \theta_3 = P_1\hat{O}P_2,$$

all assumed $< \pi$. Define the symmetric matrix $V^{\mathrm{T}}V = \begin{pmatrix} 1 & c_3 & c_2 \\ c_3 & 1 & c_1 \\ c_2 & c_1 & 1 \end{pmatrix}$ where

$$c_1 = \cos\theta_1 = \mathbf{v}_2 \cdot \mathbf{v}_3, \qquad c_2 = \cos\theta_2 = \mathbf{v}_3 \cdot \mathbf{v}_1, \qquad c_3 = \cos\theta_3 = \mathbf{v}_1 \cdot \mathbf{v}_2.$$

Introduce the basis $\mathbf{w}_i = \Delta^{-1}\mathbf{v}_{i+1} \times \mathbf{v}_{i+2}$ ($i \in \mathbb{N}$ (mod 3)), where $\Delta = \det(V^{\mathrm{T}}V)$, and let $W = (V^{-1})^{\mathrm{T}}$ be the matrix of those vectors. By construction each $\mathbf{w}_i$ is orthogonal to the plane through O, P_{i+1}, P_{i+2}. The spherical angle at P_i is the angle formed by the tangents to the equatorial arcs meeting at P_i, i.e. the dihedral angle of the planes meeting along line OP_i. The internal angles ϕ_i measure π minus the angles formed by normal lines:

$$\cos\phi_i = -\frac{\mathbf{w}_{i+1} \cdot \mathbf{w}_{i+2}}{\|\mathbf{w}_{i+1}\| \, \|\mathbf{w}_{i+2}\|}.$$

Consider the matrix

$$W^T W = \frac{1}{\Delta^2} \begin{pmatrix} 1 - c_1^2 & c_1 c_2 - c_3 & c_1 c_3 - c_2 \\ c_1 c_2 - c_3 & 1 - c_2^2 & c_2 c_3 - c_1^2 \\ c_1 c_3 - c_2 & c_2 c_3 - c_1^2 & 1 - c_3^2 \end{pmatrix}.$$

Its rows are the cross products of the columns of $V^T V$, its entries the cofactors of $V^T V$. Therefore $\|\mathbf{w}_1\| = \dfrac{1 - c_1^2}{\Delta^2} = \dfrac{\sin \theta_3}{|\Delta|}$ (and actually, $\|\mathbf{v}_1 \times \mathbf{v}_2\| = \sin \theta_3$). Furthermore $\mathbf{w}_1 \cdot \mathbf{w}_2 = \dfrac{c_1 c_2 - c_3^2}{\Delta^2}$, from which we deduce $-\cos \phi_3 = \dfrac{c_1 c_2 - c_3}{\sin \theta_1 \sin \theta_2}$, and eventually

$$\cos \theta_3 = \cos \theta_1 \cos \theta_2 + \sin \theta_1 \sin \theta_2 \cos \phi_3.$$

In conclusion we have proved

Proposition 18.25 (Spherical Cosine Law) *In a spherical triangle where C is the vertex opposing side c, we have $\cos c = \cos a \cos b + \sin a \sin b \cos \hat{C}$.* ☐

From the above relationship it's possible to deduce further trigonometry theorems.

Exercises Prove

i) the *spherical Pythagoras' theorem*: $\gamma = \pi/2 \Longrightarrow \cos c = \cos a \cos b$.
ii) The *spherical sine law*: $\dfrac{\sin \alpha}{\sin a} = \dfrac{\sin \beta}{\sin b} = \dfrac{\sin \gamma}{\sin c}$.
iii) The dual relation to the cosine law: $\cos \gamma = -\cos \alpha \cos \beta + \sin \alpha \sin \beta \cos c$.
    (Remember $\gamma \mapsto \cos \gamma$ is a bijection $[0, \pi] \to [-1, 1]$.)
iv) Two spherical triangles with $a = a', b = b', \gamma = \gamma'$ are congruent.

Before we move on, let's remind that spherical geometry is no moot matter: it has a major employ in astronomy, spaceflight (communications / weather / GPS satellites) and stereochemistry (e.g. protein folding), to name a few.

- *Hyperbolic trigonometry*

The hyperbolic analogue of Proposition 18.25 on the pseudosphere reads

Proposition (Hyperbolic Cosine Law) *A hyperbolic triangle on $\Psi^2(1)$ satisfies $\cosh c = \cosh a \cosh b - \sinh a \sinh b \cos \gamma$.*

Proof Formally, $\Psi^2(1)$ is a sphere of radius i, and using relations Exercises 13.7 i) the claim follows. This argument is grounded in solid geometry (it's a matter of complexifying everything), even though at an elementary level it might seem awkward. ☐

Corollary (Hyperbolic Pythagoras) *In a hyperbolic right triangle where $\gamma = \pi/2$ we have $\cosh c = \cosh a \cosh b$.*

Proof To simplify a little let's consider only the case $C = O, A = (x, 0), B = (0, y)$. Then $\tanh(a/2) = y, \tanh(b/2) = x, \tanh(c/2) = \sqrt{\dfrac{x^2 + y^2}{1 + x^2 y^2}}$, from which we deduce $\cosh(a) = \dfrac{1 + x^2}{1 - x^2}, \cosh(b) = \dfrac{1 + y^2}{1 - y^2}$ and $\cosh(c) = \dfrac{(1 + x^2)(1 + y^2)}{(1 - x^2)(1 - y^2)}$.

$\square$

Exercise Prove the *hyperbolic sine law*: $\dfrac{\sin \alpha}{\sinh a} = \dfrac{\sin \beta}{\sinh b} = \dfrac{\sin \gamma}{\sinh c}$.

Two final exercises to compare the three trigonometries:

Exercises

i) Let $\overset{\triangle}{ABC}$ be an isosceles right triangle with legs of length 1. Show that the hypotenuse measures

$$\arccos\left(\cos^2(1)\right), \qquad \sqrt{2}, \qquad \cosh^{-1}\left(\frac{e^2 + 2 + e^{-2}}{4}\right),$$

in $S^2, \mathscr{E}, \mathscr{H}$ respectively. Compare the three numbers. The naive moral is that the hyperbolic world stretches lengths, whereas spherical geometry compresses them.

ii) Take a small triangle (with edges $\ll r$) and use the Taylor expansion of cos, sin to compare the three cosine laws, the three sine laws and the three Pythagoras' theorems. Conclude that spherical geometry and hyperbolic geometry are asymptotically Euclidean.

Before we wrap up, we include as a curiosity three deep theorems that reflect, each in its own way, the three-fold division of geometric theories, and at the same time link ☞ metric geometry to topology to ☞ complex geometry.

Theorem (**Constant-Curvature Surfaces**) *Up to isometries, the unique connected surfaces $\Sigma \subseteq \mathbb{R}^3$ with constant curvature K are*

Σ	S^2	$\mathscr{E}$	Ψ^2
K	> 0	0	< 0

Theorem (**Gauß–Bonnet, 1848**) *Let $K(r)$ denote the curvature function of $\Sigma = S^2(r), \mathscr{E}, \Psi^2(r)$. On any triangle $\overset{\triangle}{ABC} \subset \Sigma$,*

$$\frac{1}{r^2} \int_{\overset{\triangle}{ABC}} K(r) d\mathscr{A} = \alpha + \beta + \gamma - \pi.$$

The integral can be understood as an average, and matches the excess Theorems 18.1, 18.16 and 18.22.

A related matter: let $B(r) = \{x \in \Sigma \mid \|x\|_\Sigma < r\}$ be the disc of radius r in Σ. Expanding the area in Taylor series in the variable r gives

$$\mathscr{A}\big(B(r)\big) = \pi r^2 - \frac{K(r)}{12}\pi r^4 + \text{ terms of order } \geqslant 6 \text{ in } r.$$

Hence in spherical geometry areas are smaller, in hyperbolic geometry they are bigger:

$$\mathscr{A}_{\text{spherical}} \leqslant \mathscr{A}_{\text{Euclidean}} \leqslant \mathscr{A}_{\text{hyperbolic}}.$$

The translation of all this in complex-geometric language is the paramount

Theorem (Riemann's Uniformisation Theorem) *Any 1-connected Riemann surface is conformally equivalent to either S^2 or $\mathscr{E}$ or $\mathbb{D}$.*

For the record, this is the ultimate example of result baptised with a name different from the first person who proved it: here it was Poincaré and Koebe (both in 1907).

18.5   The Euler Characteristic

The present subject befits the theory called ☞ *simplicial homology*. For conciseness we shall rely on the intuition and notions of point-set topology.

Definition A (convex) **polyhedron** $K \subseteq \mathbb{R}^3$ is the intersection of a finite number m of half-spaces:

$$K = \{x \in \mathbb{R}^3 \mid Ax \leqslant b\},$$

where $b \in \mathbb{R}^m$ and the matrix $A \in \mathbb{R}^{3 \times m}$ has rank three.

The maximal-rank condition ensures the interior of K is non-empty (since we do not want K to lie in a proper affine subspace of $\mathbb{R}^3$). The *skeleton* of K is made of points that belong to at least one boundary plane. Hence the skeleton is a finite union of F convex polygons bounding K, called **faces**. One calls **edges** the common edges of polygons and **vertices** the points where the edges meet. (Thus K is the **convex hull** of its vertices).

Definition Let F, E, V denote the number of faces, edges and vertices of K. The integer

$$\chi(K) = V - E + F$$

is the **Euler–Poincaré characteristic** of K.

The topological classification of surfaces tells K is homeomorphic to a ball in $\mathbb{R}^3$, and its skeleton to a sphere S^2. Therefore

Theorem *Every compact, convex polyhedron K has characteristic $\chi(K) = 2$.*

We wish to consider the subclass of **regular polyhedra** (a.k.a. **Platonic solids**), which are made of congruent regular polygons of n edges so that k edges meet at each vertex.

Lemma *In any regular polyhedron: $kV = 2E = nF$.*

Proof Exercise.   □

Theorem 18.26 (Classification of Platonic Polyhedra) *There exist only 5 compact, convex regular polyhedra:*

k	3	3	3	4	5
n	3	4	5	3	3
F	4	6	12	8	20
K	tetrahedron	cube	dodecahedron	octahedron	icosahedron

Proof The argument is elementary, and based on the observation that the sum of the internal angles of an n-polygon is equal to $(n-2)\pi$. We propose another classical proof. The previous lemma, together with condition $V - E + F = 2$, implies

$$\frac{1}{n} + \frac{1}{k} = \frac{1}{2} + \frac{1}{E}.$$

So first of all $n, k \geqslant 3$, but they cannot be both > 3 (since $E > 0$). Moreover, if $n = 3$ then $k < 6$, and symmetrically $k = 3 \implies n < 6$. This exhausts all possibilities, giving the instances of the table.   □

The analogue of Platonic solids in the plane are the (infinitely many) regular polygons of $E > 2$ edges. Surprisingly, in dimension higher than 3 there is a small number of so-called regular *polytopes*. Schläfli classified all possibilities: in dimension 4 there are five, and in each dimension $\geqslant 5$ only three (generalising the tetrahedron, cube and octahedron).

The Euler–Poincaré characteristic has two hip applications that we'll spend a few words on: (1) interior design and (2) cartography.

(1) As any polyhedron is topologically like S^2, Theorem 18.26 can be interpreted as a classification of regular *tessellations*, or *tilings*, of the sphere. Let's briefly mention for comparison how Euclidean and hyperbolic tilings can be classified too.

Consider a tessellation of $\mathcal{E}$ where the tiles are regular, congruent n-polygons, with k polygons touching at each vertex. On $\mathcal{E}$ this can happen only if $k\dfrac{(n-2)\pi}{n} = 2\pi$, hence $\dfrac{1}{n} + \dfrac{1}{k} = \dfrac{1}{2}$. The only solutions are $(k, n) = (3, 6), (4, 4), (6, 3)$, which produce the familiar triangular, square and hexagonal tilings of the plane.

Exercise Explain why it is possible to place 6 circles around a seventh, all of the same size, so that each circle touches at least three others.

A recent survey article about regular planar tessellations is [100]. There's much more flexibility on the hyperbolic plane: the absence of $\mathbf{B_V}$ (p. 404) guarantees that the larger the number of a polygon's edges, the smaller become internal angles. A hyperbolic polygon's internal angles add up to less than $(n-2)\pi$, so there always exists a tiling provided $\dfrac{1}{n} + \dfrac{1}{k} < \dfrac{1}{2}$, and there are infinitely many possibilities.

(2) The second application is even simpler to formulate and conceptualise [33, 58]. Consider a plane region divided in subregions, say a geopolitical map of states. The problem is to find the smallest number of colours needed to colour it, so that adjacent regions can be distinguished by different colours. Adjacent regions have a common boundary curve, so if they only touch at one point they are not considered adjacent.

Theorem 18.27 (Four-Colour Theorem) *Four colours are enough to colour any plane map.*

The proof is intricate to follow (☞ *graph theory*) and was established by Appel and Haken (1976) with a computer in 1200 h, an eternity by today's standards. Despite the successive simpler arguments, no-one has still managed to find a proof that doesn't rely on computational tools. The uncommon nature of the proof—it was the first important theorem proved by a machine—and the complexity of the verifiable part caused quite a controversy. The mathematical community still asks whether this really represents a proof, since nobody can check the billions of computations done by the computer. Furthermore, some mathematicians object to the inelegance of a proof based on a case-by-case analysis, and beauty in mathematics is very debated. For comparison, the proof of the weaker 'five-colour theorem' (Heawood 1890, Kempe 1879) is a completely theoretical three-liner. That said, computer science, engineering or any area of applied mathematics demand viable tools that could well eclipse any aesthetic ideal of proof. Besides, twenty-second century mathematics might look very different from what we practise today.

It's very tempting to make some experiments and try to produce a region that touches all the others. The typical mistake is to focus on one region, and forgetting that the remaining regions can be coloured with just three colours. In fact, if certain regions' colours are selected beforehand, it's easy to come up with a map that can't be coloured without using more than four colours. Sometimes, changing these regions' colours might solve the problem. The point is that a region should be different only from those it touches, not necessarily from the ones bordering the ones it touches. For example, in a world map Brazil and Chile don't have a common border, so they are allowed to be of the same colour.

The next question would be the colouring problem of surfaces other than the plane. The smallest number of necessary colours (the *chromatic number*) depends on the surface's Euler–Poincaré characteristic χ. Theorem 18.27 says the plane's chromatic number is 4, and one can prove that the same holds for the sphere (which has Euler characteristic $\chi = 2$, as the plane).

Theorem (Heawood 1890, Ringel–Youngs 1968) *For compact surfaces without boundary, except for the sphere and the Klein bottle, the chromatic number is $p = \left\lfloor \frac{7+\sqrt{49-24\chi}}{2} \right\rfloor$.*

For example the torus has characteristic zero, so we need at least 7 tones to colour an arbitrary map on the torus. The projective plane has $\chi = 1$ and $p = 6$. The Klein bottle ($\chi = 0$) is an exception because its chromatic number is $p = 6$ (Franklin 1934), while the formula would give the non-optimal value 7.

The Möbius strip (Fig. 18.12) has $p = 6$ as well (Tietze 1910), but the formula doesn't apply, since this surface has a boundary.

Fig. 18.12 Möbius strip

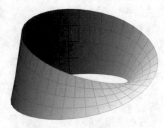

Chapter 19
Metric Spaces

In this introduction to metric notions we didn't, with a few occasional exceptions, use topology. A complete reference is [75].

Prerequisites: the proof of the Bolzano–Weierstraß Theorem 19.19 and Banach's fixed-point Theorem 19.24 rely on (little) infinitesimal calculus.

Definition A **distance** (or **metric**) on a set $X \neq \varnothing$ is a function $d \colon X \times X \to [0, +\infty)$ such that:

i) $d(x, y) = 0 \iff x = y$ (*positive definiteness*)
ii) $d(x, y) = d(y, x)$ (*symmetry*)
iii) $d(x, y) \leqslant d(x, z) + d(z, y)$ (*triangle inequality*)

for every $x, y, z \in X$. The pair (X, d) is called a **metric space**, and X is **metrised** by d.

Examples 19.1

i) The function $d(x, y) = 1 \; \forall x \neq y$ (and $d(x, x) = 0$) is a distance, called discrete or trivial.
ii) Let (X, d) be a metric space. The function $\overline{d} \colon X \times X \to \mathbb{R}$, $\overline{d}(x, y) = \min\left(1, d(x, y)\right)$ is a distance: conditions i)-ii) are obvious, so let's prove iii). Since $\overline{d} \leqslant 1$, if $\overline{d}(x, z) + \overline{d}(z, y) \geqslant 1$ there is nothing to prove. If, instead, $\overline{d}(x, z) + \overline{d}(z, y) < 1$, then $\overline{d}(x, z) = d(x, z)$, $\overline{d}(z, y) = d(z, y)$, and so $\overline{d}(x, y) \leqslant d(x, y) \leqslant d(x, z) + d(y, z) = \overline{d}(x, z) + \overline{d}(z, y)$.
iii) Let X be a finite set and define $d(A, B) = \operatorname{card}(A \triangle B)$ for any $A, B \subseteq X$, where $\triangle$ is the set-theoretical symmetric difference. Thus d metrises $\mathscr{P}(X)$.
iv) Let p be a prime number, and for $m, n \in \mathbb{Z}$ call r the largest integer such that p^{r-1} divides $m - n$. Thus $d(m, n) = \dfrac{1}{r}$, for $m \neq n$, is a distance on $\mathbb{Z}$.

© The Author(s), under exclusive license to Springer Nature Switzerland AG 2021
S. G. Chiossi, *Essential Mathematics for Undergraduates*,
https://doi.org/10.1007/978-3-030-87174-1_19

The triangle inequality is the only non-immediate property. For that, suppose $m - n = p^{r-1}k$, $n - q = p^{s-1}k'$ with p not dividing k, k'. Then $m - q = p^{t-1}k''$ for some $t \geq \min\{r, s\}$ and k'' not divided by p. Therefore

$$d(m, q) = \frac{1}{t} \leqslant \frac{1}{\min\{r, s\}} = \max\left\{\frac{1}{r}, \frac{1}{s}\right\}$$

$$= \max\{d(m, n), d(n, q)\} \leqslant d(m, n) + d(n, q).$$

v)   Consider real polynomials $p(x) = \sum_{i=0}^{n} p_i x^i$, $q(x) = \sum s j = 0^m q_j x^j \in \mathbb{R}[x]$. Due to the absolute value's properties and the additivity of the supremum, $d(p, q) = \sup\{|p_k - q_k| \mid 0 \leqslant k \leqslant \max\{m, n\}\}$ is a metric on $\mathbb{R}[x]$.

Exercises Show that

i)   Rectangular matrices $\mathbb{R}^{m \times n}$ are metrised by $d(A, B) = \mathrm{rk}(A - B)$.
    (*Hint*: the rank is the maximum number of LI columns, so it can only vanish in one case; furthermore, it's subadditive.)
ii) $d(x, y) = |\log(y/x)|$ is a distance on $\mathbb{R}^+$. What's the distance between x and x', as $x' \longrightarrow 0$?

From a geometric perspective the following metric spaces are rather relevant.

Examples 19.2

vi)   The real line $\mathbb{R}$ with $d(x, y) = |x - y|$ is a metric space: the triangle inequality holds by Proposition 7.6.
vii) For any 1-1 function $f : X \to \mathbb{R}$, $d(x, y) = |f(x) - f(y)|$ is a metric on $\mathbb{R}$.
     One of the axioms of elementary geometry is based on this, cf. p. 403.
viii) Probably the mother of all examples is $\mathbb{R}^n$ with the **Euclidean distance**

$$d_{\mathrm{Eucl}}(x, y) := \sqrt{(x_1 - y_1)^2 + \ldots + (x_n - y_n)^2},$$

where $x = (x_i)_{i=1,\ldots,n}$, $y = (y_i)_{i=1,\ldots,n}$, generalising case vi). The metric space

$$\mathbb{R}_{\mathrm{Eucl}}^n := (\mathbb{R}^n, d_{\mathrm{Eucl}})$$

is called **n-dimensional Euclidean space** (Euclidean line for $n = 1$, Euclidean plane for $n = 2$ etc.).

ix) Other distances on $\mathbb{R}^n$ are the 'taxi distance' $d_1(x, y) := \sum_{i=1}^{n} |x_i - y_i|$ and the **supremum distance**

$$d_\infty(x, y) := \sup_{i=1,\ldots,n} |x_i - y_i|.$$

x) Let $\mathbb{D} := \{z \in \mathbb{C} \mid |z| < 1\}$ denote the open unit disc in $\mathbb{C}$. The map

$$d_{\mathcal{H}}(z, z') = \tanh^{-1}\left|\frac{z - z'}{1 - \overline{z}z'}\right|, \qquad z, z' \in \mathbb{D}$$

is called **Poincaré metric**, see (18.5).

Proposition 19.3 *The Poincaré metric $d_{\mathcal{H}}$ is a distance.*

Proof Positive definiteness and symmetry are direct consequences of the modulus' properties. As for the triangle inequality, first of all we reduce to $z = 0, z' \in \mathbb{R}$ by means of a roto-translation (an isometry of the homogeneous space $\mathbb{D}$). So we must show $d_{\mathcal{H}}(0, z) \leqslant d_{\mathcal{H}}(0, u) + d_{\mathcal{H}}(u, z)$. Putting $|z| = r, |u| = s$ this becomes:

$$\tanh^{-1}(r) \leqslant \tanh^{-1}(s) + \tanh^{-1}\left|\frac{u - r}{1 - ru}\right|.$$

Now, when $r \leqslant s$ that is true because $\tanh^{-1}$ is increasing. When $r > s$ the claim is

$$\tanh^{-1}\left|\frac{s - r}{1 - rs}\right| \leqslant \tanh^{-1}\left|\frac{u - r}{1 - ru}\right|,$$

so it's enough to prove $\dfrac{s - r}{1 - rs} \leqslant \left|\dfrac{u - r}{1 - ru}\right|$. But the map $\theta \mapsto \dfrac{|r - se^{i\theta}|^2}{|1 - rse^{i\theta}|^2}$ has a minimum at $\theta = 0$, i.e. $u = s$, thus proving the claim. $\square$

xi) The upper plane $\mathcal{H}^+ := \{z \in \mathbb{C} \mid \operatorname{Im} z > 0\}$ is metrised by the hyperbolic distance (18.6).

xii) Consider $\mathbb{C}^{n+1}$ with (16.2), the standard Hermitian product $\langle w, z \rangle$, and distance d_{FS} from Definition 18.14. The quadratic coefficient in the expansion of $d_{FS}([w + t\xi], [z + t\zeta])$ in powers of t is called **Fubini–Study metric** on the complex projective space $\mathbb{CP}^n$ (and d_{FS} is its 'integrated form').

Exercises 19.4 Prove that

i) for any x, y, z, w in a metric space,

$$|d(x, y) - d(z, w)| \leqslant d(x, z) + d(y, w). \tag{19.1}$$

From that deduce the triangle inequality, and then the 'other' triangle inequality $|d(x, y) - d(z, y)| \leqslant d(x, z)$.

ii) $d_\infty(x, y) \leqslant d_{\text{Eucl}}(x, y) \leqslant d_1(x, y) \leqslant n\, d_\infty(x, y)$ for every (x, y) in $\mathbb{R}^n$.

iii) Consider a family's genealogy tree and let's measure kinship by declaring that the separation between person A and person B is 0 if $A = B$, otherwise we count the smallest number of descendant relationships connecting A, B, going once upwards and once downwards along the tree. In the scheme below: I would have 2 degrees of separation from my nan, and 3 from any of my first cousins.

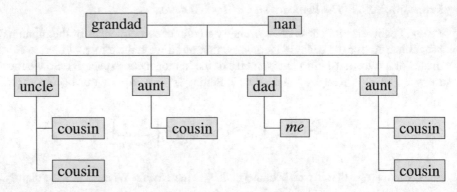

Build a more complicated family tree of your choosing and show that the notion of degree of separation satisfies the triangle inequality only in very peculiar circumstances (beware that the shortest path might not go through the nearest common ancestor). Hence it doesn't define a metric.

19.1   Metric Topology

Definition 19.5 Let S be a subset in a metric space (X, d). We call **diameter** of S the quantity

$$\text{diam}(S) := \sup \{ d(a_1, a_2) \mid a_1, a_2 \in S \}.$$

In case $\text{diam}(S) < \infty$ we say S is **bounded**, otherwise **unbounded**.

For any set Z, a map $f : Z \to (X, d)$ is **bounded** if the range $f(Z) \subseteq X$ is a bounded set.

Examples

i) Real intervals (a, b), $[a, b)$, $(a, b]$ and $[a, b]$ are bounded (in all cases diam $= b - a$), and their finite unions are bounded as well.

ii) Consider $\mathbb{R}^+$ with metric $d(a, b) = \left| a^{-1} - b^{-1} \right|$. The subset $\mathbb{N}$ is bounded (diam $\mathbb{N} = 1$), while $\{ 1/n \mid n \in \mathbb{N}^* \}$ is unbounded.

iii) More bounded real maps can be found in Example 10.10.

Exercise Show $S \subseteq \mathbb{R}$ is bounded if and only if it is contained in a bounded interval.

Proposition *Take $S \subseteq \mathbb{R}_{Eucl}$. Then* $\text{diam}(S) = \sup S - \inf S$.

Proof If $S = \varnothing$ then $\inf S = +\infty$, $\sup S = -\infty$ and hence $\operatorname{diam} \varnothing = -\infty - \infty = -\infty$.

Supposing $\sup S = +\infty$, take $b \in S$. For any $p > 0$ there exists $a \in S$ with $a > b + p$, so $\operatorname{diam} S > p$. As p is arbitrary, $\operatorname{diam} S = +\infty = \sup S - \inf S$. Similarly if $\inf S = -\infty$ (exercise).

Now assume $\sup S$, $\inf S \in \mathbb{R}$ and take $r > 0$. There exist $a, b \in S$ such that $a - r/2 < \inf S \leqslant a \leqslant b \leqslant \sup S < b + r/2$, from which $\sup S - \inf S \leqslant b - a + r \leqslant \operatorname{diam} S + r$. Therefore $\sup S - \inf S \leqslant \operatorname{diam} S$. But $\inf S \leqslant S \leqslant \sup S$, and then $\operatorname{diam} S \leqslant \sup S - \inf S$. $\qquad\square$

Exercises 19.6 Let $A \subseteq B \subseteq (X, d)$ be subsets. Show that

i) B bounded $\implies A$ bounded, and $\operatorname{diam}(A) \leqslant \operatorname{diam}(B)$;
ii) $\operatorname{dist}(x, B) \leqslant \operatorname{dist}(x, A) \leqslant \operatorname{dist}(x, B) + \operatorname{diam}(B)$ for all $x \in X$;
iii) if A, B are bounded and not disjoint, $\operatorname{diam}(A \cup B) \leqslant \operatorname{diam}(A) + \operatorname{diam}(B)$.

Definition 19.7 Let (X, d) be a metric space, $x \in X$ a point and $r \geqslant 0$ a real number. The set

$$B(x, r) = \left\{ y \in X \mid d(x, y) < r \right\}$$

is the **open ball** centred at x with radius r (for the distance d), while

$$\overline{B(x, r)} = \left\{ y \in X \mid d(x, y) \leqslant r \right\}$$

is the **closed ball** (centred at x with radius r, for the distance d).

The name comes from the Euclidean world. In contrast to the latter ambient, where balls are round, the balls in Example 19.2 xi) are hypercubes.

Examples 19.8

i) The open balls in $\mathbb{R}$ are the open intervals (a, b): the centre is the mid-point $\dfrac{b + a}{2}$, the radius the half-span $\dfrac{b - a}{2}$. Intervals of type $[a, b]$ are the closed balls.
ii) Draw the balls of Example 19.2 xi) on the plane $\mathbb{R}^2$.
iii) The relationship between $B(x, r)$ and the sphere $S(x, r) = \left\{ y \in X \mid d(x, y) = r \right\}$ is subtle and sometimes counterintuitive. Take for instance the discrete distance of Example 19.1 i). Then the open unit ball reduces to the centre: $B(x, 1) = \{x\}$, the closed unit ball is the entire space: $\overline{B(x, 1)} = X$, the unit sphere is the punctured space: $S(x, 1) = X \setminus \{x\}$.
    What are the spheres of radius $r \neq 1$?

A metric space is a topological space, since balls allow to define *neighbourhood* systems: $A \subseteq (X, d)$ would be called open if any point $a \in A$ belongs to some ball $B(a, \epsilon) \subseteq A$.

Proposition *Take $y \in B(x, r) \subseteq X$ for some $x \in X, r \in \mathbb{R}^+$. Then there exists an $\epsilon > 0$ such that $B(y, \epsilon) \subseteq B(x, r)$.*

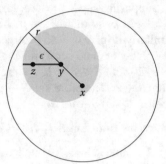

Proof Set $\epsilon = r - d(y, x) > 0$. Taking any $z \in B(y, \epsilon)$ we have $d(y, z) < \epsilon$. The triangle inequality implies $d(x, z) \leqslant d(z, y) + d(x, y) \leqslant \epsilon + d(x, y) = r$, meaning $z \in B(x, r)$. $\square$

The same argument works with any $0 < \epsilon \leqslant r - d(y, x)$.

Exercises 19.9 Prove the following statements.

i) Every open ball is contained in a closed ball, and conversely.
ii) The open ball $B(x, 1)$ at an arbitrary point x cannot contain $B(x, 2)$ as proper subset.
iii) Every ball is bounded: diam $B(x, r) = 2r$. (That's why the name 'diameter'.)
iv) $A \subseteq X$ bounded $\Longrightarrow A \subseteq B(x, r)$ for some x, with $r > \text{diam}(A)$.
v) $B(x, \epsilon) \cap B(y, \epsilon) = \varnothing$ for any $x \neq y \in X$ and any $\epsilon \leqslant d(x, y)/2$.
     Put equivalently: distinct points are contained in disjoint open balls. Such 'separation' property, indicated by $\mathbf{T_2}$, characterises **Hausdorff** topological spaces.
vi) If d is a distance, for any real number $\lambda > 0$ the function λd is a distance with the same balls as d:

$$B_d(x, r) = B_{\lambda d}(x, r/\lambda).$$

(B_d is the ball relative to the distance d.)
vii) Let d, d' be metrics such that $d(x, y) \leqslant k d'(x, y)$ for some $k > 0$ and all $x, y \in X$. Show that $B_{d'}(x, r/k) \subseteq B_d(x, r)$ for all $x \in X, r > 0$.

Definition Let (X, d) be a metric space. For any $\varnothing \neq Z \subseteq X$ the **distance to** Z is the map $\text{dist}(Z, \cdot) : X \to \mathbb{R}$

$$\text{dist}(Z, x) = \inf_{z \in Z} d(z, x).$$

This number always exists, since d is lower bounded (by 0).

Examples

i) Let's compute the distance of a point $P(x_0, y_0)$ from the line $r : ax + by + c = 0$ on the plane $\mathbb{R}^2_{\text{Eucl}}$. First, the unit vector $\vec{n} = \left(\dfrac{a}{\sqrt{a^2+b^2}}, \dfrac{b}{\sqrt{a^2+b^2}} \right)$ is normal to r. Moreover, if $Q(\xi, \eta) \in r$ the orthogonal projection of the vector $P - Q$ on $\vec{n}$ doesn't depend on the choice of Q. By elementary geometry reasons (the hypotenuse is longer than the legs: $|\cos \phi| \leqslant 1$, $\forall \phi$) we have $|\vec{n} \cdot (P - Q)| \leqslant |P - Q|$. Hence the distance equals the length of the projection:

$$\text{dist}(P, r) = |\vec{n} \cdot (P - Q)| = \frac{|a(x_0 - \xi) + b(y_0 - \eta)|}{\sqrt{a^2 + b^2}} = \frac{|ax_0 + by_0 + c|}{\sqrt{a^2 + b^2}}.$$

ii) We claim $\text{dist}(x, \mathbb{Q}) = 0 \ \forall x \in \mathbb{R}_{\text{Eucl}}$. By density, in fact, $\mathbb{Q} \cap (x - r, x + r) \neq \varnothing$ for any $r \in \mathbb{R}^+$, so $d(x, \mathbb{Q}) < r$. But this holds for any $r > 0$, so the claim follows.

More generally, for non-empty sets $Z, W \subseteq (X, d)$

$$\text{dist}(Z, W) := \inf \{ d(z, w) \mid z \in Z, w \in W \}.$$

Examples

i) $\text{dist}(\mathbb{R}, \mathbb{Q}) = 0 = \text{dist}(\mathbb{R}, \mathbb{R} \setminus \mathbb{Q})$.
ii) In $X = \mathbb{R}^2_{\text{Eucl}}$ take the y-axis $Z = \{ (0, z) \mid z \in \mathbb{R} \}$ and the graph of the hyperbola $W = \{ (w^{-1}, w) \mid w \in \mathbb{R}^* \}$. The sets Z, W are disjoint so the distance between arbitrary points never vanishes: $\| (0, z) - (1/w, w) \| > |w|^{-1} > 0$. But $\text{dist}(Z, W) = 0$.

Despite the name, though, dist is not a metric since $\text{dist}(Z, W) = 0$ if $Z \cap W \neq \varnothing$. It doesn't satisfy the triangle inequality either: take $A = [-3, -1]$, $C = [-2, 2]$, $B = [1, 3] \subseteq \mathbb{R}$, so $\text{dist}(A, B) = 2 > 0 = \text{dist}(A, C) + \text{dist}(C, B)$.

Proposition 19.10 $\text{dist}(Z, \cdot)$ *satisfies* $\big| \text{dist}(Z, x) - \text{dist}(Z, y) \big| \leqslant d(x, y)$ *for all* $x, y \in X$.

Proof Due to the obvious symmetry it suffices to show $\text{dist}(Z, y) - \text{dist}(Z, x) \leqslant d(x, y)$ for all $x, y \in X$. By definition, for any $\epsilon > 0$, there exists a $z \in Z$ such that $\text{dist}(Z, x) + \epsilon \geqslant d(x, z)$, so $\text{dist}(Z, y) \leqslant d(z, y) \leqslant d(z, x) + d(x, y) \leqslant \text{dist}(Z, x) + \epsilon + d(x, y)$. $\qquad \square$

This result proves $\text{dist}(Z, \cdot)$ is, at any rate, Lipschitz (hence continuous).

Exercise 19.11 Let $Z \subseteq X$ be non-empty. For any $x \in X$ prove

$$\text{dist}(Z, x) = \inf \{ r \in \mathbb{R} \mid B(x, r) \cap Z \neq \varnothing \}.$$

Theorem *Let* (X, d) *be a metric space,* x *a point,* $\mathscr{A} = \{ A_i \} \subseteq \mathscr{P}(X)$ *a collection of subsets with* $\bigcap \mathscr{A} \neq \varnothing$. *Then*

$$\text{dist}\left(x, \bigcup_{i \in I} A_i \right) = \inf_{i \in I} \text{dist}(x, A_i), \qquad \text{dist}\left(x, \bigcap_{i \in I} A_i \right) \geqslant \sup_{i \in I} \text{dist}(x, A_i)$$

Proof First, $A_i \subseteq \bigcup \mathscr{A}$ clearly implies dist $\left(x, \bigcup \mathscr{A}\right) \leqslant$ dist(x, A_i) for all $i \in I$, which then gives dist $\left(x, \bigcup \mathscr{A}\right) \leqslant \inf_{i \in I}$ dist(x, A_i).

For the converse, fix $r \in \mathbb{R}^+$. There exists $z \in \bigcup \mathscr{A}$ such that $d(x, z) \leqslant$ dist $\left(x, \bigcup \mathscr{A}\right) + r$, and we also have $z \in A_{i*}$ for some A_{i*}. Therefore $\inf_{i \in I}$ dist$(x, A_i) \leqslant$ dist$(x, A_{i*}) \leqslant d(x, z)$. Since r is arbitrary, dist $\left(x, \bigcup \mathscr{A}\right) \geqslant \inf_{i \in I}$ dist(x, A_i), which proves the first relation.

As for the second relation, $A_i \supseteq \bigcap \mathscr{A}$ for all i, and Exercise 19.6 allows to conclude.   □

The above inequality is typically strict: consider unit balls centred at $\pm 1 \in \mathbb{C}$

$$B_1 := \left\{ z \in \mathbb{C} \colon |z - 1| \leqslant 1 \right\}, \quad B_2 := \left\{ z \in \mathbb{C} \colon |z + 1| \leqslant 1 \right\}.$$

We have dist$(i, B_1 \cap B_2) = $ dist$(i, 0) = 1$ but dist$(i, B_1) = $ dist$(i, B_2) = \sqrt{2} - 1$.

Definition 19.12 A point $x \in (X, d)$ is called **accumulation point** for a subset $S \subseteq X$ if

$$\text{dist} \left(x, S \setminus \{x\} \right) = 0.$$

A point $x \in S$ is **isolated** if it is not an accumulation point.

Observe we are not demanding that an accumulation point belong in S. The underlying idea is that there exist in S points lying as close as we want to an accumulation point.

Examples 19.13

i) The accumulation points of $S = [0, 1) \cup \{3\} \subseteq \mathbb{R}_{\text{Eucl}}$ are all those in $[0, 1]$. The number 3 is isolated.

ii) Any real number is an accumulation point for $\mathbb{Q}$ (and $\mathbb{R}$).

iii) The set $S = \left\{ \dfrac{1}{n} \right\}_{n \in \mathbb{N}^*} \cup \{0\}$ has a unique accumulation point 0, and each $\dfrac{1}{n}$ is isolated.

iv) The graph of the map $f \colon \mathbb{R}^+ \to [-1, 1]$, $f(x) = \sin \dfrac{1}{x}$, is called **topologist's sine curve** due to its special features. It roughly looks like a sinusoidal curve that is being compressed towards the y-axis. As it approaches the origin, the number of oscillations (frequency) increases indefinitely (Fig. 19.1).

Call Z the vertical segment $\{0\} \times [-1, 1]$ and pick a point y on it. Since an arbitrary ball $B(y, r)$ will intersect graf(f) infinitely many times, by Exercise 19.11 dist$(\text{graf}(f), y) = 0$, meaning that Z is entirely made of accumulation points for the curve. See also Example 19.18 iv).

Fig. 19.1 The map
$y = \sin \dfrac{1}{x}$ for $x > 0$

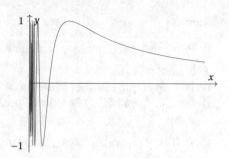

Definition A subset $S \subseteq (X, d)$ is **dense** if $\text{dist}(x, S) = 0$ for all $x \in X$. Equivalently, every point in X is an accumulation point for S. Borrowing a topological notation we'll write $\overline{S} = X$.

Remarks of topological nature 19.1

1) The above definition of metric density is *a priori* different from the order density of Lemma 6.29. Dense subsets are metrically big (the whole metric space is dense in itself: $\overline{X} = X$), but can be thought of as small/thin relative to, say, cardinality: $\mathbb{R}$ is uncountable, but several of its dense subsets are countable, like $\mathbb{Q}$ or dyadic numbers $\mathfrak{D} \subset \mathbb{Q}$, see (8.1). Regarding Theorem 8.10 we have: $\overline{\mathbb{N}} = \mathbb{N}$, $\overline{\mathbb{Z}} = \mathbb{Z}$ (since discrete), whereas $\overline{\mathbb{Q}} = \mathbb{R} = \overline{\mathbb{R} \setminus \mathbb{Q}}$.
2) With Definition 19.12 we simplified the general picture. Just for the sake of completeness, we mention that a point $x \in (X, d)$ is called

   - *external* to $S \subseteq X$ if there exists a ball $B(x, r)$ disjoint from S;
   - *internal* if there exists a ball $B(x, r)$ entirely contained in S;
   - *adherent* if it's not external: $B(x, r) \cap S \neq \varnothing$ for every ball $B(x, r)$.
     Adherent points fall in two categories:
   - *accumulation points*, if $B(x, r) \cap (S \setminus \{x\}) \neq \varnothing$ for any ball $B(x, r)$;
   - *isolated* if there is a ball $B(x, r)$ such that $B(x, r) \cap S = \{x\}$.

   The set of adherent points defines the *topological closure* $\overline{S}$ of S.

Definition Two distances d_1, d_2 on X are **equivalent** if there exist positive numbers $\lambda, \mu \in \mathbb{R}^+$ such that

$$\lambda d_1(x, y) \leqslant d_2(x, y) \leqslant \mu d_1(x, y)$$

for all $x, y \in X$.

Exercises

i) Prove that the distance $\overline{d}$ of Example 19.1 iii) and $\delta(x, y) = \dfrac{d(x, y)}{1 + d(x, y)}$ are equivalent to d.

ii) Is 'being equivalent' an equivalence relation for metrics?

iii) Let (X, d) and (Y, δ) be metric spaces, and $f : [0, +\infty)^2 \to \mathbb{R}$ a function such that

- $0 \leqslant f(c_1, c_2) \leqslant f(a_1, a_2) + f(b_1, b_2)$ whenever $c_1 \leqslant a_1 + b_1$ and $c_2 \leqslant a_2 + b_2$;
- $f(a, b) = 0$ if and only if $a = b = 0$.

Prove that $d\big((x_1, y_1), (x_2, y_2)\big) := f\big(d(x_1, x_2), \delta(y_1, y_2)\big)$ defines a distance on the Cartesian product $X \times Y$. Deduce that

$$d_1\big((x_1, y_1), (x_2, y_2)\big) = d(x_1, x_2) + \delta(y_1, y_2),$$

$$d_2\big((x_1, y_1), (x_2, y_2)\big) = \sqrt{d(x_1, x_2)^2 + \delta(y_1, y_2)^2},$$

$$d_\infty\big((x_1, y_1), (x_2, y_2)\big) = \max\left(d(x_1, x_2), \delta(y_1, y_2)\right)$$

are equivalent metrics on $\mathbb{R}^2$ (cf. Exercises 19.4).

19.2   Isometries

We return to the exquisitely geometric notion on which we based our approach to elementary geometry in Chap. 18: maps that preserve the metric structure of a space.

Definition 19.14 Let (X, d_X), (Y, d_Y) denote metric spaces. A surjective map $h \colon X \twoheadrightarrow Y$ is called an **isometry** if

$$d_Y\big(h(x), h(y)\big) = d_X(x, y) \quad \text{for every } x, y \in X,$$

i.e. if it preserves distances. If so we say (X, d_X), (Y, d_Y) are **isometric** spaces.

Corollary *An isometry is bijective, and hence 'being isometric' is an equivalence relation.*

Proof Exercise. $\square$

Consequently,

$$\mathrm{Isom}(X, d) := \big\{ h \colon (X, d) \to (X, d) \text{ isometry} \big\}$$

is called the **isometry group** of the metric space (X, d).

Examples

i) Fix $(x_0, y_0) \in \mathbb{R}^2$, $\theta \in \mathbb{R}$. The roto-translation

$$F(x, y) = (x \cos\theta + y \sin\theta + x_0, -x \sin\theta + y \cos\theta + y_0)$$

is an isometry of the Euclidean plane. Writing $z_0 = x_0 + iy_0$, we may view $F : \mathbb{C} \to \mathbb{C}$ as

$$F(z) = e^{i\theta}z + z_0,$$

that is the composite $F = T_{z_0} \circ R_\theta$ of the θ-rotation about the origin with the translation T_{z_0}. It was shown in Sect. 18.1.2 every isometry of $\mathbb{R}^2_{\text{Eucl}}$ has the form $F(z) = \pm e^{i\theta}z + z_0$ for certain θ, z_0. The group $\text{Isom}(\mathbb{R}^n_{\text{Eucl}})$ is characterised by Theorem 18.11.

iii) The isometry group of the Euclidean 2-sphere $S^2 = \{x \in \mathbb{R}^3 \mid \|x\|_{\text{Eucl}} = 1\}$ is (isomorphic to) $SO(3)$, the rotations of $\mathbb{R}^3$ fixing the origin. See (17.2) and Theorem 18.17.

iv) Proposition 15.13 proves that any isometry on a vector space that fixes the origin must me a linear map. In particular orthogonal transformations $O(n)$ and unitary transformations $U(n)$ are isometries of the standard metrics on $\mathbb{R}^n$ and $\mathbb{C}^n$ respectively.

v) Consider spherical coordinates (ϕ, ψ) on S^2, cf. (18.2), and call N, S the poles $(0, \ldots, 0, \pm 1)$. The *Mercator projection* (1569) is the map $M : S^2 \setminus \{N, S\} \to \mathbb{R}^2$:

$$M(\phi, \psi) = \left(\phi, \log\tan\left(\tfrac{\pi}{4} + \tfrac{\psi}{2}\right)\right).$$

It's not an isometry, so it cannot be used to estimate distances. Yet, it enjoys another important property: M is *conformal*, meaning it preserves angles. The images of curves on the sphere with constant azimuth (the angle formed with the parallels), called *loxodromes*), are straight lines.

It's known from ☞ *Riemannian geometry* that no map $S^2 \to \mathbb{R}^2$ can preserve distances, angles and areas at the same time. The distortions of the Earth's surface are clear if we take a look at a planispheric map: Greenland is invariably represented as being larger than both China (in reality it's four times smaller) and Africa (Greenland is about the size of Algeria). Not to mention Antarctica, which is way out of proportions. The reason for all this is that M exaggerates areas away from the equator.

ii) The **Cayley transform** (Theorem 18.21)

$$C : \mathscr{H}^+ \to \mathbb{D}, \ z \mapsto i\frac{z - i}{z + i}$$

is an isometry between the hyperbolic upper half-plane and the Poincaré disc
(☞ *complex geometry, functional analysis*). The isometries of the hyperbolic
plane are characterised in Theorem 18.24.

Remarks of topological nature, II 19.2

1) A pivotal theorem (of Nash's et al.) says that any metric space X that is paracom-
   pact or second-countable (a.k.a. a **topological manifold**) can be embedded in $\mathbb{R}^n$
   for some n. In other words there exists an injective map $(X, d) \hookrightarrow \mathbb{R}^n_{\text{Eucl}}$ that is
   isometric on its image.

2) Let $A, B \neq \varnothing$ be compact subsets in (X, d). It's an easy exercise to check
   $\sup_{a \in A} d(a, b)$ doesn't define a distance between A, B. It's not symmetric, and
   $\sup_{a \in A} d(a, b) = 0$ only implies $A \subseteq B$. A way to correct this is to symmetrise
   it, obtaining the *Hausdorff distance*

$$d_H(a, b) := \max \Big\{ \sup_{a \in A} d(a, b), \ \sup_{b \in B} d(a, b) \Big\}.$$

It can be proved that $\big| \operatorname{diam}(B) - \operatorname{diam}(A) \big| \leqslant 2 d_H(a, b)$.

19.3   Sequential Convergence

Recall that a **sequence** in a metric space (X, d) is a function $a \colon \mathbb{N} \to X$. As we
know, the domain is used as index set, so a sequence $a = (a_n)_{n \in \mathbb{N}}$ is viewed as a
subset in X.

Definition A sequence $(a_n) \subseteq (X, d)$ **converges** to the point $l \in X$ if, for every
open ball $B(l, r) \subseteq X$, there exists an $N \in \mathbb{N}$ such that $a_n \in B(l, r)$ for all $n \geqslant N$.
This is shortened as

$$a_n \longrightarrow l, \quad \text{or} \quad \lim_{n \to \infty} a_n = l.$$

A sequence is said **convergent** if it converges to some point $l \in X$, called **limit**
(point) of the sequence.

Proposition 19.15 *The limit l of a sequence (a_n) is an accumulation point:*

$$\lim_{n \to \infty} a_n = l \implies \operatorname{dist}\big((a_n) \setminus \{l\}, l\big) = 0.$$

Proof Just observe $a_n \in B(l, r)$ means $d(a_n, l) < r$.   □

Here's a fundamental property:

Proposition *If $a_n \longrightarrow l$ and $a_n \longrightarrow l'$, then $l = l'$. That is: limit points are unique.*

Proof By contradiction suppose $l \neq l'$. Exercise 19.9 gives disjoint balls centred at x and y that contain every a_n when n is large enough. Hence Proposition 19.15 is violated ⨳⨳ □

This is secretly a consequence of the fact that every metric space is Hausdorff. As a matter of fact the topology of a metric space (X, d) is defined in terms of limits, since the closure $\overline{S}$ of a subset $S \subseteq X$ (adherent points) is the union of S with its accumulation points, cf. item 2) in Remark 19.1.

Examples

i) The real sequence $a_n = \dfrac{1}{n}$ converges to $0 \in \mathbb{R}$: for any integer n larger than a chosen real number r we have $a_n < \dfrac{1}{r}$. Hence $B(0, r)$ contains the entire tail of the sequence, starting from the element $a_{\lfloor r \rfloor + 1}$.

ii) The sequence $a_n = \sin \dfrac{1}{n}$ does not converge in $\mathbb{R}$. The graph of $f : \mathbb{R}^+ \to [-1, 1]$, $f(x) = \sin(1/x)$ has infinitely many accumulation points, see Example 19.13 iv), none of which is a limit point for $\mathrm{graf}(f)$.

iii) From Example 6.10 we know that every real number is the limit of a rational sequence. This fact is the essence of the incompleteness of $\mathbb{Q}$ (see Theorem 7.8).

Proposition 19.16 *A non-decreasing real sequence (a_n) converges, and $\lim\limits_{n \to \infty} a_n = \sup a_n$.*

Proof The key point is that the sequence is real-valued, so the supremum $M := \sup a_n$ exists (either finite or $+\infty$). By definition $a_n \leqslant M$, and for any $M' < M$ there is an m such that $M' < a_m \leqslant M$. As the sequence is non-decreasing, $M' < a_n \leqslant M$ for all $n > m$. In case $M = +\infty$ the proof ends. If not, for every $\epsilon > 0$ there's an m such that $M - \epsilon < a_n < M + \epsilon$ for all $n > m$. Hence, every ball $B(M, \epsilon)$ contains the tail of the sequence starting from index m. □

Exercises

i) Adapt Proposition 19.16 and prove that a non-increasing sequence (a_n) converges to $\lim\limits_{n \to \infty} a_n = \inf a_n$.

ii) Take a real sequence (x_n) and consider the sequence $M_m := \sup\limits_{n \geqslant m} x_n$. The latter is non-increasing by Exercise 2.11, hence it has limit (the infimum). Call **limit superior** the quantity

$$\limsup_{n \to \infty} x_n := \inf_{m \in \mathbb{N}} \sup_{n \geqslant m} x_n \quad \left(= \lim_{m \to \infty} \sup_{n \geqslant m} x_n \right).$$

Show that $\limsup\limits_{n \to \infty} x_n = \inf \left\{ M \mid \forall \epsilon > 0 \ \exists N \text{ such that } x_n < M + \epsilon \ \forall n > N \right\}$.

iii) Mimicking ii), define the **limit inferior** $\liminf\limits_{n \to \infty} x_n$ and prove $\liminf\limits_{n \to \infty} x_n \leqslant \limsup\limits_{n \to \infty} x_n$.

iv) Prove

$$\limsup_{n\to\infty} x_n = \liminf_{n\to\infty} x_n = l \iff \lim_{n\to\infty} x_n = l.$$

Definition A **subsequence** (a_{k_n}) of a sequence $a \colon \mathbb{N} \to X$ is the composite map $a \circ k$ where $k \colon \mathbb{N} \to \mathbb{N}$ is an increasing map.

Lemma 19.17 *Let (a_n) be a sequence in X. If a subsequence (a_{k_n}) converges to a point $p \in X$, then p is an accumulation point for (a_n).*

Proof Let B be a ball centred at p. There is an $N \in \mathbb{N}$ such that $a_{k_n} \in B$ for all $n \geqslant N$. For every M one can find $m \in \mathbb{N}$ such that $a_{k_m} \in B$ and $k_m \geqslant M$. Therefore (a_n) accumulates at p. $\square$

Exercises Prove the following facts.

  i) If $a_n \longrightarrow l$, every subsequence (a_{k_n}) converges to l. Equivalently, for (a_n) not to converge to l it suffices to have a subsequence $(a_{k_n}) \not\longrightarrow l$.
 ii) The lim sup of a real sequence x_n is the greatest accumulation point (namely, the largest limit of all subsequences (x_{k_n})).

Examples 19.18

  i) The alternating sequence $a_n = (-1)^n$ has two accumulation points: $a_{2k+1} \longrightarrow -1 = \liminf_{n\to\infty} a_n$ and $a_{2k} \longrightarrow +1 = \limsup_{n\to\infty} a_n$. Hence it doesn't converge.
 ii) For $n > 0$ the sequence $a_n = \left\lfloor \dfrac{n}{2} \right\rfloor^{(-1)^{n+1}} = 1, \dfrac{1}{2}, 3, \dfrac{1}{4}, 5, \dfrac{1}{6}, 7, \dfrac{1}{8}, \ldots$
     accumulates at 0, because $a_{2k} = k^{-1} \longrightarrow 0$. But (a_n) doesn't converge, due to the divergent subsequence $a_{2k+1} = k \longrightarrow +\infty$. In fact, $\limsup_{n\to\infty} a_n = +\infty$ and $\liminf_{n\to\infty} a_n = 0$.
iii) The sequence $a_n = \sin(n)$ has no limit, yet $\limsup_{n\to\infty} a_n = 1$ since π is irrational.
 iv) The sequence $a_n = \sin \dfrac{1}{n}$ does not converge in $\mathbb{R}$. We've seen in Example 19.13 iv) that $f(x) = \sin(1/x)$, $x > 0$ accumulates at every point of the vertical segment $\{0\} \times [-1, 1]$. In fact, intersecting the graph with an arbitrary line $y = c \in [-1, 1]$ produces a sequence of points $P_k = \left(\dfrac{1}{\arcsin(c) + 2k\pi}, c \right)$ converging to $(0, c)$. This gives subsequences of (a_n) converging to any number in $[-1, 1]$, e.g.:

$$a_{n_k} = \sin \dfrac{1}{k\pi} \longrightarrow 0,$$

$$a_{n_h} = \sin \dfrac{1}{\pi/2 + 2h\pi} \longrightarrow 1 = \limsup_{n\to\infty} a_n,$$

$$a_{n_j} = \sin \frac{1}{-\pi/2 + 2h\pi} \longrightarrow -1 = \liminf_{n \to \infty} a_n.$$

Remark The topology of metric spaces (which we shall not address) is particularly rich: they are normal, first countable, second countable iff separable, compact iff sequentially compact.

Definition (X, d) is **sequentially compact** if every sequence has a convergent subsequence. Equivalently,[1] X is sequentially compact if every sequence admits an accumulation point.

Exercises

i) Let $a \colon \mathbb{N} \to \mathbb{R}$ be a bijection $\mathbb{N} \cong \mathbb{Q}$. Find its accumulation points.
     (*Hint*: any non-empty real interval contains an infinite number of rationals.)
ii) Prove $(0, 1) \subseteq \mathbb{R}$ is not sequentially compact.
iii) Show that every convergent sequence is bounded, but not vice versa.

The relationship between convergence and boundedness specialises if the ambient space X is the set of real numbers:

Theorem 19.19 (Bolzano–Weierstraß, 1817) *Every bounded real sequence* $(x_n) \subseteq \mathbb{R}$ *has a convergent subsequence (and as such, it accumulates at one point at least).*

Proof The classical argument is based on the **bisection method** (also used for Theorem 7.24) and requires infinitesimal calculus. In Proposition 19.23 we will prove a generalisation.

As (x_n) is bounded there exist $a_0 \leqslant b_0$ such that $(x_n) \subseteq [a_0, b_0]$. That means the pre-image of the sequence, seen as map $\mathbb{N} \to \mathbb{R}$, is $\mathbb{N} = (x_n)^{-1}([a_0, b_0])$. Call $m_0 = (a_0 + b_0)/2$ the midpoint, so

$$\mathbb{N} = (x_n)^{-1}([a_0, m_0]) \cup (x_n)^{-1}([m_0, b_0]).$$

These two sets cannot be both finite, so let's pick the infinite one (if both are infinite, choose the left one). Define

$$a_1 = \begin{cases} a_0 \text{ if } (x_n)^{-1}([a_0, m_0]) \text{ is infinite} \\ m_0 \text{ otherwise} \end{cases}$$

$$b_1 = \begin{cases} m_0 \text{ if } (x_n)^{-1}([a_0, m_0]) \text{ is infinite} \\ b_0 \text{ otherwise} \end{cases}$$

---

[1] Because metric spaces are first countable.

Then $[a_1, b_1]$ is an infinite subinterval of the above two (or the left-most one if both are infinite). By construction $a_0 \leqslant a_1 < b_1 \leqslant b_0$, the length $b_1 - a_1 = (b_0 - a_0)/2$ was halved, and $x_n \in [a_1, b_1]$ for infinitely many indices. Now we repeat the bisection n times, and prove by induction $b_n - a_n = (b_0 - a_0)/2^n$. By induction hypothesis $a_0, a_1, \ldots, a_k, b_0, b_1, \ldots, b_k$ satisfy

$$b_k - a_k = (b_0 - a_0)/2^k \ (\text{so } a_k < b_k \text{ and } \lim_{k \to \infty} a_k = \lim_{k \to \infty} b_k)$$

$x_n \in [a_k, b_k]$ for infinitely many n

(a_k) is non-decreasing, (b_k) non-increasing.

Set $m_k = (a_k + b_k)/2$, allowing us to assume $a_{k+1} = a_k$, $b_{k+1} = m_k$. The sequences (a_k), (b_k) bound (x_n). As (a_k) is monotone and upper bounded (by any b_k), it has finite limit: $a_k \longrightarrow \sup a_k =: x_0$ (from below). Then $a_0 \leqslant x_0 \leqslant b_0$, and $b_k \longrightarrow x_0$ as well (but from above). This means that for any $\epsilon > 0$ there exist $\overline{k}, \widehat{k}$ such that $x_0 - \epsilon < a_k \leqslant x_0 \leqslant b_k < x_0 + \epsilon$, for all $k \geqslant \overline{k}, \widehat{k}$. Therefore $\forall \epsilon > 0$, $[a_k, b_k] \subseteq (x_0 - \epsilon, x_0 + \epsilon)$ for any $k \geqslant \max\{\overline{k}, \widehat{k}\}$. Then infinitely many terms in (x_n) belong in a neighbourhood $[a_k, b_k)$ of x_0. Taking a point between a_0, b_0, a point between a_1, b_1 and so on we get closer and closer to x_0: this is done by defining inductively the increasing sequence

$$k_0 = 0$$
$$k_1 = \min \{m > k_0 \mid x_m \in [a_1, b_1]\}$$
$$\vdots$$
$$k_{n+1} = \min \{m > k_n \mid x_m \in [a_{n+1}, b_{n+1}]\}.$$

By construction, eventually, $x_{k_n} \longrightarrow x_0$.   □

19.4 Complete Metric Spaces

Cauchy sequences (in truth due to Cantor) generalise the notion of convergence and represent the basic tool in the analysis of metric spaces.

Definition A sequence (a_n) in (X, d) is a **Cauchy sequence** if, for every $\epsilon > 0$, there exists an $N \in \mathbb{N}$ such that $d(a_n, a_m) < \epsilon$ for all $n, m \geqslant N$. In words, there's a ball $B(x, \epsilon)$ containing a tail of a_n, starting from a certain element a_N.

Cauchy sequences are bounded: since there's only a finite number of terms $a_0, \ldots, a_{N-1}$ outside $B(x, \epsilon)$, it's possible to find a ball large enough to contain the sequence entirely: choose radius $R = \max \{\epsilon, r_0, \ldots, r_{N-1}\}$, where $r_j = d(x, a_j)$.

Examples

i) The sequence $a_n = \dfrac{1}{n}$ in $(0, 1) \subseteq \mathbb{R}$ is Cauchy, and $\lim_{n \to \infty} a_n = 0 \notin (0, 1)$.

ii) The sequence $a_n = (-1)^n$ is not Cauchy, but it's bounded.

Lemma *Every convergent sequence is a Cauchy sequence.*

Proof If l denotes the limit of (a_n), for any $\epsilon > 0$ there's an N such that $a_n \in B(l, \epsilon/2)$, for all $n \geq N$. The triangle inequality implies $d(a_n, a_m) \leq d(a_n, l) + d(l, a_m) < \epsilon$ for all $n, m \geq N$. □

Lemma 19.20 *A Cauchy sequence is convergent (say, to l) if and only if it accumulates at l.*

In particular, every Cauchy sequence in a sequentially compact metric space converges.

Proof

($\Longrightarrow$) See Proposition 19.15.

($\Longleftarrow$) Let (a_n) be a Cauchy sequence that accumulates at l. The Cauchy property implies there exists $M \in \mathbb{N}$ such that $d(a_n, a_m) < \epsilon/2$ for all $n, m \geq M$. As l is an accumulation point, there is an $N \geq M$ such that $d(l, a_N) < \epsilon/2$. The triangle inequality $d(l, a_n) \leq d(l, a_N) + d(a_N, a_n) < \epsilon$ says that for any $\epsilon > 0$ there's an $N \in \mathbb{N}$ such that $d(l, a_n) < \epsilon$ for all $n \geq N$. But if (a_n) has a convergent subsequence, by Lemma 19.17 it admits limit.

□

A pertinent application of this result is the possibility of writing any real number in decimal form, or in any base if it comes to that. Computers, for instance, codify and store data as numbers in base 2 and 8.

Proposition 19.21 (Representation in base m) *Fix an integer $m \geq 2$.*

For any $x \in \mathbb{R}^+ \cup \{0\}$ there exists a unique integer sequence $(d_i)_{i \in \mathbb{N}}$ such that

i) $d_0 = \lfloor x \rfloor$

ii) $0 \leq d_i < m$, for all i

iii) $y_0 = d_0$, $y_{n+1} = y_n + \dfrac{d_{n+1}}{m^{n+1}}$ is a Cauchy sequence, and $\lim\limits_{n \to \infty} y_n = x$.

Conversely, let $(d_i)_{i \in \mathbb{N}}$ be an integer sequence such that $0 \leq d_i < m$ for all i. Then there exists a unique real number $x \geq 0$ satisfying iii).

Proof *([92, p.157])* Fix. m, x. We shall only prove the sequence's existence, and leave the rest as exercise. Define $d_0 = \lfloor x \rfloor$. Then $x_1 := (x - d_0)m$ belongs to $[0, m)$. Set $d_1 = \lfloor x_1 \rfloor$ and iterate the division. By induction set $x_n := (x_{n-1} - d_{n-1})m$ and $d_n := \lfloor x_n \rfloor$, $n > 1$. Now

$$x = \underbrace{d_0 + \frac{d_1}{m} + \frac{d_2}{m^2} + \ldots + \frac{d_n}{m^n}}_{y_n} + \frac{x_{n+1}}{m^{n+1}}, \qquad 0 \leq x_{n+1} < m. \tag{19.2}$$

Therefore $0 \leq x - y_n < \dfrac{1}{m^n}$, so clearly $y_n \longrightarrow x$ as $n \longrightarrow +\infty$. □

Taking $m = 10$ in (19.2) recovers the decimal expansion of x. See Theorem 7.13.

Definition A metric space (X, d) is called **complete** if every Cauchy sequence converges in X.

Theorem *Every sequentially compact space is complete.*

Proof Immediate, since the statement is a mere reformulation of Lemma 19.20.   □

Theorem 19.22 *The Euclidean spaces $\mathbb{R}^n$ and $\mathbb{C}^n$ are complete, for any $n \in \mathbb{N}$.*

Proof Since $\mathbb{C}^n \cong \mathbb{R}^{2n}$, we can just treat $\mathbb{R}^n$. Let (a_n) be a Cauchy sequence in $\mathbb{R}^n$ and $N \in \mathbb{N}$ a number such that $|a_n - a_N| < 1$ for all $n \geqslant N$. Setting $R = \max\{\|a_1\|, \|a_2\|, \ldots, \|a_N\|\}$, the triangle inequality forces (a_n) to stay inside the compact ball $\{x \in \mathbb{R}^n \mid \|x\| \leqslant R + 1\}$. Then Lemma 19.20 guarantees the sequence converges.   □

It's enough the pull one point from $\mathbb{R}^n$ to destroy completeness: the sequence $a_n = n^{-1}$ in $\mathbb{R} \setminus \{0\}$ does not converge.

Remark Theoretically, the completeness of $\mathbb{R}$ allows to solve equations that do not have roots in $\mathbb{Q}$. Here's the practical version of the phenomenon. As 0 isn't a root of $x^2 = 2$, we may rewrite the equation as $x = \dfrac{1}{2}\left(x + \dfrac{2}{x}\right)$.

On one hand, this expression gives rise to a continuous fraction, akin to those on p.173.

On the other, consider the sequence $(x_n)_{n>0}$ defined by $x_1 = 4$, $x_{n+1} = \dfrac{1}{2}\left(x_n + \dfrac{2}{x_n}\right)$. We claim $x_n \longrightarrow \sqrt{2}$. For starters we show boundedness: $1 \leqslant x_n \leqslant 2$. By induction, this is true for $n = 1$, and assuming it true for n, then $1 \leqslant 2/x_n \leqslant 2$, so $1 \leqslant x_{n+1} \leqslant 2$ as well. Next we prove $x_n^2 \geqslant 2$. Still by induction, this is fine for $n = 1$, and $(x_n - 2/x_n)^2 \geqslant 0 \implies (s_n^2 + 4/s_n^2) \geqslant 4$, so

$$x_{n+1}^2 = \frac{1}{4}\left(x_n + \frac{2}{x_n}\right)^2 = \frac{1}{4}\left(s_n^2 + \frac{4}{s_n^2} + 4\right)^2 \geqslant 2.$$

Finally, (x_n) is decreasing: this is easy since $\dfrac{x_{n+1}}{x_n} = \dfrac{1}{2}\left(1 + \dfrac{2}{s_n^2}\right) \leqslant \dfrac{1}{2}(1+1) = 1$. The real line's completeness ensures (x_n) has a limit, which Example 19.4 will prove to be precisely $\sqrt{2}$.

Exercises Implement the same technique, algebraically and pictorially, using

i) the same sequence $x_{n+1} = \frac{1}{2}\left(x_n + \frac{2}{x_n}\right)$, but choosing initial point $x_1 \in (0, \sqrt{2})$;

ii) $x_{n+1} = \dfrac{2}{x_n}$ (from writing $x^2 = 2$ as $x = \dfrac{2}{x}$).

Theorem 19.22 also warrants the completeness of $\mathbb{R}$ as a consequence of the (sequential) compactness of $[0, 1]$. It's instructive to prove the converse, as well. To do so we need to show that every sequence in $[0, 1]$ has a Cauchy subsequence. First

Lemma *Let $\{A_n\}_{n \in \mathbb{N}}$ be a (countable) collection of finite sets, and $f_n \colon \mathbb{N} \to A_n$ maps. There exists an increasing rearrangement $g \colon \mathbb{N} \to \mathbb{N}$ such that $f_n(g(m)) = f_n(g(n))$ for every $m \geqslant n$.*

Proof As A_1 is finite, at least one pre-image of f_1 is infinite. Lemma 6.3 guarantees the existence of an increasing map $g_1 \colon \mathbb{N} \to \mathbb{N}$ such that $f_1 \circ g_1$ is constant. Similarly, one pre-image of $f_2 \circ g_1$ is infinite, so there is an increasing $g_2 \colon \mathbb{N} \to \mathbb{N}$ such that $f_2 \circ g_1 \circ g_2$ is constant. Induction generates a sequence of increasing maps $g_n \colon \mathbb{N} \to \mathbb{N}$ such that, for all n, $f_n \circ g_1 \circ g_2 \circ \cdots \circ g_n$ is constant. The 'diagonal' map $g \colon \mathbb{N} \to \mathbb{N}$, $g = g_1 \circ \cdots \circ g_n$ satisfies the request: for any $n < m \in \mathbb{N}$ we set $l = g_{n+1} \circ \cdots \circ g_m(m)$. Then $l \geqslant m$ and $f_n(g(m)) = f_n \circ g_1 \circ \cdots \circ g_n(l) = f_n \circ g_1 \circ \cdots \circ g_n(n) = f_n(g(n))$. $\qquad \square$

Indicate by A_n the collection of intervals

$$\left[0, \frac{1}{2^n} \right), \ \ldots, \ \left[\frac{i}{2^n}, \frac{i+1}{2^n} \right), \ \ldots, \ \left[1 - \frac{1}{2^n}, 1 \right], \quad 0 \leqslant i < 2^n.$$

Note A_n is finite, since there are 2^n such intervals. Now take a sequence (a_n) in $[0, 1]$ and choose maps $f_n \colon \mathbb{N} \to A_n$ such that $a_m \in f_n(m)$ for every m. By the lemma there is an increasing map $g \colon \mathbb{N} \to \mathbb{N}$ such that $f_n(g(n)) = f_n(g(m))$ for all $n \leqslant m$. The subsequence $(a_{g(n)})$ satisfies $\left| a_{g(n)} - a_{g(m)} \right| \leqslant 2^{-n}$ for all $n \leqslant m$, hence it is a Cauchy subsequence. $\qquad \square$

Definition A subset $T \subseteq (X, d)$ is **totally bounded** if for any $\epsilon > 0$ there exists a finite cover of T made by closed balls of radius ϵ:

$$\forall \epsilon > 0 \ \exists x_1, \ldots, x_k \in X \ : \ T \subseteq \bigcup_{i=1}^{k} \overline{B(x_i, \epsilon)}.$$

Example A totally bounded set $T \subseteq X$ is evidently bounded, because it's contained in a ball of radius $\dfrac{1}{2} \operatorname{diam} \left(\bigcup_{i=1}^{k} \overline{B(x_i, \epsilon)} \right)$.

Exercises For subsets $T \subseteq \mathbb{R}^n_{\mathrm{Eucl}}$ the converse holds, and so: totally bounded $\iff$ bounded. But this is a special feature of the standard Euclidean distance, because in general the lack of completeness prevents a bounded subset in an arbitrary metric space X from being totally bounded.

Show that the unit interval $[0, 1]$ equipped with the discrete metric (Example 19.1 1)) is bounded but not totally bounded. Using the same circle of ideas, prove that $\mathbb{N}$ with the discrete metric is bounded but not totally bounded. Note, by the way, that a Cauchy sequence of natural numbers is eventually constant, hence it converges. $\mathbb{N}$ is therefore complete for the discrete metric.

In order to continue, recall that Theorem 19.19 is essentially a rephrasing of the *Heine–Borel theorem*: $T \subseteq \mathbb{R}^n_{\text{Eucl}}$ is compact $\iff$ closed and bounded. The next result generalises this to arbitrary metric spaces:

Proposition 19.23 *A subset* $T \subseteq (X, d)$ *is sequentially compact* $\iff$ *it is complete and totally bounded.*

Proof

($\implies$) From Lemma 19.20 sequential compactness implies completeness. So there remains to prove T is totally bounded. Suppose the contrary. Then there is an $\epsilon > 0$ such that T cannot be covered by k closed balls $B_i := \overline{B(x_i, \epsilon)}$ of radius ϵ, and from them we'll construct a sequence with no convergent subsequences. Namely:

a) $\exists x_1 \in T$ such that B_1 doesn't cover T; hence
b) $\exists x_2 \notin B_1$ such that $B_1 \cup B_2$ doesn't cover T; hence
c) $\exists x_3 \notin B_1 \cup B_2$ such that $T \not\subseteq B_1 \cup B_2 \cup B_3$;

By induction we generate a sequence (x_n) of centres such that
$d(x_2, x_1) \geqslant \epsilon$ due to a),
$d(x_3, x_2) \geqslant \epsilon$, $d(x_3, x_1) \geqslant \epsilon$ due to b), etc.
In general, $d(x_m, x_n) \geqslant \epsilon$ for $m \neq n$.
Consequently the points lie ϵ-away from one another, and there cannot be any Cauchy subsequence (i.e. convergent, by Lemma 19.20). But T is sequentially compact ⚡.

($\impliedby$) Let $(x_n) \subseteq T$ be a sequence. The idea is extracting from it a Cauchy subsequence (which, by completeness, has to converge). Let $B_1(1) \cup \ldots \cup B_k(1) \supset T$ be a finite covering by unit balls. Since there cannot be a finite number of elements x_n in each ball, we call x_{1_n} the subsequence which always belongs in $B_1(1)$. Then $d(x_{1_n}, x_{1_m}) < 2$.

Now take a finite cover of balls with radius $1/2$, and as before call x_{2_n} the subsequence entirely contained in $B_2(1/2)$. By induction we get

$$x_{1_1}, \quad x_{1_2}, \quad x_{1_3}, \quad \ldots \quad \text{lie in a ball of radius } 1;$$
$$x_{2_1}, \quad x_{2_2}, \quad x_{2_3}, \quad \ldots \quad \text{lie in a ball of radius } 1/2;$$
$$x_{3_1}, \quad x_{3_2}, \quad x_{3_3}, \quad \ldots \quad \text{lie in a ball of radius } 1/3;$$
$$\vdots \qquad \ddots \qquad \vdots$$

Each row x_{σ_n} is a subsequence of row $x_{\sigma_{n-1}}$ above it, and as we descend the distances decrease. The diagonal subsequence (x_{k_k}) is Cauchy, because $d(x_{\sigma_n}, x_{\sigma_m}) \leqslant 2/k$ for any $n, m \geqslant k$.   □

This result is employed, amongst other things, to show that the unit ball in an infinite-dimensional normed vector space is not compact. Very naively, imagine $\mathbb{R}^\infty$ as infinitely many copies of $\mathbb{R}$: the vectors $\mathbf{e}_j = (0, \cdots, 0, 1, 0, \cdots)$, $j \in \mathbb{N}$ are ostensibly ON, so each one is 'far' enough from the others so that the sequence $(\mathbf{e}_j)$ does not converge.

Exercise Fill in the gaps in the proof of Theorem 3.18.

The theory of complete metric spaces has two cornerstones: *Baire's category theorem* and *Banach's fixed-point theorem*. As the former would lead us astray we'll discuss the latter, and for that

Definition 19.24 Let (X, d) be a metric space and $0 < \gamma < 1$ a real number. A map $f \colon X \to X$ is called **contraction** if

$$d\big(f(x), f(y)\big) \leqslant \gamma \, d(x, y)$$

for all $x, y \in X$.

Examples

i) The map $f \colon [1, 5] \to [1, 5]$, $f(x) = \dfrac{x}{2} + \dfrac{1}{x}$ is a contraction for the Euclidean metric, with $\gamma = 1/2$. The point $x = \sqrt{2}$ is fixed under f. See Fig. 19.2.
   The constant is not optimal, since the restriction $f \colon [\sqrt{2}, 3/2] \to [\sqrt{2}, 3/2]$ is a contraction for $\gamma = 1/18$.

ii) In $\mathbb{R}^n_{\text{Eucl}}$ a contraction is a kind of Lipschitz function (Definition 10.11).

iii) The *Schwarz–Pick lemma* states that any holomorphic map f of the disc $\mathbb{D} \subseteq \mathbb{C}$ is a contraction for the Poincaré metric

$$d_{\mathscr{H}}(f(z), f(z')) \leqslant d_{\mathscr{H}}(z, z'), \qquad z, z' \in \mathbb{D}.$$

Fig. 19.2 Babylonian method for solving $x^2 = 2$

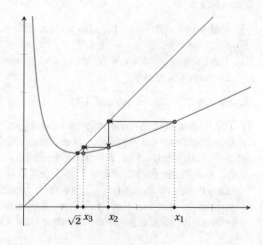

$\sqrt{2}$ x_3 x_2 x_1

Equality occurs precisely when $f \in \mathrm{Aut}(\mathbb{D})$ (☞ *complex analysis*).

Theorem 19.25 (Contraction Theorem/Banach's Fixed-Point Theorem, 1922)
Let (X, d) be complete. Then any contraction $f: X \to X$ has a unique fixed point $z \in X: f(z) = z$.

Proof (Existence) Pick $x_1 \in X$ and define $x_{n+1} = f(x_n)$ recursively. Then $d(x_{r+1}, x_r) \leqslant \gamma^{r-1} d(x_2, x_1)$ for all $r > 1$. Repeatedly using the triangle inequality we obtain

$$d(x_m, x_n) \leqslant d(x_m, x_{m-1}) + d(x_{m-1}, x_{m-2}) + \ldots + d(x_{n+1}, x_n)$$

$$\leqslant (\gamma^{m-2} + \gamma^{m-3} + \ldots + \gamma^{n-1}) d(x_2, x_1)$$

$$= \gamma^{n-1} \frac{1 - \gamma^{m-n}}{1 - \gamma} d(x_1, x_2) < \frac{\gamma^{n-1}}{1 - \gamma} d(x_1, x_2) \longrightarrow 0 \quad \text{as } n \longrightarrow +\infty$$

for all $m > n$. Therefore (x_n) is Cauchy, and by completeness it converges to, say, $p \in X$. As f is continuous at p, $f(x_n) \longrightarrow f(p)$, but $f(x_n) = x_{n+1} \longrightarrow p$, and hence p is a fixed point.

(Uniqueness) Supposing $f(p) = p$, $f(q) = q$, from $d(p, q) = d(f(p), f(q)) \leqslant \gamma d(p, q)$ we deduce $d(q, p) = 0$. □

The argument also shows that for every $x, y \in X$, the sequence $\big(d(f^n(x), f^n(y))\big)$ converges to 0, and $\{f^n(x)\}$ is Cauchy. Since any iterate f^n is still a contraction, then $f^n(x) \longrightarrow z$ for any $x \in X$.

Theorem 19.25 is but one version of a general statement for closed subspaces of Banach spaces. It's used in many situations, in particular for

- the proof of the *inverse-function theorem*,
- the existence of solutions of many differential and integral equations,
- *Newton's method* for detecting roots of differentiable functions.

Exercises

i) Prove $\cos: [0, 1] \to [0, 1]$ is a contraction.
ii) Show $f: (0, 1/4) \to (0, 1/4)$, $f(x) = x^2$ is a contraction with no fixed points.
iii) Find a map $f: \mathbb{R} \to \mathbb{R}$ without fixed points that satisfies $\big|f(x) - f(y)\big| < |x - y|$ for every $x, y \in \mathbb{R}$.

Remark of topological nature, III

1) The notion of sequence may be generalised to topological spaces X that are not first countable (in particular, non-metrisable) in the following way.

   Call *directed set* a partially ordered non-empty set $(I, \leqslant)$ such that, for all $i, j \in I$, there exists an $h \in I$ such that $h \geqslant i, j$. Examples include $(\mathbb{N}, \leqslant)$, finite subsets of any poset $(A, \subseteq)$, and any totally ordered set. A *net*, or *generalised sequence*, is a function $f: I \to X$ defined on a directed set. The concept is intimately related to that of a filter (see Definition 5.9), and in fact it can be proved that both define the same notion of convergence. Nets are employed to prove that any finite-dimensional subspace in a topological vector space is closed.

2) Lest we think of sequences only as list of numbers, bear in mind that a good deal of ☞ *real, complex, functional, harmonic analysis* is about sequences, and hence series, of functions. In ☞ *metric geometry* many problems are centred around sequences of metric spaces $\mathscr{M} = \{(M_1, d_1),\ (M_2, d_2),\ \dots\}$. The correct notion of convergence for sequences of compact metric spaces comes from *Gromov–Hausdorff theory*, which places on $\mathscr{M}$ the metric

$$d_{GH}(X, Y) := \inf \{ d_H (f(X), g(Y)) \mid f : X \to M,\ g : Y \to M \text{ are} $$
$$\text{isometric embeddings in some } M \in \mathscr{M}\}.$$

(d_H was defined in Remark 19.2.) This *Gromov–Hausdorff distance* is a means to measure how far two compact metric spaces X, Y are from being isometric.

19.5 Normed Vector Spaces

A large class of metric spaces arises in linear algebra, where certain vector spaces are metrised in a natural way. Together with their infinite-dimensional cousins they are the primary focus of ☞ *functional analysis*.

Definition A **norm** on an $\mathbb{R}$-vector space X is non-negative map $N : X \to [0, +\infty)$ such that:

a) $N(\lambda \mathbf{x}) = |\lambda| N(\mathbf{x})$ (*homogeneity*)
b) $N(\mathbf{x} + \mathbf{y}) \leqslant N(\mathbf{x}) + N(\mathbf{y})$ (*subadditivity/triangle inequality*)
c) $N(\mathbf{x}) = 0 \Longrightarrow \mathbf{x} = \mathbf{0}$ (*non-singularity*)

for all $\lambda \in \mathbb{R}$, $\mathbf{x}, \mathbf{y} \in X$. The pair (X, N) is called a real **normed (vector) space**.

Property c) says N is positive definite, and so it is customary to indicate $N(\mathbf{x}) = \|\mathbf{x}\|$ in analogy to the examples below:

Examples Take $\mathbf{x} = (x_1, \dots, x_n) \in \mathbb{R}^n$, $\mathbf{z} = (z_1, \dots, z_n) \in \mathbb{C}^n$ and define

i) the **Euclidean norm** on $\mathbb{R}^n$: $\|\mathbf{x}\|_{\text{Eucl}} := \sqrt{\sum_i x_i^2}$;

ii) the **Hermitian norm** on $\mathbb{C}^n$: $\|\mathbf{z}\|_{\text{Herm}} := \sqrt{\sum_i z_i \overline{z_i}}$;

iii) the **supremum norm** on $\mathbb{R}^n$: $\|\mathbf{x}\|_\infty := \sup_{i=1,\dots,n} |x_i|$.

Exercises

i) Identify $\mathbb{C}^n = \mathbb{R}^{2n} = \mathbb{R}^n \times \mathbb{R}^n$ and show $\|\cdot\|_{\text{Herm}} = \|\cdot\|_{\text{Eucl}} = \|\cdot\|_2$.
ii) For any $\mathbf{x}$ in a real normed space $(X, \|\cdot\|)$, prove that $\|\mathbf{x}\| = \inf_{t \in \mathbb{R}^*} \{ |t|^{-1} : \|t\mathbf{x}\| \leqslant 1 \}$.

These examples suggest the following

Proposition *A normed space* (X, N) *is metrisable.*

Proof Define

$$d(\mathbf{x}, \mathbf{y}) := N(\mathbf{x} - \mathbf{y}).$$

This is a metric, called **distance induced by/associated with the norm**. Property c) implies d is positive definite, a) guarantees the symmetry

$$d(\mathbf{x}, \mathbf{y}) = N(\mathbf{x} - \mathbf{y}) = N(-(\mathbf{y} - \mathbf{x})) = |-1| N(\mathbf{y} - \mathbf{x}) = d(\mathbf{y}, \mathbf{x}),$$

and b) is the triangle inequality.   □

The structure of a general metric space is simpler that the one of a normed space, since the former does not possess linear operations (addition and multiplication by scalars).

Examples

− the Euclidean and Hermitian norms are both associated with $d_{\mathrm{Eucl}}(\mathbf{z}, \mathbf{w}) := \|\mathbf{z} - \mathbf{w}\|$;
− the supremum norm $\|\cdot\|_\infty$ is induced by the distance d_∞ of Examples 19.2.

When we consider (X, N) as metric space, implicitly we are referring to the norm-induced metric.

Corollary *The distance d induced by the norm N is homogeneous and translation-invariant:*

$$\begin{aligned} d(\lambda\mathbf{x}, \lambda\mathbf{y}) &= |\lambda|\, d(\mathbf{x}, \mathbf{y}) \\ d(\mathbf{x}, \mathbf{y}) &= d(\mathbf{x} + \mathbf{z}, \mathbf{y} + \mathbf{z}) \end{aligned} \tag{19.3}$$

for any $\mathbf{x}, \mathbf{y}, \mathbf{z} \in X, \lambda \in \mathbb{R}.$

Proof Exercise.   □

Conversely,

Proposition *Let X be a vector space equipped with a distance function satisfying* (19.3). *Then*

$$N(\mathbf{x}) := d(\mathbf{x}, \mathbf{0})$$

is a norm on X.

Proof N is positive definite since $d(\mathbf{x}, \mathbf{0}) = 0 \iff \mathbf{x} - \mathbf{0} = \mathbf{0}$. It's homogeneous because d is invariant under translations and homogeneous: $N(\lambda \mathbf{x}) = d(\lambda \mathbf{x}, \mathbf{0}) = |\lambda| d(\mathbf{x}, \mathbf{0}) = |\lambda| N(\mathbf{x})$. At last, the triangle inequality $N(\mathbf{x} + \mathbf{y}) = d(\mathbf{x} + \mathbf{y}, \mathbf{0}) = d(\mathbf{x}, -\mathbf{y}) \leqslant d(\mathbf{x}, \mathbf{0}) + d(-\mathbf{y}, \mathbf{0}) = d(\mathbf{x}, \mathbf{0}) + d(\mathbf{y}, \mathbf{0}) = N(\mathbf{x}) + N(\mathbf{y})$ guarantees subadditivity. □

Corollary 19.26 *An inner-product space (X, ϕ) is metrisable.*

Proof The quadratic form associated with ϕ is a norm $\|\mathbf{w}\| := \sqrt{\phi(\mathbf{w}, \mathbf{w})}$, see Sect. 15.5. □

Examples

i) The standard dot product (15.8) on $\mathbb{R}^n$ induces the Euclidean norm/distance.
ii) The Hermitian product (16.2) on $\mathbb{C}^n$ induces the Euclidean distance.
iii) The space of real square matrices End $(\mathbb{R}^n)$ is metrised by

$$d(A, B) := \operatorname{tr}(A^T B).$$

In turn this induces the **matrix norm** $\|A\|_M := \sqrt{\operatorname{tr}(A^T A)}$, which satisfies

$$\|A\mathbf{v}\| \leqslant \|A\|_M \|\mathbf{v}\|, \qquad \mathbf{v} \in \mathbb{R}^n$$
$$\|AB\|_M \leqslant \|A\|_M \|B\|_M$$

Exploiting this one proves that an $n \times n$ orthogonal matrix P satisfies $\|P\|_M = \sqrt{n} < \infty$. Therefore O$(n)$ is a bounded set (hence compact, as End $(\mathbb{R}^n) \cong \mathbb{R}^{n^2}$ is complete).

Exercises Reword Definitions 19.14 and 19.7 and prove:

i) balls in $(X, \|\cdot\|)$ can be characterised as

$$B(\mathbf{x}, r) = \{\mathbf{y} \in X \mid \|\mathbf{x} - \mathbf{y}\| < r\}.$$

The *frontier* of $B(\mathbf{x}, r)$ now coincides with the sphere $\partial B(\mathbf{x}, r) = S(\mathbf{x}, r) := \{\mathbf{y} \in X \mid \|\mathbf{x} - \mathbf{y}\| = r\}$. This not the case in general metric spaces, see Examples 19.8.
ii) A mapping $\phi : (X, \|\cdot\|_X) \to (Y, \|\cdot\|_Y)$ is an isometry if and only if it preserves norms:

$$\|\phi(\mathbf{x})\|_Y = \|\mathbf{x}\|_X \quad \text{for any } \mathbf{x} \in X.$$

Remark The other major source of complete metric spaces is ☞ *functional analysis*. A complete normed space is called a **Banach space**, and a Banach space whose norm is induced by an inner product is a **Hilbert space**.

   For instance $\mathbb{R}^n_{\text{Eucl}}$ is a Hilbert space, by Theorem 19.22. More exciting examples have infinite dimension: take a real sequence $a = (a_n)_{n \in \mathbb{N}}$ and define its **L^2 norm**

$$\|a\|_{L^2} := \left(\sum_{n=1}^{\infty} a_n^2 \right)^{1/2}. \tag{19.4}$$

The space of L^2-bounded real sequences $(a_n)_{n\in\mathbb{N}}$ is indicated by

$$\ell^2(\mathbb{R}) := \left\{ a \in \mathbb{R}^{\mathbb{N}} \mid \|a\|_{L^2} < \infty \right\}.$$

Exercise Using that $(x+y)^2 + (x-y)^2 = 2(x^2+y^2)$ for any $x, y \in \mathbb{R}$, show $\ell^2(\mathbb{R})$ is a normed vector space.

The norm of (19.4) induces the distance $d_{L^2}(a, b) := \|a - b\|_{L^2}$, and

Theorem $\left(\ell^2(\mathbb{R}), d_{L^2} \right)$ *is a separable metric space.*

Proof That $d(a, b)$ is positive definite and symmetric is immediate. For $(a_n), (b_n) \in \ell^2(\mathbb{R})$ the triangle inequality

$$\left(\sum_{n=1}^{N} (a_n - b_n)^2 \right)^{1/2} \leqslant \left(\sum_{n=1}^{N} a_n^2 \right)^{1/2} + \left(\sum_{n=1}^{N} b_n^2 \right)^{1/2}$$

results in $\|a - b\|_{L^2} \leqslant \|a\|_{L^2} + \|b\|_{L^2}$ as $N \to \infty$.

(Separability) Consider the set of rational sequences $(q_1, q_2, \ldots, q_N, 0, 0, \ldots)$ that stabilise to zero:

$$Q := \left\{ q = (q_i) \in \ell^2(\mathbb{R}) \mid \exists N > 0 \colon q_i = 0 \; \forall i \geqslant N \right\}.$$

This is countable because card $Q = \aleph_0^N = \aleph_0$. Moreover, for any $a = (a_n) \in \ell^2(\mathbb{R})$ and $\epsilon > 0$ there exist $N > 0$ and $q_1, \ldots, q_N \in \mathbb{Q}$ such that $\sum_{n>N} a_n^2 < \epsilon$ and $(a_n - q_n)^2 < \epsilon/N$, for any $n \leqslant N$. Therefore the sequence $q \in Q$ fulfils

$$\|a - q\|_{L^2}^2 = \sum_{n \leqslant N} (a_n - q_n)^2 + \sum_{n>N} a_n^2 < 2\epsilon,$$

proving that Q is dense in $\ell^2(\mathbb{R})$. $\qquad\square$

Exercise If $S = \ell^2(\mathbb{R})$ is viewed as subset of the space of real sequences $\mathbb{R}^{\mathbb{N}}$, what's the difference between Definition 19.5 and saying $\|a\|_{L^2} < \infty$?

Whilst the relationship between norms and distances is straightforward, the relationship between norms and inner products is subtler. We have seen in Corollary 19.26, or (15.6), that a positive definite bilinear form defines a norm forthwith (the square root of the associated quadratic form). The converse does not always hold. There exist normed spaces whose norm is not induced by any inner product. Fix $1 \leqslant p \leqslant \infty$ and define the **L^p norm** of sequence $a = (a_i)_{i \in \mathbb{N}}$ by

$$\|a\|_{L^p} = \left(\sum_{i=0}^{\infty} a_i^p \right)^{1/p}.$$

When $p \neq 2$ it can be proved that $\ell^p(\mathbb{R}) := \left\{ (a_n)_{n \in \mathbb{N}} \mid \|a\|_{L^p} < \infty \right\}$ does not admit inner products inducing the L^p norm. In this sense the space $\ell^2(\mathbb{R})$ is special.

Exercises Show that

i) the metric space $\left(\ell^2(\mathbb{R}), d_{L^2} \right)$ is complete, hence a Banach space.
ii) Let $\ell^\infty(\mathbb{R})$ denote the space of real sequences (a_n) that are bounded for

$$d_\infty\big((a_n), (b_n)\big) = \sup_{n \in \mathbb{N}} |a_n - b_n|.$$

Then $\ell^\infty(\mathbb{R})$ is a Banach space. (In contrast to ℓ^2, though, ℓ^∞ is not separable.)

The universe of Banach spaces provides us with sophisticated examples of distances (☞ *complex and functional analysis*). For example, take the hyperbolic disc $(\mathbb{D}, d_{\mathcal{H}})$ of Example 19.2 and the open unit ball $B = B(0, 1)$ in some complex Banach space. Then

$$d(x, y) = \sup \left\{ d_{\mathcal{H}}\left(f(x), f(y) \right) \mid f \colon B \to \mathbb{D} \text{ holomorphic} \right\}$$

is called *Carathéodory metric* on B [61].

19.6 Completions ☕

In this final section we'll prove the existence and uniqueness of canonical complete metric spaces. To familiarise ourselves with the circle of ideas let's start with the prototypical incomplete space, the rational numbers. Any real number r is the limit of a rational Cauchy sequence—essentially due to the density of $\mathbb{Q}$ in $\mathbb{R}$—as we saw in Example 6.10. Hence the set

$$\mathscr{C}(\mathbb{Q}) := \left\{ q \colon \mathbb{N} \to \mathbb{Q} \text{ Cauchy sequence} \right\}$$

determines $\mathbb{R}$. But there are infinitely many $(2^{\aleph_0})$ sequences having the same limit r, so we declare two Cauchy sequences equivalent, $(q_n) \sim (q'_n) \in \mathscr{C}(\mathbb{Q})$, whenever the zig-zag sequence $q_1, q'_1, q_2, q'_2, \ldots, q_k, q'_k, \ldots$ converges. We claim that

$$\mathbb{R} \cong \mathscr{C}(\mathbb{Q})/\sim$$

In fact, it's immediate to check $\mathscr{C}(\mathbb{Q})/\sim$ is an ordered field, and not so hard to show it is complete. Therefore Theorem 7.11 says it must be isomorphic to the real line. The isomorphism, by the way, forces $\mathscr{C}(\mathbb{Q})/\sim$ to be metrisable.

The above incarnation of real numbers as equivalence classes of rational Cauchy sequences was independently discovered by Méray (1869–1872) and Cantor (1872–1883). Relative to Dedekind cuts, its disadvantage is the excess of technicalities required when working with cosets rather than just sequences. The advantage, on the other hand, it that it can be used in a general context to embed metric spaces in complete spaces.

Definition Let (X, d) be a metric space. A **(metric) completion** of (X, d) is a pair formed by a complete metric space $(\widehat{X}, \hat{d})$ and an isometry $\Phi \colon X \to \widehat{X}$ with dense image.

Examples

i) The Euclidean inclusions $\mathbb{Q} \hookrightarrow \mathbb{R}$ and $(0, 1) \hookrightarrow [0, 1]$ are completions.

ii) If p is a prime number, consider the *p-adic metric* $d_p(x, y) = p^{-n}$ on $\mathbb{Q}$, where n is the largest integer such that $x - y \in p^n \mathbb{Z}$. The corresponding completion $\mathbb{Q}_p$ defines *p-adic numbers* (☞ *number theory*). When $p = 2$ we recover dyadic numbers (8.1). Triadic numbers were implicitly used in Example 3.20.

Lemma 19.27 *Let* $(a_n), (b_n)$ *denote Cauchy sequences in* (X, d). *The limit* $\lim\limits_{n \to \infty} d(a_n, b_n)$ *exists and is finite.*

Proof The sequence of distances $\{d(a_n, b_n)\}_{n \in \mathbb{N}}$ is Cauchy because inequality (19.1) implies

$$|d(a_n, b_n) - d(a_m, b_m)| \leqslant d(a_n, a_m) + d(b_n, b_m).$$

As $\mathbb{R}$ is complete, the limit is non-negative. $\square$

We can construct a completion for (X, d) explicitly, as quotient of the space of Cauchy sequences $\mathscr{C}(X, d)$ (this will be Theorem 19.29). Introduce the equivalence relation on $\mathscr{C}(X, d)$

$$(a_n) \sim (b_n) \overset{\text{def}^n}{\Longleftrightarrow} \lim_{n \to \infty} d(a_n, b_n) = 0.$$

and set

$$\widehat{X} := \mathscr{C}(X, d)\big/\sim\,.$$

For any $a \in X$ let $\Phi(a) \in \widehat{X}$ indicate the coset of the constant sequence $a, a, a, \ldots$. In this way $[a_n] = \Phi(a)$ if and only if $\lim_{n\to\infty} a_n = a$, and we identify $a \in X$ with the equivalence class of sequences that converge to a. The canonical inclusion

$$\Phi \colon X \hookrightarrow \widehat{X}$$

becomes onto (bijective) if and only if (X, d) is already complete. Lemma 19.27 guarantees the function

$$\hat{d} \colon \widehat{X} \times \widehat{X} \to \mathbb{R}, \qquad \hat{d}\big([a_n], [b_n]\big) := \lim_{n\to\infty} d(a_n, b_n) \tag{19.5}$$

is well defined.

Remark Beware that $\hat{d}$ is positive, but not positive definite: there exist distinct Cauchy sequences at zero distance.

Another example of such a 'pseudo-metric' is the *Kobayashi metric* of ☞ *complex geometry*, namely the biggest distance $d \colon \mathbb{C}^n \times \mathbb{C}^n \to \mathbb{R}$ such that $d\big(f(x), f(y)\big) \leqslant d_{\mathscr{H}}(x, y)$ for any holomorphic map $f \colon \mathbb{D} \to \mathbb{C}^n$. Restricted to the unit ball $B(0, 1) \subseteq \mathbb{C}^n$, it goes under the name of *Bergman metric* [61].

We stress that $\widehat{X}$ depends on X and on d as well. Furthermore, the construction makes explicit use of the completeness of $\mathbb{R}$, which means that completing $\mathbb{Q}$ necessitates a slightly special treatment.

Lemma 19.28 *If $(Q, d\big|_Q) \subseteq (X, d)$ is dense, the inclusion $\widehat{Q} \hookrightarrow \widehat{X}$ is a 1-1 correspondence.*

Proof First, $\mathscr{C}(Q, d\big|_Q) \subseteq \mathscr{C}(X, d)$. Since the relation $\sim$ restricts compatibly, we have $\widehat{Q} \subseteq \widehat{X}$, thus proving injectivity.

Now we claim every Cauchy sequence (x_n) in X is Cauchy in Q. For any n pick $q_n \in Q$ such that $d(q_n, x_n) \leqslant 2^{-n}$. Then (19.1) implies $d(q_n, q_m) \leqslant d(x_n, x_m) + 2^{-N}$ for every $n, m > N$, showing that (q_n) is Cauchy. And clearly $[q_n] = [x_n]$. $\square$

Theorem 19.29 (Existence of Completions) *The function $\hat{d}$ of (19.5) is a distance on $\widehat{X}$, and the canonical inclusion $\Phi \colon (X, d) \hookrightarrow (\widehat{X}, \hat{d})$ is a completion.*

Proof By definition $\hat{d}$ is non-negative, and $\hat{d}\big([a_n], [b_n]\big) = 0$ if and only if $[a_n] = [b_n]$. The symmetry of $\hat{d}$ is also evident. Now take $(a_n), (b_n), (c_n)$ Cauchy, and compute:

$$\hat{d}\big([a_n], [b_n]\big) = \lim_{n\to\infty} d(a_n, b_n) \leqslant \lim_{n\to\infty} \big(d(a_n, c_n) + d(c_n, b_n)\big)$$

$$\leqslant \lim_{n\to\infty} d(a_n, c_n) + \lim_{n\to\infty} d(c_n, b_n) = \hat{d}\big([a_n], [c_n]\big) + \hat{d}\big([c_n], [b_n]\big).$$

Moreover, $\hat{d}(\Phi(a), \Phi(b)) = \lim_{n\to\infty} d(a, b) = d(a, b)$ for any $a, b \in X$, which proves Φ is an isometry.

Density: if (a_n) is Cauchy, for any $\epsilon > 0$ there is an N such that $d(a_N, a_n) \leqslant \epsilon$ for any $n \geqslant N$. Taking $b = a_N \in X$ gives $\hat{d}([a_n], \Phi(b)) \leqslant \epsilon$ for all $\epsilon > 0$, $[a_n] \in \widehat{X}$.

Only completeness is missing: Φ is a bijective isometry by Lemma 19.28, and $\Phi(X)$ is dense in $\widehat{X}$. Therefore the composite map $\widehat{X} \to \widehat{\Phi(X)} \to \widehat{\widehat{X}}$ is bijective. But this is the same as saying $\widehat{X}$ is complete.   □

Truth be told, all completions of (X, d) are isometric:

Theorem (Uniqueness of Completions) *Let*
$\Phi\colon (X, d) \to (Y, \delta),\ \Phi'\colon (X, d) \to (Y', \delta')$
be completions of the same metric space
(X, d). Then there exists a unique isometry
$h\colon (Y, \delta) \to (Y', \delta')$ such that

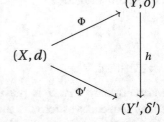

$$\Phi' = h \circ \Phi.$$

Proof The uniqueness of h follows from the density of $\Phi(X)$ in Y. Regarding the existence, without loss of generality we assume $Y = \widehat{X}$, as in Theorem 19.29. As the isometry $\Phi'\colon X \to Y'$ has dense image, it induces an invertible mapping $\widehat{\Phi'}\colon \widehat{X} \to \widehat{Y'}$. But (Y', δ') is complete, so $\Phi\colon Y' \to \widehat{Y'}$ is an invertible isometry. Then $h = \Phi^{-1} \circ \widehat{\Phi'}$ is the map we seek.   □

Immediately, then,

Corollary 19.30 (Universality) *Let (X, d) be a metric space and (Y, δ) a complete metric space. Every isometry $h\colon (X, d) \to (Y, \delta)$ factors through the completion $\Phi\colon (X, d) \to (\widehat{X}, \hat{d})$.*

In other terms, there exists a unique isometry $\hat{h}\colon (\widehat{X}, \hat{d}) \to (Y, \delta)$ such that $\hat{h} \circ \Phi = h$.

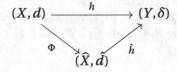

Examples (without proof)

i) [75, p.124]. Take the ring $\mathbb{R}[x]$ of real polynomials and define on it

$$d\big(p(x), q(x)\big) = \inf_{n\in\mathbb{N}} \left\{ \frac{1}{2^n} \ \Big|\ x^n \text{ divides } p(x) - q(x) \right\}.$$

This is a distance, and the completion of $(\mathbb{R}[x], d)$ is canonically isomorphic to the ring of formal power series $\mathbb{R}[\![x]\!]$.

ii) Let $B(X, \mathbb{R})$ denote the set of continuous and bounded maps $f : X \to \mathbb{R}$, equipped with

$$d_\infty(f, g) := \sup_{x \in X} |f(x) - g(x)| \, .$$

$$\Psi_x : X \to \mathbb{R}, \quad \Psi_x(y) = d(x, y) - d(x_0, y)$$

is an isometry. The closure $(\overline{\Psi(X)}, d_\infty)$ is the completion of (X, d).

Appendices

Etymologies

A quick glance at any maths book reveals the need of a wealth of symbols, calligraphic variations ($\mathscr{G}$, $\mathbb{G}$, $\mathfrak{g}$, **g** etc.) and a zoo of accents and diacritics (a′, â, ă, ã, $\vec{a}$, $\overset{*}{a}$, $\underset{*}{a}$, ...). Letter-like symbols include ∀ (for all), ∃ (exists), ℵ (aleph), ∂ (del), ∇ (nabla), ∪ (union), $\mathfrak{S}$ (cyclic sum) and many others. Furthermore, as Roman letters are not enough, learning the Greek alphabet is a must:

Letter		Name
A	α	alpha
B	β	beta
Γ	γ	gamma
Δ	δ	delta
E	ϵ, ε	epsilon
Z	ζ	zeta
H	η	eta
Θ	θ, ϑ	theta
I	ι	iota
K	κ	kappa
Λ	λ	lambda
M	μ	mu

Letter		Name
N	ν	nu
Ξ	ξ	xi
O	o	omicron
Π	π	pi
P	ρ	rho
Σ	σ	sigma
T	τ	tau
Y	υ	upsilon
Φ	ϕ, φ	phi
X	χ	chi
Ψ	ψ	psi
Ω	ω	omega

Just for fun, next comes a selection of terms with (Ancient) Greek root. Knowing how to analyse the language proficiently is an added value, and polishes its use. In this respect etymology never lets us down.

© The Author(s), under exclusive license to Springer Nature Switzerland AG 2021
S. G. Chiossi, *Essential Mathematics for Undergraduates*,
https://doi.org/10.1007/978-3-030-87174-1

Word	Origin	Literal meaning
glossary	γλῶσσα	tongue
analysis	ἀνάλυσις	dissolution, decomposition
physics	φύσική (ἐπιστήμη)	(knowledge of) nature
topology	τόπος + λόγος	study of places
category	κατηγορία	predicate, category
dynamics	δύναμις	force, power
kinematics	κινέω	to set in motion
chaos	χάος	big empty space
mechanics	μηκανή	contraption, tool
-graphy, graph	γράφω	to write
lexicographic	λεξίς + γράφω	speech, word + to write
analogous	ἀνά + λόγος	similar + aspect
program(me)	πρόγραμμα	written public notice
-morphic, -morphism	μορφή	shape
holo-	ὅλός	whole, entire
homo-	ὁμός	similar, analogous
mono-	μόνος	solitary, unique, only
epi-	ἐπί	above
iso-	ἴσος	equal
endo-	ἔνδον	internal
auto-	αὐτός	same, itself
crypto-	κρυπτός	hidden
chiral	χείρ	hand
isotropic	ἴσος + τρόπος	of the same style
poly-	πολύς	many
polynomial	πολύς + νομός	many + term
polygon	πολύς + γωνία	many + angle
pentagon	πέντε + γωνία	5 + angle
hexagon	ἕξ + γωνία	6 + angle
polyhedron	πολύς + ἕδρα	many + seat, place
tetrahedron	τετράς + ἕδρα	4 + place
octahedron	ὀκτώ + ἕδρα	8 + place
dodecahedron	δώδεκα + ἕδρα	12 + place
icosahedron	εἴκοσι + ἕδρα	20 + place
logic	λογική (τέχνη)	(art of) reasoning
ontologic	ὦν + λόγος	that which speaks of existence
eterologic	ἕτερος + λόγος	that which speaks of the other
axiom	ἀξίωμα	that which is appropriate
lemma	λῆμμα (λαμβάνω)	that which is supposed (to take, assume)

(continued)

Word	Origin	Literal meaning
theorem	θεώρημα (θεωρέω)	that which is to be proved (to look, examine)
theory	θεωρία	that which is considered, thought
arithmetics	' ἀριθμητικὴ (τέχνη) (ἀριθμός)	(art of) counting (number)
ellipse	ἔλλέιπσις (ἐλλέιπω)	that which is excluded (to exclude)
parabola	ά + βάλλω	to throw side-by-side
parallel	ά + ἄλληλος	aligned with one another
hyperbola	ὑπέρ + βάλλω	to throw beyond
logarithm	λόγος + ἀριθμός	computation + number
orthogonal	ορθός + γωνία	upright + angle
diagonal	διαγοναλ (διά + γωνία)	from angle to angle (across + angle)
trichotomy	τριχα + τέμνω	thrice + to slice
-metry, metric	μέτρον	measurement
symmetry	σύν + μέτρον	with the same measure
geometry	γῆ + μέτρον	measurement of Earth
characteristic	χαρακτήρ	stamp, distinction sign
thesis	θέσις	statement, claim
antithesis	ἀντί + θέσις	opposite claim
hypothesis	ὑπό + θέσις	that which lies underneath (basis, assumption)
hypotenuse	ὑπότείνω	to stretch under
antipodal	ἀντί + πούς	opposite to the foot
tautology	ταῦτός + λόγος	of the same meaning
syllogism	συλλογισμός	conclusion, inference
atom	ἄτομος (ἀ- + τέμνω)	indivisible (non + divide)
monotone	μονότονος	unchanged, stable
stereographic	στερεός + γράφω	solid (three-dimensional) + to write
canonical	κανών	measuring tool, standard
periodic	περίοδος	time lapse
harmonic	ἀρμονία	harmony
cycle	κύκλος	circle, round
cycloid	κύκλος + εἶδος	of circular shape
trochoid	τροχός + εἶδος	of round appearance
sphere	σφαῖρα	ball, globe
loxodrome	λοξός + δρόμος	oblique path
lemniscate	λημνίσχος	ribbon

Exercise Trace back to the etymology the meaning of: aperiodic (asymmetric, acyclic, ...), homology, isotonic, homotopy, autologic, homography, isometry, homothety, dodecagon, homogeneous, synthesis, holomorphic, trigonometry, cyclotomic, monomial, parameter.

The interesting book [68] uncovers the origin of lots of words, be it Latin (equivalence, annihilator, function, domain, sequence, corollary, (ir)rational, absolute, abscissa, projection, addendum, transcendental, tractrix, data, ergo, circa, exponent, equation, (in)equality, surface, tangent, *i.e., a priori, a posteriori, q.e.d., a fortiori, ad libitum, ad hoc, modus ponens/tollens, vice versa, mutatis mutandis, et al., etc.*), Arabic (sine, algebra, algorithm, . . .) or other.

References

1. E. Abbena, A. Gray, S. Salamon, *Modern Differential Geometry of Curves and Surfaces with Mathematica* (Chapman & Hall/CRC, Boca Raton, 2006)
2. I. Agricola, Th. Friedrich, *Elementary Geometry* (AMS, Providence, 2008)
3. I. Agricola, Th. Friedrich, *Global Analysis. Differential Forms in Analysis, Geometry and Physics* (AMS, Providence, 2002)
4. M. Aigner, G.M. Ziegler, *Proofs from the Book* (Springer, Berlin, 1998)
5. J. Arndt, C. Haenel, π *Unleashed* (Springer, Berlin, 2001)
6. P. Beckmann, *A History of Pi*, 3rd edn. (Dorset Press, New York, 1989)
7. E.G. Beltrametti, G. Cassinelli, *The Logic of Quantum Mechanics, Encyclopedia of Mathematics and Its Applications*, vol. 15 (Addison-Wesley, Reading, 1981)
8. P. Bencerraf, H. Putnam, *Philosophy of Mathematics, Selected Readings* (Cambridge University Press, Cambridge, 1964)
9. M. Berger, *Geometry Revealed, a Jacob's Ladder to Modern Higher Geometry* (Springer, Berlin, 2010)
10. G.D. Birkhoff, A set of postulates for plane geometry. Ann. Math. **33**(2), 329–345 (1932)
11. G. Birkhoff, M.K. Bennett, Felix Klein and his 'Erlanger Programm', in *History and Philosophy of Modern Mathematics*, ed. by W. Aspray, P. Kitcher (University of Minnesota Press, Minneapolis, 1988), pp. 145–176
12. A. Bogomolny, Pigeonhole principle from interactive mathematics miscellany and puzzles. Available http://www.cut-the-knot.org/do_you_know/pigeon.shtml
13. F. Borceux, *Handbook of Categorical Algebra*, vol. 1 (Cambridge University Press, Cambridge, 1994)
14. P. Borwein, The amazing number π. Nieuw Arch. Wisk. **1**, 254–258 (2000)
15. P. Borwein, S. Choi, B. Rooney, A. Weirathmüller (eds.), *The Riemann Hypothesis, CMS Books in Mathematics* (Springer, Berlin, 2008)
16. J.M. Borwein, R.M. Corless, Gamma and factorial in the monthly. Am. Math. Mon. **121**, 1 (2017)
17. C.B. Boyer, *A History of Mathematics* (Wiley, New York, 1968)
18. D.C. Brody, L.P. Hughston, Geometric quantum mechanics. J. Geom. Phys. **38**, 19–53 (2001)
19. D. Castellanos, The ubiquitous π. Math. Mag. **61**, 67–98 (part I), 148–163 (part II) (1988)
20. R. Courant, H. Robbins, *What Is Mathematics? An Elementary Approach to Ideas and Methods* (Oxford University Press, Oxford, 1978)
21. M. Davis (ed.), *The Undecidable* (Raven Press, New York, 1964)
22. P.J. Davis, R. Hersch, *The Mathematical Experience* (Birkhäuser, Boston, 1981)

23. R. Dedekind, Stetigkeit und irrationale Zahlen (1872). In Complete mathematical works, available at https://gdz.sub.uni-goettingen.de/id/PPN235685380

24. R. Dedekind, Was sind und was sollen die Zahlen?, Vieweg (1888). Available at https://archive.org/details/wassindundwasso00dedegoog

25. R. Dedekind, *Vorlesungen über Zahlentheorie*, 4th ed. (1894). Available at https://archive.org/details/vorlesungenber00lejeuoft/page/n8/mode/2up

26. J.J. Duistermaat, J.A.C. Kolk, *Lie Groups* (Springer, Berlin, 2000)

27. M. Erné, J. Koslowski, A. Melton, G.E. Strecker, A primer on Galois connections, in *Proceedings of the 1991 Summer Conference on General Topology and Applications*, vol. 704 (Annals New York Acad. Sci., New York, 1993), pp. 103–125

28. B. Fine, G. Rosenberger, *The Fundamental Theorem of Algebra* (Springer, Berlin, 1997)

29. P. Flajolet, R. Sedgewick, *Analytic Combinatorics* (Cambridge University Press, Cambridge, 2009)

30. H.G. Forder, *The Foundations of Euclidean Geometry* (Dover, New York, 1958)

31. L. Fortnow, *The Golden Ticket: P, NP, and the Search for the Impossible* (Princeton University Press, Princeton, 2013)

32. T. Franzén, *Gödel's Theorem. An Incomplete Guide to Its Use and Abuse* (A.K. Peters, Wellesley, 2005)

33. R. Fritsch, G. Fritsch, *The Four-Color Theorem: History, Topological Foundations, and Idea of Proof* (Springer, Berlin, 1998)

34. W. Fulton, J. Harris, *Representation Theory. A First Course* (Springer, Berlin, 1991)

35. H. Furstenberg, On the infinitude of primes. AMS Mon. **62**, 353 (1955)

36. E.M. García-Caballero, S.G. Moreno, M.P. Prophet, The golden ratio and Viète's formula. Teach. Math. Comput. Sci. **12**, 43–54 (2014)

37. M. Gardner, The Bells: versatile numbers that can count partitions of a set, primes and even rhymes. Sci. Am. **238**, 24–30 (1978)

38. L. Gatto, *Generic Linear Recurrent Sequences and Related Topics* (Publicações Matemáticas do IMPA, Rio de Janeiro, 2015)

39. B.R. Gelbaum, J.M.H. Olmsted, *Counterexamples in Analysis* (Dover, Mineola, 1964)

40. R. Goldblatt, *The Mathematics of Modality*. CSLI Lecture Notes, vol. 43 (Stanford University Press, Stanford, 1993)

41. D. Goldrei, *Classic Set Theory* (Chapman and Hall, Boca Raton, 1996)

42. S. Gomes da Silva, *Equivalências e consequências da hipótese do contínuo em análise e topologia*. Mini course at UFBA (2015)

43. R.K. Guy, *Unsolved Problems in Number Theory* (Springer, Berlin, 2004)

44. J. Hadamard, *An Essay on the Psychology of Invention in the Mathematical Field* (Princeton University Press, Princeton, 1945)

45. P.R. Halmos, S. Givant, *Introduction to Boolean Algebras* (Springer, Berlin, 2009)

46. G.H. Hardy, *A Mathematician's Apology* (Cambridge University Press, Cambridge, 1967)

47. R. Hersch, *What Is Mathematics, Really?* (Oxford University Press, Oxford, 1997)

48. A. Herschfeld, On infinite radicals. Am. Math. Mon. **42**(7), 419–429 (1935)

49. D.R. Hofstadter, *Gödel, Escher, Bach: An Eternal Golden Braid* (Basic Books, New York, 1979)

50. P. Howard, J.E. Rubin, *Consequences of the Axiom of Choice*. Math. Surveys and Monographs, vol. 59 (AMS, Providence, 1998)

51. J.E. Humphreys, *Introduction to Lie Algebras and Representation Theory* (Springer, Berlin, 1972)

52. R.S. Irving, *Integers, Polynomials, and Rings* (Springer, Berlin, 2004)

53. L. Ji, A. Papadopoulos (eds.), *Sophus Lie and Felix Klein: The Erlangen Program and Its Impact in Mathematics and Physics*. IRMA Lectures in Mathematics and Theoretical Physics, vol. 23 (EMS, Zürich, 2015)

54. P.T. Johnstone, *Stone Spaces* (Cambridge University Press, Cambridge, 1986)

55. J.M. Kantor, Hilbert's problems and their sequels. Math. Intell. **18**(1), 21–30 (1996)

56. O. Karpenkov, *Geometry of Continued Fractions* (Springer, Berlin, 2013)

57. J.L. Kelley, *General Topology* (Van Nostrand, Princeton, 1955)
58. L.C. Kinsey, *Topology of Surfaces* (Springer, Berlin, 1993)
59. P. Kitcher, *The Nature of Mathematical Knowledge* (Oxford University Press, Oxford, 1983)
60. M. Kline, *Mathematical Thought from Ancient to Modern Times* (Oxford University Press, Oxford, 1972)
61. S. Kobayashi, *Hyperbolic Manifolds and Holomorphic Mappings, an Introduction*, 2nd ed. (World Sci., Singapore, 2005)
62. D. Koukoulopoulos, *The Distribution of Prime Numbers* (AMS, Providence, 2019)
63. A.W. Knapp, *Basic Algebra* (Birkhäuser, Boston, 2006)
64. S.G. Krantz, The axiom of choice, in *Handbook of Logic and Proof Techniques for Computer Science* (Springer, Berlin, 2002), pp. 121–126
65. T. Kuhn, *The Structure of Scientific Revolutions* (University of Chicago Press, Chicago, 1962)
66. S. Lang, *Introduction to Transcendental Numbers* (Addison-Wesley, Reading, 1966)
67. I. Lakatos, *Proofs and Refutations* (Cambridge University Press, Cambridge, 1976)
68. A. Lo Bello, *Origins of Mathematical Words: A Comprehensive Dictionary of Latin, Greek, and Arabic Roots* (Johns Hopkins University Press, Baltimore, 2013)
69. L. Lovász, J. Pelikán, K. Vesztergombi, *Matemática discreta* (SBM, Rio de Janeiro, 2003)
70. M. Manetti, *Topology* (Springer, Berlin, 2015)
71. V. Moretti, *Spectral Theory and Quantum Mechanics. Mathematical Foundations of Quantum Theories, Symmetries and Introduction to the Algebraic Formulation*, 2nd ed. (Springer, Cham, 2017)
72. G.J. Murphy, *C*-algebras and Operator Theory* (Academic, Boston, 1990)
73. S. Müller-Stach (ed.), Richard Dedekind: *Stetigkeit und Irrationale Zahlen* (Vieweg Verlag, Braunschweig, 1872) and *Was sind und was sollen die Zahlen?* (Vieweg Verlag, Braunschweig, 1888). *Klassische Texte in der Mathematik* (Springer, Berlin, 2017)
74. P. Odifreddi, *Il dio della logica* (Longanesi, Milano, 2018)
75. M. Ó Searcóid, *Metric Spaces* (Springer, Berlin, 2007)
76. F.W.J. Olver, D.W. Lozier, R.F. Boisvert, C.W. Clark, *NIST Handbook of Mathematical Functions* (Cambridge University Press, Cambridge, 2010)
77. G. Pólya, *Mathematical Discovery: On Understanding, Learning and Teaching Problem Solving* (Wiley, New York, 1981)
78. B. Poonen, Undecidable problems: a sampler, in *Interpreting Gödel. Critical Essays*, ed. by J. Kennedy (Cambridge University Press, Cambridge, 2014), pp. 211–241
79. H.A. Priestley, *Introduction to Complex Analysis*, 2nd ed. (Oxford University Press, Oxford, 2003)
80. H. Putnam, Models and reality. J. Symb. Logic **45**(3), 464–482 (1980)
81. W.V. Quine, *Mathematical Logic*, revised ed. (Harvard University Press, Cambridge, 1981)
82. K.R. Rebman, The pigeonhole principle (what it is, how it works, and how it applies to map coloring). Two-Year Coll. Math. J. **10**, 3–13 (1979)
83. M. Reid, B. Szendrői, *Geometry and Topology* (Cambridge University Press, Cambridge, 2005)
84. P. Ribenboim, *The New Book of Prime Number Records* (Springer, Berlin, 1996)
85. B. Russell, *Introduction to Mathematical Philosophy* (Dover, New York, 1919)
86. S.M. Salamon, Differential equations and discrete mathematics. Lecture Notes, University of Oxford, 1996
87. R. Sikorski, *Boolean Algebras*. Ergebnisse der Mathematik und ihrer Grenzgebiete, 2. Folge (Springer, Berlin, 1969)
88. M. Spivak, *Calculus* (W.A. Benjamin, New York, 1967)
89. R.P. Stanley, *Enumerative Combinatorics*, vols. 1 and 2 (Cambridge University Press, Cambridge, 1997, 1999)
90. J. Stillwell, *The Four Pillars of Geometry* (Springer, Berlin, 2005)
91. J. Stillwell, *Mathematics and Its History* (Springer, Berlin, 2010)
92. R.R. Stoll, *Set Theory and Logic* (Dover, New York, 1979)
93. M.H. Stone, The representation of Boolean algebras. Bull. AMS **44**, 807–816 (1938)

94. C. Tapp, *An den Grenzen des Endlichen: Das Hilbertprogramm im Kontext von Formalismus und Finitismus* (Springer, Berlin, 2013)
95. G. Tourlakis, *Lectures in Logic and Set Theory*, vols. 1 and 2 (Cambridge University Press, Cambridge, 2003)
96. G. Vitali, *Sul problema della misura dei gruppi di punti di una retta* (Tipografia Gamberini e Parmeggiani, Bologna, 1905), pp. 231–235
97. H. Weyl, Invariants. Duke Math. J. **5**(3), 489–502 (1939)
98. H.S. Wilf, *Generatingfunctionology* (Academic, San Diego, 1994)
99. E. Wigner, *The Unreasonable Effectiveness of Mathematics in the Natural Sciences* (Wiley, New York, 1960)
100. C. Zong, Can you pave the plane with identical tiles? Notices AMS **67**(5), 635–646 (2020)
101. Clay Mathematics Institute.
102. MacTutor History of Mathematics archive.
103. Mathematics Genealogy Project.

These reading suggestions, like any other, should be weighed against Samuel Johnson's recommendation that "*a man ought to read just as inclination leads him, for what he reads as a task will do him little good*". It goes without saying that the advice applies (especially) to the present book.

Author Index

Below, for reference, are the names cited throughout. Readers interested in historical and biographical aspects might want to look at the websites [102, 103], the texts [17, 46, 60, 91] and the cult book [49].

Niels H. Abel (1802–1829)
Wilhelm F. Ackermann (1896–1962)
Leonard Adleman (1945–) 💻
Pavel S. Alexandrov (1896–1982)
André-Marie Ampère (1775–1836)
Roger Apéry (1916–1994)
Kenneth I. Appel (1932–2013)
Jean-Robert Argand (1768–1822)
Archimedes (Ἀρχιμήδης 287–212 aC)
Emil Artin (1898–1962)
Michael F. Atiyah (1929–2019) 🌑🌑🌑
René-Louis Baire (1874–1932)
Stefan Banach (1892–1945)
Hyman Bass (1932–)
Eric T. Bell (1883–1960)
Eugenio Beltrami (1835–1900)
Stefan Bergman (1895–1977)
Paul I. Bernays (1888–1977)
Jacob Bernoulli (1655–1705)
Felix Bernstein (1878–1956)
Friedrich W. Bessel (1784–1846)
Bryan J. Birch (1931–) 🌑
George D. Birkhoff (1884–1944)
Garrett Birkhoff (1911–1996)
János Bolyai (1802–1860)
Bernard Bolzano (1781–1848)
Rafael Bombelli (~1526–1572)
Pierre O. Bonnet (1819–1892)
George Boole (1815–1864)
F.É.J. Émile Borel (1871–1956)

Nathaniel Bowditch (1773–1838)
Luitzen E.J. Brouwer (1881–1966)
François G.R. Bruhat (1929–2007)
Cesare Burali-Forti (1861–1931)
Jost Bürgi (1552–1632)
Georg F.L.Ph. Cantor (1845–1918)
Constantin Carathéodory (1873–1950)
Gerolamo Cardano (1501–1576)
Élie J. Cartan (1869–1951)
Henri P. Cartan (1904–2008) 🌑
René Descartes (1596–1650)
Giovanni D. Cassini (1625–1712)
Eugène Catalan (1814–1894)
Augustin-Louis Cauchy (1789–1857)
Arthur Cayley (1821–1895) 🌑
Eduard Čech (1893–1960)
Claude Chevalley (1909–1984)
André-Louis Cholesky (1875–1918)
Alonzo Church (1903–1995)
Paul J. Cohen (1934–2007) 🌑
Nicolaus Copernicus (Mikołaj Kopernik, 1473–1543)
Charles-Augustin de Coulomb (1736–1806)
Gabriel Cramer (1704–1752)
Haskell B. Curry (1900–1982)
J.W. Richard Dedekind (1831–1916)
Jean F.A. Delsarte (1903–1968)
Augustus De Morgan (1806–1871)
Leonard E. Dickson (1874–1954)
Jean A.E. Dieudonné (1906–1992)

Lorenzo Mascheroni (1750–1800)
Yurĭ V. Matiyasevich (1947–)
James C. Maxwell (1831–1879)
Charles Méray (1835–1911)
Gerardus Mercator (Gerhard Kremer, 1512–1594)
Marin Mersenne (1588–1648)
Robert L. Mills (1927–1999)
Hermann Minkowski (1864–1909)
August F. Möbius (1790–1868)
Abraham de Moivre (1667–1754)
Eliakim H. Moore (1862–1932)
Anthony P. Morse (1911–1984)
H.C. Marston Morse (1892–1977)
Werner Nahm (1949–)
Nakayama Tadashi (1912–1964)
John Napier (1550–1617)

John F. Nash Jr. (1928–2015)
Claude-Louis Navier (1785–1836)
John von Neumann (1903–1957)
Isaac Newton (1643–1727)
A. Emmy Noether (1882–1935)
Marc-Antoine Parseval (1755–1836)
Blaise Pascal (1623–1662)

Wolfgang E. Pauli (1900–1958)
Giuseppe Peano (1858–1932)
Benjamin O. Peirce (1809–1880)
Charles S. Peirce (1839–1914)
Georg A. Pick (1859–1942)
Pythagoras of Samos (Πυθαγόρας ο σάμιος ~570–495 aC)
Plato (Πλάτων ~428–348 aC)
John Playfair (1748–1819)
J. Henri Poincaré (1854–1912)
Siméon D. Poisson (1781–1840)
L.A.C. René de Possel (1905–1974)
Emil L. Post (1897–1954)
Mojżesz Presburger (1904–1943)
Georges de Rham (1903–1990)
Henry G. Rice (1920–2003)
G.F. Bernhard Riemann (1826–1866)
Frigyes Riesz (1880–1956)
Gerhard Ringel (1919–2008)
Ronald L. Rivest (1947–)
John B. Rosser Sr. (1907–1989)
Gian-Carlo Rota (1932–1999)
Walter Rudin (1921–2010)
Paolo Ruffini (1765–1822)

Bertrand A.W. Russell (1872–1970)
Stephen H. Schanuel (1934–2014)
Ludwig Schläfli (1814–1895)
Erhard Schmidt (1876–1959)

Theodor Schneider (1911–1988)
Lowell Schoenfeld (1920–2002)
F.W.K. Ernst Schröder (1841–1902)
Erwin R.J.A. Schrödinger (1887–1961)
Friedrich H. Schur (1856–1932)
Laurent-Moïse Schwartz (1915–2002)
K. Hermann A. Schwarz (1843–1921)

Jean-Pierre Serre (1926–)
Adi Shamir (1952–)
Saharon Shelah (1945–)
Carl L. Siegel (1896–1981)
Wacław F. Sierpiński (1882–1969)
Thoralf A. Skolem (1887–1963)
Robert M. Solovay (1938–)
Michael D. Spivak (1940–)
Simon Stevin (1548–1620)
Michael Stifel (1487–1567)
George G. Stokes (1819–1903)
Otto Stolz (1842–1905)
Marshall H. Stone (1903–1989)
Ch.H. Eduard Study (1862–1930)
H. Peter F. Swinnerton-Dyer (1927–2018)
James J. Sylvester (1814–1897)
Thales of Miletus (Θαλῆς ὁ μιλήσιος ~624–546 aC)
Alfred Tarski (1901–1983)
Niccolò F. Tartaglia (1500–1557)
Terence C.-S. Tao (1975–)
Clifford H. Taubes (1954–)
Brook Taylor (1685–1731)

René F. Thom (1923–2002)
Heinrich F.F. Tietze (1880–1964)
Andreĭ N. Tikhonov (1906–1993)
John A. Todd (1908–1994)
Ehrenfried W. von Tschirnhaus (1651–1708)
Alan M. Turing (1912–1954)
Pavel S. Urysohn (1898–1924)
Charles J. de la Vallée-Poussin (1866–1962)
Alexandre-Théophile Vandermonde (1735–1796)
Oswald Veblen (1880–1960)
John Venn (1834–1923)
Giuseppe Vitali (1875–1932)
Joseph H.M. Wedderburn (1882–1948)
Karl Th.W. Weierstraß (1815–1897)
André Weil (1906–1998)
Caspar Wessel (1745–1818)
Hermann K.H. Weyl (1885–1955)
J. Henry C. Whitehead (1904–1960)
Andrew J. Wiles (1953–)

Legend

Fields Medal   Steele Prize

Abel Prize   Nobel Prize in Physics

Wolf Prize in Mathematics   Nobel Prize in Literature

De Morgan Medal   Turing Award

Breakthrough Prize in Mathematics

Index

© The Author(s), under exclusive license to Springer Nature Switzerland AG 2021
S. G. Chiossi, *Essential Mathematics for Undergraduates*,
https://doi.org/10.1007/978-3-030-87174-1

Printed in the United States
by Baker & Taylor Publisher Services